Towards Green Hydrogen Generation

Scrivener Publishing
100 Cummings Center, Suite 541J
Beverly, MA 01915-6106

Advances in Hydrogen Production and Storage

Series Editors: Mehmet Sankir and Nurdan Demirci Sankir

Scope: Energy is one of the most important issues for humankind. Increasing energy demand, regional limitations, and serious environmental effects of the conventional energy sources provide the urgent need for new, clean, and sustainable energy.

Advances in Hydrogen Production and Storage emphasizes the basics of renewable energy and storage as well as the cutting-edge technologies employed for these applications. The series focuses mainly on hydrogen generation, photoelectrochemical solar cells, fuel cells and flow batteries.

Publishers at Scrivener
Martin Scrivener (martin@scrivenerpublishing.com)
Phillip Carmical (pcarmical@scrivenerpublishing.com)

Towards Green Hydrogen Generation

Edited by

Mehmet Sankir

Department of Materials Science and Nanotechnology Engineering, TOBB University of Economics and Technology, Ankara, Turkey

and

Nurdan Demirci Sankir

Department of Materials Science and Nanotechnology Engineering, TOBB University of Economics and Technology, Ankara, Turkey

Scrivener Publishing

WILEY

This edition first published 2024 by John Wiley & Sons, Inc., 111 River Street, Hoboken, NJ 07030, USA
and Scrivener Publishing LLC, 100 Cummings Center, Suite 541J, Beverly, MA 01915, USA

For more information about Scrivener publications please visit www.scrivenerpublishing.com.

Wiley Global Headquarters
111 River Street, Hoboken, NJ 07030, USA

For details of our global editorial offices, customer services, and more information about Wiley products visit us at www.wiley.com.

Library of Congress Cataloging-in-Publication Data

ISBN 978-1-394-23408-0

Cover image: Pixabay.Com
Cover design by Russell Richardson

Set in size of 11pt and Minion Pro by Manila Typesetting Company, Makati, Philippines

Printed in the USA

10 9 8 7 6 5 4 3 2 1

Contents

Preface

The world is undergoing a new period of upheaval. This process started with the COVID-19 pandemic, and now there is talk of "new" distruptions in many areas. Perhaps the most prominent of these is the emergence of new export items such as hydrogen for countries.

In December 2022, Japan and Saudi Arabia signed a partnership agreement on carbon capture, carbon economy, clean hydrogen, and ammonia fuels. To meet this agreement, there is a need for economically attractive changes in the production, distribution, storage, and consumption of energy. With these changes, it is possible to create new jobs, increase the number workers in existing jobs, and thus create a large economy. In short, while fulfilling an important function, such as reducing emissions and decarbonizing energy, it is also possible to create a large market with the establishment of a hydrogen ecosystem.

The realization of the 28 countries' decision to keep global warming at or below two degrees Celsius, as stated in the Paris Climate Agreement, and the achievement of minimizing emissions can only be realized through the establishment of a hydrogen ecosystem. This requires the gradual utilization of carbon-free resources to generate power for everyday life and industry. This is why hydrogen is so important. Therefore, hydrogen technologies come to the fore in the context of achieving net zero emissions. There is virtually no other way to meet the 2053 net zero emissions commitment. Hydrogen is a rare gas in nature and is often found together with natural gas. In summary, while hydrogen is the most abundant element in the known universe, molecular hydrogen is very rare in nature and needs to be produced—and produced in large quantities, if we are serious about a Green Deal.

This book has been organized into three parts to introduce and discuss these crucial topics. Part I discusses the Green Deal and the current state and challenges encountered in the industrialization of green hydrogen production, as well as related politics. Also, it describes a gradual shift in the approach of hydrogen production technologies from non-renewable to renewable. On the other hand, Part II is devoted to carbon capturing and hydrogen. Biomass mass waste-to-hydrogen conversion and related learning models, life cycle analysis, and metal oxide-based carbon capture technologies are all addressed in this chapter. The last part of the book was designed to present all features of the green hydrogen generation. PEM water electrolysis and other electrolyzers, wind-driven hydrogen production, and bifunctional electrocatalysts-driven hybrid water splitting are introduced and thoroughly discussed.

As a result, despite the uncertainty created by the pandemic, efforts to create social impact on climate change, clean and reliable energy sources, energy-efficient buildings, sustainable and smart mobility, biodiversity and ecosystems, zero pollution, environmentally friendly environments, and a sustainable industry have gained momentum. Therefore, a new geopolitical order is envisaged where energy production, distribution, storage, and sectors that increase carbon footprint are reconfigured. In short, a new economic order in which the carbon footprint will be monitored is desired. As mentioned above, the Green Deal will have a significant impact not only on the European Union countries, but also on all global players that produce, distribute, and dominate the energy market, as well as on stakeholders, such as countries that produce steel, cement, aluminum, and carbon-intensive outputs. This volume covers also these issues by presenting the relations and opportunities between green hydrogen and related sectors.

In this book, readers will find a multidisciplinary approach that enlightens all the features of green hydrogen in the book. It has been edited for both undergraduate and graduate students, and it will be a valuable reference text in the area of green hydrogen generation and other diverse fields. A wide readership will find this volume quite useful, from professionals in chemistry, physics materials science, chemical engineering, and energy engineering, to those in the fields of economics, political science and international relations.

We are deeply grateful to everyone who helped with this book and greatly appreciate the dedicated support and valuable assistance rendered by Martin Scrivener and the Scrivener Publishing team during its publication.

The Editors
Mehmet Sankir and Nurdan Demirci Sankir
Department of Materials Science and Nanotechnology Engineering
TOBB University of Economics and Technology
Ankara, Turkey
July 2024

Part 1

HYDROGEN AND THE GREEN DEAL

1

Decarbonizing the Industry with Green Hydrogen

Cigdem Tuc Altaf[1], Orcun Demir[2], Tuluhan Olcayto Colak[3], Emine Karagöz[3], Mehmet Kurt[3], Nurdan Demirci Sankir[1,3†] and Mehmet Sankir[1,3*]

[1]Department of Materials Science and Nanotechnology Engineering, TOBB University of Economics and Technology, Sogutozu, Ankara, Turkey
[2]Political Science and International Relations, TOBB University of Economics and Technology, Sogutozu, Ankara, Turkey
[3]Micro and Nanotechnology Graduate Program, TOBB University of Economics and Technology, Sogutozu, Ankara, Turkey

Abstract
Perseverance to reduce global greenhouse gas emissions and backward economic and ecological issues of excessive use of carbon-based fuels is more preponderant than any other environmental-related policies. In this respect, hydrogen (H_2) appears to be a revolutionary in decarbonizing the common and energy-intensive industrial applications along with its use in transportation. Nations agree to explore a diverse range of innovations to integrate green H_2 into the common industrial plants. Energy-intensive industrial sectors emitting large amounts of carbon dioxide (CO_2) annually need widespread analysis and investigations to reform the adaptation to carbon-neutral resources. This chapter plunges into these energy-intensive, yet crucial for human needs, applications such as ammonia (NH_3), iron/steel, cement, and oil refining. These sectors can be assisted with green H_2 as an energy source with its high energy density and as a potential carbon-neutral feedstock for the chemical industry.

***Keywords*:** Hydrogen energy, green hydrogen, climate change, hydrogen economy, industrial decarbonizaiton, H_2 strategies

**Corresponding author*: msankir@etu.edu.tr
†*Corresponding author*: nsankir@etu.edu.tr

Mehmet Sankir and Nurdan Demirci Sankir (eds.) Towards Green Hydrogen Generation, (3–48)

1.1 Introduction

In the gradual transition of the fossil-intensive energy landscape toward a green structure, renewable energy sources (RES) face challenges, particularly in two aspects compared to fossil sources. First, there is an intermittency problem with RES in response to the demand for an uninterrupted energy supply. Despite recent improvements in fields such as storage and smart grid technologies, this issue remains a significant shortcoming. The second challenge is that power generation for hard-to-electrify sectors, which involve processes with very high-temperature combustion, can only be achieved using fossil sources. In this respect, H_2 has become a complementary source of future projections with its potential to fill the gap in the thermal and renewable power generation mix [1]. Therefore, H_2 appears to be a problem solver in decarbonizing hard-to-electrify sectors and storing electricity generated from renewables. Additionally, it is a candidate that could replace fossil sources in chemical production and long-haul transportation [2].

There is a global H_2 initiative inflation, although, in current economic and technological conditions, H_2 is still a niche that is not cheap. Since the 1970s, the H_2 economy has been seen as an initiative with significant breakthrough potential. Although this interest has continued since the early 2000s, problems related to some technological components (such as detonability) and the high cost of economic infrastructure such as storage have created a deflation in governments' initiatives on H_2. According to the IEA, global government spending on H_2 investments reduced by 35% between 2008 and 2018 [3]. However, the idea that the time had come for the real acceleration of H_2 became increasingly commonplace. Striving greenhouse gas emissions reduction universally against the challenges of climate change, which is a much more driving factor than past technological and economic initiatives, is today's impetus to increase H_2 deployment. As a matter of fact, many companies have made a splash with serious investments in addition to the national strategy plans of the countries. According to the World Energy Council [4], twelve governments and the EU published national H_2 strategies in 2021, while according to the IEA Global H_2 Review, this number reached 41 by 2023 [5].

Each national plan has different priorities within countries' own technological and logistical capacity projections. Countries are exploring a diverse range of policy instruments and innovative measures to incorporate H_2 into their energy systems. In other words, future expectations for H_2 are not just technological and economic. How countries allocate these developments within the framework of their national goals also fills the issue with political question marks. Forecasts regarding imports and exports and expectations for the realization of green H_2 potential are new puzzle pieces of the energy landscape. In addition, transportation as well as production of H_2 contain new opportunities and challenge potentials in parallel with global political developments.

van Renssen argues that developments in the global H_2 market may lead to new political parties and competition and reshape geopolitical relations and alliances [2]. The risk that may lead the system to a conflictual orientation was emphasized, despite the differentiation of actors in a structure like the oil age. Although serious targets continue to spread globally, this study carries the possibility of securitizing the turmoil regarding H_2, as it is still in its infancy in terms of technology and economy. Moreover, its global share is predicted to be around 25% even in the most ambitious target [6]. In that regard, authors, who are concerned about problems similar to the political consequences of oil in the H_2 structure, recommend that institutionalization be done through international governance mechanisms [2].

Scott and Powells, on the other hand, question the inclusion of H_2 in daily practices at the individual level and open up space for further studies in terms of social sciences with justifications such as heating, poverty, and place attachments [7]. In addition, Sadik-Zada [1] claims that global developments and national plans for H_2 will widen the inequality between the global north and south. The author shows that developing countries do not have a comparative advantage in realizing their high potential due to their lack of access to finance to realize electrolysis and infrastructure investments for H_2, which are not very cheap at present [1]. According to the research, as the measure of renewables in the energy flock increases and a strong natural gas pipeline structure is established, the economic and technological developments of countries will also provide a political comparative advantage.

Pflugmann and de Blasio [8] predict that if H_2 settles at scale, it will have a shape similar to today's global natural gas structure. According to them, countries will play roles in the H_2 landscape that will be formed within the framework of potential infrastructure and resource equipment. In this context, they argue that the future substantiality of geopolitically reserves-poor countries in Europe and Southeast Asia will not differ from current realities [8]. In general, it seems that political perspectives beyond economic and technological developments contain opportunities and risks regarding what national plans will bring. In this respect, it is necessary to focus on the national strategy road maps of countries with critical roles. The sectoral distributions of these road maps, especially within the framework of countries' import and export orientations, also present their preferences in the versatile deployment areas of H_2.

1.2 Intersections of National H_2 Strategies

H_2 investments gained momentum after 2020, as the number of countries announcing their national strategies increased and RES costs decreased. This increase accelerated as national road maps set targets before the increase as of 2020. Japan was the first country to institutionalize the H_2 target (Figure 1.1).

Japan's "Basic H_2 Strategy" is based on two key areas where it faces challenges. The first is the fragility caused by import dependence on fossil fuels, especially in the East. The second is their commitment to greenhouse gasses. Japan does not deny the share of H_2 in reducing greenhouse gas emissions, which it has promised to reduce by 26% by 2030 and 80% by 2050, by meeting its demand [9]. In this context, Japan has announced a commitment of \$1.5 billion to improve its distribution infrastructure and produce green H_2. Thus, H_2 also has an important place in attempts to reduce import dependency in economic and technological targets. Japan predicts that H_2, as an energy storage medium, will help stabilize fluctuations in energy costs caused by its own imports. In June 2023, Japan reaffirmed its dedication to utilizing H_2 and NH_3 in the power sector through a revision of its 2017 strategy. In the 6th Strategic Energy Plan [9], Japan emphasizes

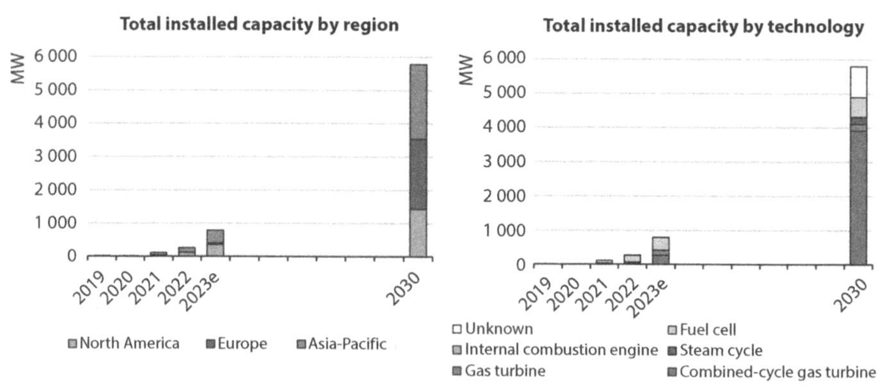

Figure 1.1 Capacity in power generation using H_2 and NH_3 by region, historical, and from announced projects, 2019–2030 [5].

that H_2 and NH_3 will contribute 1% to electricity production by 2030. In this context, capacity auctions started as of October 2023 [10]. In January 2022, the first liquefied H_2 shipment was made from Japan to Australia. Thus, Japan became a pioneer in the field of transportation [11]. In addition, the United States is currently investing heavily in fuel-cell vehicles. "Road Map to a US H_2 Economy" (2020) highlights the potential for H_2 to competitively meet 14% of US energy demand by 2050 and millions of new jobs will be created. However, although this road map is supported by the government, it does not constitute a national action. This road map, prepared jointly by important companies investing in R&D for the H_2 economy, is a policy recommendation.

On the contrary, South Korea is actively promoting the utilization of fuel cells in buildings through decentralized systems. While the country announces goals that endorse centralized production research, there is a significant increase in the number of fuel cell electric vehicles (FCEVs) in transportation. South Korea has introduced its H_2 Economy Roadmap, emphasizing leadership in fuel cells for both automotive applications and expansive immobile fuel cells for power sources, leveraging the significant presence of the automotive sector in South Korea. In this framework, South Korea aims to produce more than 6 million FCEVs by 2040 [12]. China, as the world's leading H_2 promoter, generates more than 20 million tons of H_2 annually, which accounts for nearly one-third of the global production. However, the majority of China's H_2 is derived from coal [13]. Nevertheless, China has set ambitious green H_2 targets for 2050. Anticipating a rise in H_2 demand to 45 million tons by 2040, with green H_2 comprising 40%, the goal is to further increase it to 60 million tons by 2050, with a green H_2 rate of 75% [14].

On the other side of the world, Europe, both the EU and its member states, has been paying attention to H_2 for a long time. Natural gas dependence on Russia has always raised questions, and H_2 discussions gained momentum with the annexation of Crimea in 2014. However, H_2 discussions gained momentum after technological and economic developments and the deep impact of the Ukrainian War on the European energy market. As the norm distributor of the green economy, the EU prioritizes green in H_2 but also plans to use blue H_2 as a transition period. In this context, it is committed to establishing 40 GW electrolysis capacity by 2030 (EU, 2021). Furthermore, the EU has revealed plans to establish an import logistic network, incorporating an extra 40 GW sourced from neighboring countries through eastern and southern, including Ukraine and North African lands [15]. At this point, although Europe has very strong targets for a carbon-neutral economy, it is extremely crucial to foresee that its own resources alone will not be sufficient to achieve this. Even though Europe is very eager in production, its green H_2 production potential will be limited, so it is developing policies to ensure safe imports after the transition with the help of blue H_2. As shown in Figure 1.2, European stakeholders have a low potential in respect of production. Although the use of existing natural gas pipelines is poised for new technological developments, it has a significant transportation potential.

Countries in the Middle East and North Africa (MENA) region, known for their significant oil and gas production, are now showing an increased interest in H_2 projects. Given their pivotal roles in the current global energy landscape, largely dependent on fossil fuels, these states rely heavily on revenue from fossil fuel exports. Consequently, the ongoing global shift toward RES poses a substantial existential challenge to these countries' domestic stability. In light of this, they are considering strategies to address the potential adverse macroeconomic effects and reduced geopolitical influence in a future "decarbonize" world [15].

With abundant potential in renewable energy and carbon capture and storage (CCS), oil and gas producers in the MENA could become major exporters of green hydrogen. However, despite this promising outlook, the hydrogen ambitions of MENA countries might encounter obstacles owing to the significant water scarcity in the region. The expected exacerbation of water stress in MENA, intensified by climate change, introduces

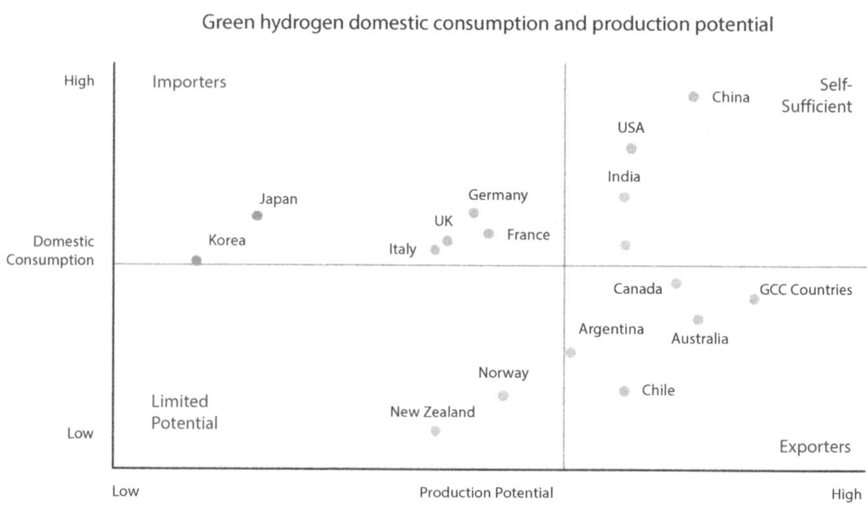

Figure 1.2 Analyzing the potential for domestic consumption and production of green hydrogen in specific countries and making comparisons [16].

complexity to their efforts in the field of hydrogen production. Addressing the water deficit, MENA states might integrate hydrogen projects with desalination plans, as seen in Neom. This approach aims not only to tackle water scarcity but also to boost MENA's desalination capacity, currently comprising almost half of the global capacity. Plans for hydrogen projects have been announced by countries like Saudi Arabia, the United Arab Emirates (UAE), and Oman. Notably, in July 2020, a substantial venture involving Air Products, Saudi ACWA, and Neom was unveiled, with plans for commissioning in 2025. Valued at $5 billion, this project aims to establish the world's largest green hydrogen and green NH_3 plant [17]. Gulf economies, which are institutionally close to the export culture, are trying to strengthen their financial capacity with their small demands and the quest to protect and diversify their exporter structures in the changing world order.

In the Americas, Chile and Canada aim to create a supply structure that will meet domestic demand before considering exports in the medium term [4]. Chile's renewable potential is above the average of the American continent. Therefore, it wishes to eliminate the fragility in the structure of imported NH_3 by pioneering green H_2. Conversely, Canada, in addition to being a major hydrocarbon exporter, has short-term transportation targets. In this sense, like Japan and Spain, Canada is also working on H_2 passenger trains [5]. In general terms, the national plans of the countries that play a leading role in this path are both remarkable and guiding. Within the framework of these plans, the potential for a hopeful, albeit still controversial, H_2 economy attracts high interest. In this context, a significant part of H_2's undoubted benefits for decarbonization is to industry. Carbon-intensive sectors have a significant share in greenhouse gas emissions.

1.3 Decarbonization of Carbon-Intensive Sectors

Carbon-intensive industrial sectors, with demand for temperatures greater than 1000°C in production, require comprehensive restructuring in the transition to a carbon-neutral economy. In this restructuring, carbon-intensive sectors such as steel, cement, chemicals, glass, paper, and refinery will need a very challenging transition planning in terms of both high energy demand and the direct use of fossil fuels as raw materials. H_2 offers two contributions to these sectors: First, it can be used to provide high-temperature process heat and replace fossil fuels. Second, it can be used as a potential carbon-neutral feedstock for the chemical industry [18]. Nevertheless, it appears challenging to circumvent the carbon dioxide emissions generated by these industries, coupled with the processing of raw materials during production. High infrastructure investments and cost-competitive optimizations are essential for H_2 to contribute to the transition of carbon-intensive sectors to carbon-neutral economies. In this respect, CCS and carbon capture and utilization (CCU) are presented as remarkable solutions in cost and efficiency projections.

1.3.1 NH_3 Industry

NH_3 is important as a means of H_2 carrier and storage and will play a marginal role in H_2 demand after 2030. It is considered much safer in three respects: First, it greatly reduces the possibility of explosion with its high autoignition temperature feature. Second, it has low condensation pressure, and last, it has low gas density [19]. The NH_3 supply chain in the energy sector is expected to be operational by 2030. According to IEA, NH_3 demand will increase by 40% by 2050 [20]. Australia, the United Kingdom, and Germany pioneered research and development programs to use NH_3 for production and subsequent export by other means. For 80% domestically produced NH_3, Japan plans to triple its production in 2030 [21]. As of 2020, the Asia Pacific region stands as the predominant force in the NH_3 market, contributing to 47% of the global NH_3 production. Notably, China alone constitutes 30% of the worldwide production. Despite experiencing rapid growth at a rate of 4% annually between 1990 and 2010, China's production has slowed down to 1% per year since 2010, a pace below the global average growth rate. Asia Pacific, while witnessing a slight reduction in its share to 42% of global production by 2050, retains its position as the primary NH_3-producing region. The production in China is predicted to peak within the next 5 to 10 years and subsequently decrease by almost 35% by 2050. This decline is attributed to economic restructuring, steering the country away from heavy industry toward more high-value industries and sectors. Despite the reduction, China remains the leading producer in 2050, contributing 16% to global production. Concurrently, India's production experiences more than a twofold increase, reaching 15% of global production [10].

To sum up, it is critical that investments in their transformation, especially in the cement and ISI splint sectors, are implemented in the short term. Likewise, although serious plans are made in petrochemicals and NH_3, collaborations and incentives toward global goals are required. It is essential that new technologies and investments have competitive power to accelerate investments and achieve targets. In this respect, IEA emphasizes the importance of directly stimulating investment in R&D studies and draws attention to taking regulatory actions. Efficiency efforts on the consumption side are also important. It is necessary to reduce reliance on single-use plastics and improve waste management. Additionally, projects to raise consumer awareness should be encouraged. In short, the importance of circular economy and resource efficiency is increasing. Although short-term efficiency targets are important, a new design is required in the whole industrial system in line with zero-emission targets.

1.3.2 Cement Industry

The cement industry is one of the most carbon-intensive sectors. The cement industry, having expanded to about 25 times its size since the 1950s, grew by 73% from 2005 to 2013 [22]. It stabilized in the following years, reaching a plateau appearance [23]. It is responsible for approximately 5% of global greenhouse gas emissions, almost the same as the steel sector. According to IEA (2022), direct CO_2 emission intensity in cement production has generally remained stable over the past 5 years and has not tended to decrease. The lack of acceleration for the cement sector in the period after the Paris Agreement requires taking more urgent steps for the 2030 targets. On the other hand, it will not be easy to achieve this acceleration in terms of both decreasing costs and maturing long-term investments. By 2030, the sector is targeted to achieve 4% reductions in annual CO_2 intensity to align with the Net Zero Emissions (NZE) Scenario by 2050. Factors that play a key role in achieving this goal include reducing the clinker/cement ratio, continuous energy efficiency improvements, adoption of low-carbon fuels, material efficiency improvements, and innovative technologies such as CCS.

In terms of CO_2 emissions, one of the most critical processes is the calcination process in the cement industry. Due to the nature of this process, CO_2 is released by processing limestone. In other words, regardless of what you use as the heating medium in the calcination process, the process itself is the severe cause of CO_2 emissions. Fifty percent of the CO_2 emissions in production are released from raw materials, 40% from fuel, and the remaining 10% from electricity and other processes [24]. Therefore, in addition to the use of green H_2 as a fuel, in a carbon-neutral economy, the CO_2 in cement production must complete the process with CCS and reduce/prevent emissions [25]. On the cost side, with today's technology, flue gas CO_2 capture offers an infrastructure that significantly increases the cost. According to Nhuchhen *et al.*, the clinker production cost increases by 72% with CCS technology [25]. In addition to this risk, the intermittency problem that may arise due to the provision of green H_2 from modern renewable sources presents a risky supply mechanism for a clinker plant that must operate 24 hours a day [25]. According to Farfan *et al.*, only certain stages of the cement production process can be 100% carbon-free. Green H_2 and other renewable resources can be used in the transportation and heat treatment process of cement. Heat treatment can even be done directly with solar energy. Yet, emissions from the material cannot be prevented as long as limestone is the raw material [24].

Bartlett and Krupnick (2020) [16] also emphasize that heat reliability is a challenge for the use of intermittent renewable resources in the heat treatment process. However, according to Bartlett and Krupnick, who argue that H_2 can be cost-competitive as a fuel when it drops below certain costs, CCUS is a more cost-friendly method. But, CCUS also has its drawbacks. The most important of these is the low concentration of CO_2 in the combustion gasses compared to the almost pure CO_2 released during the calcination process. In this respect, the authors propose a hybrid model and see the combined use of both H_2 and CCUS as the method that will reduce emissions most strongly and effectively [16].

On the other hand, studies on raw materials in the cement industry have produced some innovative products aimed at reducing carbon emissions. According to a report by an Austrian think tank named *Beyond Zero Emission*, geopolymer cement has a lower carbon emission potential than Portland and has the advantage of lower setup costs [26]. In this context, one of the most successful R&D studies is CO_2 SUICOM produced in Japan. This product, called "carbon negative concrete", does not emit CO_2 in its production and even CO_2 is used as raw material [27].

Within the framework of both negative and positive developments, the national plans of the countries also include certain plans for the cement sector. In cement production, China is, by far, the leader in production, covering approximately 52% of the global share by 2022. After that, India ranks second with a share of nearly 10%. Thereafter, (3) Vietnam, (4) United States, (5) Turkey, and (6) Brazil are listed as the largest cement producers [28]. Between 2011 and 2013, China produced more cement than the United States produced in 100 years [24]. Chen *et al.* (2023) [29] show that the transition in the cement sector under China's Nationally Determined Contributions (NDCs) is limited and not sufficient. They highly recommend using CCUS to reduce CO_2 emissions and air pollution, which has become a serious threat. In November 2022, the Carbon Peak Implementation Plan for the Building Materials Industry was initiated in China. Targeting the cement and lime sectors, the plan aims to achieve the peak of carbon emissions before 2030. This will be accomplished through the ongoing development and utilization of low-carbon materials and technology, a decrease in energy consumption, and a shift toward clean and sustainable energy sources [30].

The EU, on the other hand, announced its Industrial Plan within the scope of the Green Deal initiative in February 2023 [5]. The main purpose of this plan is to facilitate, accelerate, and harmonize incentives while preserving the competitiveness of the EU. It is taking detailed measures specifically designed to support the transition in energy-intensive sectors in line with the goal of a net zero economy. In the United States, the Inflation Reduction Act announced in August 2022 sets a budget of $5.8 billion for industrial decarbonization to support a reduction of approximately 200 metric tons (Mt) of CO_2. Within the framework of the law, the United States aims to reduce CO_2 emissions by 17% by 2030. New CCS applications and investments for the 45th quarter tax credit are also supported. Low embodied carbon appropriation incentives, in particular, have the potential to be a driving force in the regulation of concrete and cement [31].

From the perspective of the Middle East, the cement targets of the Gulf countries, which have low carbon emissions in parallel with their production, are also shaped in line with the H_2 transition targets. Although the UAE does not have a clear plan for cement, there is an emphasis on the use of H_2. Industry accounts for 50% of total energy demand. Even though we are talking about a relatively low demand, the rate is high. Therefore, H_2 incentives in the cement sector will be significant in reducing CO_2 emissions [32]. In addition to this, Saudi Arabia also emphasizes the industrial contribution of H_2 within the circular carbon economy and NEOM green H_2 projects [32]. Industry has a very low emission compared to energy. In this context, Saudi Arabia does not set a clear industry-specific target and rather pioneers H_2 value chain opportunities.

In general terms, the cement sector is promising in terms of long-term targets for CO_2 emissions, both in terms of raw materials and technologies for combustion energy. The broader impact on CO_2 emissions in terms of investment and technological costs will be seen in the long term. Short-term policies and initiatives are critical for the rapid implementation of investments that will provide long-term benefits and for promising technologies to become cost-competitive.

1.3.3 Iron and Steel Industry (ISI)

The iron and steel industry (ISI) is one of the largest sectors contributing to global GHG emissions, as much as the cement industry. ISI, which is responsible for one-third of global industrial CO_2 emission and 7% to 9% of total global CO_2 emission, is one of the most conventional industrial structures in terms of its technological lock-ins and commercial path dependency [8, 33, 34]. In the ISI, which is an important part of global industrial production with huge investments, targets for CO_2 reduction are divided into short- and long-term targets. For a large-scale transition, the steel industry is resilient and slow-moving [33]. Considering the investment, project, construction, and production processes, short-term emission reduction targets for iron and steel production facilities, which are long-term projects, focus on energy efficiency [35]. Long-term targets, on the other hand, emphasize the potential for a much more structured and effective emission reduction target by changing the fuel type(s) in production and crude steelmaking route(s) with emerging technologies [35]. Accordingly, reduction targets in countries are modeled with these short and long-term targets.

Two methods predominate in current technologies in steelmaking: the blast furnace–basic oxygen furnace (BF-BOF) route and the scrap–electric arc furnace (scrap-EAF). Since scrap-EAF offers a recycling model, increasing the share of the scrap-EAF route will reduce carbon emissions in production. But, BF-BOF is still the dominant model in all global steelmaking. The technological model that is aimed to provide the most ambitious reduction in carbon emissions is H_2-based direct reduction (H-DR) [8]. However, a remarkable cost increase is calculated with H-DR route steelmaking [36]. In their study focusing on Germany, Otto *et al.* [35] claim that it is possible to reduce CO_2 emissions by 47% to 95% by coupling EAF with H-DR. Based on this, the potential for countries to realize their ambitious goals with rapid investments and collaborations has a positive outlook. Yet, iron and steel investments are both cost-prone and long-term projects. The investment potential of the sector will be directly proportional to the cost analysis of technological developments.

Total steel production in the world decreased by 4.2% in 2022 compared to the previous year [37]. China, the world's largest producer, supplies approximately 54% of total global steel production by 2022. The largest producers after China are India, Japan, the United States, and Russia, respectively (see Table 1.1). The carbon emissions reduction target at ISI is critical in line with the climate targets of the largest manufacturers. According to Ren *et al.*, it calculates that China can only reduce GHG emissions in the steel industry by 15% to 20% through energy efficiency [8]. However, China's target of peaking carbon emissions by 2030 is a difficult target to achieve with efficiency alone.

The general trend in the world, including the top 10 largest steel producers, is that steel production will decline. The share of developed countries in steel production in general is 25%, but this rate is predicted to decrease to

Table 1.1 Crude steel production by countries [34].

Ranking	Crude steel production(million metric tons), 2022	
1	China	1018
2	India	154,06
3	Japan	89,2
4	United States	80,5
5	Russia	71,5
6	South Korea	65,9
7	Turkey	35,1
8	Germany	36,8
9	Brazil	34
10	Iran	30,06

20% in the next decade. On the other hand, the increase in steel production in developing economies such as Iran, India, Malaysia, and Indonesia is noteworthy. Although China is the largest producer, it has reached saturation point in the field of infrastructure and construction and is developing in less energy-intensive sectors. In this respect, there is a decline in the steel industry compared to developing countries. China's share in global steel production is expected to decrease to 35% by 2050. On the other hand, India's share in global steel production is predicted to reach 17% in 2050 [38]. India also emphasizes the importance of scrap in steelmaking toward its goal of halving CO_2 intensity by 2030. In this sense, India is planning to 70% increase ferrous scrap by 2030 [39].

The most important emphasis on developing countries is limited by geographical conditions. According to Bararzadeh Ledari *et al.*, each country has different technological capabilities and goals in different geographical conditions [40]. Considering the cost criteria for Green Steel production, countries with fossil resources investing in blue steel production will chart a much more progressive path. On the other hand, they emphasize that green steel production is more cost-driven for countries dependent on imports for fossil resource supply. They also point out the potential for brown steel production through biomass production to be suitable for developing countries with large agricultural areas [40].

As one of the historical leaders of steel production technology, Japan is also a leader in energy efficiency and one of the pioneers of R&D expenditures. In December 2020, the government of Japan took the lead in crafting the "Green Growth Strategy Through Achieving Carbon Neutrality"

(COURSE50) by 2050 [41]. This comprehensive strategy outlined key initiatives, including the integration of the CO_2 Ultimate Reduction System for Cool Earth 50 (COURSE50) and H_2 reduction steelmaking. These initiatives were strategically placed within the "process chart" of the growth strategy for the H_2 industry, identified as one of the 14 priority fields in the overall strategy [27].

Steelmaking in the European Union plans to equip around 15% of blast furnaces with carbon capture systems by 2050. The innovative "smelting reduction/basic oxygen furnace" (SR-BOF) with CCUS dates back to the 1950s but very few projects have reached a pilot level; the route is aimed to be phased in to replace old blast furnaces from the late 2020s [42]. These routes require large-scale CO_2 transport and storage infrastructure along with industrial facilities in other sectors (e.g., cement). CCS faces acceptance challenges in many countries in the region. Germany and Austria, for example, have introduced partial or temporary bans on large-scale geological CO_2 storage in the past decade. However, in some countries, it has been identified as a strategic priority at various times (the Netherlands, the United Kingdom, and Norway are notable examples, and even the German government recently revisited CCS). Regional coordination of CO_2 transport infrastructure and the promise of possible offshore storage could ease ongoing concerns about the technology [38].

In addition, The Green Deal Industrial Plan, published by the EU in February 2023, covers not only cement but also the entire carbon-intensive sector. In this respect, it is also an important stage for ISI. In April 2023, with the approval of the implementation of the carbon border adjustment mechanism (CBAM), free emissions trading system (ETS) allowances for steelmakers will end by 2035. Some other initiatives, such as the European H_2 Bank or H_2 pipeline corridors, also support this uptake of low-carbon H_2 in steel production [5]. One of the main debates in the European Union is about whether it will be more costly to supply H_2 through production or import.

At this stage, considering the transportation and supplier risks of H_2, Lopez *et al.* suggest that the supply chain to be established with local production will be cost-competitive around Europe [43]. According to Lopez *et al.*, although H_2 import seems costly due to transportation costs, it can make a positive contribution to the development of the global south by supporting projects. They even emphasize the possibility of moving green steel production to North Africa through collaborations and technology transfer. In 2013, a low-carbon steel road map called EUROFER set 2050 targets for European Union members with 56 BOFs by 2023. In this context, the EU emphasized the importance of H-DR technology until the middle of the century.

Among EU countries, the Swedish steel industry, which has a deep-rooted technological path dependency, is also a notable case. H_2 Breakthrough Ironmaking Technology (HYBRIT) is a groundbreaking R&D program developed in Sweden. HYBRIT, as a DR initiative, aims to replace coal with H_2. HYBRIT technology has achieved ambitious results on the properties and quality of directly reduced sponge iron (H-DRI) with the produced H_2. Test results show that the direct reduction of iron ore using H_2 offers a superior product that provides ease of processing, transportation, and storage. It also shows that it almost completely eliminates CO_2 emissions in the mitigation process [44]. Karakaya *et al.*, in their research among stakeholders in three different fields (one steel producer, one electricity generation company, and one mining company), claimed that the stakeholders were very eager for the goal of developing a strong technology [33]. In their study within the framework of multi-level perspective (MLP), they concluded that incumbent regime agents strongly promote H_2-based DR technology.

Lastly, Brazil, as the ninth largest steel producer, is a unique case with the use of biomass in ISI. Yet, electricity, charcoal, and coal have the dominant share in steel production. Like other countries, short-term efficiency is one of the main goals in Brazil. Accelerating cost reductions of niche technologies is also critical for Brazil. Otherwise, investment targets in new technologies will fall behind. According to Hebeda *et al.* (2023) [45], for the 88% emission reduction target, investments in DR technology in Brazil should be implemented rapidly in 2029 and reach 21% in 2050, with the contribution of biomass technologies.

To summarize, ISI-related carbon reduction projects and potentials in both developed and developing countries are increasing significantly. The increase in steel production in developing countries makes investments and research on emission reduction in steel production of these countries more important. The ISI technology has a lock-in structure and is resistant to technological transitions. It is difficult for countries to achieve ambitious emission reduction targets if rapid steps are not taken within the framework of cost analysis and externalities that will affect the cost.

1.3.4 Petrochemical Industry

The petrochemical industry has a production process that relies on fossil fuels and is responsible for significant amounts of CO_2 emissions. The sector is dependent on fossil resources in three areas: production, use, and end of life. It is the third largest sector in industrial CO_2 emissions after cement and ISI [13]. Production in the sector is expected to quadruple by 2060, but emissions reduction targets for 2050 pose a dramatic challenge. Saygin and Gielen predicted that energy and feedstock costs will increase by 35% for full decarbonization in the sector in line with 2050 targets [13].

The primary demand for pure H_2, which is mostly used in NH_3 manufacturing and oil refining, was 72 million Mt of H_2. Additionally, 18 Mt of H_2 was combined with other gasses and used to produce methanol and direct reduction iron (DRI) steel [10]. In addition, plastic production has also increased dramatically by more than 20 times in the past 50 years and reached 390.7 million Mt in 2021 [46]. In this manner, it is crucial to increase short-term efficiency by focusing on recycling plastics. Kullmann *et al.* in this sense, emphasize defossilization by focusing on the deployment of renewable raw materials in the chemical industry [47].

It should not be ignored that the climate change implications of petroleum product processing extend beyond the industry's own borders. The usage of petroleum products has a spillover impact that spreads to many aspects of life and increases CO_2 emissions. In this context, green H_2 should quickly gain traction in the sector in accordance with CCS, circular economy, and zero-emission goals. Cost-cutting R&D studies are crucial in this cost-sensitive shift. However, Saygin and Gielen argue that the competitive market limits global cooperation. They highlight the significance of government incentives for carbon leakage. However, because a fundamental transformation is necessary, the existence of stakeholders who wish to produce cheaply in the market without risking cost rises with a "wild" competition strategy threatens the entire transition [13].

Examining the European chemical industry, Bazzanella *et al.* (2017) find that while many chemical products are carbon-based, it is inherently impossible to decarbonize them [48]. They argue that carbon should be created from renewable sources to make the chemical industry more sustainable. The European chemical sector will consume around 600 TWh of energy in 2020 alone. According to the "maximum" scenario in their study, in which they generated numerous scenarios for the transformation of the chemical industry by 2050, they project that the European chemical industry's energy demand in 2050 will be 2.5 times greater than in 2015. The conversion of methanol to olefins stands out as a critical technology for a sustainable European chemical sector. This strategy incorporates a transformation in which 200 million net consumers will be achieved by 2050 and metric tons of CO_2 will be produced through methanol, which will assist this sector in meeting carbon reduction targets. They argue that carbon should be created from renewable sources to make the chemical industry more sustainable. The European chemical sector will consume around 600 TWh of energy in 2020 alone. According to the "maximum" scenario in their study, in which they generated numerous scenarios for the transformation of the chemical industry by 2050, they project that the European chemical industry's energy demand in 2050 will be 2.5 times greater than in 2015. The conversion of methanol to olefins stands out as a critical technology for a sustainable European chemical sector. This strategy incorporates a transformation in which 200 million net consumers will be achieved by 2050 and metric tons of CO_2 will be produced through methanol, which will assist this sector in meeting carbon reduction targets [48]. Kullmann *et al.* underline the importance of national goals. In this regard, they created a national energy system model tailored exclusively for Germany. They discovered that when the chemical sector transitions from fossil to renewable fuels, H_2 demand will climb by 40%. They say that the methanol-to-olefins (MTO) pathway, which stands out for sustainability in research on alternative raw materials in the production of polymers and plastics in the petrochemical industry, could have a vital role in defossilizing the German chemical industry [47]. In addition, non-energy demand is an important CO_2 catalyst of the petrochemical sector, not as fuel but as raw material. Non-energy consumption accounts for approximately 15% of industrial CO_2 emissions worldwide. Kullmann *et al.* claim that demand for the chemical industry is expected to increase, although there has been a slight decline for three years [47, 48]. Yet, Hall *et al.* argue that although petroleum products can compete as chemical raw materials in non-energy demand, their competitiveness is low due to the additional costs of processing fossil fuels. As part of refinery production, H_2 is an input for converting crude oil into refined products and desulfurization [49].

1.4 Developments in Industrial Use of Green H_2

1.4.1 Decarbonizing the NH_3 Production by Green H_2

NH_3 has a vital role in the global food supply as it is widely used to produce agricultural fertilizers. Haber-Bosch (HB) production is the main route for NH_3 production at an industrial rate. In the industrial plants for NH_3 production, H_2 and nitrogen gasses react at high temperatures and pressures (20,000 kPa and 450°C) with the use of a catalyst to form anhydrous NH_3 [50]. Currently, the NH_3 production through the HB process is considered a "brown" route since it relies on H_2 produced from methane through the steam methane reforming (SMR) method. Thus, NH_3 synthesis is considered the most carbon-intensive process in the chemical industry since the SMR process results in approximately 90% of the CO_2. Reducing the CO_2 amount produced during the NH_3 manufacturing process, which is around 1.8% of global carbon dioxide emissions, is a highly important step that will be taken toward net-zero targets by 2050 [51, 52]. In fact, brown NH_3 production consumes intense energy (8 MWh of energy/tons of NH_3). It should be noticed that the highest portion of the energy use and nearly 90% of the carbon release are from the step of H_2 production. As we can consider the NH_3 production from the SMR process as brown, the low carbon and 100% renewable NH_3 can be named "blue" and "green" NH_3, respectively. Blue NH_3 is produced from the H_2 obtained by SMR with CCS, while green NH_3 is produced from the H_2 obtained through electrolysis of water. In this context, the production of green NH_3 is achieved when the process is 100% renewable and carbon-free. The best alternative to lower carbon emissions when producing NH_3 is to supply H_2 with low-grade carbon (blue) or carbon-free (green). Green NH_3 synthesis from renewable H_2 obtained by water electrolysis and its utilization pathways is described in Figure 1.3 [53]. Another alternative is the synthesis of hybrid-green NH_3 by modernizing the existing conventional plants employing electrolyzers to replace partly the fossil-based H_2 feedstock with a green source.

Green NH_3 production plants are constructed from renewable sources such as solar photovoltaic (PV) or wind (or a combination of wind-solar PV), water electrolyzers, air-separation unit (ASU) to collect nitrogen from the air, and the HB loop for process conditions such as high temperature

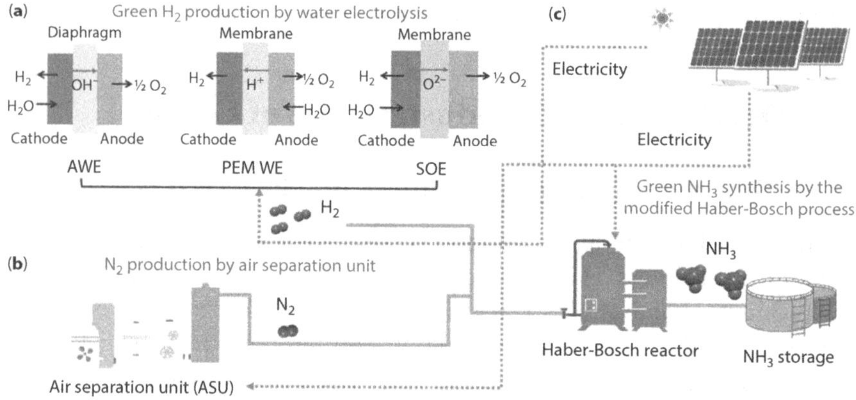

Figure 1.3 Green NH_3 synthesis and utilization pathways [53]. (a) All three types of water electrolyzers powered with renewables to split water. (b) ASU unit powered with renewables for nitrogen supply. (c) Electricity obtained from solar PV units.

and pressure [54]. The technological developments enabled a great variety of innovations for green H_2 such as photocatalytic, solar thermochemical, and photoelectrochemical (PEC) processes; biological conversion; and electrolysis [55]. However, apart from all, only water electrolysis is valid on an industrial scale [53].

As for electrolytic units, a solid oxide electrolyzer (SOEC), proton exchange membrane electrolyzer (PEM-EC), and alkaline water electrolyzer (AWE) can be used. All types of electrolyzers are built with an anode, cathode, electrolyte, power source, and sometimes a membrane. SOEC electrolyzers can operate at high temperatures (800°C and 1000°C). Recently, Topsoe and First NH_3 companies signed a launch company agreement to build a plant that allows high-temperature SOEC to produce green NH_3 at an industrial scale [56]. A 500 MW of capacity is planned to operate in the world's first commercial-scale green NH_3 plants at sites in northern Germany and the southwestern United States by 2025. AWE electrolyzer employs potassium hydroxide as an electrolyte to prevent acid corrosion but consumes more electricity than PEM-EC. Among all, PEM-EC is the most considered technology in the industrial scale of NH_3 production due to its compact design, the high purity of H_2 production, and high dynamics [57, 58]. The projects on green NH_3 synthesis using H_2 produced from water electrolysis have recently been marketed by major companies such as Fertiberia, Siemens, Air Products, OCP, and Krupp [53, 59–61]. For instance, the facility for Europe's largest plant to combine green H_2 production from electrolyzers for industrial NH_3 production is developed by two companies in Spain: Fertiberia which is the fertilizer manufacturer, and Iberdrola which is the energy organization. The plant comprises 20-MW PEM electrolyzers to produce 3000 tons of green H_2 annually for renewable NH_3 production [61]. On the scientific area side, the research on the economic feasibility of all types of electrolyzers has been going on to find the most practical and economical way for water electrolysis from renewables [62–64]. A theoretical study approved that SOEC is potentially suitable for being an electrolyzer for green NH_3 production in terms of economic viewpoint [62]. However, in another work, it was suggested that the PEM-EC–type electrolyzer can be considered when the renewable energy is fluctuating to supply better conditions for the electrolysis process [65].

1.4.1.1 Green H_2 Production and Storage for NH_3 Plants

In green NH_3 production plants, energy is supplied by solar or wind turbines to power electrolyzers. There are numerous case studies for the feasibility of either wind turbines or solar power and their hybrid power systems for the plants to be constructed all around the world [65–68]. For instance, Saudi Arabia has launched a large-scale project to obtain green NH_3 by solar and wind power, with a goal of annually producing 1 million Mt of green NH_3 by 2025 [69]. The Japanese government is planning to fuel local production and large-quantity imports of green NH_3 by fully converting all coal-fired power plants to NH_3 by 2050 [70]. As a small-scale NH_3 conversion, a wind turbine of a Minnesota farm already supplies power for a chemical plant that generates green NH_3 for both fertilizing the land and powering the tractor and heats the barns for drying grains. It was reported that the system reduces the farm's carbon footprint by 90% [71]. Australia aims to become a pioneering green NH_3 exporter with a large-budget project involving integrating a 900-MW solar project with existing hydropower and port facilities and electrolyzers to achieve renewable H_2 and N NH_3 production.

Liquid H_2, compressed H_2, and cry-compressed H_2 are the most preferred H_2 storage technologies [72]. For large-scale industrial applications, it is more convenient to use compressed H_2 storage technologies, even though it can cause a loss of energy [73]. Apart from industrial requirements for NH_3, it has a great potential for being an energy carrier as it offers H_2 gas upon decomposition [74, 75]. Besides, NH_3 storage consumes less energy as compared with cryogenic conditions of liquid H_2 storage and it is less flammable. NH_3 formed by the HB process can be stored in a single tank on the plant for large volumes, underground in drilled shafts or salt caverns. This would significantly reduce the cost of concentrated solar power with storage. However, the decomposition reaction of NH_3 could reach equilibrium conversions of 99.7% at 500°C [76]. The produced H_2 contains unreacted NH_3 impurity and requires an additional purification step which affects the overall cost of the process. Currently, the technological growth in NH_3 as an H_2 carrier is less mature than in NH_3 production technologies. In addition, most scientific reports on NH_3 decomposition to H_2 utilized high-cost Ru-based catalysts. Therefore, it is most urgent to decrease the catalyst cost by replacing it with more abundant Fe-, Co-, or Ni-based catalysts to enable the industrial process [77]. Table 1.2 displays the total energy input requirements for five scenarios of H_2 synthesis, storage, transportation, and NH_3 conversion/cracking to store renewable electricity [75].

1.4.1.2 Designing Catalysts for NH_3 Synthesis

Nitrogen is a decidedly inert gas composed of strictly bound σ and π electrons that must be weakened for an effective conversion from nitrogen

Table 1.2 The energy investment for storage routes [75].

	Situation 1	*Situation 2*	*Situation 3*	*Situation 4*	*Situation 5*
Production 1	Electrolysis	Electrolysis	Electrolysis	Electrolysis	Electrolysis
Production 2		HB process	HB process		
Storage and transport	Compression	Liquefaction (7 days)	Liquefaction (182 days)	Liquefaction (7 days)	Liquefaction (182 days)
Conversion		Cracking	Cracking		
MWh required pert-H_2	39.7–52.7	63.7–78.3	65.1–79.75	50.5–63.5	73.5–86.5
Overall return on energy investment	63–84%	42.5– 52.3%	41.8–51%	52.5–66%	38.5%–45%

to NH_3. The kinetics of the reaction are sluggish and require high energy to break N≡N triple bonds. Therefore, the HB process is highly energy-intensive (30 GJ per ton of NH_3) due to the requirements of high temperatures and pressures [78]. Consequently, the necessity of high-pressure compressors and reactors results in substantial capital costs in addition to the high-cost energy needed to achieve synthesis conditions.

A proper catalyst is required for NH_3 production through the HB process. The industrial HB process utilizes a magnetite iron (Fe_3O_4) catalyst with the addition of minor quantities of alkali and alkaline earth oxides (K_2O, CaO, MgO, Al_2O_3, and SiO_2). These oxides reflect as promoters to perfect the catalytic action by stabilizing activated Fe and enhancing the surface area [79]. In addition, a combination of theoretical and experimental investigation carried out by Huo *et al.* proved that the reaction equilibrium favors the NH_3 product in the presence of a K_2O promoter by stabilizing high-index and more active Fe facets [80]. In the 1970s, British Petroleum (BP) and Kellogg commercialized a graphitized carbon–supported Ru catalyst with Cs_2O and BaO promoters for NH_3 forming *via* the Kellogg Advanced NH_3 Process (KAAP) [81]. In the following years, apart from the Fe catalyst, the ruthenium-based (Ru) catalyst was deeply investigated with alkali promotes [82]. Thus, Cs_2O-promoted and MgO-supported Ru is considered as a standard catalyst in the current analyses. In recent years, encouraging studies on heterogeneous catalysts for NH_3 synthesis under milder conditions have been presented to overcome the sluggish kinetics of nitrogen conversion [83, 84]. The scientific studies showed that higher

NH_3 production rates were observed with lanthanide oxides (Sm_2O_3) and zeolites-supported Ru catalysts as compared to calcium or alumina oxide supports [85, 86]. Consequently, an appropriate catalyst having superior selectivity, activity, and minimal consumption can ease the harsh reaction requirements and thus reduce the overall cost. Therefore, laboratory-scale experimental findings are very important to link an industrial-scale catalyst. However, the activity is not the major point in selecting a catalyst. Because poison tolerance in case natural gas is the widespread H_2 source, handling regeneration and, most importantly, the catalyst cost are the major concerns for the industrial HB process.

1.4.2 Green H_2 in Cement Industry

The concrete industry is well known for its substantial environmental impact on Earth since billions of tons of concrete are produced globally each year, requiring massive amounts of materials. The vast amount of CO_2 emissions is caused by cement manufacturing, specifically Portland cement. The environmental issues along with the energy needs, water consumption, and the accumulation of building and demolition trash contribute to the widespread notion that concrete is not particularly environmentally friendly or consistent with the ideals of sustainable development [87].

China is the foremost cement producer in the world with a contribution of 51.17% of global cement output in 2022. India (9.23%) comes as the second important manufacturer, followed by Vietnam (2.95%), the United States (2.28%), Turkey (1.75%), Iran (1.56%), Indonesia (1.54%), Brazil (1.51%), Russian (1.46%), Saudi Arabia (1.26%), Egypt (1.25%), Korea (1.25%), Mexico (1.20%), and Japan (1.16%) [88, 89]. Portland cement is a critical and dominant material in the construction sector. It is made of burning limestone (roughly 80%) and clay (20%) by grinding clinker with around 5% gypsum (calcium sulfate). Cement clinker is fabricated by calcination of limestone (calcium source), clay, or sand (silica and alumina source) at around 1450°C to form CaO and CO_2 [90]. Nearly, 50% of CO_2 gas releases are directly overcome by this calcination, and 40% of emissions emerge from fuel combustion in rotary kilns. Additionally, significant indirect emissions (10%) come from quarrying and product hauling [91].

$$3CaCO_3+SiO_2 \rightarrow \Delta+Ca_3SiO_5+3CO_2$$

Clinker is crushed to form a fine powder with water addition to make Portland cement. Calcium silicate hydrate hydraulic types of cement include Portland cement. Hydraulic cement is mostly calcium compounds that hydrate quickly [92, 93]. The prominent emission of carbon in cement formation arises from the two sources of anthropogenic emissions of CO_2 which are decomposition of carbonate and fossil fuel oxidation. The creation of the main elements of Portland cement causes a large quantity of CO_2 released during the calcination [94, 95]. The energy necessary to acquire sintering temperatures (about 1450°C) and overall procedures entail the combustion of considerable quantities of fossil fuels directly (generation of thermal energy) and indirectly (through the production of electricity). The emissions originating from the thermal process to obtain the clinker add a further 60 to 90% of CO_2 to the progression emissions [95, 96].

According to the net zero tracker of the World Economic Forum, producing cement is responsible for 6% of the world's CO_2 release. Emissions only dropped 0.3% and 0.58 tons of CO_2 was released per ton of cement. The fuel mix is 92% fossil fuel, and the sector is expected to grow 1.5 times by 2050. Low and reduced emission productions account for less than 1% each respectively. The global cement formation annually is forecasted to rise from around 2540 Mt in 2006 to between 3680 Mt and 4380 Mt by 2050 [97]. This noteworthy rise in cement fabrication coincides with an increase in the cement industry's overall energy consumption and CO_2 emissions. In contrast to most other industries, the cement industry does not rely heavily on fuel usage to generate CO_2 emissions. Calcination is responsible for more than half of the CO_2 produced during cement manufacturing, or around 540 kg CO_2 per ton of clinker [98]. The remaining CO_2 from cement manufacture comes from burning fuel for calcination. Kilns heat to 1400°C to 1500°C for calcination. Energy accounts for 30% to 50% of cement production costs. Each ton of cement uses 100 to 110 kWh of power. Electricity use accounts for 5% of cement industry CO_2 emissions. This figure could be less than 1% or more than 10%, depending on the energy source and its efficiency in the local electrical mix [99]. The "Intergovernmental Panel on Climate Change's Fifth Assessment Report" established that CO_2 emissions by industrial applications were mainly initiated by the manufacturing of materials. Cement fabrication accounts for approximately 12% to 15% of total industrial energy consumption in terms of the technique of energy and resource [100–103]. In addition, the cement sector generated 136 tons of NO_x, 4833 tons of SO_2, 183 tons of hazardous dioxins and furans, and 320 kg of mercury in 2010. Then again, 1 ton of Portland cement creates about 1 ton of greenhouse gas, and each ton of

Portland cement clinker emits 1.5 to 10 kg of NO_x which is a potent greenhouse gas formed by the combustion of fossil fuel and biomass, as well as the production of nitric acid [104, 105]. Dust pollution from cement manufacture reduces vision and air quality. As asserted by "the Centers for Disease Control and Prevention", drained dust can pollute water and harm humans and animals. Wastewater discharge into the atmosphere pollutes rivers and groundwater. In expanding populations, construction, urbanization, and land use degrade soil [106, 107].

The problems with cement production and the waste from construction and demolition can be solved by cement recycling. This approach is in line with the circular economy idea put out by the EU since it uses closed-loop recycling. If Europe wants to meet the green deal criteria, it may be necessary to use recycled cement (RC) in place of Portland clinker, either totally or partially [108–112]. Avoiding the use of virgin materials and the disposal of debris from building and demolition projects are not the only advantages of recycling cement. Between half and three-quarters of the CO_2 released from contemporary kilns originates from the calcination reaction [113, 114]. The theoretically calculated minimum energy involved for the calcination is assessed to be about 1760 MJ/ton clinker, although the best-performing plants utilize around 2800 MJ/ton of clinker [115, 116]. The considerable disparity in energy consumption (more than 40%) is due to heat losses that occur throughout the process, with a big portion of this loss occurring during the drying of the raw meal (usually 900 MJ/ton). It is feasible to restore the hydration characteristics of recycled hydrated cement and produce cement without emitting CO_2 during calcination or subjecting the material to high sintering temperatures [95, 117].

Green concrete may be made in several methods, one of which is by using types of cement made of calcium aluminate (CA) or calcium sulfoaluminate (CSA). CA developed in the early 20th century; monocalcium aluminate and trace amounts of other compounds make up this cement type [118]. Concrete used with this cement instead of ordinary Portland cement (OPC) is considered "green" as it reduces carbon emissions. In its original form, CA cement served just for concrete protection from sulfate attack. Nonetheless, in places where sulfate resistance is not a problem, it has been utilized for buildings because of its sustainable benefit. In addition to being more resistant to abrasion and the alkali-silica reaction, certain cement types also offer other durability benefits. Plus, they are great for specific kinds of building because of how quickly their strength increases. Although OPC may help achieve green concrete, its restricted utilization is due to certain other cement kinds being more costly. It is worth noting that fly ash has been evaluated as a binder capable of increasing the compressive strength and hardness of cement composites [119].

The use of numerous solid wastes as aggregate in concrete forming produces green concrete by eliminating the need to extract natural aggregate process. In addition, employing these solid wastes as aggregate in concrete will help to conserve natural aggregate sources, manage trash, and reduce the overall embodied energy and carbon of concrete. Several solid wastes have been identified as aggregate in concrete. Plastic [120], construction and demolition wastes [121], glass [122], agricultural wastes [123], slag, and ceramics [124] are some of these wastes. Nevertheless, contingent on the replacement level and physical qualities of the aggregate, the final effect of each form of waste utilized as aggregate varies. Glass, for example, has been shown to improve freeze-thaw resistance and surface scaling of concrete when used as a partial replacement of fine aggregate in the range of 7.5% to 25% [125]. The use of glass as an aggregate at higher levels has been observed to result in a reduction in slump and mechanical properties [126].

It is valuable to conduct concrete production in an energy-efficient and environmentally friendly technology [127, 128]. Polyethylene, polypropylene, and other polymer matrices have been investigated for green composites. Polyethylene is mostly used as aggregate, while polypropylene is mostly used as concrete reinforcement. In a study reported by Rumšys *et al.*, polyethylene had a minor effect on the workability and air permeability of concrete [129]. The study also indicated that increasing polyethylene lowered concrete strength and shrinkage. Other investigations have found polyethylene and polypropylene to be suitable for concrete manufacture [130–132]. Green concrete is made by using solid wastes as aggregate instead of mining and treating natural aggregates. These solid wastes can also be used as concrete aggregate to conserve natural aggregate sources, manage trash, and reduce concrete's embodied energy and carbon. Concrete aggregates include various solid wastes [133].

Because the cement business is material-, energy,- and carbon emission–heavy, increasing studies have been directed worldwide to reduce greenhouse gas emissions from cement producers. We are currently experiencing a global energy crisis, which should meet faster action for clean energy transitions to meet the target of net zero emissions by 2050 [134, 135]. A study has integrated a cement manufacturing system that combines the dehydrogenation of NH_3 and the combustion of H_2 to produce the thermal energy required for clinker formation. Several studies on cement manufacture utilizing alternative fuels and process modeling for reducing

CO_2 emissions from cement plants have been done [136, 137]. Previous research has primarily concentrated on energy efficiency improvements, CCS (carbon capture and storage) technology, biomass, and waste material utilization using primarily Aspen plus–based simulation, and optimization for environmental performance improvement in the cement industry [138]. Overall, there is a need to optimize the architecture of the rotary kiln process by the use of green H_2 to decarbonize the cement manufacturing process [139, 140].

Lafarge Holcim indicated in its 2022 climate report that it is evaluating H_2 for two key applications: replacing fossil fuels and supporting CCS (using captured CO_2). The company is now working on two H_2 projects: Hyscale100 and Go4Zero, in which it is upgrading its kilns to permit oxy-fuel combustion, allowing for more efficient CO_2 absorption and the production of synthetic fuel through H_2 reactions. Heidelberg Materials stated its intention to decarbonize its cement sites in its 2021 sustainability report by increasing the use of biomass fuels, exploring the usage of H_2, and boosting the electrification of its operations. Hanson UK, its UK-based business, has conducted H_2 testing at two of its three locations. The Ribblesdale site uses a net zero fuel mix as part of a world-first H_2 technology demonstration, while the Padeswood cement site will be part of the UK's Track 1 Industry Cluster HyNet, where CO_2 emissions will be absorbed and stored. CEMEX indicated in its 2021 integrated report that it had set yearly CO_2 roadmaps to reach its 2030 sustainability commitments. These roadmaps involve growing the use of alternate fuels and clean energy, as well as leveraging energy efficiency enhancers such as H_2. CEMEX, in collaboration with HiiRoc, has developed a new H_2 injection method to accelerate the deployment of H_2 in all its cement plants [141].

The utilization of H_2, besides alternative fuels, can influence the decrease in the fuel-related CO_2 emissions of the cement plants. By using the H_2 as a fuel, only water is out. Another benefit of H_2 in cement production is its high efficiency compared to other industries, as no process-related compression of the gas is required. From this perspective, it is very important if the H_2 comes from carbon-neutral sources. Currently, H_2 utilization is still insignificant in the cement industry. In 2022, the largest known kiln test was performed within the Fuel Switch project (BEIS, Mineral Products Association) in the United Kingdom, where a kiln was operated with a net-carbon-zero fuel mix at the primary burner [142, 143].

The fuel mix employed contained approximately 40% H_2 in the thermal energy used in the kiln primary burner. The short-term test lasting a few hours revealed that the practical execution of H_2 utilization in kiln firing is feasible. Because H_2 has superior ignition qualities and a higher calorific value, it draws the flame closer to the burner tip and raises the temperature of the core flame. Furthermore, an H_2 flame emits far less radiation than a solid fuel flame, which should reduce heat transfer to the clinker bed. However, by increasing the H_2 injection velocity, the flame position shift and temperature increase can be avoided. The total thermal energy consumption appears to be unaltered; however, the oxygen demand falls as the H_2 replacement rate increases. When compared to coal, H_2 requires less air to burn, resulting in a higher oxygen level at the kiln inlet. In the Global Cement and Concrete Association (GCCA) roadmap 2050 future scenario, an average contribution of 10% H_2 of total thermal energy consumption of the cement sector can be assumed, and fuel-based CO_2 emissions will drop by roughly 10%. Long-term tests are still needed to investigate the effect of H_2 on the clinker-burning process as well as long-term equipment corrosion over time. H_2 can be delivered to a cement mill in tiny quantities by truck or train. If bigger volumes are necessary, a pipeline or on-site electrolysis is required. H_2 is created *via* water electrolysis in this assessment. With an electrolyzer efficiency of 75%, the electrolysis consumes 44 MWh of electric energy to produce 1 ton of H_2. Its manufacture, particularly the power consumed, must be fully carbon-neutral. H_2 has a higher mass-based calorific value than oil or coke (120 MJ/kg), but a far lower volume-based calorific value (10.8 MJ/m^3). Because H_2 is transported in gaseous or liquid form, extremely high pressure (500 to 1000 bars) must be provided to allow H_2 to be transported in trucks or railway trains. Because H_2 is injected into the cement kiln burner at near atmospheric pressure, the cement plant must install pressure reduction valves. When H_2 is created on-site using water electrolysis, just a tiny amount of depressurization is required. This minimizes overall power usage. Currently, most cement plants do not consider H_2 to be a feasible choice due to a lack of access and/or affordability, although green H_2 prices are predicted to settle at 1.5 €/kg (12.5 €/GJ) in 2050 [144, 145]. Green H_2 produced locally at the facility using an electrolyzer will indirectly electrify clinker production by giving heat energy *via* electricity. Future electrolyzer performance should improve. New capillary electrolysis technologies may be more efficient. The facility can employ O_2 from a neighboring H_2 electrolyzer to enhance the flame for clinker formation or upgrade alternate fuel combustion in the main burner [145, 146]. Industrial decarbonization is a major energy transition challenge, especially for cement manufacturing. Combined efforts of both local and international stakeholders are required to solve these challenges to achieve sustainable, resilient, and decarbonized cement production.

1.4.3 Green H_2 in Iron and Steel Production

H_2 steel manufacture is commonly known as "carbonless steel." This procedure uses H_2 in the iron ore reduction reaction. In this example, the reaction produces water vapor as a byproduct, which greatly reduces greenhouse gas emissions. H_2 is considered the "energy source of the future," as well as the most environmentally friendly and appealing. This is critical to the development of H_2 direct reduction technology and the steel industry's energy transformation. This section describes his current H_2 production technology and green H_2 production development in the steel sector [147]. The production of H_2-based green steel, which started with the European steel industry launched the ULCOS Project in 2004, continues today. it continues with the agreement on green steel production with the giant BWM company in 2022 [40, 148].

At present, H_2 iron production expertise generally comprises the following:

- H_2 plasma reduction (HPSR)
- H_2 new flash
- Iron production technology (FIT)
- H_2 BF iron production (HBFI)
- Axial furnace direct H_2 reduction (H-DR) technologies [40]

Aside from the benefits of employing H_2 in steel manufacturing, research has been undertaken into some of the drawbacks. Özgün *et al.* performed EDS analysis and H_2 measurements on steels manufactured using typical metallurgical processes, direct reduction with H_2, and H_2 plasma melting procedures [149]. When the results of the tests were evaluated, it was established that the amount of H_2 in the steel was less than in the steels manufactured using H_2 [150].

Steel is the most significant material in industries such as automotive and construction.

While the global steel sector consumes 5% of worldwide energy expenditure, it produces 7% of global greenhouse gas emissions. A ton of crude steel produced globally emits around 1.9 tons of CO_2 [151, 152]. On average, each ton of raw steel produced globally emits 1.9 tons of CO_2. We plan to reduce emissions by 2030 to maintain the global economy. The solution to this is to reduce fossil fuel production by around 6% [153, 154]. Marocco *et al.* created a technique in which they could make H_2 using natural gas from the grid, as greenhouse gas emissions are one of the most significant repercussions of the steel sector. According to the modeling results, the H_2-powered system is less expensive and more efficient in terms of greenhouse gas emissions than standard steel production processes [155]. Another study by Wang *et al.* investigated the economics of steel manufacture using various energy sources such as wind, sun, and H_2 (Figure 1.4). According to the data, wind energy is 20% more expensive than solar energy, and the cost of steel production is significantly lower with solar energy plus H_2 [156]. Steel is made of low-carbon iron. In order for iron to become steel, Fe_2O_3 or Fe_3O_4 must be reduced in the system [157, 158].

Bhaskar *et al.* investigated the greenhouse gas emissions and costs of H_2 directly reduced steel materials using Monte Carlo simulation. According to the results, the steel material created with H_2 is both more inexpensive and more ecologically friendly than conventional production methods [36]. The most common ways of steel manufacture are (i) integrated BF-BOF, (ii) electric arc furnaces (EAF), and (iii) direct reduced iron (DRI). All three production processes described above require different raw materials

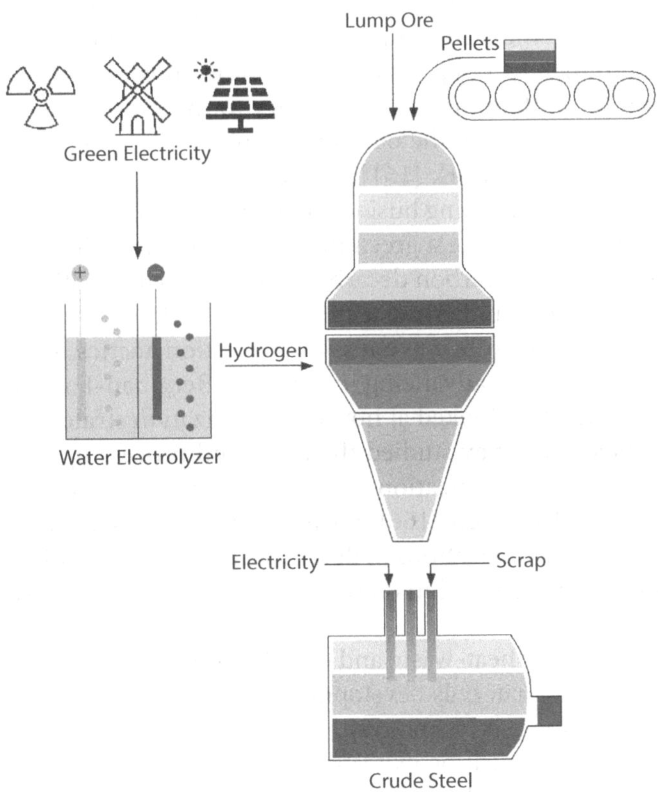

Figure 1.4 Steel production method with H_2 [156].

and energy sources and emit varied amounts of greenhouse gasses. At 2000°C, BF interacts with a reducing agent such as coke to generate pig iron. Electric arc furnaces may also recycle steel at high temperatures. The DRI process is like the BF method, except steel is made by heating iron ore pellets to 1000°C. H_2 serves as a reductant in the DRI process. This significantly cuts greenhouse gas emissions [157, 159].

1.4.4 Green H_2 in Oil Refining

The oil refining process, as the world's largest industrial application, produces raw materials and fuel for the transportation, chemical, and heating industries. Because of rapid population expansion and urbanization, the need for energy resources has increased. Today, fossil fuels represent 84% of the world's energy needs, with oil accounting for 33%. Oil is a nonrenewable resource that releases poisonous gasses. In recent years, worldwide organizations have focused on reducing dangerous emissions and researching ways to improve refining technology. However, the research focuses on direct emission reduction, and the connections between the petroleum products trading network and other business sites are overlooked [160–165]. Therefore, it is necessary to monitor the direction and volume of air pollutants in the oil refinery and to comprehensively examine the commercial network [161]. There have been recent decarbonization studies in the oil refining business that are based on sustainability and dependability concepts. For a green future, it is vital to reduce greenhouse gas emissions, increase carbon decarbonization activities with H_2 from the oil extraction process, and create technologies [165].

Recent research studies on emission reduction tactics and technical solutions are critical for advancing integration. Bergman-Fonte *et al.* published a case study that looked at how decarbonization strategies affect the oil refining business. They studied the reuse and processing of renewable feeds within the decarbonization targets of oil refineries and found that carbon risks are decreased [166]. Griffiths *et al.* assessed the economic, social, and political implications of decarbonizing the oil refining sector by categorizing carbon removal in oil refining into six categories [163, 167].

- Recovery for heat-waste and over-bottom pressure
- Innovative materials developments
- New equipment technology
- Process optimization expertise
- Optimization technology by intelligent system scheduling
- Energy-saving technology for circulating water system

Bergero *et al.* selected a Canadian example for the decarbonization of the oil sector, and a global change analysis model (GCAM) was developed and examined. In this context, the topic is addressed with three representative low-carbon extraction and global direct capture (DAC) for the oil sands system. As a result of the study, it was emphasized that the application of reduced carbon extraction technologies in oil sands will drop the amount of emissions and help to keep significant production amounts [168]. Jia *et al.* investigated carbon emissions reduction during the conversion of paraffin-based petroleum derivatives, using Fushun Petrochemical as an example. A modeling framework was developed to investigate the energy-CO_2 emission relationship at the process level. This modeling begins by presenting the likelihood that may be applied to representative and similar investigations. Second, an accounting approach for energy consumption and CO_2 release in paraffin-based oil refining is described. Third, based on the results, energy-saving and emission-lessening methods as well as strategic imputation of paraffin-based oil refining industries are proposed [169].

Carbon reduction methods and technology are required to mitigate the climatic and global impact caused by petrochemical refineries. In the oil refining business, H_2 is primarily utilized to break the long carbon chain of crude oil and reduce the sulfur level of diesel. The recent surge in petitions for diesel fuel has prompted an increase in demand for H_2. Green H_2 can play a critical role in meeting global carbon reduction targets and is a crucial actor in carbon reduction in the oil refining industry. Incidentally, there are various H_2 decarbonization solutions for a clean future with zero emissions. Choosing sustainable and safe technology has a crucial role [165, 170]. Carbon-reduction measures are unlikely to be long-term solutions unless they achieve zero or low carbon emissions. There are numerous H_2 production systems, including steam reforming, gasification, partial oxidation, methane pyrolysis, autothermal reforming, fermentation, electrolysis, biomass-derived liquid reforming, microbial electrolysis cells, photoelectrochemical, and photobiological. These methods can be divided into four categories on the basis of the leading energy source used to transform material resources to H_2. These are electrochemical, photochemical, thermochemical, and biological. Alternative waste sources, such as oil and gas, provide raw materials for H_2 synthesis [167].

Fetanat *et al.* discussed an appropriate technology for decarbonization in the oil and gas segments of Iran regarding sustainability and steadfastness. The selection of the most suitable technology was examined based on H_2 energy. As a result of the study, it was revealed that the green H_2 alternative obtained with renewable energy (solar) is the most appropriate method for decarbonizing oil refineries in Abadan, Iran [165]. Moradpoor *et al.* studied the influence of several methods of supplying wind energy to electrolysis, considering alternate supply pathways. This study used the Finnish example, and a techno-economic analysis of green power production was undertaken. Green H_2 analysis has been researched using various electric sensors and electricity supply systems. In Finland, how to generate energy from wind to replace natural gas H_2 in the oil refining business was discussed. In this study, many wind energy scenarios were developed. Furthermore, the green H_2 cost and its sensitivity to the influencing causes were investigated. Large-scale electrolysis is a rapidly evolving field and the necessity for a comprehensive examination of analyses as new facts on these technologies is underlined [170].

Griffiths and others investigated all industries and industrial processes that promote H_2 decarbonization. This study analyzed the technological, political, social, and economic factors that will influence H_2 acceptability from a sociotechnical standpoint. Examining it at a social and technological conceptual level covers not only the materials and energy utilized for H_2 production but also the full manufacturing technology, intermediate goods, and end-use applications [167]. Nikhil *et al.* studied green energy as a fuel for the decarbonization of the industrial sector of the southeastern region of India. This study conducted research on green H_2 to meet energy demands by 2050. The study emphasized the importance of H_2 production based on electrolysis and the enhanced need for these supplies. The research examined solar power as the fundamental source to reach green H_2 and envisioned the installation of 400 W of solar panels by 2050. The study concluded that stored H_2 can be used safely for 7 years. It is also predicted that the worldwide temperature rise can be reduced by ~0.0038°C as a result of removing carbon from the industrial sector in the southeastern

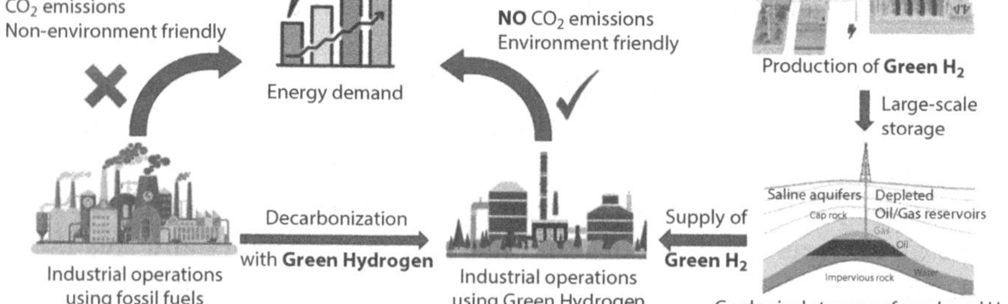

Figure 1.5 Decarbonization of oil refining process [171].

region of India (Figure 1.5) [171]. Pivetta *et al.* investigated solutions for decarbonizing energy derived from renewable and fossil fuels (diesel) used in the port system. Because more than 90% of the industrial port area (IPA) is located near the city center, environmental and health concerns have been raised, emphasizing the significance of decarbonization. In this study, IPA discovered in northern Italy was investigated. The most cost-effective energy conversion and storage solutions have been investigated to reduce carbon emissions [172].

1.5 Conclusions

Today, the main economies are focusing on green H_2, along with international energy partnerships to guarantee long-term imports. The new opportunities for industrial development come forward with the increasing demand, especially for countries that are rich in RESs. By investing in renewable power projects and electrolyzers, energy-intensive steel or base chemicals investments, NH_3 production, and the cement industry would be integrated with green H_2 replacing fossil fuels. Henceforth, many significant corporations have revealed their plans to stop carbon footprint soon. However, there is still a requirement for comprehensive analyses for upscaling of renewable potential to meet the electricity demand on a global scale. A significant amount of renewable energy will be necessary to produce green H_2 to replace fossil fuels, as well as clean electricity production to be used for powering industrial plants including road transport, heating, and cooling. Green H_2 is still at high expensive, as compared to H_2 produced from fossil fuels. However, we can expect to have cheaper green H_2 soon, as the average cost of renewable power is rapidly decreasing at a global scale with the help of technological innovations and efficiency improvements to reduce the cost of electrolyzers. The major renewable pathways in which green H_2 can induce industrial advancement are solar PV plants and wind farms, geothermal and hydropower projects, and the use of biomass. These renewable power sources are drawing massive investments for the conversion into green H_2 in electrolyzers. The produced green H_2 can be simply kept and promptly used in some industrial developments. In addition, it is easier to store and transport in the form of privileged energy-density compounds (such as methanol or NH_3) after chemical conversion. Consequently, it is expected that green H_2 will become cost-competitive by the mid-2030s by estimations of IRENA.

References

1. Sadik-Zada, E.R., Political Economy of Green H2 Rollout: A Global Perspective. *Sustainability*, 13, 13464, 2021, https://doi.org/10.3390/su132313464.
2. van Renssen, S., The H2 solution? *Nat. Clim. Change*, 10, 799–801, 2020.
3. IEA, *The Future of Hydrogen*, IEA, Paris, 2019, https://www.iea.org/reports/the-future-of-hydrogen, Licence: CC BY 4.0.
4. WEC (World Energy Council), *Working Paper: H2 on the Horizon: National H2 Strategies*, World Energy Council, London, 2021.
5. IEA, *Global Hydrogen Review 2023*, IEA, Paris, 2023, https://www.iea.org/reports/global-hydrogen-review-2023, Licence: CC BY 4.0.
6. Bloomberg NEF (BNEF), *Hydrogen economy outlook key messages, Bloomberg New Energy Finance*, March 2020, https://www.nwhydrogenalliance.co.uk/news-stories/report-hydrogen-economy-outlook-bloombergnef/#:~:text=Hydrogen%20Economy%20Outlook%2C%20a%20new,industry%20-%20at%20a%20manageable%20cost. Accessed 15 Jan 2024.
7. Scott, M. and Powells, G., Towards a new social science research agenda for H2 transitions: Social practices, energy justice, and place attachment. *Energy Res. Soc. Sci.*, 61, 101346, 2020, https://doi.org/10.1016/j.erss.2019.101346.
8. Pflugmann, F. and De Blasio, N., The geopolitics of renewable hydrogen in low-carbon energy markets. *Geopolit., Hist. Int. Relat.*, 12, 1, 9–44, 1 Jan 2020.
9. M.E.T.I., *Outline of the 6th Strategic Energy Plan*, Minist. Econ. Trade Ind, Tokyo, Japan, 2021, https://policy.asiapacificenergy.org. Accessed 24 Jan 2024.
10. IEA, *Global H2 Review 2023*, OECD, Paris, 2023.
11. H2Council, Toward a new era of H2 energy: Suiso Frontier built by Japan's Kawasaki Heavy Industries, Brussels, Belgium, 2022, https://H2council.com. Accessed 24 Jan 2024.
12. Kan, S., South Korea's H2 Strategy and Industrial Perspectives, 2018, https://www.ifri.org/en/publications/editoriaux-de-lifri/edito-energie/south-koreas-hydrogen-strategy-and-industrial. Accessed 4 May 2024.
13. Saygin, D. and Gielen, D., Zero-emission pathway for the global chemical and petrochemical sector. *Energies*, 14, 3772, 2021, https://doi.org/10.3390/en14133772.
14. Schiavo, M. and Nietvel, K., How H2 Can Fuel The Energy Transition, S&P Global Ratings, 2020, https://www.spglobal.com/ratings/en/research/articles/201119-how-H2-can-fuel-the-energy-transition-11740867. Accessed 24 Jan 2024.
15. Noussan, M., Raimondi, P.P., Scita, R., Hafner, M., The role of green and blue H2 in the energy transition—a technological and geopolitical perspective. *Sustain.*, 13, 1–26, 2021.

16. Bartlett, J. and Krupnick, A., Decarbonized hydrogen in the US power and industrial sectors: Identifying and incentivizing opportunities to lower emissions. *Resources for the Future (RFF)*, 20–25, 2020.
17. Seznec, J.F. and Mosis, S., The ACWA Power–Air Products joint venture for green H2: A new Saudi energy policy? *Atlantic Council*, 2020. https://www.atlanticcouncil.org/blogs/energysource/the-acwa-power-air-products-joint-venture-for-green-H2-a-new-saudi-energy-policy/. Accessed 24 Jan 2024.
18. Egerer, J., Farhang-Damghani, N., Grimm, V., Runge, P., The Industry Transformation from Fossil Fuels to H2 will reorganize Value Chains: Big Picture and Case Studies for Germany. *SSRN Electron. J.*, 358, 122485, 2023, https://doi.org/10.2139/ssrn.4390325.
19. Aziz, M., Wijayanta, A.T., Nandiyanto, A.B.D., Ammonia as Effective H2 Storage: A Review on Production, Storage and Utilization. *Energies*, 13, 3062, 2020, https://doi.org/10.3390/en13123062.
20. International Energy Agency, Global H2 Review. *Glob. Hydrog. Rev.*, 2021, 223, 2021.
21. Reuters, *Japan aims to boost Ammonia fuel demand to 3 mln tonnes a year by 2030*, Reuters, Tokyo, Japan, 2021, reuters.com, https://www.reuters.com. Accessed 24 Jan 2024.
22. Kajaste, R. and Hurme, M., Cement industry greenhouse gas emissions – management options and abatement cost. *J. Clean. Prod.*, 112, 4041–4052, 2016. https://doi.org/10.1016/j.jclepro.2015.07.055.
23. Statisca, *Production volume of cement worldwide from 1995 to 2022*, Statisca Res. Dep., New York, NY, USA, 2023, https://www.statista.com. Accessed 24 Jan 2024.
24. Farfan, J., Fasihi, M., Breyer, C., Trends in the global cement industry and opportunities for long-term sustainable CCU potential for Power-to-X. *J. Clean. Prod.*, 217, 821–835, 2019, https://doi.org/10.1016/j.jclepro.2019.01.226.
25. Nhuchhen, D.R., Sit, S.P., Layzell, D.B., Decarbonization of cement production in a H2 economy. *Appl. Energy*, 317, 119180, 2022, https://doi.org/10.1016/j.apenergy.2022.119180.
26. Lord, M., Zero Carbon Industry Plan: Rethinking Cement, Beyond Zero Emissions Incorporated, East Melbourne, VIC, Australia, 2017.
27. The Government of Japan, Carbon-Negative Concrete: A Game Changer for a Sustainable Future, 2022, https://www.japan.go.jp/kizuna/2022/08/carbon-negative_concrete.html. Accessed 24 Jan 2024.
28. Cement Production by Country 2024, World Popul. Rev., Lancester, Pennsylvania, USA, 2024.
29. Chen, P., Zhang, S., Meng, J. *et al.*, Technological Solutions to China's Carbon Neutrality in the Steel and Cement Sectors. *Earth's Future*, 11, e2022EF003255, 2023, https://doi.org/10.1029/2022EF003255.

30. Global Times, China issues regulations for construction-material industry, targeting sector's emissions peak by 2030, 2022 , https://www.globaltimes.cn/page/202211/1278980.shtml.
31. The US GSA, Low Embodied Carbon Concrete Requirements, 2022, www.gsa.gov. https://www.gsa.gov/. Accessed 24 Jan 2024.
32. Zaiter, I., Ramadan, M., Bouabid, A. *et al.*, Potential utilization of H2 in the UAE's industrial sector. *Energy*, 280, 128108, 2023, https://doi.org/10.1016/j.energy.2023.128108.
33. Karakaya, E., Nuur, C., Assbring, L., Potential transitions in the iron and steel industry in Sweden: Towards a H2-based future? *J. Clean. Prod.*, 195, 651–663, 2018, https://doi.org/10.1016/j.jclepro.2018.05.142.
34. Worldsteel, Climate change and the production of iron and steel: Transforming steel production, Brussels, Belgium, 2021.
35. Otto, A., Robinius, M., Grube, T. *et al.*, Power-to-Steel: Reducing CO2 through the Integration of Renewable Energy and H2 into the German Steel Industry. *Energies*, 10, 451, 2017, https://doi.org/10.3390/en10040451.
36. Bhaskar, A., Assadi, M., Somehsaraei, H.N., Decarbonization of the iron and steel industry with direct reduction of iron ore with green H2. *Energies*, 13, 1–23, 2020, https://doi.org/10.3390/en13030758.
37. Laguna, J.C., Meinke-Hubeny, F., Duerinck, J. *et al.*, Carbon-free steel production: cost reduction options and usage of existing gas infrastructure, European Union, Brussels, 2021.
38. IEA, Iron and Steel Technology Roadmap, 2020, iea.org., https://www.iea.org/reports/iron-and-steel-technology-roadmap. Accessed 18 Jan 2024.
39. SteelMint, India's ferrous scrap generation seen rising 70% to 45 mnt by 2030, 2023, steelmint.com, https://www.steelmint.com/insights. Accessed 24 Jan 2024.
40. Bararzadeh Ledari, M., Khajehpour, H., Akbarnavasi, H., Edalati, S., Greening steel industry by H2: Lessons learned for the developing world. *Int. J. H2 Energy*, 48, 36623–36649, 2023, https://doi.org/10.1016/j.ijhydene.2023.06.058.
41. COURSE50, Development of high-Strength coke production and waste heat recovery technologies for Course 50 process, 2022, course50.com, https://www.course50.com/en/technology/technology03/. Accessed 24 Jan 2024.
42. Engel, E., Schenk, J., Skorianz, M., Sustainable Steelmaking — A Strategic Evaluation of the Future Potential of H2 in the Steel Industry, in: *AISTech2020 Proceedings of the Iron and Steel Technology Conference*, AIST, pp. 372–381, 2020.
43. Lopez, G., Galimova, T., Fasihi, M. *et al.*, Towards defossilised steel: Supply chain options for a green European steel industry. *Energy*, 273, 127236, 2023, https://doi.org/10.1016/j.energy.2023.127236.
44. HYBRIT New research shows H2-reduced iron has superior properties, hybritdevelopment.se, https://www.hybritdevelopment.se/en/hybrit-new-re-

search-shows-H2-reduced-iron-has-superior-properties, 2022. Accessed 24 Jan 2024.

45. Hebeda, O., Guimarães, B.S., Cretton-Souza, G. *et al.*, Pathways for deep decarbonization of the Brazilian iron and steel industry. *J. Clean. Prod.*, 401, 136675, 2023, https://doi.org/10.1016/j.jclepro.2023.136675.
46. Islam, M.S., Islam, A.R.M.T., Ismail, Z. *et al.*, Effects of microplastic and heavy metals on coral reefs: A new window for analytical research. *Heliyon*, 9, e22692, 2023.
47. Kullmann, F., Linßen, J., Stolten, D., The role of H2 for the defossilization of the German chemical industry. *Int. J. H2 Energy*, 48, 38936–38952, 2023, https://doi.org/10.1016/j.ijhydene.2023.04.191.
48. Bazzanella, A.M. and Ausfelder, F., Low carbon energy and feedstock for the European chemical industry, p. 168, Eur. Chem. Ind. Counc., DECHEMA, Frankfurt, Germany, 2017.
49. Hall, W., Spencer, T., Renjith, G., Dayal, S., The potential role of H2 in India, p. 49, SustainablefinanceHsbcCom, New Delhi, India, 2020.
50. Rafiqul, I., Weber, C., Lehmann, B., Voss, A., Energy efficiency improvements in Ammonia production - Perspectives and uncertainties. *Energy*, 30, 2487–2504, 2005, https://doi.org/10.1016/j.energy.2004.12.004.
51. Bird, F., Clarke, A., Davies, P., Surkovic, E., Ammonia: zero-carbon fertiliser, fuel and energy store, Royal Society, London, UK, 2020.
52. Suryanto, B.H.R., Kang, C.S.M., Wang, D. *et al.*, Rational Electrode–Electrolyte Design for Efficient Ammonia Electrosynthesis under Ambient Conditions. *ACS Energy Lett.*, 3, 1219–1224, 2018, https://doi.org/10.1021/acsenergylett.8b00487.
53. Lee, B., Winter, L.R., Lee, H. *et al.*, Pathways to a Green Ammonia Future. *ACS Energy Lett.*, 7, 3032–3038, 2022, https://doi.org/10.1021/acsenergylett.2c01615.
54. Rouwenhorst, K.H.R., Van der Ham, A.G.J., Mul, G., Kersten, S.R.A., Islanded Ammonia power systems: Technology review & conceptual process design. *Renew. Sustain. Energy Rev.*, 114, 109339, 2019, https://doi.org/10.1016/j.rser.2019.109339.
55. Müller, M., Pfeifer, M., Holtz, D., Müller, K., Comparison of green Ammonia and green H2 pathways in terms of energy efficiency. *Fuel*, 357, 129843, 2024, https://doi.org/10.1016/j.fuel.2023.129843.
56. Frøhlke, U., Topsoe and First Ammonia Launch Zero Emission Ammonia Production With the World'S Largest Reservation of Electrolyzer Capacity, Hydrogen Central, 2022. https://hydrogen-central.com/topsoe-first-ammonia-launch-zero-emission-ammonia-production-worlds-largest-reservation-electrolyzer-capacity/. Accessed 4 May 2024.
57. Altaf, C.T., Çolak, T.O., Kumtepe, A. *et al.*, H2 Production and Polymer Electrode Membrane (PEM) Fuel Cells for Electrical Vehicles, in: *H2 Electrical Vehicles*, pp. 149–198, 2023.

58. Kumar, S.S. and Himabindu, V., H2 production by PEM water electrolysis – A review. *Mater. Sci. Energy Technol.*, 2, 442–454, 2019, https://doi.org/10.1016/j.mset.2019.03.002.
59. Air Products, *Air Products, ACWA Power and NEOM Sign Agreement for $5 Billion Production Facility in NEOM Powered by Renewable Energy for Production and Export of Green H2 to Global Markets*, pp. 1–4, Air Prod Chem Inc, Lehigh Valley, Pennsylvania, USA, 2020, Press release.
60. Siemens, Green' ammonia is the key to meeting the twin challenges of the 21st century, https://new.siemens.com/uk/en/company/topic-areas/sustainable-energy/green-ammonia.html, Accessed July-11-2020.
61. A. Scott, Spanish to make fertilizer from green hydrogen, *C&EN Glob Enterp*, 2020. https://cen.acs.org/environment/Spanish-make-fertilizer-green-hydrogen/98/i30 (accessed May 04, 2024).
62. Lee, B., Lim, D., Lee, H., Lim, H., Which water electrolysis technology is appropriate?: Critical insights of potential water electrolysis for green ammonia production. *Renew. Sustain. Energy Rev.*, 143, 110963, 2021, https://doi.org/10.1016/j.rser.2021.110963.
63. Campion, N., Nami, H., Swisher, P.R. *et al.*, Techno-economic assessment of green Ammonia production with different wind and solar potentials. *Renew. Sustain. Energy Rev.*, 173, 113057, 2023, https://doi.org/10.1016/j.rser.2022.113057.
64. Pagani, G., Hajimolana, Y., Acar, C., Green H2 for ammonia production – A case for the Netherlands. *Int. J. H2 Energy*, 52, 418–432, 2024, https://doi.org/10.1016/j. ijhydene.2023.06.309.
65. Vinardell, S., Nicolas, P., Sastre, A.M. *et al.*, Sustainability Assessment of Green Ammonia Production To Promote Industrial Decarbonization in Spain. *ACS Sustain. Chem. Eng.*, 11, 15975–15983, 2023, https://doi.org/10.1021/acssuschemeng.3c04694.
66. Fasihi, M., Weiss, R., Savolainen, J., Breyer, C., Global potential of green ammonia based on hybrid PV-wind power plants. *Appl. Energy*, 294, 116170, 2021, https://doi.org/10.1016/j.apenergy.2020.116170.
67. Sagel, V.N., Rouwenhorst, K.H.R., Faria, J.A., Leveraging Green Ammonia for Resilient and Cost-Competitive Islanded Electricity Generation from Hybrid Solar Photovoltaic–Wind Farms: A Case Study in South Africa. *Energy Fuels*, 37, 14383–14392, 2023, https://doi.org/10.1021/acs.energyfuels.3c01950.
68. Jaszczur, M., Hassan, Q., Sameen, A.Z. *et al.*, Massive Green H2 Production Using Solar and Wind Energy: Comparison between Europe and the Middle East. *Energies*, 16, 5445, 2023.
69. Brown, T., Saudi Arabia to export renewable energy using green ammonia, Ashburn, Virginia, USA, 2020. https://ammoniaenergy.org/articles/saudi-arabia-to-export-renewable-energy-using-green-ammonia/. Accessed 4 May 2024.
70. Brown T. Japan's Road Map for Fuel Ammonia Fuel Ammonia in Japan's Green Growth Strategy, Ashburn, VA, 2024.

71. Crowley, S., Regional, S., Commission, D., Specialist, C., Green H2 and ammonia: Implications for Minnesota and beyond, Southwest, MN, 2024, https://www.cleanenergyresourceteams.org/story/green-hydrogen-and-ammonia-implications-minnesota-and-beyond. Accessed 4 May 2024.
72. Sankir, N.D. and Sankir, M., *H2 Storage Technologies*, Wiley, Scrivener Publishing, Salem, 2018.
73. Ma, N., Zhao, W., Wang, W. *et al.*, Large scale of green H2 storage: Opportunities and challenges. *Int. J. H2 Energy*, 50, 379–396, 2024, https://doi.org/10.1016/j.ijhydene.2023.09.021.
74. Wang, B., Li, T., Gong, F. *et al.*, Ammonia as a green energy carrier: Electrochemical synthesis and direct Ammonia fuel cell - a comprehensive review. *Fuel Process. Technol.*, 235, 107380, 2022, https://doi.org/10.1016/j.fuproc.2022.107380.
75. Hsain, Z., Chu, A., Serpell, O. *et al.*, Ammonia's Role in a Net-Zero H2 Economy, Philadelphia, Pennsylvania, 2023.
76. Cechetto, V., Di Felice, L., Medrano, J.A. *et al.*, H2 production via Ammonia decomposition in a catalytic membrane reactor. *Fuel Process. Technol.*, 216, 106772, 2021, https://doi.org/10.1016/j.fuproc.2021.106772.
77. Morlanés, N., Katikaneni, S.P., Paglieri, S.N. *et al.*, A technological roadmap to the Ammonia energy economy: Current state and missing technologies. *Chem. Eng. J.*, 408, 127310, 2021, https://doi.org/10.1016/j.cej.2020.127310.
78. Wang, L., Xia, M., Wang, H. *et al.*, Greening Ammonia toward the Solar Ammonia Refinery. *Joule*, 2, 1055–1074, 2018, https://doi.org/10.1016/j.joule.2018.04.017.
79. Wang, Q., Guo, J., Chen, P., The impact of alkali and alkaline earth metals on green Ammonia synthesis. *Chem.*, 7, 3203–3220, 2021, https://doi.org/10.1016/j.chempr.2021.08.021.
80. Huo, C.-F., Wu, B.-S., Gao, P. *et al.*, The Mechanism of Potassium Promoter: Enhancing the Stability of Active Surfaces. *Angew. Chem. Int. Ed.*, 50, 7403–7406, 2011, https://doi.org/10.1002/anie.201007484.
81. Brown, D.E., Edmonds, T., Joyner, R.W. *et al.*, The Genesis and Development of the Commercial BP Doubly Promoted Catalyst for Ammonia Synthesis. *Catal. Lett.*, 144, 545–552, 2014, https://doi.org/10.1007/s10562-014-1226-4.
82. Aika K., Role of alkali promoter in ammonia synthesis over ruthenium catalysts—Effect on reaction mechanism. *Catal. Today*, 286, 14–20, 2017, https://doi.org/10.1016/j.cattod.2016.08.012.
83. Ravi, M. and Makepeace, J.W., Facilitating green ammonia manufacture under milder conditions: What do heterogeneous catalyst formulations have to offer? *Chem. Sci.*, 13, 890–908, 2022, https://doi.org/10.1039/d1sc04734e.
84. Stoltze, P. and Nørskov, J.K., An interpretation of the high-pressure kinetics of ammonia synthesis based on a microscopic model. *J. Catal.*, 110, 1–10, 1988, https://doi.org/10.1016/0021-9517(88)90291-6.

85. Niwa, Y. and Aika, K., The Effect of Lanthanide Oxides as a Support for Ruthenium Catalysts in Ammonia Synthesis. *J. Catal.*, 162, 138–142, 1996, https://doi.org/10.1006/jcat.1996.0268.
86. Fishel, C.T., Davis, R.J., Garces, J.M., Ammonia Synthesis Catalyzed by Ruthenium Supported on Basic Zeolites. *J. Catal.*, 163, 148–157, 1996, https://doi.org/10.1006/jcat.1996.0314.
87. IEA, Global cement production in the Net Zero Scenario, 2010–2030, IEA, Paris, 2023. https://www.iea.org/data-and-statistics/charts/global-cement-production-in-the-net-zero-scenario-2010-2030-5260, Licence: CC BY 4.0.
88. Xiang, Q., Pan, H., Ma, X. *et al.*, Impacts of energy-saving and emission-reduction on sustainability of cement production. *Renew. Sustain. Energy Rev.*, 191, 114089, 2024, https://doi.org/10.1016/j.rser.2023.114089.
89. Shen, W., Liu, Y., Yan, B. *et al.*, Cement industry of China: Driving force, environment impact and sustainable development. *Renew. Sustain. Energy Rev.*, 75, 618–628, 2017, https://doi.org/10.1016/j.rser.2016.11.033.
90. Phair, J.W., Green chemistry for sustainable cement production and use. *Green Chem.*, 8, 763–780, 2006, https://doi.org/10.1039/B603997A.
91. Bosoaga, A., Masek, O., Oakey, J.E., CO2 Capture Technologies for Cement Industry. *Energy Proc.*, 1, 133–140, 2009, https://doi.org/10.1016/j.egypro.2009.01.020.
92. Gartner, E.M. and Macphee, D.E., A physico-chemical basis for novel cementitious binders. *Cem. Concr. Res.*, 41, 736–749, 2011, https://doi.org/10.1016/j.cemconres.2011.03.006.
93. Naqi, A. and Jang, J.G., Recent Progress in Green Cement Technology Utilizing Low-Carbon Emission Fuels and Raw Materials: A Review. *Sustainability*, 11, 537, 2019, https://doi.org/10.3390/su11020537.
94. Boden, T.A., Marland, G., Andres, R.J., Global, Regional, and National Fossil-Fuel CO2 Emissions, 1751-2007 (Version 2010). Environmental System Science Data Infrastructure for a Virtual Ecosystem (ESS-DIVE) (United States); Carbon Dioxide Information Analysis Center (CDIAC), Oak Ridge National Laboratory (ORNL), Oak Ridge, TN (United States), 2010.
95. Sousa, V. and Bogas, J.A., Comparison of energy consumption and carbon emissions from clinker and recycled cement production. *J. Clean. Prod.*, 306, 127277, 2021, https://doi.org/10.1016/j.jclepro.2021.127277.
96. Poponi, D., Bryant, T., Burnard, K. *et al.*, Energy Technology Perspectives 2016: Towards Sustainable Urban Energy Systems, International Energy Agency, Paris, France, 2016.
97. WEF, Cement industry net-zero tracker. In: Net-Zero Ind. Tracker 2023, Geneva, Switzerland, 2023, https://www.weforum.org/publications/net-zero-industry-tracker-2023/in-full/cement-industry-net-zero-tracker/. Accessed 24 Jan 2024.
98. WBCSD, Cement Sustainability Initiative: Cement Industry Energy and CO2 Performance "Getting the Numbers Right," Geneva and Washington

DC: World Business Council on Sustainable Development, 2009, https://www.wbcsd.org/. Accessed 24 Jan 2024.

99. Hasanbeigi, A., Price, L., Lin, E., Emerging energy-efficiency and CO2 emission-reduction technologies for cement and concrete production: A technical review. *Renew. Sustain. Energy Rev.*, 16, 6220–6238, 2012, https://doi.org/10.1016/j.rser.2012.07.019.
100. Change IC, others, AR5 Climate Change 2014: Mitigation of Climate Change, Contrib Work Gr III to fifth Assess Rep Intergov panel Clim Chang. 1454, 147, 2014.
101. Madlool, N.A., Saidur, R., Hossain, M.S., Rahim, N.A., A critical review on energy use and savings in the cement industries. *Renew. Sustain. Energy Rev.*, 15, 2042–2060, 2011, https://doi.org/10.1016/j.rser.2011.01.005.
102. Shen, W., Cao, L., Li, Q. *et al.*, Quantifying CO2 emissions from China's cement industry. *Renew. Sustain. Energy Rev.*, 50, 1004–1012, 2015, https://doi.org/10.1016/j.rser.2015.05.031.
103. Wi, K., Lee, H.-S., Lim, S. *et al.*, Use of an agricultural by-product, nano sized Palm Oil Fuel Ash as a supplementary cementitious material. *Constr .Build. Mater.*, 183, 139–149, 2018, https://doi.org/10.1016/j.conbuildmat.2018.06.156.
104. Malhotra, V.M., Role of supplementary cementing materials in reducing greenhouse gas emissions, in: *Concr. Technol. a Sustain. Dev. 21st century*, vol. 5, p. 6, 2000.
105. Mishra, S. and Siddiqui, N.A., A review on environmental and health impacts of cement manufacturing emissions. *Int. J. Geol. Agric. Environ. Sci.*, 2, 26–31, 2014.
106. Ding, L., Chen, K., Cheng, S., Wang, X., Water ecological carrying capacity of urban lakes in the context of rapid urbanization: A case study of East Lake in Wuhan. *Phys. Chem. Earth, Parts A/B/C,* 89–90, 104–113, 2015, https://doi.org/10.1016/j.pce.2015.08.004.
107. Mohamad, N., Muthusamy, K., Embong, R. *et al.*, Environmental impact of cement production and Solutions: A review. *Mater. Today Proc.*, 48, 741–746, 2022, https://doi.org/10.1016/j.matpr.2021.02.212.
108. Müller, C., Reiners, J., Palm, S., Closing the loop: What type of concrete re-use is the most sustainable option, Eur. Cem. Res. Acad, Duesseldorf, Germany, 2015.
109. Cembureau, T.E., Cementing the European Grean Deal—Reaching Climate Neutrality Along the Cement and Concrete Value Chain by 2050, Brussels, Belgium, 2021.
110. Commission E, *Communication from the Commission to the European Parliament, the Council, the European Economic and Social Committee and the Committee of the Regions Youth Opportunities Initiative*, European Commission, Brussels, Belgium, 2011.
111. EEA, Construction and Demolition waste: Challenges and Opportunities in a Circular Economy, p. 8, Eur Union, Copenhagen, Denmark, 2020.

112. Carriço, A., Bogas, J.A., Guedes, M., Thermoactivated cementitious materials – A review. *Constr. Build. Mater.*, 250, 118873, 2020, https://doi.org/10.1016/j.conbuildmat.2020.118873.
113. IEA, Technology Roadmap - Low-Carbon Transition in the Cement Industry, IEA, Paris, 2018, https://www.iea.org/reports/technology-roadmap-low-carbon-transition-in-the-cement-industry.
114. Müller, N., Harnisch, J., Mesguen, J.P., How to turn around the trend of cement related emissions in the developing world, WWF—Lafarge Conserv Partnersh Gland Switz, Nurnberg, Germany, 2008.
115. Summerbell, D.L., Barlow, C.Y., Cullen, J.M., Potential reduction of carbon emissions by performance improvement: A cement industry case study. *J. Clean. Prod.*, 135, 1327–1339, 2016, https://doi.org/10.1016/j.jclepro.2016.06.155.
116. IEA, Tracking Industrial Energy Efficiency and CO2 Emissions, IEA, Paris, 2007, https://www.iea.org/reports/tracking-industrial-energy-efficiency-and-co2-emissions.
117. Meyer, C., The greening of the concrete industry. *Cem. Concr. Compos.*, 31, 601–605, 2009, https://doi.org/10.1016/j.cemconcomp.2008.12.010.
118. Scrivener, K.L. and Capmas, A., Calcium Aluminate Cements, in: *Lea's Chemistry of Cement and Concrete*, Fourth Edition, P.C. Hewlett (Ed.), pp. 713–782, Butterworth-Heinemann, Oxford, 1998.
119. Golewski, G.L., Green concrete composite incorporating fly ash with high strength and fracture toughness. *J. Clean. Prod.*, 172, 218–226, 2018, https://doi.org/10.1016/j.jclepro.2017.10.065.
120. Araghi, H.J., Nikbin, I.M., Rahimi Reskati, S. *et al.*, An experimental investigation on the erosion resistance of concrete containing various PET particles percentages against sulfuric acid attack. *Constr. Build. Mater.*, 77, 461–471, 2015, https://doi.org/10.1016/j.conbuildmat.2014.12.037.
121. Rao, A., Jha, K.N., Misra, S., Use of aggregates from recycled construction and demolition waste in concrete. *Resour. Conserv. Recycl.*, 50, 71–81, 2007, https://doi.org/10.1016/j.resconrec.2006.05.010.
122. Aliabdo, A.A., Abd Elmoaty, A.E.M., Aboshama, A.Y., Utilization of waste glass powder in the production of cement and concrete. *Constr. Build. Mater.*, 124, 866–877, 2016, https://doi.org/10.1016/j.conbuildmat.2016.08.016.
123. Olanipekun, E.A., Olusola, K.O., Ata, O., A comparative study of concrete properties using coconut shell and palm kernel shell as coarse aggregates. *Build. Environ.*, 41, 297–301, 2006, https://doi.org/10.1016/j.buildenv.2005.01.029.
124. Awoyera, P., Gobinath, R., Haripriya, S., Kulandaisami, P., New Light Weight Mortar for Structural Application: Assessment of Porosity, Strength and Morphology Properties, in: *International Conference on Emerging Trends in Engineering (ICETE)*, S.C. Satapathy, K.S. Raju, K. Molugaram, *et al.* (Eds.), pp. 59–65, Springer International Publishing, Cham, 2020.

125. Ismail, Z.Z. and AL-Hashmi, E.A., Recycling of waste glass as a partial replacement for fine aggregate in concrete. *Waste Manage.*, 29, 655–659, 2009, https://doi.org/10.1016/j.wasman.2008.08.012.
126. Imbabi, M.S., Carrigan, C., McKenna, S., Trends and developments in green cement and concrete technology. *Int. J. Sustain. Built Environ.*, 1, 194–216, 2012, https://doi.org/10.1016/j.ijsbe.2013.05.001.
127. Radonjanin, V., Malešev, M., Marinković, S., Al Malty, A.E.S., Green recycled aggregate concrete. *Constr. Build. Mater.*, 47, 1503–1511, 2013, https://doi.org/10.1016/j.conbuildmat.2013.06.076.
128. Zéhil, G.-P. and Assaad, J.J., Feasibility of concrete mixtures containing cross-linked polyethylene waste materials. *Constr. Build. Mater.*, 226, 1–10, 2019, https://doi.org/10.1016/j.conbuildmat.2019.07.285.
129. Rumšys, D., Bačinskas, D., Spudulis, E., Meškėnas, A., Comparison of Material Properties of Lightweight Concrete with Recycled Polyethylene and Expanded Clay Aggregates. *Proc. Eng.*, 172, 937–944, 2017, https://doi.org/10.1016/j.proeng.2017.02.105.
130. Qin, Y., Zhang, X., Chai, J. *et al.*, Experimental study of compressive behavior of polypropylene-fiber-reinforced and polypropylene-fiber-fabric-reinforced concrete. *Constr. Build. Mater.*, 194, 216–225, 2019, https://doi.org/10.1016/j.conbuildmat.2018.11.042.
131. Thorneycroft, J., Orr, J., Savoikar, P., Ball, R.J., Performance of structural concrete with recycled plastic waste as a partial replacement for sand. *Constr. Build. Mater.*, 161, 63–69, 2018, https://doi.org/10.1016/j.conbuildmat.2017.11.127.
132. Awoyera, P. and Adesina, A., A critical review on application of alkali activated slag as a sustainable composite binder. *Case Stud. Constr. Mater.*, 11, e00268, 2019, https://doi.org/10.1016/j.cscm.2019.e00268.
133. Sivakrishna, A., Adesina, A., Awoyera, P.O., Rajesh Kumar, K., Green concrete: A review of recent developments. *Mater. Today Proc.*, 27, 54–58, 2020, https://doi.org/10.1016/j.matpr.2019.08.202.
134. IEA (2022), World Energy Outlook 2022, IEA, Paris https://www.iea.org/reports/world-energy-outlook-2022,
135. Pisciotta, M., Pilorgé, H., Feldmann, J. *et al.*, Current state of industrial heating and opportunities for decarbonization. *Prog. Energy Combust. Sci.*, 91, 100982, 2022, https://doi.org/10.1016/j.pecs.2021.100982.
136. Okoji, A.I., Anozie, A.N., Omoleye, J.A. *et al.*, Energetic assessment of a precalcining rotary kiln in a cement plant using process simulator and neural networks. *Alexandria Eng. J.*, 61, 5097–5109, 2022, https://doi.org/10.1016/j.aej.2021.10.010.
137. Juangsa, F.B., Cezeliano, A.S., Darmanto, P.S., Aziz, M., Thermodynamic analysis of H2 utilization as alternative fuel in cement production. *South Afr. J. Chem. Eng.*, 42, 23–31, 2022, https://doi.org/10.1016/j.sajce.2022.07.003.
138. Zhang, Y., Cao, S.-X., Shao, S. *et al.*, Aspen Plus-based simulation of a cement calciner and optimization analysis of air pollutants emission.

Clean Technol. Environ. Policy, 13, 459–468, 2011, https://doi.org/10.1007/s10098-010-0328-y.

139. Thapa, B.S., Neupane, B., Yang, H., Lee, Y.-H., Green H2 potentials from surplus hydro energy in Nepal. *Int. J. H2 Energy*, 46, 22256–22267, 2021, https://doi.org/10.1016/j.ijhydene.2021.04.096.
140. Jibran, J.A. and Mahat, C., Application of green H2 for decarbonization of cement manufacturing process: A technical review. *J. Phys. Conf. Ser.*, 2629, 12027, 2023, https://doi.org/10.1088/1742-6596/2629/1/012027.
141. Det Norske Veritas, Why is the cement industry labelled hard-to-abate?, 2023, https://www.dnv.com/article/why-is-the-cement-industry-labelled-hard-to-abate--241192/. Accessed 24 Jan 2024.
142. Association MP, others, Options for switching UK cement production sites to near zero CO 2 emission fuel: Technical and financial feasibility, London, England, 2019.
143. Progressive Energy Ltd, Industrial Fuel Switching competition phase 3: successful projects, 2022, https://www.gov.uk/government/publications/industrial-fuel-switching-to-low-carbon-alternatives/industrial-fuel-switching-demonstration-successful-projects-phase-3. Accessed 24 Jan 2024.
144. Reduction GHC, Scaling up electrolysers to meet the 1.5 C climate goal, Int Renew Energy Agency, Abu Dhabi, 2020.
145. Schneider, M., Hoenig, V., Ruppert, J., Rickert, J., The cement plant of tomorrow. *Cem. Concr. Res.*, 173, 107290, 2023, https://doi.org/10.1016/j.cemconres.2023.107290.
146. Hodges, A., Hoang, A.L., Tsekouras, G. *et al.*, A high-performance capillary-fed electrolysis cell promises more cost-competitive renewable H2. *Nat. Commun.*, 13, 1304, 2022, https://doi.org/10.1038/s41467-022-28953-x.
147. Lee, H., Decarbonization strategies for steel production with uncertainty in H2 direct reduction. *Energy*, 283, 129057, 2023, https://doi.org/10.1016/j.energy.2023.129057.
148. Vogl, V. and Åhman, M., What is green steel?-Towards a strategic decision tool for decarbonising EU steel, pp. 24–28, ESTAD Proc [P532], Lund, Sweden, 2019.
149. Özgün, Ö, Lu, X., Ma, Y., Raabe, D., How much H2 is in green steel? *NPJ Mater. Degrad.*, 7, 78, 2023, https://doi.org/10.1038/s41529-023-00397-8.
150. Moritz, M., Seidenberg, J.R., Siska, M. *et al.*, A Path to Sustainability: Green H2 Based Production of Steel and Ammonia, 2020.
151. Muslemani, H., Liang, X., Kaesehage, K. *et al.*, Opportunities and challenges for decarbonizing steel production by creating markets for 'green steel' products. *J. Clean. Prod.*, 315, 128127, 2021, https://doi.org/10.1016/j.jclepro.2021.128127.
152. Bhaskar, A., Assadi, M., Somehsaraei, H.N., Decarbonization of the Iron and Steel Industry with Direct Reduction of Iron Ore with Green H2. *Energies*, 13, 758, 2020, https://doi.org/10.3390/en13030758.

153. Mahat, C., Jibran, J.A., Sharma, N., Thapa, B., Challenges and prospects of steel production using green H2 in Nepal. *J. Phys. Conf. Ser.*, 2629, 2023, https://doi.org/10.1088/1742-6596/2629/1/012026.
154. Devlin, A., Kossen, J., Goldie-Jones, H., Yang, A., Global green H2-based steel opportunities surrounding high quality renewable energy and iron ore deposits. *Nat. Commun.*, 14, Article number 2578, 2023, https://doi.org/10.1038/s41467-023-38123-2.
155. Marocco, P., Gandiglio, M., Audisio, D., Santarelli, M., Assessment of the role of H2 to produce high-temperature heat in the steel industry. *J. Clean. Prod*, 388, 135969, 2023, https://doi.org/10.1016/j.jclepro.2023.135969.
156. Wang, R.R., Zhao, Y.Q., Babich, A. *et al.*, H2 direct reduction (H-DR) in steel industry—An overview of challenges and opportunities. *J. Clean. Prod.*, 329, 129797, 2021, https://doi.org/10.1016/j.jclepro.2021.129797.
157. Levesque, C., Green New Steel: Opportunities, Insight, and Barriers to Green Hydrogen Use in U.S. Steel Production. Climate Science and Policy, UC San Diego, 2022. Retrieved from https://escholarship.org/uc/item/85r1z0w5. Accesed 4 May 2024.
158. Trinca, A., Patrizi, D., Verdone, N. *et al.*, Toward green steel: Modeling and environmental economic analysis of iron direct reduction with different reducing gases. *J. Clean. Prod.*, 427, 139081, 2023, https://doi.org/10.1016/j.jclepro.2023.139081.
159. Toktarova, A., Göransson, L., Johnsson, F., Design of clean steel production with H2: Impact of electricity system composition. *Energies*, 14, 24, 8349, 2021, https://doi.org/10.3390/en14248349.
160. Groysman, A., Process Units in Oil Refineries and Petrochemical Plants, in: *Corrosion Problems and Solutions in Oil Refining and Petrochemical Industry*, pp. 1–7, Springer International Publishing, Cham, 2017.
161. Wu, J., Jia, Y., Cheng, M., Xia, X., A complex network perspective on embodiment of air pollutants from global oil refining industry. *Sci. Total Environ.*, 824, 153740, 2022, https://doi.org/10.1016/j.scitotenv.2022.153740.
162. Gregory, J. and Geels, F.W., Unfolding low-carbon reorientation in a declining industry: A contextual analysis of changing company strategies in UK oil refining (1990–2023). *Energy Res. Soc Sci.*, 107, 103345, 2024, https://doi.org/10.1016/j.erss.2023.103345.
163. Griffiths, S., Sovacool, B.K., Kim, J. *et al.*, Decarbonizing the oil refining industry: A systematic review of sociotechnical systems, technological innovations, and policy options. *Energy Res. Soc Sci.*, 89, 102542, 2022, https://doi.org/10.1016/j.erss.2022.102542.
164. Kapustin, V. and Rudin, M., *Chemistry and Technology of Oil Refining*, Begell House Inc, Danbury, Connecticut, USA, 2015.
165. Fetanat, A. and Tayebi, M., Sustainability and reliability-based H2 technologies prioritization for decarbonization in the oil refining industry: A decision support system under single-valued neutrosophic set. *Int. J. H2 Energy*, 52, 765–786, 2024, https://doi.org/10.1016/j.ijhydene.2023.08.229.

166. Bergman-Fonte, C., da Silva, G.N., Império, M. *et al.*, Repurposing, co-processing and greenhouse gas mitigation – The Brazilian refining sector under deep decarbonization scenarios: A case study using integrated assessment modeling. *Energy*, 282, 128435, 2023, https://doi.org/10.1016/j.energy.2023.128435.
167. Griffiths, S., Sovacool, B.K., Kim, J. *et al.*, Industrial decarbonization via H2: A critical and systematic review of developments, socio-technical systems and policy options. *Energy Res. Soc Sci.*, 80, 102208, 2021, https://doi.org/10.1016/j.erss.2021.102208.
168. Bergero, C., Binsted, M., Younis, O. *et al.*, Technology, technology, technology: An integrated assessment of deep decarbonization pathways for the Canadian oil sands. *Energy Strateg. Rev.*, 41, 100804, 2022, https://doi.org/10.1016/j.esr.2022.100804.
169. Jia, F.-R., Jing, W.-T., Liu, G.-X. *et al.*, Paraffin-based crude oil refining process unit-level energy consumption and CO2 emissions in China. *J. Clean. Prod.*, 255, 120347, 2020, https://doi.org/10.1016/j.jclepro.2020.120347.
170. Moradpoor, I., Syri, S., Santasalo-Aarnio, A., Green H2 production for oil refining – Finnish case. *Renew. Sustain. Energy Rev.*, 175, 113159, 2023, https://doi.org/10.1016/j.rser.2023.113159.
171. Nikhil, P., Prakash, G.B., Kavyasri, K., Sivasankar, P., Analysis on CO2 emissions, green H2 requirement and geo-storage potential of H2 for decarbonization of industrial operations in southeastern coast of India. *Int. J. H2 Energy*, 52, 1507–1521, 2024, https://doi.org/10.1016/j.ijhydene.2023.08.235.
172. Pivetta, D., Volpato, G., Carraro, G. *et al.*, Optimal decarbonization strategies for an industrial port area by using H2 as energy carrier. *Int. J. H2 Energy*, 52, 1084–1103, 2024, https://doi.org/10.1016/j.ijhydene.2023.07.008.

2

Conventional to Renewable Hydrogen Production: A Paradigm Shift Toward Sustainable Energy

Kriti Shrivasava[1] and Ankur Jain[1,2*]

[1]*Centre for Renewable Energy & Storage, Suresh Gyan Vihar University, Jaipur, Rajasthan, India*
[2]*Natural Science Centre for Basic Research & Development, Hiroshima University, Hiroshima, Japan*

Abstract
Renewable or green hydrogen is one key component that has the potential to displace fossil fuels as the primary source of energy for the world, but for this to occur, dependable strategies for its carbon-free generation must be devised. Hydrogen is acknowledged as a potential fuel due to its use as an energy carrier, a storage medium, in fuel cells, and as a fuel and offers carbon-free solutions. It is anticipated to be the future global energy vector that will significantly improve the air quality in the Earth's atmosphere. The major concerns in hydrogen production are the process economy and sustainability. Currently, most of the hydrogen production is fossil fuel based by employing steam reforming technique. Researchers and policymakers have given green hydrogen production a lot of consideration as a primary requirement for it to become a clean and low-carbon energy source. This book chapter presents a gradual shift in the approach of hydrogen production technologies from nonrenewable to renewable. It also investigates major green hydrogen production techniques by renewable feedstocks and alternative methods such as hydrogen production from water electrolysis, biological, photochemical, and biomass utilization. A comparative assessment in terms of economy, environmental sustainability, and technological readiness is also discussed.

Keywords: Green hydrogen, water electrolysis, biomass gasification, biohydrogen

**Corresponding author*: ankur.j.ankur@gmail.com

Mehmet Sankir and Nurdan Demirci Sankir (eds.) Towards Green Hydrogen Generation, (49–86)

2.1 Introduction

Hydrogen is among the most available elements on Earth. It is a colorless and odorless gas, uncommon on Earth in its free form. It has the capability to combine chemically with other elements to form a wide range of compounds such as hydrocarbon, alcohol, and water. Naturally, it is present in the biomass as one of the basic elements in the structure of plants and animals. For hydrogen production from water and carbon resources, energy is required. Therefore, hydrogen is generally not considered as an energy source rather, it can be produced as an energy source [1]. Hydrogen can be obtained utilizing domestic resources, i.e., coal, biomass, natural gas, and other renewable resources including solar, wind, geothermal, or hydroelectric energy. Therefore, hydrogen is nominated as an important energy carrier to ensure future energy security due to the availability of diverse domestic energy sources [2].

Hydrogen is commonly utilized in huge amounts in petroleum, automobile vehicles, and pharmaceuticals, and is used as an alternative to fuel, and chemical industries. As a fuel, hydrogen has the maximum combustion energy discharge per unit of mass (286 kJ mol^{-1}) and is extremely clean as the only byproduct is water. Consequently, it is one of the highly potential gasses that is considered as an energy source. In the future, H_2 might be utilized to power our homes, offices, and transportation vehicles and aircraft [1]. Hydrogen is a desirable alternative for long-term energy management due to the emerging need for environment and sustainability. The use of hydrogen for fuel cells in motorized transport can greatly reduce the emission of greenhouse gasses from their internal combustion engines. Therefore, hydrogen is advocated as the most efficient, cost-competitive, renewable, and potentially zero-discharge fuel for today's rising energy demands.

Researchers and policymakers have given hydrogen a lot of consideration as a clean energy carrier for future low-carbon power generation [3]. Although hydrogen is nontoxic, lighter than air, and quickly disperses when released, there are several safety issues pertaining to its commercial use. Its quick dispersion in the event of a leak, which if left unattended and the gas accumulates in a small area, can eventually catch fire and explode. To ensure the safe use of hydrogen, it is essential to have a complete understanding of its properties, design safety features in hydrogen systems, and provide training on safe hydrogen handling and storing procedures [4].

Figure 2.1 depicts the hydrogen value chain which consists of three principal phases starting from production to the applications. The first

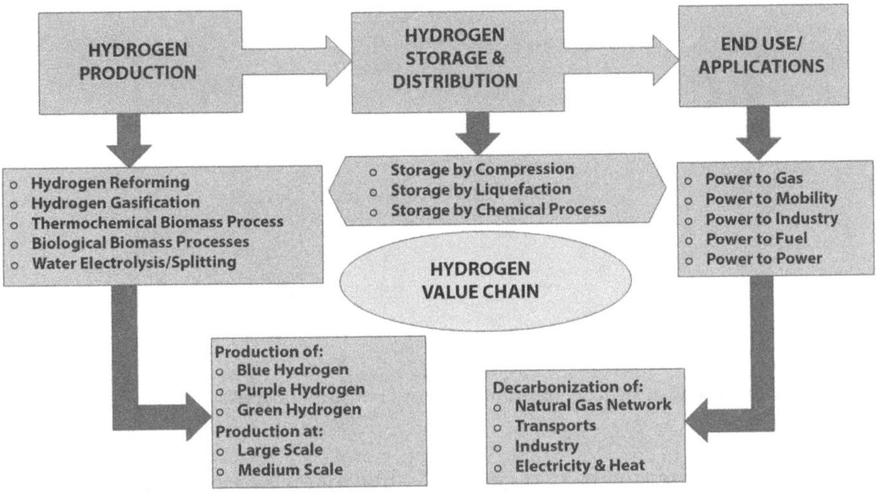

Figure 2.1 Hydrogen value chain [5].

stage is the production of hydrogen, which includes a variety of traditional and modern technologies, both being commercially utilized and currently under development or research stage. It can be produced on a big, medium, or small scale, but it should ideally be close to the point of consumption to reduce transportation costs and risks. The second stage involves storing and distributing hydrogen, which entails risk and safety considerations because it operates at extremely low temperatures and/or high pressures, necessitating the use of tanks with greater thicknesses to ensure adequate insulation. Generally, the most convenient storage solution includes hydrogen as compressed gas, cryogenically frozen gas, liquefied, and chemical storage.

Applications for end usage in the industrial and mobility/transport sectors find chemical storage as the most suitable. Hydrogen and natural gas combinations can be used in stationary home and commercial applications to provide heat and energy [5]. The status of hydrogen production (Figure 2.2) reflects approximately 49% of the hydrogen production from natural gas, primarily *via* steam methane reforming. The rest of the hydrogen is produced from oil (around 29%), which is generally consumed in hydroprocessing applications in petroleum refineries, from coal (18%) consumed for the manufacture of ammonia, and the remaining 4% *via* water electrolysis. Some other commonly used methods are catalytic steam reforming (SR) of natural gas, the partial oxidation (POX) of petroleum, steam iron treatment, and the gasification of coal. Additionally, industrial production

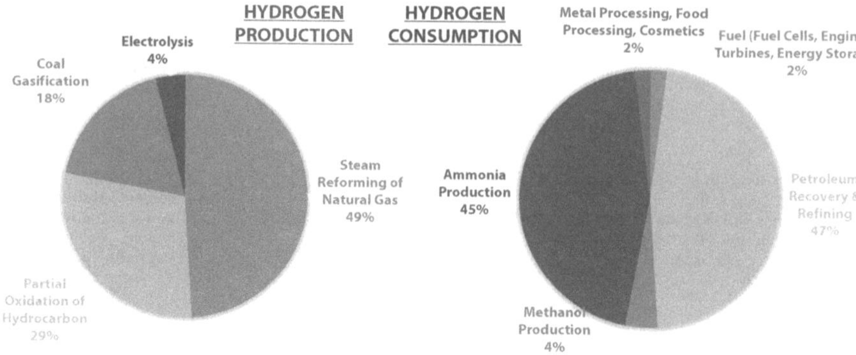

Figure 2.2 Overview of hydrogen production and consumption [6].

of some of the chemicals also generates hydrogen as a byproduct like the production of counter-chlorine from chlor-alkaline, the generation of light gasses in the refining process of crude oil, coke production from coal in the coke ovens, and chemical hydrogenation processes in margarine industry. Hydrogen consumption (Figure 2.2) includes the industrial sectors like petroleum recovery and refining, methanol and ammonia production, metal processing, food processing, cosmetics as a fuel in fuel cells, engines and turbines, and other energy storage applications [6, 7].

2.2 Hydrogen Production Technologies

Nowadays, several techniques are available to produce hydrogen from different renewable and nonrenewable sources, with a variation in cost of production, level of CO_2 emission, and formation of byproducts (Figure 2.3). Currently, a significant proportion of the world's hydrogen is still produced by converting fossil fuels, such as natural gas, into steam, which produces a lot of carbon dioxide emissions. Hydrogen production from fossil fuel–based resources is comparatively cost-effective as compared to nuclear and renewable resources due to the high efficiency of the process and well-established technology. However, such production on a large scale causes high carbon emissions and elevated energy consumption which are still appalling challenges. A significant overhaul of the energy system is necessary to reduce global CO_2 emission output and the risk of climate change (greenhouse effect). A long-term solution to lower CO_2 emissions

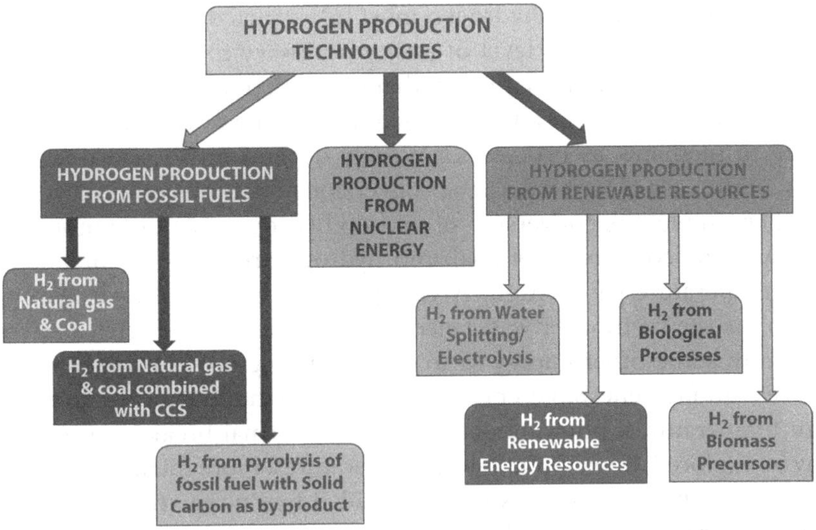

Figure 2.3 Hydrogen production and its color spectrum [9].

is the use of hydrogen as an energy carrier. However, due to its high manufacturing costs, hydrogen is currently not as competitive as other energy sources [8].

However, process economy and developing an environmentally sustainable synthesis technique are the main considerations and difficulties in producing hydrogen [10]. Other dominant factors related to hydrogen production are the overall availability of sources and the availability of uninterrupted energy supply, as both factors determine the cost of hydrogen production. Nuclear energy is the only other form of energy that is permanent and abundant, while the rest are restricted to geographical areas and climatic circumstances. Solar energy storage's intermittent nature might be a flexible concern for energy storage, but it increases production costs because of supplementary components. Consequently, nuclear energy is also considered a promising alternative for affordable and clean hydrogen generation because of its stability, surplus quantity, and portability [2] despite its own disadvantages and associated risk factors. However, recent scientific evidence suggest the utility of cost-effective hydrogen synthesis methods from cheap and abundant biomass, biological fermentation, photochemical, and electrolysis. These methods could feasibly replace current hydrogen production practices, guaranteeing clean hydrogen production for all future needs.

To produce hydrogen, hydrogen must be extracted and isolated as separate molecules with the level of purity necessary for a particular application. The present dominating methane production strategy only makes sense if the energy is initially contained in methane or can be easily transferred to methane, as processes naturally depend on the starting point. Thus, in the case of fossil fuels, the conversion of natural gas into hydrogen is rather simple, the conversion of oil into hydrogen is a little more complex, and the conversion of coal into hydrogen requires the first step of high-temperature gasification. A very popular method of hydrogen production is the method of water electrolysis, which uses electricity generated from some other energy sources. Depending on their type, biomass sources among renewable energy sources demand extra monitoring. At high temperatures, photoinduced or direct thermal breakdown of water may be explored; however, at lower temperatures, more intricate, multistep schemes are needed, such as those provided by nuclear reactor steam or concentrating solar power plants [11].

2.3 Production of Blue Hydrogen

Hydrogen-containing materials derived from fossil fuels, such as gasoline, hydrocarbons, methanol, or ethanol, are transformed into a hydrogen-rich gas stream *via* fossil fuel processing techniques. Methane (i.e., natural gas) is widely used as a commercial precursor for hydrogen production. This conventional hydrogen production route generates a major fraction of hydrogen for global hydrogen demand and to make this process more environment-friendly, such conventional hydrogen production facilities are being integrated with other technologies like carbon capture and storage (CCS) or carbon capture and utilization (CCU) which help in reducing the CO_2 emissions [12].

Blue hydrogen is produced from hydrocarbon fuels and their products, i.e., methane, LPG, gasoline, methanol, ethanol, coal, oil, naphtha, natural gas, and heavy oil fractions by several chemical processes (Figure 2.4). The three major industrial technologies available are (i) SR, (ii) POX, and (iii) autothermal reforming (ATR). Apart from these, coal gasification and cracking of heavy oil fractions are also feasible processes for commercial H_2 production. Since these technologies generate huge carbon monoxide (CO), they integrate one or more chemical reactors in the subsequent step to efficiently convert CO into carbon dioxide (CO_2) by using water-gas shift (WGS) and preferential oxidation (PrOx) or methanation reactions [10]. Most fossil fuels contain a certain amount of sulfur, which needs prior

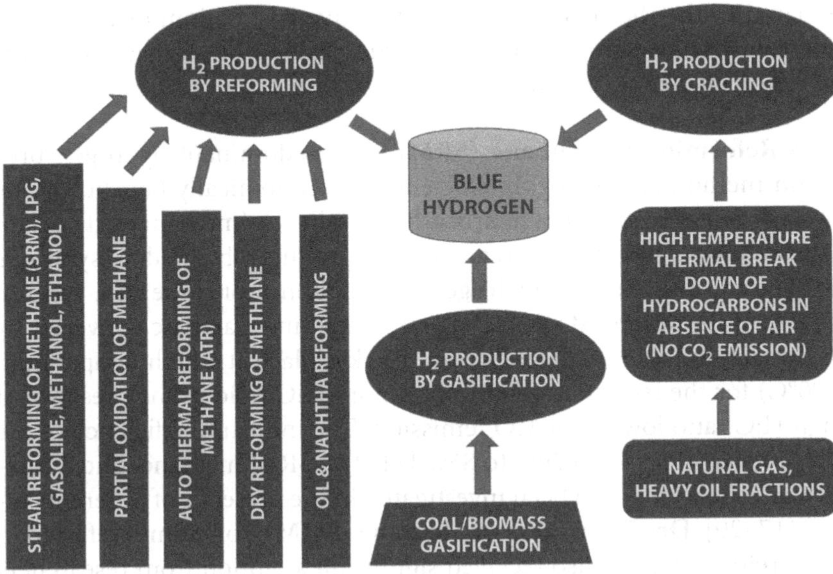

Figure 2.4 Production of blue hydrogen.

removal and needs to be considered in the future planning of a hydrogen-based economy.

Hydrogen production from hydrocarbon fuels can be done by SR, POX, and ATR. Their major disadvantages are the high production of greenhouse gasses, the requirement of high temperature, complex operation formation of soot, and low hydrogen production [13]. Therefore, for hydrogen production, these methods should be integrated with CCS techniques as steam methane reforming is a CO_2 intensive process that is currently used globally for the major portion of hydrogen production.

Reforming

Reforming is a thermal or catalytic refining process in which the hydrocarbon molecules are rearranged to obtain new products with desired characteristics. A significant portion of the traditional hydrogen is generated from fossil fuels and, more specifically, natural gas reforming and methane POX. ATR of methane, dry reforming of methane, and oil/naphtha reforming are some other reforming techniques for hydrogen production. Generally, the process begins with the desulfurization of precursor material to separate the sulfur content, especially H_2S [14]. The sulfur is recovered using the Claus process in which gaseous hydrogen sulfide chemically reacts with oxygen to separate sulfur. The catalytic process is highly efficient as compared to the thermal process where activated titanium or aluminum oxide catalyst is employed in the catalytic Claus process that boosts the conversion efficiency of sulfur.

Steam Reforming of Methane (SRM) is a predominant hydrogen production method as it is highly efficient and economically beneficial. The first stage is the treatment of hydrocarbon feedstock (methane) with steam in the tubular catalytic bed reactor. The product obtained is synthesis gas (syngas), a mixture of hydrogen and carbon monoxide in a ratio of 3:1 [15]. In the second stage, it is again treated in a catalytic converter by WGH reaction to give CO and steam. It takes place at a high temperature (>500°C) for the feedstock with an optimum H:C ratio as the presence of a high H:C ratio lowers the CO_2 emission. The industrial efficiency of the SRM process varies from 70% to 85% [11, 16]. SR of methanol and a mixture of alcohols have also been investigated in the presence of different catalysts [17–20]. **Dry Reforming of Methane (DRM)**, sometimes referred to as CO_2 reforming, has also received significant attention from researchers as it consumes two greenhouse gasses (i.e., CO_2 and CH_4) to produce clean hydrogen fuel. Another advantage is the unit H_2:CO ratio, due to which the syngas can be directly used in F-T synthesis for manufacturing value-added products. As dry reforming is an endothermic reaction, it occurs at high temperatures, although the reaction temperature can be reduced in the presence of a suitable catalyst [21–23]. **Partial Oxidation of Methane (POM)** is an exothermic process, mainly utilized in hydrogen synthesis for automobile fuel cells and other commercially viable purposes. The process is less efficient as compared to SRM and costly due to the consumption of large amounts of pure oxygen. For catalytic POM, Ni- and Rh-based catalysts are typically used; however, employing Rh is not economically feasible. Due to its quick kinetics and exothermic nature, catalytic POM (CPOM) is an energy-efficient method for producing syngas. Additionally, this has stopped the use of large reactors and a lot of superheated steam. Another benefit of POM is the generation of syngas with a 2:1 H_2:CO ratio, allowing for direct use in Fischer-Tropsch or methanol synthesis without the need for additional modifications. According to reports, the POM process has an efficiency range of 60% to 75% [24–26]. ATR is another technique for hydrogen production, combining POX and adiabatic SR, and does not require external energy to operate. The ATR process is comparatively simpler and more economical than SRM. The additional advantage of the ATR process over SRM is its rapid ease of start and close down while yielding a larger quantity of hydrogen than POM. Moreover, it has been observed that by adjusting the oxygen-to-methane and steam-to-carbon ratio, the coke formation can be minimized. The ATR process due to its encouraging product gas composition for F-T synthesis, comparative superior efficiency, and improved economy is expected to be the promising alternative technology for the 'gas to liquid' (GTL) fuel industry. The efficiency of the ATR process varies between 65% and 75% [23, 27]. Hydrogen can be produced in an environmentally friendly and sustainable method through SR of bio-oil using different catalytic systems nickel, rhodium, or ruthenium. In catalytic naphtha-reforming processes, the reactors used today are either tubular or spherical, and the feedstock may flow in axial or radial direction; additionally, to improve the performance of the whole process, thus increasing the output and reducing waste, all new units are to be designed based on continuous catalyst regeneration (CCR) reformers [28–30].

A. Hydrogen Production by Cracking

Hydrocarbon compounds are cracked when heated to extremely high temperatures in an oxygen-free atmosphere, releasing hydrogen and carbon. This reaction takes place in the presence of a catalyst. No emission of greenhouse gasses is the main favorable feature of this process. Ten percent of the natural gas flow is used for the energy input needed in this endothermic reaction. Plasma reformers are a technology created for this purpose; they operate at about 2000°C, have the advantage of being able to reform all hydrocarbons, including heavy oil fractions, and break down organic materials in an air- and oxygen-free environment. As a main byproduct, hydrogen is obtained by thermal dissociation, without the creation of carbon dioxide. As the carbon byproduct has a granular structure, it can be easily separated from hydrogen [7]. Methane cracking is another promising option for producing hydrogen due to minimal CO/CO_2 emissions, and it is encouraged due to environmental concerns. Metallic and carbonaceous catalysts have been extensively studied for efficient methane cracking. Traditional catalytic methane cracking and newly created solar cracking reactors are two creative ways to increase the process sustainability [31]. Ammonia is a promising and popular material as a direct source of green hydrogen production because it does not carry carbon, and its total hydrogen content is also higher than other fuels and can be suitably converted to hydrogen [32, 33]. Ammonia Cracking by electrochemical, photocatalysis, and thermochemical processes is attracting global scientific attention.

B. Hydrogen Production by Gasification

Gasification is a commercially used technology that uses the thermal POX of both coal and biomass to produce syngas. A series of thermochemical processes are used for POX (Table 2.1). Another common technique is coal gasification which includes an initial phase of air separation to remove oxygen from the air and deliver it to the gasifier. For air separation, one of the three extensively employed techniques—membrane separation, cryogenic air separation, or pressure swing adsorption—is used. Coal pyrolysis comes after the air separation process. To gasify coal, the pyrolysis stage needs two inputs: steam and oxygen [34].

Several coal-based gasification systems have been investigated to evaluate hydrogen and electricity production [36]. Implementing carbon dioxide sequestration has led to a large decrease in GHG emissions, lowering this energy source's worldwide carbon footprint. But, it also has enhanced the other environmental effect indicators like the abiotic resource depletion index, eutrophication potential index, and acidification potential index [37]. The technology of coal gasification by supercritical water has a great prospect because it generates hydrogen-rich gas products in an efficient and environmentally clean manner [38]. To produce hydrogen fuel from coal and have a practical application in the energy industry, plasma gasification is suggested as a feasible solution. This is because the plasma gasification method generates more hydrogen than other gasification processes, releases fewer solid wastes (slag, ash, tar, etc.), and is more environmentally friendly in general [39].

Table 2.1 Chemical reactions during the gasification process [35].

Type of reaction	**Chemical equation**	**Enthalpy change ΔH (MJ/kmol)**
Combustion reaction	$C + 1/2O_2 \rightarrow CO$	**-111**
Combustion reaction	$C + O_2 \rightarrow CO_2$	**-283**
Boudouard reaction	$C + CO_2 \leftrightarrows 2CO$	**+172**
Methanation reaction	$C + 2H_2 \leftrightarrows CH_4$	**-75**
Water gas reaction	$C + H_2O \leftrightarrows CO + H_2$	**+131**
Water gas shift reaction	$CO + H_2O \leftrightarrows CO_2 + H_2$	**-41**
Steam methane reforming reaction	$CH_4 + H_2O \leftrightarrows CO + 3H_2$	**+206**
Dry reforming reaction	$CH_4 + CO_2 \leftrightarrows 2CO + 2H_2$	**+260**
Methanation reaction	$2C + 2H_2O \leftrightarrows CH_4 + CO2$	**+103**

Production of Purple Hydrogen

Purple hydrogen is produced by using nuclear energy. Hydrogen produced from the high temperature of nuclear power plant reactors is known as thermochemical hydrogen (Figure 2.5). In such types of power plants, nuclear fission of uranium atoms generates energy because nuclear fission reactions are controllable inside the nuclear reactors. This process does not emit greenhouse gasses, and therefore, nuclear power is popularly recognized as a source of economic and zero-carbon electricity for the grid, but this power cannot be directly utilized for transportation due to technological limitations. The thermal energy generated in nuclear reactors can be used for several purposes in addition to electricity generation. This thermal energy can be directly used for hydrogen production *via* different processes such as thermochemical water splitting (TWS), SR, or water electrolysis. Indirectly, electricity generated by this thermal energy is used in the fuel cells for hydrogen production [12].

Purple hydrogen could be a feasible way to carry this nuclear energy to cut down global transportation costs [40]. Nuclear energy can also be used for water splitting which results in hydrogen production through different chemical routes like sulfur-iodine, vanadium-chlorine, hybrid sulfur, and copper-chlorine. These chemical routes take the input of mainly water and heat and create an internal cycle of several intermediate species to generate hydrogen and oxygen as final products without any gaseous emission into the atmosphere. Four-step magnesium chlorine cycle was also investigated as an efficient hydrogen production system in a nuclear-powered integrated plant [41]. The copper-chlorine (Cu-Cl) cycle was studied for TWS linked to nuclear energy and any other available and desired renewable sources [42]. A viable method for producing hydrogen is made possible by advanced electrolysis technology, which follows reliable chemistry being operated inside the reactors, with relatively lower temperature requirements and sufficient efficiency levels. Another practical and eco-friendly method of commercial hydrogen production is commercially established water electrolysis technology. This is especially important for all countries facing an energy crisis since they must reduce their reliance on imported

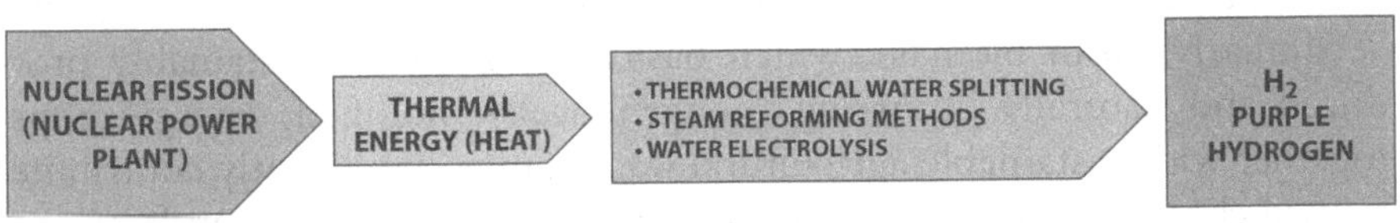

Figure 2.5 Production of purple hydrogen.

fossil fuels and chemical feedstocks like natural gas. Such nuclear hydrogen and allied power systems can strengthen renewable energy resources during times of unavailability when they are unable to meet a substantial share of the larger extent of the global energy supply chain [42].

Disadvantages associated with the utilization of nuclear energy are mainly the long-lasting radioactive waste generated during the process of nuclear fission of uranium. The main environmental issue is related to the safe and long-lasting disposal of radioactive wastes. One of the suggested solutions to this problem is the conversion of long half-life radioactive elements into short half-life elements. This strategy can lessen the scope of this problem, although it will not solve the problem of nuclear waste completely [43]. The usage of renewable energy is a key component of the entire decarbonization and green hydrogen initiatives, which also call for the acquisition of energy conversion technology like solar PV cells, wind turbines, and batteries. However, the low-carbon hydrogen production scenario must also consider the contribution of the domestic nuclear power sector as well as renewable resources. It will expand the role of nuclear power in serving and decarbonizing industries other than electricity, especially those where electrification alone would not reduce GHG emissions, and will help in diversifying the available sources of green hydrogen [44].

Production of Green Hydrogen

Green hydrogen can be produced in two ways: first by the water splitting using green energy (i.e., from renewable energy resources) and second from the biological sources by using biotic or abiotic agents. Hydrogen production can be achieved by converting solar photovoltaic (PV), wind energy, hydropower, waste biomass, geothermal, and sometimes even tidal energy (Figure 2.6) into sustainable green electricity by combining them with currently available eco-friendly technologies. When such green electricity is used for water electrolysis, green hydrogen is produced. In this way, hydrogen generation can be used as a method to store intermittently available renewable energy. Therefore, hydrogen may act as both the energy carrier and energy storage medium to meet the large-scale commercial energy demand of transportation (electric vehicles, shipping, and aircraft) [45, 46] and other industrial sectors with minimal environmental impact.

Similarly, clean biomass, which basically comprises sustainably produced and naturally occurring (nongenetically modified) energy crops together with sustainable agricultural wastes, may significantly contribute toward cheap hydrogen production for upcoming power demands. Clean biomass is a well-known source of renewable energy that can be converted

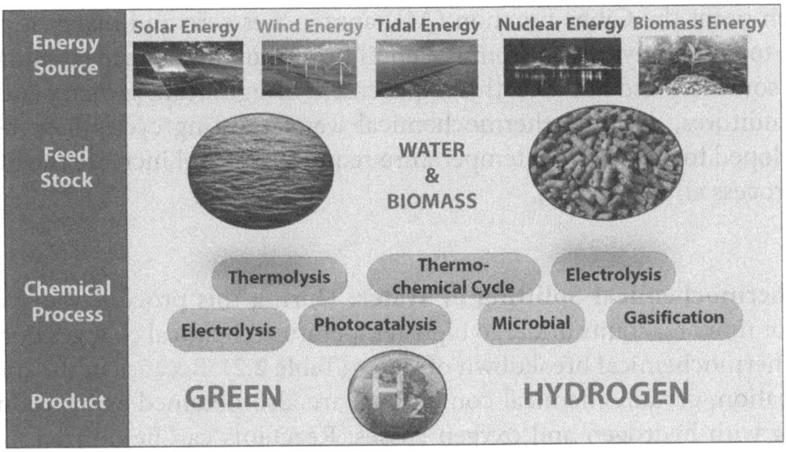

Figure 2.6 Production of green hydrogen.

into heat, electricity, and other fuels by conversion into a gas by gasification and extracting hydrogen. It also maintains a natural carbon cycle which is extracted from the environment during plant growth and released during the process of hydrogen generation. Reforestation and replanting are required to maintain a hydrogen source based on renewable biomass [3, 6, 47, 48].

A. Hydrogen Production from Water Using Renewable Energy

Hydrogen can be produced from water by several methods like simple water thermolysis, electrolysis, photocatalytic water splitting, or TWS. Nuclear, solar, and wind energy are examples of renewable energy sources that can be utilized to conduct these procedures. Specifically, the energy produced by these renewable sources—whether thermal, radiant, or electrical—can be utilized to produce hydrogen [13, 49].

a. Thermal Splitting of Water (Thermolysis): Thermolysis is the term for the process of synthesizing hydrogen by the thermal breaking of water at high temperatures. With this technique, the vapor molecules start to break down as if they will break down into hydrogen and oxygen gasses when the temperature is raised to 1400°C or higher. To produce hydrogen suitable for commercial application, the temperature needs to be increased to between 2500°C and 3000°C [7]. Usually, water does not break down until it reaches extremely high temperatures—typically above 2500°C—at which point the Gibbs function (ΔG) approaches zero and makes it possible to extract hydrogen from the equilibrium mixture. Since sustainable heat sources could not meet this requirement of significant primary energy expenditures, different thermochemical water-splitting cycles have been developed to decrease the temperature requirement and increase the overall process efficiency [50].

b. Thermochemical Splitting of Water: During this process, water and one or more reactants undergo high-temperature chemical changes during the thermochemical breakdown of water (Table 2.2). Because of the transformation, certain chemical compounds are also obtained as byproducts along with hydrogen and oxygen gasses. Reactants can be utilized more than once in these modifications. This thermochemical breakdown of water typically involves three steps: generation of hydrogen, oxygen, and recycled materials. The requirement of high temperature can be met by

Table 2.2 Comparative analysis of electrolysis methods for hydrogen production [7, 50].

Thermochemical cycles	Maximum operating temperature (°C)
Cerium-chlorine (Ce-Cl)	850
Hybrid copper-chloride (Cu-Cl)	550
Hybrid copper sulfate (Cu-SO_4)	850
Iron-chlorine (Fe-Cl)	925
Hybrid chlorine (CI_2)	850
Magnesium-iodine (Mg-I)	600
Vanadium-chlorine (V-Cl)	850
Bromo-calcium-iron (Br-Ca-Fe)	725
Sulfur-iodine (S-I)	850
Zinc oxide-Zn (ZnO/Zn)	1727
Hybrid oxides (Fe_3O_4/FeO)	1627
Hybrid oxides (Fe_2O_3/Fe_3O_4)	1327

solar or nuclear energy by developing different solar collectors and parabolic reflectors, e.g., dish, trough, and tower systems [51, 52].

Currently, Mg-Cl and Cu-Cl thermochemical cycles are considered the most promising low-temperature cycles due to no GHG emission into the environment and low-temperature requirements of up to 550°C. It is discovered that as solar light intensity rises, so do the energy efficiency and rates of hydrogen production. The cost of hydrogen production is determined by several factors like cost, availability, and toxicity of the constituent elements and chemicals, material separation, and corrosion issues. A solar thermal water splitting (STWS) system works in two stages: initially, at high temperatures produced by focused sunlight, metal oxide releases oxygen. Then, reduced material is exposed to H_2O and re-oxidized to the initial oxidation state [53]. If STWS is to become economically feasible, optimal STWS materials and appropriate reactor designs must be developed to build new STWS reactors [54].

The two-step thermochemical cycles can be further divided into two types: volatile and nonvolatile. In volatile redox cycles, the metal oxide converts into a gas phase after thermal reduction, while in the nonvolatile cycles, metal oxides remain in the solid phase. Because the cycles containing volatile products have higher entropy, they are thermodynamically more advantageous and involve more reduction. To prevent recombination between the O_2 and the volatile metal, which would reduce the process's overall energy efficiency, the volatile products must be quenched. Because of this, most of the investigations are based on nonvolatile cycles, particularly those involving ferrites or another spinel. For the reduction stage, these cycles necessitate extremely high temperatures, between 1400°C and 1500°C. Nonstoichiometric oxides, on the other hand, can serve as possible building blocks for solar thermal hydrogen generation at lower operating temperatures. These nonstoichiometric oxides have rapid oxidation and reduction kinetics, favorable oxidation thermodynamics, stability at high temperatures, and a considerable degree of reduction at moderate temperatures. Perovskite oxides have lately been suggested as possible materials in this regard [55]. A new class of oxides—poly-cation oxides (PCOs)—was reported recently which contains multiple cations and is able to thermochemically split water in a two-stage process to give hydrogen (H_2) and oxygen (O_2). Increased hydrogen yield was observed FeMgCoNiOx ($x \approx 1.2$) with a reduction in the working temperatures from 1300°C to 1100°C. PCOs with complex cationic compositions offer new opportunities to fundamentally investigate redox thermochemistry together with scalable H_2 production [56]. A study based on high-throughput density functional theory (HT-DFT) evaluated 11,000 cubic and distorted perovskite

ABO_3 compounds based on thermodynamic stability and energy of oxygen vacancy formation for a thermodynamically favored two-step TWS process and revealed 139 materials as potential new candidates for TWS application [57], e.g., cubic perovskite SrTi0.5Mn0.5O3-δ (STM55) during cyclic operations resulted in a 2:1 yield of H_2/O_2 [58].

Solar thermochemical cycles have excellent potential to become a commercially viable process for future green hydrogen production. Specific technological elements need scientific attention due to their influence on overall efficiency and economy, including particle heating techniques, the suitability, and characteristics of redox materials, techniques to achieve low oxygen pressure in the reduction chamber, and the significance of implementing heat recovery [59].

c. Electrolysis of Water: Using a few electrodes submerged in water and electricity flowing through them to convert the aqueous solution into oxygen and hydrogen, this technique is thought to be the easiest way to create hydrogen. It can be categorized by reaction conditions as solid oxide electrolysis, electrolyte alkaline, proton exchange membrane, and microbial electrolysis cells (MECs) (Figure 2.7).

This process was developed to manufacture hydrogen at room temperature with an efficiency of 56% to 73%. The alkaline electrolysis process is cost-effective and generally employed and studied, while the solid oxide electrolysis, even being the most electrically efficient process, is still under development due to issues like heat cycling, chromium migration, seals,

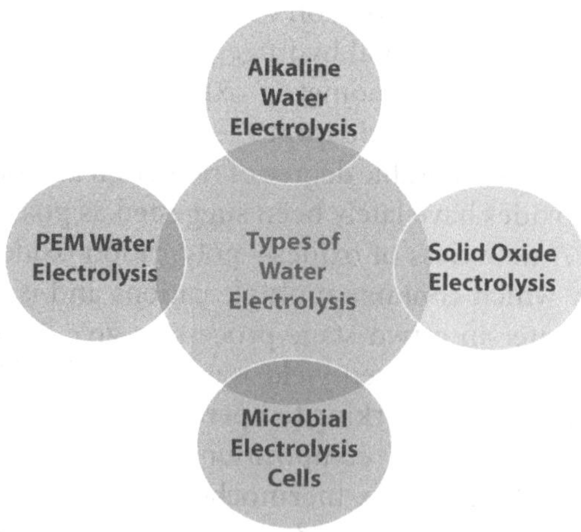

Figure 2.7 Types of water electrolysis [60].

and corrosion. Table 2.3 presents a comparative analysis of electrolysis methods for hydrogen production.

The technology of alkaline water electrolysis is well-established, low-cost of production, and suitable for industrial-scale hydrogen production up to the quantity of 1000 m^3/hour. Major drawbacks are numerous parts of the device, corrosion, sluggish startup times, and difficult maintenance. PEM hydrogen production technology is much more feasible due to quick startup times, little corrosion, easy maintenance, and fewer parts. But, the hydrogen production can reach only up to 400 m^3/hour. A major factor

Table 2.3 Comparative analysis of electrolysis methods for hydrogen production [61].

Specification	**Alkaline**	**PEM**	**SOE**
Technology maturity	State of the art	Demonstration	R & D
Cell temperature (°C)	60–80	50–80	900–1000
Cell pressure (bar)	<30	<30	<30
Current density (A/cm²)	0.2–0.4	0.6–2.0	0.3–1.0
Cell voltage (V)	1.8–2.4	1.8–2.2	0.95–1.3
Power density (W/cm²)	Up to 1.0	Up to 4.4	-
Voltage efficiency (%)	62–82	67–82	81–86
Specific energy consumption (kWh/Nm³)	4.5–7.0	4.5–7.5	2.5–3.5
Partial load range (%)	20–40	0–10	-
Cell area (m²)	<4	<300	-
Hydrogen production (Nm³/hour)	<760	<30	-
Life time (hour)	<90,000	<20,000	<40,000
System life time (year)	20–30	10–20	-
Purity of hydrogen produced (%)	>99.8	99.999	-
Cold start-up time (min.)	15	<15	>60

limiting the advancement of PEM hydrogen-generating technology is the high manufacturing cost [62].

Hydrogen production is greatly aided by the development of water-splitting cells as an effective energy conversion and storage technology but it accounts for only 4% of the world's hydrogen production because of the slow reaction kinetics of OER and HER due to high overpotentials, suppressing the energy efficiency of water electrolysis. Effective catalysts are required to reduce the overpotential of both the hydrogen evolution reaction (HER) and oxygen evolution reaction (OER) and to increase energy efficiency. However, it is anticipated that the recent upsurge in interest in lattice oxygen engineering and nanostructure will lead to new developments in the preparation of active, stable, and low-cost OER and HER for the widespread commercialization of water-splitting–based hydrogen production [63]. Some of the ongoing difficulties in producing hydrogen using low-cost non-precious catalysts are the limited information available on the activity of non-precious catalytic processes in alkaline solutions, and corrosion for carbon-intensive catalysts.

Using MEC technology, organic materials, such as renewable biomass and wastewater, may produce hydrogen. Electrical energy is transformed into chemical energy in MECs [64]. Hydrogen is produced from the complex organic molecules by using electric current. On the anode, microorganisms oxidize the substrate and produce CO_2, protons, and electrons. Protons travel to the cathode *via* an electrolyte, or proton conducting membrane, while electrons move through an external circuit to the cathode side. On the combination of protons and electrons, hydrogen is produced. This process requires a small amount of external voltage (0.2 V to 1.0 V) as compared to water electrolysis [65, 66]. However, this MEC technology is still under development due to several challenges related to the hydrogen production rate, high internal resistance, electrode materials, and complicated design. Enhancing the hydrogen production rate and lowering the energy input are the main challenges of MEC technology. The primary hurdles for MEC technology are increasing the pace of hydrogen production while reducing the energy input. One major factor that directly affects the rate of hydrogen and current production in MECs is the reactor architecture. Another important component of upscaling is the reactor architecture. Although membranes were a feature of traditional MEC designs, current studies have indicated that membrane-free designs can result in high rates of production and hydrogen recovery. Since then, several membrane-free reactors have been developed. MECs can generate a high yield of hydrogen from various organic materials under relatively mild circumstances, circumventing thermodynamic constraints. Comparing this method to direct water electrolysis, the electric energy cost for producing hydrogen is significantly lower. MECs can sustain several energetically unfavorable biological and chemical reactions in addition to producing hydrogen. Because of its special benefit, MECs have found use in a wide range of alternative applications, including biosensors, chemical synthesis, resource recovery, stubborn pollution remediation, and bioelectrochemical research platforms [67–69]. MECs are evolving into a flexible platform technology that provides a fresh approach to emerging environmental problems pertaining to resource and energy recovery as well as waste stream management.

Anion exchange membrane (AEM) water electrolysis is a technique that uses electricity and an AEM to produce hydrogen. The use of inexpensive transition metal catalysts in favor of more expensive noble metal electrocatalysts is one of the main benefits of AEM water electrolysis [66, 70]. Since AEM electrolysis is still in the primary development stage, it is difficult to be utilized for the generation of economically viable hydrogen. The newest emerging technology, AEM water electrolysis, was developed to address the shortcomings of PEM and alkane water electrolysis. The main issue with this technology, though, is its lack of dependability [71]. For low-cost AEM electrolysis, a novel, ultrathin AEM is suggested. Thin membranes have two specific advantages in terms of decreased ohmic resistance and decreased mass transfer resistance. Because the ion transfer channel is shorter in a membrane electrode assembly (MEA) with a thinner membrane, hydroxide ion transfer will be enhanced [72].

d. Photoelectrolysis: In general, photolysis occurs when visible light energy is absorbed with the aid of certain photocatalysts and used to break down water into hydrogen and oxygen. Solar energy is absorbed through certain semiconducting materials in photoelectrolysis, and water splitting takes place like an electrolysis process [73]. When a photon strikes the semiconductor anode surface having energy greater than or equal to the semiconductor's bandgap, an electron-hole pair is generated and separated by the electric field between the semiconductor and the electrolyte. The electrons travel in an external circuit to the cathode and interact with H^+ to generate H_2, while the holes stay at the anode and split water into H^+ that travels through the electrolyte to the cathode and O_2 that remains with water [50]. Several studies have concentrated on creating noble metal–free catalysts for hydrogen synthesis, with a particular emphasis on materials based on nickel, to reduce the overpotential and energy expenses. Several studies have concentrated on creating noble metal–free catalysts for hydrogen synthesis, with a particular emphasis on materials based on nickel, to reduce the overpotential and energy expenses [74]. The most important factor in the HER and OER is the concentration of anion ionomer in the catalyst layers [75].

B. Hydrogen Production from Biological Processes

An environmentally acceptable and sustainable method of clean energy production from renewable resources is through biological hydrogen evolution. Photosynthetic or fermentative organisms are primarily in charge of biological hydrogen generation activities. Direct biophotolysis, indirect biophotolysis, photo-fermentation, dark fermentation, combinations of these processes (e.g., integrating dark and photo-fermentation), and the WGS reaction are biological ways to create hydrogen [76]. Cyanobacteria are a prominent source of prospective cell factories for producing hydrogen among a variety of biological systems. These microbes have great promise for producing biological photohydrogen. Photosynthesis is how cyanobacteria grow. They are primarily composed of different carotenoids and chlorophyll, which are used for photoprotection and light harvesting. In addition to producing chlorophyll a, most of them also feature distinctive pigments known as phycobilin, which serve as auxiliary pigments during photosynthesis. Cyanobacteria use hydrogenase or nitrogenase to generate hydrogen gas [77].

Biological hydrogen production processes can be classified as follows:

- Biophotolysis of water by algae and cyanobacteria.
- Photodecomposition of organic compounds by photosynthetic bacteria (PSB).
- Fermentative hydrogen production from waste organic compounds.
- Hybrid systems using photosynthetic and fermentative bacteria.

Direct biophotolysis offers a practical approach for producing hydrogen from water and solar energy, although being constrained by its relatively low hydrogen productivity. A biomimetic system for water photolysis that can separate water into distinct gas streams and get around the inherent incompatibility of simultaneous hydrogen and oxygen evolution may be developed because of new discoveries and technological advancements in the fields of hydrogen enzymes, electron carriers, biomaterials, and nanotechnology. Even with the maximum energy productivity of 0.38 kJ/L/hour, indirect biophotolysis is still far behind the other biofuels. For instance, the rate of ethanol fermentation is 15 to 30 kJ/L/hour. By reusing the energy contained in fermentative products like acetic acid, biohydrogen's comparatively poor energy productivity can be increased [78]. Cyanothece sp. ATCC 51142 is a nitrogen-fixing, unicellular cyanobacterium that is a promising strain with an amazing capacity to produce massive amounts of H_2. The cyanobacterium fixes atmospheric nitrogen (N_2) biologically into ammonia (NH_3) while simultaneously generating H_2 as a byproduct in an anaerobic environment. A unique two-stage chemostat photobioreactor (PBR) system was designed and developed to handle the incompatible requirements between the sequential H_2 production of the cyanobacteria and its growth. The system operated continuously for 30 days without experiencing any performance losses, and it subsequently showed a remarkable improvement in H_2 production, with a yield that was more than 6.4 times higher than that of a single-stage batch system [79–81]. Because dark fermentation of microalgae may produce hydrogen consecutively and is not dependent on energy from natural sunlight, it is often superior to photo-fermentation, biophotolysis, and microbial electrolysis to produce biohydrogen. The most potent enzymes that produce hydrogen are [Fe-Fe]-hydrogenases, which are notably 1000 times more active than nitrogenases and around 10 to 100 times more active than [Ni-Fe]-hydrogenases [82, 83]. The sensitivity of the process to oxygen (O_2), the main inhibitor of the hydrogenase enzyme responsible for H_2 generation, is one of the major obstacles toward the creation of hydrogen (H_2) by biophotolysis. Extending the algal cells' life for biophotolysis together with improving algal growth parameters like pH, temperature, light intensity, and cell density can either directly or indirectly stimulate enzyme activity for increased H_2 generation. Analysis has been done on several O_2-regulating methods, including mineral deprivation, oxygen-reducing/scavengers, coculturing bacteria and microalgae, and adding antioxidants and nanoparticles [84].

Purple non-sulfur bacteria (PNSB) can be used on a range of organic substrates and are able to function over a wide spectrum of light wavelengths. The technique of cell immobilization has improved this procedure [85]. Apple waste was taken as a carbon source for biohydrogen production from high-efficient PSB HAU-M1. The effect of initial pH, light intensity, temperature, and material-to-liquid ratio on hydrogen production by HAU-M1 was evaluated using both single-factor design and response surface methodology (RSM). The optimal conditions were an initial pH of 7.14, light intensity of 3029.67 lx, temperature of 30.46°C, and material-to-liquid ratio of 0.21, respectively. A maximum specific hydrogen yield (SHY) of 111.85±1 ml H_2/g TS was obtained under the optimal conditions [86]. A bio-photoreactor with a transparent glass flat panel with polymethyl methacrylate (PMMA) grid columnar for enhanced biofilm growth with *Rhodopseudomonas palustris* GCA009 was developed and tested at 590-nm incident light. When hydrogen production by continuous photo-fermentative process from glucose was investigated using this novel reactor, at the light intensity of 210 W/m^2, feed substrate concentration of 56.0 mmol/L, and crossflow velocity of 1.68×10^{-6} m/s, a maximum hydrogen production rate of 32.6 mmol/L/day (3.56 mmol/m^2/hour), hydrogen yield of 1.15 mol H_2/mol glucose and light conversion efficiency of 5.34% was achieved [87]. Similarly, another technique of dark-photo circular baffled reactor (DP-CBR) was introduced to operate at ambient temperatures (21 ± 10°C). It gave a maximum hydrogen yield (HY) of 0.4 L/g COD with COD removal of 82%, and organic-N removal of 95% at a hydraulic retention time of 24 hours and initial pH of 6.5. It was found that by increasing HRT, reactor efficiency can be maintained at ambient temperature, and higher hydrogen yield was obtained by the protein hydrolysis in the dark fermentation [88]. An intimately coupled inorganic-biohybrid with enhanced hydrogen-producing activity was constructed by utilizing self-assembled selenide/sulfide semiconductor nanoparticles for *E. coli* as a light harvester. This hybrid demonstrated a 2.6-fold increase in hydrogen production rate compared to those containing extracellular nanoparticles and attained a light energy conversion efficiency of up to 27.6% because of overcoming the transmembrane electron transfer barrier. Because metabolites like lactate effectively scavenge photo-excited holes and produce very little reactive oxygen species, the addition of photocatalysis did not reduce the survival of the bacteria [89]. Through the coculturing of PSB and acidogenic microflora, a hybrid method was developed and assessed to overcome induced fatty acid inhibition during the dark-fermentative hydrogen (H_2) production process. In comparison to the dark-fermentation system (DFS), the augmented dark-photo fermentation system (ADPFS) demonstrated a 40% increase in cumulative H_2 production (CHP; 250 ml) and a 10% improvement in COD removal efficiency. Coculturing contributed to a 40% reduction in VFA accumulation, indicating the functional involvement of photosynthetic organisms in lowering fatty acid concentrations in relation to increased H_2 production [90].

C. Hydrogen Production from Agricultural Waste

The biological production of hydrogen can be characterized as environmentally safe and economically viable because it allows you to recycle waste to get high-value-added products. Agricultural waste can be defined as waste solids and liquids discharged by farm products processing and agricultural production, livestock and poultry breeding, and rural residents. Since agricultural waste like crop residues, livestock waste, and food waste are mainly composed of complex organic substrates, they can be biologically degraded by complex microbial ecosystems *via* the process of dark fermentation to produce hydrogen [91]. Management of huge quantities of solid organic waste in the agro-industrial complex is a major concern in which most of the waste is generated on livestock farms (56%) and during crop production (35.6%). Centralized biogas plants can be suggested as a potential solution for efficient processing of such carbon-rich organic waste to produce biofuels. Furthermore, hydrogen production can be achieved by SR of biogas as a cheap precursor. During a feasibility study, palladium membrane is used in a reforming reactor for the simultaneous production and separation of hydrogen but pure hydrogen from biogas can also be obtained by dry methane reforming in a fluidized bed reactor. A cogeneration system to produce not only hydrogen but also energy is possible by integrating the biogas SR system and the organic Rankine cycle. Decentralized production of hydrogen is not economically feasible due to high production costs. It is important to develop a larger-scale production technology that allows high yield at low cost [92].

To produce biorenewable hydrogen, agricultural waste of cow manure (CM) and stems of weed Amaranthus *retroflexus* L. (AR) were pyrolyzed in a ratio of 1:1, 2:1, and 4:1. Thermogravimetric analysis at a heating rate of 10°C/min in the temperature range from 40°C to 1000°C generated 36.95% pyrolysis liquid, 24.99% syngas, and 38.06% biochar, and the maximum hydrogen concentration of 21.17% with CM to AR ratio of 4:1 [93]. Such Biohydrogen production has some major benefits like the following:

- Cost-effective waste management.
- Capital generation from agricultural food production *via* its conversion into energy.
- Reduction of CO_2 emissions due to the utilization of sustainable biomass.
- Replacement of fossil fuels by biofuels.

However, a major challenge toward the development of biohydrogen production technology is the high cost of the feedstock because the common substrates for hydrogen dark fermentation typically require sucrose or starch. In addition to solid agricultural waste, wastewater containing high concentrations of carbohydrates, generated from textile industries, agricultural processes, and food industries, can also be preferred for economic reasons [94]. Harvesting of economically valuable crop products also generates agricultural waste and residues, like stover, straw, cobs, stalks, peelings, and bagasse. Such type of waste basically contains fibrous materials generally composed of cellulose, hemicellulose, and lignin formed by photosynthesis, and organic waste liquid rich in organic matter, i.e., livestock and poultry manure that can be converted to hydrogen (Figure 2.8) [95]. Hydrogen potentials range between less than 1 ml of H_2 per gram of dry matter for complex lignocellulosic residues and 240 ml of H_2 per gram of dry matter for purified polymers such as starch. To increase hydrogen potential, many solutions have been proposed like substrate pretreatment, microbial consortium selection, and process parameter optimization for example substrate pretreatment resulted in higher hydrogen yields up to 150 ml of hydrogen for each gram of pretreated rice straw [96].

Similarly, other agricultural residues including almond shells, walnut shells, barley straws, canola stalks, rice straws, and wheat straws were converted into a hydrogen-rich gas *via* gasification in supercritical water media in a preset condition. With the highest hydrogen content of 6.5 wt%, a maximum hydrogen yield of 8.38 mmol/g was observed for barley straw. Canola stalk with the highest C/H ratio showed the highest total gas yield of 25.3 mmol/g. Higher carbon-to-hydrogen ratio and lower oxygen percentage in the feedstocks generally resulted in higher total gas and CO_2 yields and lower hydrogen yields. However, lignin content in the feedstocks was found to be inversely proportional to total gas yields and cellulose content directly enhanced the total gas yield [97]. The enhancing effect of calcium oxide (CaO) addition on steam gasification of different typical Chinese agricultural biomass wastes on the H_2 production was demonstrated in an atmospheric fluidized bed reactor under different conditions and it indicated the dual role of CaO, which acts as both a CO_2 absorbent as well as a catalyst in the gasification process. When added, it significantly increased the concentration and yield of H_2 and at the same time decreased the CO

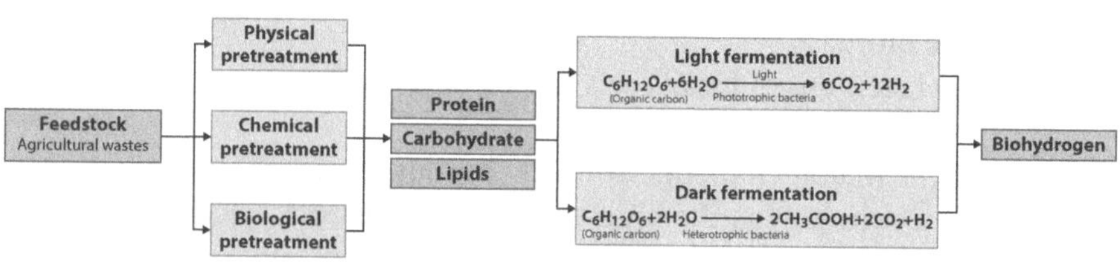

Figure 2.8 Biohydrogen production from agricultural waste (reprinted from [100] with the permission of Elsevier).

and CO_2 content in the product gas [98]. In another study, tea waste was gasified at 450°C, 650°C, and 850°C at 1-atm pressure in an updraft gasifier using potassium carbonate as a catalyst and air as a gasification agent. The effect of catalyst ratio (catalyst/biomass), temperature, and reaction time on product gas distribution determines the composition of gaseous products [99]. When grass powder was used as a carbon substrate, maximum activities of crude enzyme complex generated were employed for hydrolysis of untreated and mild acid pretreated rice husk. Generated maximum amount of reducing sugar released from enzyme-treated rice husk was 485 mg/g of the substrate which when finally used for hydrogen production by Clostridium beijerinckii, maximum cumulative H_2 production and H_2 yield were 237.97 mL and 2.93 mmoL H_2/g of reducing sugar [100].

Thermogravimetric analysis of sorghum stalk, soybean stalk, cotton stalk, corn stover, and corncob revealed that the activation energy of each fiber gets changed along with the conversion rate; the average activation energy for most of the selected fibers was close to each other, from 189.1 to 200.0 kJ mol^{-1}, except corncob. Corncob had the lowest residue content and activation energy [101]. The effect of temperature, biomass composition, steam/biomass mole ratio, and effect of CO_2 absorbent on the pyrolysis gasification of different waste biomass using CaO as the CO_2 absorbent showed that mainly biomass composition affects the H_2 and char production. The palm oil fiber produced the highest amount of H_2, and the palm oil shell generated the maximum amount of char. When steam was added, it significantly enhanced H_2 and CO production and reduced the production of solid char [102]. One study to explore the potential of biomass-to-hydrogen production took place in Kazakhstan, in which a gasification process was utilized to obtain green hydrogen. The total amount of green hydrogen produced from agricultural waste is 432 tonnes/km^2/year and from livestock waste 5262×10^{-6} tonnes/km^2/year which is equal to 856 tonnes/year. The top three regions with H_2 production were Turkestan, Almaty, and East Kazakhstan regions [103].

Conclusion

To lessen our reliance on fossil fuels and protect the environment from their harmful impacts, a great deal of research is being done on clean hydrogen production technology. Hydrogen is predicted to be a front-runner contender fuel for the new H_2 economy [2]. Hydrogen produced using fossil fuels is more affordable than hydrogen produced using nuclear or renewable energy sources. Nonetheless, there are still serious obstacles to overcome in the fossil fuel-based hydrogen generation process, including excessive carbon emissions and high energy usage. Nevertheless, current studies have demonstrated that cost-effective hydrogen synthesis can be produced *via* biomass, biological fermentation, photochemical, and electrolysis processes. These procedures have the potential to be a viable substitute for traditional hydrogen production technologies, thereby ensuring the generation of clean hydrogen in the future (Figure 2.9). The cost of producing hydrogen is dependent on several other important elements, including the availability of a source and a continuous energy supply. Solar energy is abundant and always dominates nuclear energy in terms of output; the remaining energy sources are restricted to areas and climates. The sporadic nature of solar energy storage makes it a modifiable worry for energy storage; yet, additional sections result in higher production costs. Consequently, nuclear energy is thought to be a promising substitute for the affordable and clean production of hydrogen because of its stability, excess quantity, and portability. Since hydrogen storage is the main problem that needs to be addressed to switch to hydrogen fuel, this presents another barrier.

The commercialization of renewable energy-driven hydrogen generation techniques, market and infrastructure development, commercially feasible product development, and the establishment of a sustainable hydrogen economy all require more research. Remote locations far from

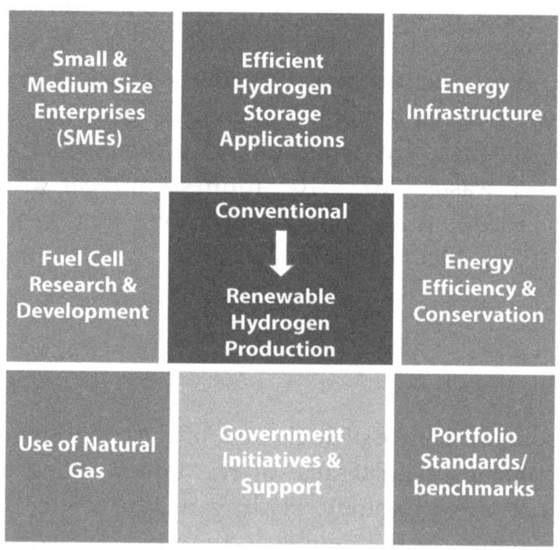

Figure 2.9 Decisive factors for the transition of conventional to renewable hydrogen production [48].

the grid can also use off-grid water electrolysis systems powered by solar or wind energy to meet their electrical needs. By using off-grid offshore wind energy to produce clean fuel (hydrogen/ammonia) for the marine industry, hydrogen can also play a significant part in the decarbonization of the maritime sector. Ammonia is a convenient form of blue hydrogen that can be stored and utilized for international trade. Ammonia is also easily broken into hydrogen on-site. By using carbon dioxide removal techniques, the commercial implementation of such systems can also aid in reaching almost net-zero CO_2 emissions.

The literature presents findings from multiple research that indicate that systems based on renewable energy are more ecologically friendly than those based on fossil fuels for producing hydrogen. For it to be used extensively, the cost of producing hydrogen from renewable energy must still be lowered. Hydrogen production technique based on biomass has some advantages over other renewable energy-based hydrogen production systems, both in terms of economics and environmental impact. In particular, the environmental impact of producing hydrogen through biomass electrolysis is less than that of the other methods under evaluation conditions like those of other studies, suggesting that this method could be a viable option for an energy-efficient and environmentally friendly method of producing hydrogen energy [49].

The research and development community is now investigating two significant methods for sunlight-driven "green" hydrogen generation: water-splitting *via* solar thermochemical hydrogen (STCH) and photoelectrochemical (PEC). While the products of both technologies—hydrogen and sunlight—are the same, they utilize different feedstocks. A recent techno-economic analysis (TEA) of the PEC and STCH water-splitting methods demonstrated the potential for producing hydrogen at a reasonable cost with the use of renewable energy sources. Although the technical readiness level (TRL) of the direct solar-to-hydrogen technology paths, such as PEC and STCH, is comparatively lower than LTE due to the proven longevity and scale of the systems, these pathways have some distinct advantages. Because they can bypass electrical transmission lines and losses, direct solar-to-hydrogen paths, for instance, can be beneficial in areas lacking dependable electrical grid infrastructures [71].

For breakthrough technology to be commercialized, procurement, tax breaks, and government incentives are essential. Another method that the government supports the adoption of new technology is through standards and regulations. Today, government monitoring and laws are tied to technological breakthroughs that speed up communications and slow down global warming. It is necessary to include small- and medium-sized businesses (SMEs) in the hydrogen deployment, research, and development process. SMEs' research, development, and use of new, cutting-edge technologies have been one of the main "economic engines" driving growth in recent decades. Federal and state governments should put more of an emphasis on fuel cell research and development. The usage of natural gas must be viewed as only a transition fuel to produce hydrogen to expedite the timely development and implementation of the hydrogen economy. The important thing is to make sure that the costs associated with reforming natural gas are paired with the electrolysis of hydrogen from renewable energy sources. If significant investments were made only in the infrastructure of the central electric grid and natural gas or liquefied natural gas, instead of allocating the necessary funds to lowering the costs spent on renewable production of hydrogen, it would be a financial and environmental disaster that would delay this paradigmatic revolution for thirty to fifty years.

The usual notation that gasoline and, as of late, LNG will be a component of the shift to a hydrogen economy needs to be taken into consideration. An important obstacle to the globalization of the hydrogen economy is the world's reliance on coal and oil. The energy, transportation, and information technology infrastructures, which make up the sixth issue of concern, are at the center of the entire conversation. The integration of energy efficiency and conservation is necessary for a sustainable transition toward hydrogen in the future. Reduced demand and addressing security concerns can be achieved most effectively through energy efficiency and conservation. When producing hydrogen, renewable energy sources must be given top priority. Undoubtedly, one of the most important elements in "pushing" or establishing a market for new clean technology is the capacity to establish standards or benchmarks that can be forced or voluntary. Standards for renewable portfolios should place a strong emphasis on collaboration between the public and commercial sectors.

References

1. Pal, D.B., Singh, A., Bhatnagar, A., A review on biomass based hydrogen production technologies. *Int. J. Hydrogen Energy*, 47, 3, Elsevier Ltd, 1461–1480, Jan. 08, 2022, doi: 10.1016/j.ijhydene.2021.10.124.
2. Yusuf, M., Alnarabiji, M.S., Abdullah, B., Clean Hydrogen Production Technologies, in: *Advances in Sustainable Energy*, pp. 159–170, Springer International Publishing, US, 2021, doi: 10.1007/978-3-030-74406-9_5.

3. Yu, M., Wang, K., Vredenburg, H., Insights into low-carbon hydrogen production methods: Green, blue and aqua hydrogen. *Int. J. Hydrogen Energy*, 46, 41, 21261–21273, Jun. 2021, doi: 10.1016/j.ijhydene.2021.04.016.
4. Dawood, F., Anda, M., Shafiullah, G.M., Hydrogen production for energy: An overview. *Int. J. Hydrogen Energy*, 45, 7, Elsevier Ltd, 3847–3869, Feb. 07, 2020, doi: 10.1016/j.ijhydene.2019.12.059.
5. Vidas, L. and Castro, R., Recent developments on hydrogen production technologies: State-of-the-art review with a focus on green-electrolysis. *Appl. Sci. (Switzerland)*, 11, 23, 11363, MDPI, Dec. 01, 2021, doi: 10.3390/app112311363.
6. Kumar, S.S. and Lim, H., An overview of water electrolysis technologies for green hydrogen production. *Energy Rep.*, 8, 13793–13813, Nov. 2022, doi: 10.1016/J.EGYR.2022.10.127.
7. Kayfeci, M., Keçebaş, A., Bayat, M., Hydrogen production, in: *Solar Hydrogen Production: Processes, Systems and Technologies*, pp. 45–83, Elsevier, Academic Press, US, 2019, doi: 10.1016/B978-0-12-814853-2.00003-5.
8. Momirlan, M. and Veziroglu, T., Recent directions of world hydrogen production. *Renew. Sustain. Energy Rev.*, 3, 219–231, 1999.
9. Ajanovic, A., Sayer, M., Haas, R., The economics and the environmental benignity of different colors of hydrogen. *Int. J. Hydrogen Energy*, 47, 57, 24136–24154, Jul. 2022, doi: 10.1016/j.ijhydene.2022.02.094.
10. Kalamaras, C.M. and Efstathiou, A.M., Hydrogen Production Technologies: Current State and Future Developments. *Conference Papers in Energy*, vol. 2013, pp. 1–9, Jun. 2013, doi: 10.1155/2013/690627.
11. Sørensen, B. and Spazzafumo, G., Hydrogen, in: *Hydrogen and Fuel Cells*, pp. 5–105, Elsevier, Academic Press, US, 2018, doi: 10.1016/B978-0-08-100708-2.00002-3.
12. Ishaq, H., Dincer, I., Crawford, C., A review on hydrogen production and utilization: Challenges and opportunities. *Int. J. Hydrogen Energy*, 47, 62, 26238–26264, 2021, doi: 10.1016/j.ijhydene.2021.11.149.
13. Li, S., Kang, Q., Baeyens, J., Zhang, H.L., Deng, Y.M., Hydrogen Production: State of Technology. *IOP Conf. Ser.: Earth Environ. Sci.*, Institute of Physics Publishing, vol. 544, no. 1, p. 012011, Jul. 2020, doi: 10.1088/1755-1315/544/1/012011.
14. C., Y.-Y., Q., F., L.-S., Kalyani, J., Mo-Doped FeS Mediated H2 Production from H2S via an *In Situ* Cyclic Sulfur Looping Scheme. *ACS Sustain. Chem. Eng.*, 9, 33, 11204–11211, 2021.
15. Ming, Q., Healey, T., Allen, L., Irving, P., Steam reforming of hydrocarbon fuels. *Catal. Today*, 77, 1–2, 51–64, Dec. 2002, doi: 10.1016/S0920-5861(02)00232-8.
16. Shirasaki, Y. *et al.*, Development of membrane reformer system for highly efficient hydrogen production from natural gas. *Int. J. Hydrogen Energy*, 34, 10, 4482–4487, May 2009, doi: 10.1016/j.ijhydene.2008.08.056.

17. Lindström, B. and Pettersson, L.J., Hydrogen generation by steam reforming of methanol over copper-based catalysts for fuel cell applications. *Int. J. Hydrogen Energy*, 26, 9, 923–933, Sep. 2001, doi: 10.1016/S0360-3199(01)00034-9.
18. Bepari, S. and Kuila, D., Steam reforming of methanol, ethanol and glycerol over nickel-based catalysts-A review. *Int. J. Hydrogen Energy*, 45, 36, 18090–18113, Jul. 2020, doi: 10.1016/J.IJHYDENE.2019.08.003.
19. Chin, Y.H., Dagle, R., Hu, J., Dohnalkova, A.C., Wang, Y., Steam reforming of methanol over highly active Pd/ZnO catalyst. *Catal. Today*, 77, 1–2, 79–88, Dec. 2002, doi: 10.1016/S0920-5861(02)00234-1.
20. Palo, D.R., Dagle, R.A., Holladay, J.D., Methanol Steam Reforming for Hydrogen Production. *Chem. Rev.*, 107, 10, 3992–4021, Sep. 2007.
21. Yusuf, M., Farooqi, A.S., Keong, L.K., Hellgardt, K., Abdullah, B., Contemporary trends in composite Ni-based catalysts for CO2 reforming of methane. *Chem. Eng. Sci.*, 229, 116072, Jan. 2021, doi: 10.1016/j.ces.2020.116072.
22. Abdullah, B., Abd Ghani, N.A., Vo, D.V.N., Recent advances in dry reforming of methane over Ni-based catalysts. *J. Cleaner Prod.*, 162, Elsevier Ltd, 170–185, Sep. 20, 2017, doi: 10.1016/j.jclepro.2017.05.176.
23. Lavoie, J.M., Review on dry reforming of methane, a potentially more environmentally-friendly approach to the increasing natural gas exploitation. *Front. Chem.*, 2, 81, NOV, Frontiers Media S. A, 2014, doi: 10.3389/fchem.2014.00081.
24. Alvarez-Galvan, C. *et al.*, Partial oxidation of methane to syngas over nickel-based catalysts: Influence of support type, addition of rhodium, and preparation method. *Front. Chem.*, 7, 104, MAR, 2019, doi: 10.3389/fchem.2019.00104.
25. Semelsberger, T.A., Brown, L.F., Borup, R.L., Inbody, M.A., Equilibrium products from autothermal processes for generating hydrogen-rich fuel-cell feeds. *Int. J. Hydrogen Energy*, 29, 10, 1047–1064, Aug. 2004, doi: 10.1016/S0360-3199(03)00214-3.
26. Krummenacher, J.J., West, K.N., Schmidt, L.D., Catalytic partial oxidation of higher hydrocarbons at millisecond contact times: Decane, hexadecane, and diesel fuel. *J. Catal.*, 215, 2, 332–343, Apr. 2003, doi: 10.1016/S0021-9517(03)00011-3.
27. Wilhelm, D.J., Simbeck, D.R., Karp, A.D., Dickenson, R.L., Syngas production for gas-to-liquids applications: technologies, issues and outlook q, *Fuel Process. Technol. (FPT)*, 71, 1–3, 139–148, 2001, [Online], Available: www.elsevier.comrlocaterfuproc.
28. Rahimpour, M.R., Jafari, M., Iranshahi, D., Progress in catalytic naphtha reforming process: A review. *Appl. Energy*, 109, Elsevier Ltd, 79–93, 2013, doi: 10.1016/j.apenergy.2013.03.080.

29. Trane, R., Dahl, S., Skjøth-Rasmussen, M.S., Jensen, A.D., Catalytic steam reforming of bio-oil. *Int. J. Hydrogen Energy*, 37, 8, Elsevier Ltd, 6447–6472, 2012, doi: 10.1016/j.ijhydene.2012.01.023.
30. Iranshahi, D., Pourazadi, E., Paymooni, K., Rahimpour, M.R., Jahanmiri, A., Moghtaderi, B., A dynamic membrane reactor concept for naphtha reforming, considering radial-flow patterns for both sweeping gas and reacting materials. *Chem. Eng. J.*, 178, 264–275, Dec. 2011, doi: 10.1016/j.cej.2011.08.005.
31. Msheik, M., Rodat, S., Abanades, S., Methane Cracking for Hydrogen Production: A Review of Catalytic and Molten Media Pyrolysis. *Energies (Basel)*, 14, 11, 3107–3107, May 2021.
32. Chung, K.H., Park, Y.K., Jung, S.C., Hydrogen production by cracking of ammonium hydroxide using liquid-phase plasma on the modified TiO2 photocatalysts. *Int. J. Hydrogen Energy*, 47, 98, 41631–41639, Dec. 2022, doi: 10.1016/J.IJHYDENE.2022.06.032.
33. Yousefi Rizi, H.A. and Shin, D., Green Hydrogen Production Technologies from Ammonia Cracking. *Energies (Basel)*, 15, 21, 8246–8246, Nov. 2022.
34. Al-Zareer, M., Dincer, I., Rosen, M.A., Analysis and assessment of a hydrogen production plant consisting of coal gasification, thermochemical water decomposition and hydrogen compression systems. *Energy Convers. Manage.*, 157, 600–618, Feb. 2018, doi: 10.1016/J.ENCONMAN.2017.11.047.
35. Moghadam, R.A., Yusup, S., Uemura, Y., Chin, B.L.F., Lam, H.L., al Shoaibi, A., Syngas production from palm kernel shell and polyethylene waste blend in fluidized bed catalytic steam co-gasification process. *Energy*, 75, 40–44, Oct. 2014, doi: 10.1016/j.energy.2014.04.062.
36. Seyitoglu, S.S., Dincer, I., Kilicarslan, A., Energy and exergy analyses of hydrogen production by coal gasification. *Int. J. Hydrogen Energy*, 42, 4, 2592–2600, Jan. 2017, doi: 10.1016/J.IJHYDENE.2016.08.228.
37. Burchart, D., Gazda-Grzywacz, M., Grzywacz, P., Burmistrz, P., Zarębska, K., Life Cycle Assessment of Hydrogen Production from Coal Gasification as an Alternative Transport Fuel. *Energies (Basel)*, 16, 1, 383, Jan. 2023, doi: 10.3390/en16010383.
38. Sun, J., Feng, H., Xu, J., Jin, H., Guo, L., Investigation of the conversion mechanism for hydrogen production by coal gasification in supercritical water. *Int. J. Hydrogen Energy*, 46, 17, 10205–10215, Mar. 2021, doi: 10.1016/J.IJHYDENE.2020.12.130.
39. Midilli, A., Kucuk, H., Topal, M.E., Akbulut, U., Dincer, I., A comprehensive review on hydrogen production from coal gasification: Challenges and Opportunities. *Int. J. Hydrogen Energy*, 46, 50, 25385–25412, Jul. 2021, doi: 10.1016/J.IJHYDENE.2021.05.088.
40. Lewis, D., Hydrogen and its relationship with nuclear energy. *Prog. Nucl. Energy*, 50, 2–6, 394–401, Mar. 2008, doi: 10.1016/J.PNUCENE.2007.11.072.

41. Ozcan, H. and Dincer, I., Thermodynamic modeling of a nuclear energy based integrated system for hydrogen production and liquefaction. *Comput. Chem. Eng.*, 90, 234–246, Jul. 2016, doi: 10.1016/J.COMPCHEMENG.2016.04.015.
42. Orhan, M.F., Dincer, I., Rosen, M.A., Kanoglu, M., Integrated hydrogen production options based on renewable and nuclear energy sources. *Renew. Sustain. Energy Rev.*, 16, 8, 6059–6082, Oct. 2012, doi: 10.1016/J.RSER.2012.06.008.
43. Nowotny, J. *et al.*, Towards sustainable energy. Generation of hydrogen fuel using nuclear energy. *Int. J. Hydrogen Energy*, 41, 30, 12812–12825, Aug. 2016, doi: 10.1016/J.IJHYDENE.2016.05.054.
44. Bhattacharyya, R., Singh, K.K., Bhanja, K., Grover, R.B., Leveraging nuclear power-to-green hydrogen production potential in India: A country perspective. *Int. J. Energy Res.*, 46, 13, 18901–18918, Oct. 2022, doi: 10.1002/er.8348.
45. Atilhan, S., Park, S., El-Halwagi, M.M., Atilhan, M., Moore, M., Nielsen, R.B., Green hydrogen as an alternative fuel for the shipping industry. *Curr. Opin. Chem. Eng.*, 31, 100668, Mar. 2021, doi: 10.1016/J.COCHE.2020.100668.
46. Hoelzen, J., Silberhorn, D., Zill, T., Bensmann, B., Hanke-Rauschenbach, R., Hydrogen-powered aviation and its reliance on green hydrogen infrastructure – Review and research gaps. *Int. J. Hydrogen Energy*, 47, 5, 3108–3130, Jan. 2022, doi: 10.1016/J.IJHYDENE.2021.10.239.
47. Agaton, C.B., Batac, K., II, Reyes, E.M., Prospects and challenges for green hydrogen production and utilization in the Philippines. *Int. J. Hydrogen Energy*, 47, 41, 17859–17870, May 2022, doi: 10.1016/J.IJHYDENE.2022.04.101.
48. Oliveira, A.M., Beswick, R.R., Yan, Y., A green hydrogen economy for a renewable energy society. *Curr. Opin. Chem. Eng.*, 33, 100701, Sep. 2021, doi: 10.1016/J.COCHE.2021.100701.
49. Wang, M., Wang, G., Sun, Z., Zhang, Y., Xu, D., Review of renewable energy-based hydrogen production processes for sustainable energy innovation. *Global Energy Interconnect.*, 2, 5, 436–443, Oct. 2019, doi: 10.1016/j.gloei.2019.11.019.
50. Nikolaidis, P. and Poullikkas, A., A comparative overview of hydrogen production processes. *Renew. Sustain. Energy Rev.*, 67, 597–611, Jan. 2017, doi: 10.1016/J.RSER.2016.09.044.
51. Mehrpooya, M. and Habibi, R., A review on hydrogen production thermochemical water-splitting cycles. *J. Clean. Prod.*, 275, 123836, Dec. 2020, doi: 10.1016/J.JCLEPRO.2020.123836.
52. Li, X., Sun, X., Song, Q., Yang, Z., Wang, H., Duan, Y., A critical review on integrated system design of solar thermochemical water-splitting cycle for hydrogen production. *Int. J. Hydrogen Energy*, 47, 79, 33619–33642, Sep. 2022, doi: 10.1016/J.IJHYDENE.2022.07.249.
53. Muhich, C.L., Ehrhart, B.D., Al-Shankiti, I., Ward, B.J., Musgrave, C.B., Weimer, A.W., A review and perspective of efficient hydrogen generation via solar thermal water splitting. *WIREs Energy Environ.*, 5, 3, 261–287, May 2016.

54. Abanades, S., Metal oxides applied to thermochemical water-splitting for hydrogen production using concentrated solar energy. *Chem. Eng.*, 3, 3, MDPI AG, 1–28, Sep. 01, 2019, doi: 10.3390/chemengineering3030063.
55. Orfila, M., Linares, M., Molina, R., Botas, J. Á., Sanz, R., Marugán, J., Perovskite materials for hydrogen production by thermochemical water splitting. *Int. J. Hydrogen Energy*, 41, 42, 19329–19338, Nov. 2016, doi: 10.1016/J.IJHYDENE.2016.07.041.
56. Zhai, S. *et al.*, The use of poly-cation oxides to lower the temperature of two-step thermochemical water splitting. *Energy Environ. Sci.*, 11, 8, 2172–2178, Jun. 2018.
57. Emery, A.A., Saal, J.E., Kirklin, S., Hegde, V., II, Wolverton, C., High-Throughput Computational Screening of Perovskites for Thermochemical Water Splitting Applications. *Chem. Mater.*, 28, 16, 5621–5634, Jun. 2016.
58. Qian, X., He, J., Mastronardo, E., Baldassarri, B., Wolverton, C., Haile, S.M., Favorable Redox Thermodynamics of SrTi0.5Mn0.5O3–δ in Solar Thermochemical Water Splitting. *Chem. Mater.*, 32, 21, 9335–9346, Oct. 2020.
59. Budama, V.K., Rincon Duarte, J.P., Roeb, M., Sattler, C., Potential of solar thermochemical water-splitting cycles: A review. *Sol. Energy*, 249, 353–366, Jan. 2023, doi: 10.1016/J.SOLENER.2022.11.001.
60. Kumar, S.S. and Himabindu, V., Hydrogen production by PEM water electrolysis – A review. *Mater. Sci. Energy Technol.*, 2, 3, 442–454, Dec. 2019, doi: 10.1016/J.MSET.2019.03.002.
61. Bhandari, R., Trudewind, C.A., Zapp, P., Life cycle assessment of hydrogen production via electrolysis - A review. *J. Cleaner Prod.*, 85, Elsevier Ltd, 151–163, Dec. 15, 2014, doi: 10.1016/j.jclepro.2013.07.048.
62. Guo, Y., Li, G., Zhou, J., Liu, Y., Comparison between hydrogen production by alkaline water electrolysis and hydrogen production by PEM electrolysis. *IOP Conf. Ser.: Earth Environ. Sci.*, 371, 4, 042022, Institute of Physics Publishing, Dec. 2019, doi: 10.1088/1755-1315/371/4/042022.
63. Wang, S., Lu, A., Zhong, C.J., Hydrogen production from water electrolysis: role of catalysts. *Nano Convergence*, 8, 1, Korea Nano Technology Research Society, Dec. 01, 2021, doi: 10.1186/s40580-021-00254-x.
64. Zhang, Y. and Angelidaki, I., Microbial electrolysis cells turning to be versatile technology: Recent advances and future challenges. *Water Res.*, 56, 11–25, Jun. 2014, doi: 10.1016/J.WATRES.2014.02.031.
65. Ndayisenga, F. *et al.*, Microbial electrohydrogenesis cell and dark fermentation integrated system enhances biohydrogen production from lignocellulosic agricultural wastes: Substrate pretreatment towards optimization. *Renew. Sustain. Energy Rev.*, 145, 111078, Jul. 2021, doi: 10.1016/J.RSER.2021.111078.
66. Miller, H.A. *et al.*, Green hydrogen from anion exchange membrane water electrolysis: a review of recent developments in critical materials and

operating conditions. *Sustain. Energy Fuels*, 4, 5, 2114–2133, 2020, doi: 10.1039/C9SE01240K.

67. Hua, T., Li, S., Li, F., Zhou, Q., Ondon, B.S., Microbial electrolysis cell as an emerging versatile technology: a review on its potential application, advance and challenge. *J. Chem. Technol. Biotechnol.*, 94, 6, 1697–1711, Jun. 2019, doi: 10.1002/jctb.5898.
68. Kadier, A., Simayi, Y., Abdeshahian, P., Azman, N.F., Chandrasekhar, K., Kalil, M.S., A comprehensive review of microbial electrolysis cells (MEC) reactor designs and configurations for sustainable hydrogen gas production. *Alexandria Eng. J.*, 55, 1, 427–443, Mar. 2016, doi: 10.1016/J.AEJ.2015.10.008.
69. Kadier, A. *et al.*, Recent advances and emerging challenges in microbial electrolysis cells (MECs) for microbial production of hydrogen and value-added chemicals. *Renew. Sustain. Energy Rev.*, 61, 501–525, Aug. 2016, doi: 10.1016/J.RSER.2016.04.017.
70. Vincent, I. and Bessarabov, D., Low cost hydrogen production by anion exchange membrane electrolysis: A review. *Renew. Sustain. Energy Rev.*, 81, 1690–1704, Jan. 2018, doi: 10.1016/J.RSER.2017.05.258.
71. Cheng, W.H., De La Calle, A., Atwater, H.A., Stechel, E.B., Xiang, C., Hydrogen from Sunlight and Water: A Side-by-Side Comparison between Photoelectrochemical and Solar Thermochemical Water-Splitting. *ACS Energy Lett.*, 6, 9, 3096–3113, Aug. 2021.
72. Vincent, I., Kruger, A., Bessarabov, D., Hydrogen Production by water Electrolysis with an Ultrathin Anion-exchange membrane (AEM). *Int. J. Electrochem. Sci.*, 13, 12, 11347–11358, Dec. 2018, doi: 10.20964/2018.12.84.
73. Dincer, I. and Acar, C., Review and evaluation of hydrogen production methods for better sustainability. *Int. J. Hydrogen Energy*, 40, 34, 11094–11111, Aug. 2014, doi: 10.1016/j.ijhydene.2014.12.035.
74. Rothschild, A. and Dotan, H., Beating the Efficiency of Photovoltaics-Powered Electrolysis with Tandem Cell Photoelectrolysis. *ACS Energy Lett.*, 2, 1, 45–51, Jan. 2017, doi: 10.1021/acsenergylett.6b00610.
75. Faid, A.Y., Xie, L., Barnett, A.O., Seland, F., Kirk, D., Sunde, S., Effect of anion exchange ionomer content on electrode performance in AEM water electrolysis. *Int. J. Hydrogen Energy*, 45, 53, 28272–28284, Oct. 2020, doi: 10.1016/J.IJHYDENE.2020.07.202.
76. Azwar, M.Y., Hussain, M.A., Abdul-Wahab, A.K., Development of biohydrogen production by photobiological, fermentation and electrochemical processes: A review. *Renew. Sustain. Energy Rev.*, 31, 158–173, Mar. 2014, doi: 10.1016/J.RSER.2013.11.022.
77. Ghiasian, M., Biophotolysis-Based Hydrogen Production by Cyanobacteria, in: *Prospects of Renewable Bioprocessing in Future Energy Systems*. Biofuel and Biorefinery Technologies, vol. 10, pp. 161–184, Springer, Cham, 2019, doi: 10.1007/978-3-030-14463-0_5.
78. Yu, J., Yu, J., Takahashi, P., Biophotolysis-based Hydrogen Production by Cyanobacteria and Green Microalgae, *Communicating current research*

and educational topics and trends in applied microbiology, 1, 79–89, 2007, [Online], Available: https://www.researchgate.net/publication/242768941.
79. Kaushik, A. and Sharma, M., Exploiting Biohydrogen Pathways of Cyanobacteria and Green Algae: An Industrial Production Approach, in: *Biohydrogen Production: Sustainability of Current Technology and Future Perspective*, pp. 97–113, Springer India, New Delhi, 2017, doi: 10.1007/978-81-322-3577-4_5.
80. Show, K.Y., Yan, Y.G., Lee, D.J., Biohydrogen production from algae: Perspectives, challenges, and prospects, in: *Biomass, Biofuels, Biochemicals: Biofuels from Algae*, Second Edition, pp. 325–343, Jan. 2019.
81. Rather, A.H., Srivastav, A.K., Abdul, A., A Study on Biohydrogen Production based on Biophotolysis from Cyanobacteria. *Ann. Rom. Soc. Cell Biol.*, 25, 6, 12500-12509, 2021, [Online], Available: http://annalsofrscb.ro.
82. Li, S., Li, F., Zhu, X., Liao, Q., Chang, J.S., Ho, S.H., Biohydrogen production from microalgae for environmental sustainability. *Chemosphere*, 291, 132717, Mar. 2022, doi: 10.1016/J.CHEMOSPHERE.2021.132717.
83. Nagarajan, D., Lee, D.J., Kondo, A., Chang, J.S., Recent insights into biohydrogen production by microalgae – From biophotolysis to dark fermentation. *Bioresour. Technol.*, 227, 373–387, Mar. 2017, doi: 10.1016/J.BIORTECH.2016.12.104.
84. Javed, M.A. *et al.*, The role of oxygen regulation and algal growth parameters in hydrogen production via biophotolysis. *J. Environ. Chem. Eng.*, 10, 1, 107003, Feb. 2022, doi: 10.1016/J.JECE.2021.107003.
85. Sagir, E. and Alipour, S., Photofermentative hydrogen production by immobilized photosynthetic bacteria: Current perspectives and challenges. *Renew. Sustain. Energy Rev.*, 141, 110796, May 2021, doi: 10.1016/J.RSER.2021.110796.
86. Lu, C. *et al.*, Bio-hydrogen production from apple waste by photosynthetic bacteria HAU-M1. *Int. J. Hydrogen Energy*, 41, 31, 13399–13407, Aug. 2016, doi: 10.1016/J.IJHYDENE.2016.06.101.
87. Wang, Y., Tahir, N., Cao, W., Zhang, Q., Lee, D.J., Grid columnar flat panel photobioreactor with immobilized photosynthetic bacteria for continuous photofermentative hydrogen production. *Bioresour. Technol.*, 291, 121806, Nov. 2019, doi: 10.1016/J.BIORTECH.2019.121806.
88. Meky, N., Ibrahim, M.G., Fujii, M., Elreedy, A., Tawfik, A., Integrated dark-photo fermentative hydrogen production from synthetic gelatinaceous wastewater via cost-effective hybrid reactor at ambient temperature. *Energy Convers. Manage.*, 203, 112250, Jan. 2020, doi: 10.1016/J.ENCONMAN.2019.112250.
89. Cui, S. *et al.*, Light-assisted fermentative hydrogen production in an intimately-coupled inorganic-bio hybrid with self-assembled nanoparticles. *Chem. Eng. J.*, 428, 131254, Jan. 2022, doi: 10.1016/J.CEJ.2021.131254.
90. Chandra, R. and Venkata Mohan, S., Enhanced bio-hydrogenesis by co-culturing photosynthetic bacteria with acidogenic process: Augmented

dark-photo fermentative hybrid system to regulate volatile fatty acid inhibition. *Int. J. Hydrogen Energy*, 39, 14, 7604–7615, May 2014, doi: 10.1016/J.IJHYDENE.2014.01.196.

91. Guo, X.M., Trably, E., Latrille, E., Carrre, H., Steyer, J.P., Hydrogen production from agricultural waste by dark fermentation: A review. *Int. J. Hydrogen Energy*, 35, 19, 10660–10673, Oct. 2010, doi: 10.1016/J.IJHYDENE.2010.03.008.
92. Karaeva, J.V., Hydrogen production at centralized utilization of agricultural waste. *Int. J. Hydrogen Energy*, 46, 69, 34089–34096, Oct. 2021, doi: 10.1016/J.IJHYDENE.2021.08.004.
93. Karaeva, J.V. *et al.*, CO-PYROLYSIS of agricultural waste and estimation of the applicability of pyrolysis in the integrated technology of biorenewable hydrogen production. *Int. J. Hydrogen Energy*, 47, 23, 11787–11798, Mar. 2022, doi: 10.1016/J.IJHYDENE.2022.02.057.
94. Li, Y.C. *et al.*, Techno-economic evaluation of biohydrogen production from wastewater and agricultural waste. *Int. J. Hydrogen Energy*, 37, 20, 15704–15710, Oct. 2012, doi: 10.1016/J.IJHYDENE.2012.05.043.
95. Zheng, Y. *et al.*, A review on biological recycling in agricultural waste-based biohydrogen production: Recent developments. *Bioresour. Technol.*, 347, 126595, Mar. 2022, doi: 10.1016/J.BIORTECH.2021.126595.
96. Chatellard, L., Marone, A., Carrère, H., Trably, E., Trends and Challenges in Biohydrogen Production from Agricultural Waste, in: *Biohydrogen Production: Sustainability of Current Technology and Future Perspective*, pp. 69–95, Springer India, New Delhi, 2017, doi: 10.1007/978-81-322-3577-4_4.
97. Salimi, M., Safari, F., Tavasoli, A., Shakeri, A., Hydrothermal gasification of different agricultural wastes in supercritical water media for hydrogen production: a comparative study. *Int. J. Ind. Chem.*, 7, 3, 277–285, Sep. 2016, doi: 10.1007/s40090-016-0091-y.
98. Li, B., Yang, H., Wei, L., Shao, J., Wang, X., Chen, H., Hydrogen production from agricultural biomass wastes gasification in a fluidized bed with calcium oxide enhancing. *Int. J. Hydrogen Energy*, 42, 8, 4832–4839, Feb. 2017, doi: 10.1016/J.IJHYDENE.2017.01.138.
99. Ayas, N. and Esen, T., Hydrogen production from tea waste. *Int. J. Hydrogen Energy*, 41, 19, 8067–8072, May 2016, doi: 10.1016/J.IJHYDENE.2015.09.156.
100. Saratale, G.D. *et al.*, Cellulolytic Enzymes Production by Utilizing Agricultural Wastes Under Solid State Fermentation and its Application for Biohydrogen Production. *Appl. Biochem. Biotechnol.*, 174, 8, 2801–2817, Dec. 2014, doi: 10.1007/s12010-014-1227-1.
101. Zhang, Z. *et al.*, Thermo-physical properties of pretreated agricultural residues for bio-hydrogen production using thermo-gravimetric analysis. *Int. J. Hydrogen Energy*, 41, 10, 5234–5242, Mar. 2016, doi: 10.1016/J.IJHYDENE.2016.01.079.

102. Isha, R. and Williams, P.T., Pyrolysis-gasification of agriculture biomass wastes for hydrogen production. *J. Energy Inst.*, 84, 2, 80–87, May 2011, doi: 10.1179/014426011X12968328625432.
103. Tleubergenova, A., Han, B.C., Meng, X.Z., Assessment of biomass-based green hydrogen production potential in Kazakhstan. *Int. J. Hydrogen Energy*, 49, 349–355, Aug. 2023, doi: 10.1016/J.IJHYDENE.2023.08.197.

3

Toward Green Hydrogen Generation

Kexin Yin, Yinglong Wang*, Xiaoying Zhang, Yusen Chen and Jianhui Zhong

College of Chemical Engineering, Qingdao University of Science and Technology, Qingdao, People's Republic of China

Abstract

Green hydrogen production is a novel method that prioritizes environmental sustainability in the production of hydrogen. In contrast to traditional gray and blue hydrogen production, green hydrogen significantly reduces carbon dioxide emissions and decreases reliance on fossil fuels. This process involves utilizing renewable energy sources such as solar and wind power to drive the electrolysis of water, resulting in the generation of hydrogen. The hydrogen produced in this manner is extensively applied in energy, transportation, and chemical industries. In addition, it is deemed a pivotal solution for reducing carbon emissions and mitigating greenhouse gasses. This chapter outlines the process description and working principle of various green hydrogen production processes, compares and analyzes various advanced hydrogen production technologies, and clarifies their respective advantages and limitations. The existing bottlenecks and barriers pertaining to hydrogen production equipment, technology, and economics are explored deeply. Finally, the current state and challenges encountered in the industrialization of green hydrogen production are summarized, thereby offering guidance for its future development. With ongoing technological advancements, green hydrogen production is expected to assume a greater role and propel the advancement and application of clean energy.

***Keywords*:** Traditional hydrogen production, green hydrogen production, carbon neutrality, sustainable hydrogen technology

**Corresponding author*: wangyinglong@qust.edu.cn

Mehmet Sankir and Nurdan Demirci Sankir (eds.) Towards Green Hydrogen Generation, (87–120)

List of Acronyms

PSA	pressure swing adsorption
AWE	alkaline water electrolysis
PEM	proton exchange membrane
SOWE	solid oxide water electrolysis
FFHP	fossil fuel hydrogen production
HPWE	hydrogen production by water electrolysis
BHP	biomass hydrogen production
NEHP	nuclear energy hydrogen production
HPS	hydrogen production from seawater
HPNG	hydrogen production from natural gas
HPP	hydrogen production from petroleum
THPT	thermochemical hydrogen production technology
PDMHP	photothermal decomposition method for hydrogen production

3.1 Introduction

In recent years, with the continuous acceleration of the process of social industrialization and the rapid growth of economic development, people's demand for energy has gradually increased. The energy consumption in the world today is still dominated by nonrenewable fossil fuels, and the world's energy consumption is facing a huge crisis and challenge. Energy is an important thrust of social development, and the development and utilization of fossil energy have greatly promoted the change of production mode and the development of human society. However, the reserves of fossil energy are limited, and the combustion process causes certain environmental pollution, which seriously restricts the realization of the goals of "carbon peak" and "carbon neutrality". Hydrogen energy, an important clean energy source, is burned to produce water and does not produce any pollutants. At present, the development of the economy is a key measure to achieving a "carbon peak" and helping "carbon neutrality", which has important strategic significance. The development of the hydrogen economy mainly consists of three parts: (1) developing green hydrogen production; (2) hydrogen energy network, the development of pipeline transportation, regional centers, or field use; (3) in terms of application, the development of hydrogen energy transportation, industrial demonstrations, and community heating tests. As the source of the "Hydrogen Economy", how to achieve green large-scale hydrogen production has become the top priority of research. Figure 3.1 shows the different routes of driving energy (i.e., thermal and electrical) from clean energy sources that can be utilized in integrated combined heat and power systems for hydrogen production through pure or hybrid thermochemical cycles.

Hydrogen energy is regarded as one of the most ideal renewable green energies in the current society for its many advantages [2–4]: (1) Hydrogen sources are relatively extensive; (2) hydrogen is a green and clean energy source; (3) the combustion heat value of hydrogen is very high and the thermal conductivity is very good; (4) hydrogen can be stored and transported. At present, mature hydrogen production methods include fossil fuel hydrogen production (FFHP), hydrogen production by water electrolysis (HPWE), solar water decomposition hydrogen production, biomass hydrogen production (BHP), nuclear energy hydrogen production (NEHP), and hydrogen production from seawater technology (HPS) [5]. The cost of hydrogen production from fossil fuels is relatively low and the

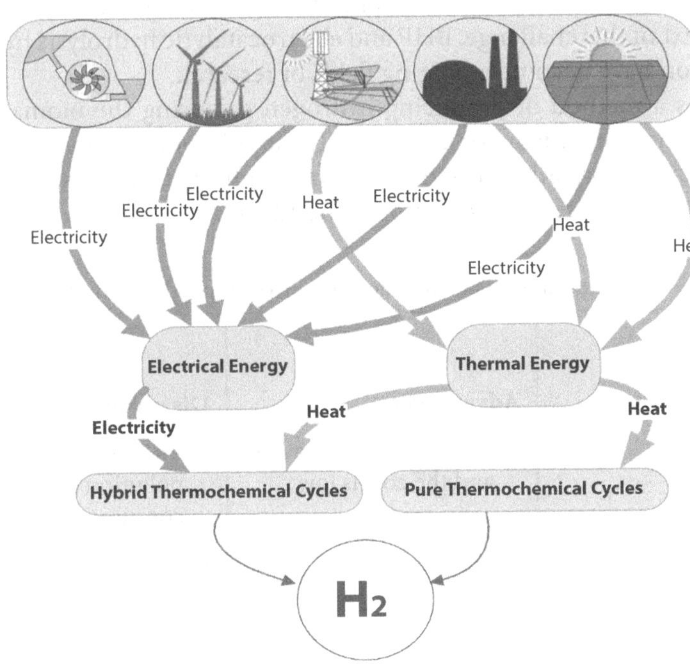

Figure 3.1 Pathways for carbon-neutral hydrogen production using nonfossil-based sources [1].

technology has been mature. The process of hydrogen production from fossil fuels produces a large amount of greenhouse gasses such as carbon dioxide, which has a bad impact on the environment. The cost of HPWE is relatively low, which only produces oxygen and hydrogen without the production of other pollutants, but it needs to consume a lot of electric energy. The BHP technology is not fully mature and needs further research before large-scale application. Hydrogen production from nuclear energy has the advantages of no greenhouse gas production, water as raw material, high efficiency, and large scale, and is an important solution for large quantities of hydrogen supply in the future. HPS has the advantage of multiple energy utilization, but the composition of seawater is very complex, the electrolyzer used may be different, and the cost is high. Table 3.1 shows the advantages and disadvantages of various hydrogen production methods. At present, the relatively mature hydrogen production methods mainly have the following three kinds [6–8].

Traditional FFHP methods usually involve natural gas steam reforming or coal gasification processes, which can produce large amounts of hydrogen, but with high carbon emissions and environmental pollution. Greenhouse gas emissions have a serious impact on climate change, so reducing carbon emissions has become an urgent task on a global scale. In the context of this challenge, BHP and electrocatalytic hydrolysis hydrogen production have become emerging fields of research.

BHP is a method of producing hydrogen gas using the biomass reactions of microorganisms or enzymes. These microorganisms are usually able to convert organic waste or wastewater into hydrogen gas under low temperature and pressure conditions. The advantage of BHP is its relatively

Table 3.1 Advantages and disadvantages of hydrogen production methods.

Hydrogen production technology	**Advantages**	**Disadvantages**
FFHP	The technology is mature and the cost is low	Environmental pollution and energy consumption
BHP	Environmentally friendly	Low hydrogen production efficiency
Water electrolysis hydrogen production	Green environmental protection and wide source of raw materials	Electric energy consumption

low energy consumption while also reducing the cost of waste disposal. One potential application of this approach is to treat organic waste and convert it into useful energy, thus achieving the double benefit of waste reuse.

HPWE is a method of producing hydrogen by electrolyzing water molecules. This process involves passing an electric current through water, splitting it into hydrogen and oxygen. One of the key advantages is that the HPWE can use renewable energy sources, such as solar or wind power, as a power source, thus eliminating greenhouse gas emissions. In addition, this method can achieve efficient energy conversion, providing an efficient way to produce hydrogen.

While these emerging technologies have many advantages, they also face some challenges. One challenge for hydrogen production from biomass is the optimization of microbial growth conditions and hydrogen production efficiency to increase productivity. HPWE requires efficient catalytic materials and electrolytic equipment to ensure economic viability.

In summary, hydrogen production is an area of intense interest, gaining attention for its potential sustainability and environmental properties. Traditional methods of producing hydrogen from fossil fuels pose serious environmental problems, so emerging technologies such as BHP and electrocatalytic hydrolysis hydrogen production offer new hope for future hydrogen production. By studying and comparing these methods in depth, they can provide important insights into clean, sustainable hydrogen production and hopefully contribute to solving global energy problems.

3.2 Traditional Hydrogen Production Technology

3.2.1 Coal-to-Hydrogen

Coal-to-hydrogen is also known as coal gasification to hydrogen, which is composed of four parts: coal steam conversion to gas, gas purification, gas conversion, and pressure swing adsorption (PSA) purification of hydrogen. Coal or coke reacts with water steam at high temperatures to generate gasses mainly containing hydrogen, carbon monoxide (CO), and carbon dioxide (CO_2). After cooling, dust removal, and desulfurization, the gas is mixed with water steam for a conversion reaction, most of the carbon monoxide is converted into hydrogen and carbon dioxide (CO_2), becoming the conversion gas, and then the conversion gas through the PSA process, to obtain high purity hydrogen. The process flow chart of coal-to-hydrogen is shown in Figure 3.2.

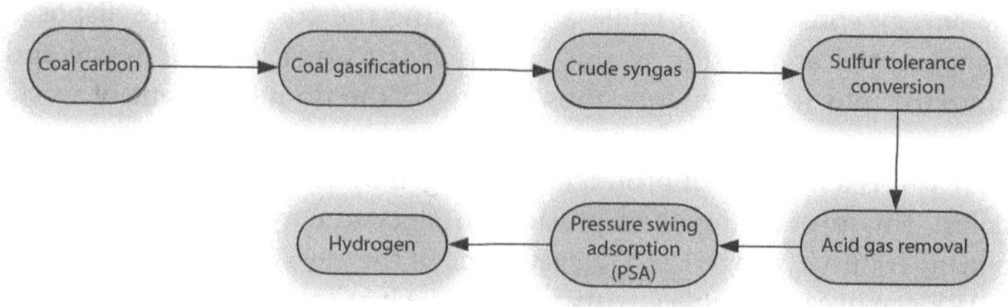

Figure 3.2 Simple flow chart of coal to hydrogen.

In the gasification process, coal uses water, air, or oxygen as a gasification agent to convert into syngas [9]. The advantages of coal gasification for hydrogen production are as follows: on the one hand, syngas obtained from coal gasification can be properly applied to combined cycle technology (IGCC) to improve power generation efficiency and comprehensive utilization efficiency of coal [10]. On the other hand, through the steam conversion reaction of syngas, CO_2 and H_2 are enriched in the product gas, which is conducive to low-energy CO_2 capture and high-quality hydrogen production. However, coal-to-hydrogen also has corresponding problems, coal-to-hydrogen process is complex and needs to carry out complex gas purification and treatment processes to remove harmful substances and impurities, which will increase equipment and operating costs, low energy efficiency, coal-to-hydrogen process energy conversion efficiency is relatively low, usually only about 50% to 60%. This means that the energy of coal cannot be fully converted into hydrogen energy, and there will be a certain amount of energy waste. Coal gasification produces large amounts of carbon dioxide (CO_2), which is one of the main greenhouse gasses and contributes significantly to climate change. Coal-to-hydrogen technology will therefore lead to a large amount of carbon emissions and exacerbate the global greenhouse effect, and measures need to be taken to capture, utilize, and store carbon dioxide to reduce the environmental impact [11].

3.2.2 Hydrogen Production from Natural Gas

3.2.2.1 Hydrogen Production from Natural Gas (HPNG)

Hydrogen production from natural gas (HPNG) includes the main reaction of the following two processes:

(1) Natural gas and water react to produce carbon monoxide and hydrogen at a high temperature of 100°C and under the condition of nickel oxide catalysis, the reaction formula is as follows:

$$CH_4+H_2O \rightarrow H_2+CO+Q$$

(2) Carbon monoxide and water react at 430°C and under the condition of ferric oxide catalyst to produce carbon dioxide and hydrogen, the reaction formula is as follows:

$$H_2O+CO \rightarrow CO_2+H_2+Q$$

3.2.2.2 Natural Gas Hydrogen Production Process

The main processes of natural gas production include atmospheric and vacuum distillation, catalytic cracking, catalytic reforming, and aromatics production, At the same time, it includes natural gas extraction, gathering, and purification. Under certain pressure, high temperatures, and catalysts, the alkane and water vapor in natural gas react chemically. The converted gas is heated by the waste heat boiler and enters the converter furnace to convert CO into H_2 and CO_2. After heat transfer, condensation, and water separation, the gas is sequentially passed through the adsorption tower equipped with three specific adsorbents through program control, and N_2, CO, CH_4, and CO_2 are extracted by PSA. The hydrogen production process from natural gas is shown in Figure 3.3.

3.2.2.3 Advantages and Disadvantages of HPNG

Compared with other hydrogen production methods, HPNG has higher production efficiency and can produce hydrogen quickly and on a large scale, adapting to the fast-paced and efficient hydrogen production methods in today's society. The raw material cost of HPNG is relatively low, the production equipment is relatively simple, and it is cheaper than other hydrogen production methods. Convenient storage and transportation: The hydrogen produced by HPNG can be compressed and stored, with relatively low storage costs. It can also be transported through pipelines for convenient and safe storage and transportation. Disadvantages include climate change issues. Natural gas contains methane, which is a potent

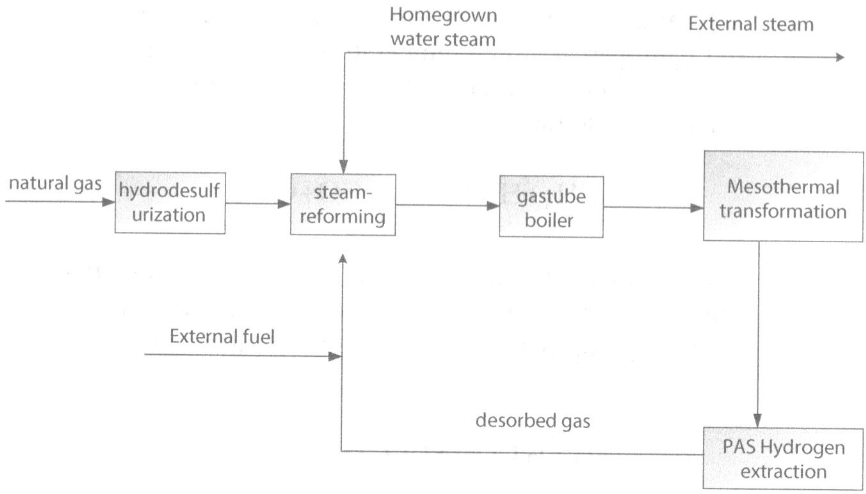

Figure 3.3 Hydrogen production process from natural gas.

greenhouse gas. Its exhaust emissions can lead to climate change and exacerbate global warming. Environmental pollution: Harmful gasses such as carbon dioxide are also produced in the process of HPNG. The pollution caused to the environment has also become a major challenge for HPNG. Energy security: Natural gas is a nonrenewable energy source that relies on it as a raw material for hydrogen production. Once the natural gas supply is interrupted, it will affect the production of hydrogen, thereby affecting the development of the entire industry.

3.2.3 Hydrogen Production from Petroleum

Hydrogen production from petroleum (HPP) is a method of producing hydrogen from petroleum as a raw material, which can be divided into surface operation and underground operation. Ground operations include refining hydrocarbons in specialized furnaces or processing plants to produce hydrogen gas. Underground operations rely on *in situ* processes that occur within the reservoir [12], such as gasification, combustion, and biodegradation to produce hydrogen gas (Figure 3.4). Ground operations mainly include raw material preparation, raw material pretreatment, steam reforming or partial oxidation, and other components [13–15]. Through the treatment of different devices, the separated hydrogen gas is purified,

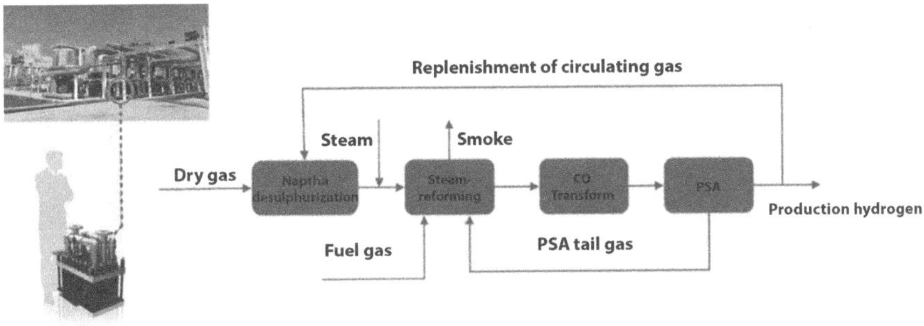

Figure 3.4 Diagram of *in situ* gasification process for hydrogen production [12].

and impurities and residual carbon oxides are removed, stored, or transported to the target location as needed.

HPP is an energy-intensive process that requires a large amount of oil as raw material and consumes a large amount of electricity or natural gas to facilitate the reaction, which puts pressure on the demand for energy resources. In addition, its high carbon emissions have a significant impact on the environment, which has gradually been restricted in recent years.

3.3 Green Hydrogen Production Technology

3.3.1 Electrolysis of Water to Produce Hydrogen

3.3.1.1 Alkaline Water Electrolysis for Hydrogen Production

The principle of hydrogen production by alkaline electrolytic water is through an electrolytic cell, which is usually composed of two electrodes, and the tank is injected with an alkaline electrolyte solution [16]. Common alkaline electrolyte solutions are sodium hydroxide (NaOH) or potassium hydroxide (KOH). Alkaline water electrolysis (AWE) for hydrogen production is characterized by a relatively low capital cost, as it avoids expensive materials and superior long-term durability. AWE process for hydrogen production is shown in Figure 3.5. Alkaline electrolytic water hydrogen production can be coupled with electrochemical refineries to achieve the production of high-value-added chemicals, further reducing the cost of hydrogen production. However, the problem of low current density faced by AWE for hydrogen production has hindered its commercialization,

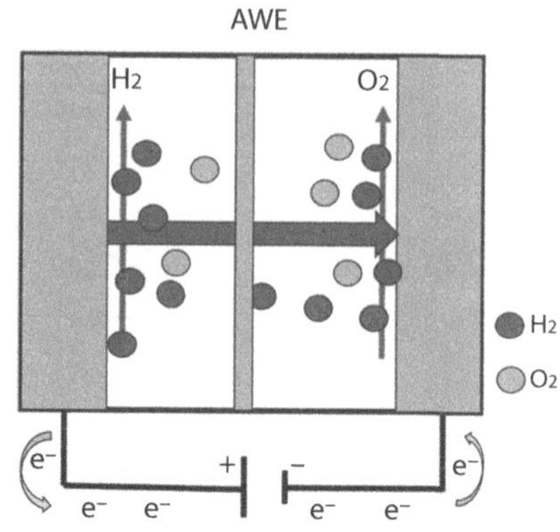

Figure 3.5 AWE process for hydrogen production.

which can be solved by optimizing the electrocatalyst and upgrading the electrolytic system [17].

In alkaline electrolytic water, the reaction of water decomposition into hydrogen and oxygen is as follows:

$$2H_2O(1) \rightarrow 2H_2{+}O_2(g)$$

Green hydrogen is generated from renewable energy sources, and hydrogen is obtained by electrolysis of water. Hydrogen production by electrolytic water decomposes water molecules into hydrogen and oxygen through an electrochemical process under the action of direct current and precipitates them in the negative and anode, respectively. In an electrolytic cell, when the positive electrode of the external DC power supply is connected to the anode and the negative electrode is connected to the cathode, the water molecules in the alkaline electrolyte solution will react under the influence of electrolysis. In this reaction, hydrogen gas is generated on the cathode, oxygen is generated on the anode, and the resulting hydrogen gas can be collected externally. It should be noted that in the process of alkaline electrolytic water hydrogen production, the cathode and anode need to conduct electricity through the ions in the electrolyte solution to maintain the electrolytic reaction. At the same time, it is important to control the voltage and current applied to the electrolyzer to ensure efficient hydrogen production [18]. AWE hydrogen production has mature technology, relatively low cost, efficient energy conversion, extensive adaptation, and sustainability. However, there are also some disadvantages that cannot be ignored. When producing hydrogen on a large scale, it needs to consume a lot of electricity, and its energy cost is relatively high. In order to improve the reaction efficiency, AWE technology usually requires the use of expensive and rare catalysts, which increases the cost of hydrogen production. The technique uses alkaline electrolyte solutions such as sodium hydroxide or potassium hydroxide. This requires proper management and disposal of alkaline substances to avoid environmental and health risks [19].

3.3.1.2 Proton Exchange Membrane Electrolysis of Water to Produce Hydrogen

Proton exchange membrane (PEM) water electrolysis is considered the most promising technology for generating green hydrogen gas from water and intermittent renewable energy sources [20–22]. In addition, PEM water electrolysis also has multiple advantages such as compact system design, high working current density, high hydrogen purity, higher energy efficiency, and fast response when combined with renewable energy [23–25]. However, the use of expensive electrocatalysts and battery components has led to expensive and limited commercial applications. PEM water electrolysis typically yields ultrapure hydrogen and oxygen at lower temperatures (30°C to 80°C) and higher current densities (2 to 4 A/cm^{-2}) [26]. In comparison to AWE and AEM water electrolysis, PEM water electrolysis has a few advantages, including high energy efficiency, good real-world loads, small system design, and quick response times when combined with renewable energy sources [27, 28]. Furthermore, the absence of corrosive liquid electrolytes makes PEM water electrolysis a dependable method. With minimal performance loss, PEM water electrolysis currently has a maximum stability of 100,000 hours [29]. Water electrolysis, however, only makes up 2% of the world's green hydrogen production due to the high cost of hydrogen production, which is mostly due to the usage of expensive ingredients in water electrolysis [30]. Therefore, significant advancements are required to reduce the price of producing green hydrogen. The working principle of PEM water electrolysis is shown in Figure 3.6.

Green hydrogen is obtained through a process of hydroelectrolysis combined with renewable energy sources such as solar, wind, geothermal, and hydroelectric power [31]. When paired with renewable energy sources, PEM water electrolysis has several advantages over other water electrolysis technologies, including a greater operational current density, high energy efficiency, a compact system design, and quick dynamic reaction. Sadly,

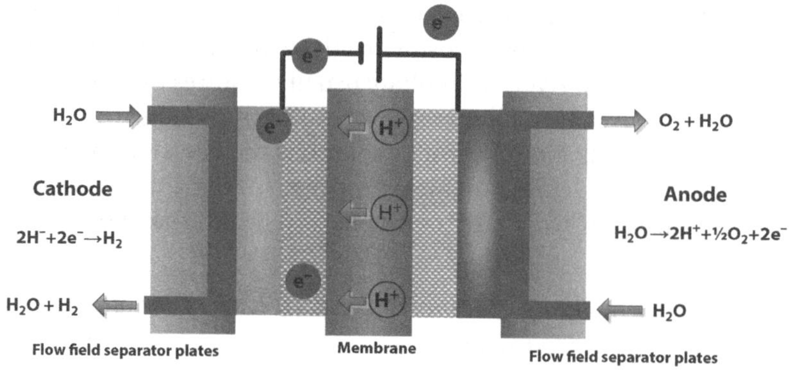

Figure 3.6 Schematic illustration of the working principle of PEM water electrolysis [32].

compared to blue hydrogen (made by conventional natural gas reforming and coal gasification), green hydrogen is created using PEM water electrolysis at a higher cost. This is because PEM water electrolysis uses expensive components, has a short running duration, and has a high cost of renewable electricity. Therefore, PEM electrolysis technology still needs to be further developed.

3.3.1.3 *Solid Oxide Electrolysis of Water for Hydrogen Production*

Solid oxide electrolysis to produce hydrogen water is a process that utilizes solid oxide electrolyte materials for water decomposition reactions to obtain hydrogen and oxygen. The principle is as follows:

1. Choose appropriate solid oxide electrolyte materials: Solid oxide electrolyte materials are usually materials with good ion conductivity, such as ceramic materials in solid oxide fuel cells.
2. Creating ion transport channels: The ion transport channels in solid oxide electrolyte materials are achieved through the gaps and defects in the intermediate phase of the material. By controlling current and temperature conditions, oxygen anions can be moved in solid oxide electrolyte materials.
3. It should be noted that the electrolysis of solid oxide to produce hydrogen water is a high-temperature electrolysis process that often needs to be carried out at high temperatures (usually between 700°C and 1000°C). In addition, the technology of solid oxide electrolysis for hydrogen water

production is still in the exploration and development stage, and further research and improvement are needed to improve efficiency and reduce costs.

Solid oxide electrolysis of water for hydrogen production is an emerging technology that uses solid oxides as electrolytes to decompose water molecules into hydrogen and oxygen under high-temperature conditions [33, 34]. The high-temperature solid oxide electrolysis of water for hydrogen production is shown in Figure 3.7. This technology has many potentials and can play an important role in energy transformation.

First, solid oxide electrolysis of water for hydrogen production technology can efficiently generate hydrogen gas. Compared to traditional water electrolysis methods, solid oxide electrolysis of water for hydrogen production technology is not limited by the electrolysis kinetics of water and can achieve efficient hydrogen production in high-temperature environments. According to research, the efficiency of solid oxide electrolysis technology for hydrogen production can exceed 80%, far higher than traditional methods.

Second, the solid oxide electrolysis technology for hydrogen production from water has good stability and corrosion resistance. Solid oxides as

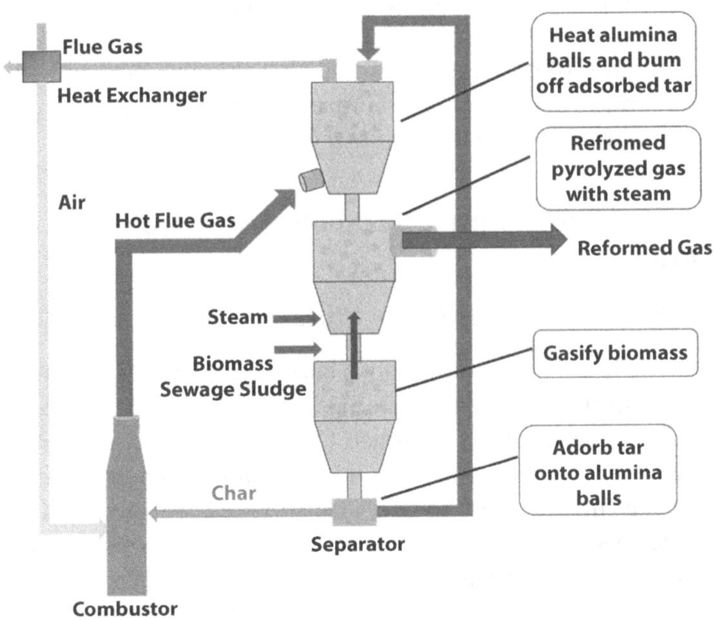

Figure 3.7 Schematic diagram of high-temperature SOEC principle.

electrolytes have good mechanical and chemical stability, which can operate stably for a long time in high-temperature environments and reduce equipment maintenance costs.

In addition, the solid oxide electrolysis of water for hydrogen production technology also has the potential for comprehensive energy utilization. High-temperature thermal energy can be used as an energy source for the solid oxide electrolysis of water to produce hydrogen, converting waste heat energy into valuable hydrogen gas. This is of great significance in energy transformation and can effectively improve energy utilization efficiency.

However, the technology of solid oxide electrolysis of water for hydrogen production still faces some challenges in commercial applications. The first issue is the cost of hydrogen production, as high-temperature operations require a large amount of energy, resulting in higher hydrogen production costs. To solve this problem, it is necessary to further reduce energy consumption and improve electrolysis efficiency. Second, the separation, storage, and utilization of hydrogen and oxygen require the development of mature hydrogen storage and utilization technologies. Last, the long-term stability of the solid oxide electrolysis cell needs further improvement to extend the equipment life [35–37].

Overall, the solid oxide electrolysis of water for hydrogen production technology, as an efficient and stable hydrogen production method, has broad application prospects. With the continuous progress of technology and the reduction of costs, it is believed that this technology will play an increasingly important role in the future energy transformation and contribute to the sustainable energy industry.

3.3.2 Solar Water Splitting Technology for Hydrogen Production

3.3.2.1 Photocatalytic Hydrogen Process

Photocatalytic hydrogen production is based on the photolysis water reaction principle of photocatalytic materials, which converts light energy into chemical energy to produce hydrogen [38]. Photocatalytic hydrogen production requires the use of suitable photocatalytic materials, commonly used materials include titanium dioxide (TiO_2), barium indium oxide (In_2O_3), zinc oxide (ZnO), silver (Ag), etc. [39, 40]. These materials have a special electronic structure that can absorb light energy and produce excited state electrons and holes when exposed to light. When light shines on the photocatalytic material, the electrons in the material absorb the energy of the photons and transition from the ground state to the excited state energy level. These excited electrons have high energy and can be used to catalyze the photolysis of water [41]. On the surface of photocatalytic materials, excited electrons and holes can participate in the photolysis water reaction (Figure 3.8). The excited electrons can reduce the oxygen atom in the water molecule, reducing it to oxygen (O_2). The holes can oxidize the hydrogen atoms in the water molecules to produce hydrogen (H_2) [42]. In the process of photocatalysis, light excites electrons on the surface of the catalyst to form a REDOX reaction. Through this reaction, electrons and holes are transported in the photocatalytic material to participate in the reduction and oxidation reactions, respectively, resulting in the photolysis of water to produce hydrogen. In order to carry out the photopreserving catalytic hydrogen production reaction, a certain electron conductivity and ion conductivity are required. A conductive agent or electrolyte is

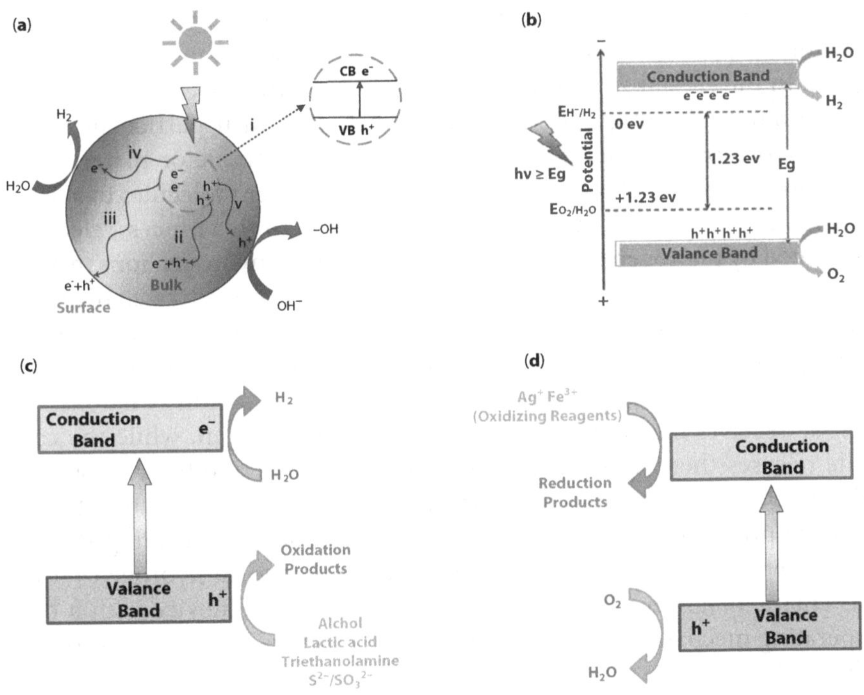

Figure 3.8 (a) Schematic diagram of the basic principle of semiconductor photocatalytic water decomposition for hydrogen production. (b) Schematic diagram of semiconductor photocatalytic water band decomposition. (c) Introduction of sacrificial agent system photolysis aquatic hydrogen half-reaction. (d) Photolysis of aquatic oxygen-half reactions after the introduction of sacrificial agents.

often used on the surface of the photocatalytic material to provide a transport path for electrons and ions and promote the progress of the photolytic water reaction.

In general, the technology is a form of clean energy production that uses renewable solar energy to photolysis water to produce hydrogen. Solar energy is the main driving force of photocatalytic hydrogen production, which has abundant resources and renewability. Compared with traditional fossil energy, photocatalytic hydrogen production can reduce environmental pollution and carbon emissions [43]. It is environmentally friendly and renewable, which is of great significance for the development of clean energy in the future. By using solar energy for hydrogen production, it can reduce the dependence on limited mineral resources and achieve sustainable development of energy.

3.3.2.2 Photoelectric Chemical Process of Hydrogen

In the photoelectrochemical process, the photocatalytic effect of semiconductors is used to convert light energy into photogenerated electron-hole pairs. In the crystal structure of semiconductors, these electron-hole pairs can promote the migration of carriers, which in turn participate in the decomposition of water [44]. In solar cells, semiconductor materials are used to change the photogenerated potential of water molecules to achieve the reduction and oxidation of water molecules. In this process, electrons activated by photogenerated potential interact with hydrogen ions in aqueous solutions to form a new catalytic system. The holes activated by photoelectric potential react with water molecules to form oxygen and hydrogen ions [45]. In the electrolysis process, the electrolyzer uses the action of applied voltage to accelerate the reduction and oxidation reactions. The anode receives electrons and reduces them to hydrogen, while the cathode releases them and oxidizes them to oxygen. In addition, in order to increase the ion concentration in the electrolyte, a suitable electrolyte must also be provided for the electrolysis reaction. Electrons leading through the outer circuit to the cathode react with ions to form hydrogen [46]. A schematic diagram of a typical photoelectrochemical device with a basic operating mechanism for splitting water to produce hydrogen is shown in Figure 3.9.

Photoelectric catalytic hydrogen production technology uses the photoelectric effect and a catalyst to convert solar energy into chemical energy and promotes the photolysis reaction of water molecules under light to produce hydrogen. Compared with traditional hydrogen production methods such as pyrolysis and electrolysis, photoelectric catalytic hydrogen production

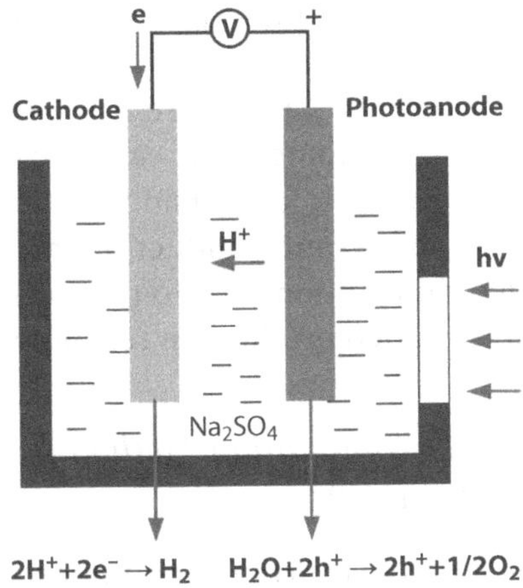

Figure 3.9 Schematic diagram of a typical photoelectrochemical device and its basic operating mechanism of hydrogen production by water cracking.

technology has the following advantages: (1) The energy band structure of photoelectron catalytic materials and the presence of surface catalyst can improve the photoelectric conversion efficiency, make the photolysis reaction of water molecules efficient, and thus improve the hydrogen production efficiency; (2) the photoelectric catalytic hydrogen production process does not produce other harmful gasses, has high purity, and is in line with the characteristics of clean energy; (3) photoelectric catalytic hydrogen production does not require high temperature, high pressure, and other conditions, and the reaction speed is fast, suitable for distributed hydrogen production; (4) photoelectric catalytic hydrogen production does not involve explosive or flammable gasses and is highly safe; (5) photoelectric catalytic materials can also produce other useful byproducts during the hydrogen production process, such as oxygen or other chemicals for versatility. However, there are still some problems and challenges in the current photoelectric catalytic hydrogen production technology: (1) The utilization efficiency of solar energy by current photoelectric catalytic materials needs to be improved, and the photoelectric conversion efficiency is relatively low; (2) the cost of some photoelectric catalytic materials is high and restricts its promotion in large-scale applications; (3) photoelectric catalytic materials are not stable enough, are easily affected by the environment and reaction conditions, and have a short life span; (4) hydrogen is a high-energy, low-density gas that requires storage and transportation, and it is still a challenge; (5) photoelectric catalytic hydrogen production technology is still in the research and development stage and requires a lot of research and experimental verification. The technology is not mature enough.

In response to these problems, scientific research institutions and enterprises are working to improve photoelectric catalytic hydrogen production technology, improve photoelectric conversion efficiency, reduce material costs, enhance material stability, and conduct research and exploration in hydrogen storage, system integration, etc., to promote the photoelectrocatalysis development of hydrogen production technology.

3.3.2.3 Photothermal Decomposition Method for Hydrogen Production (PDMHP)

The PDMHP from solar water is currently an important branch of renewable energy hydrogen technology. Photothermal decomposition hydrogen technology is a method of using high temperatures to decompose water into hydrogen and oxygen. The basic principle is to concentrate sunlight or other high-temperature energy in a small area through a concentrated reflection or spotlight system, forming a high-temperature area (Figure 3.10). In high-temperature regions, specific reaction materials serve as catalysts to promote the decomposition reaction of water, breaking down water molecules into hydrogen and oxygen [47]. This process requires precise control of reaction temperature, reaction time, and selection of reaction materials to ensure efficient water decomposition reactions.

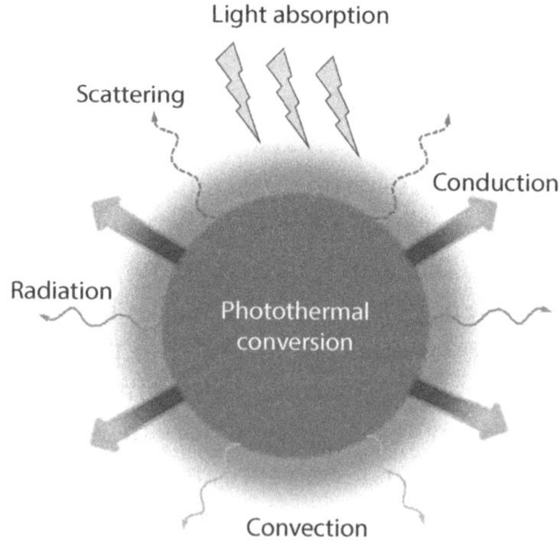

Figure 3.10 Schematic diagram of photothermal conversion.

The advantages of photothermal decomposition technology for hydrogen production include efficiency and renewability [48]. By utilizing high-temperature or concentrated thermal energy as the driving force, the photothermal decomposition method can achieve efficient water decomposition reactions, resulting in the production of a large amount of hydrogen gas. In addition, photothermal decomposition belongs to renewable energy hydrogen technology, which can fully utilize renewable energy such as solar energy, reduce dependence on limited resources, and achieve carbon neutrality and sustainable development.

However, there are also some challenges and limitations in the photothermal decomposition technology for hydrogen production. First, achieving a stable supply of high-temperature energy is a key issue, especially in very hot environments or when the energy supply is insufficient at night. Second, the catalytic activity and stability of the reaction materials have a significant impact on the entire reaction process [49], and further research and improvement are needed. In addition, the cost of hydrogen production through photothermal decomposition is relatively high, including energy and catalyst costs, which limits its promotion in commercial applications.

In terms of domestic and international research, photothermal decomposition technology for hydrogen production has received widespread attention. Many research institutions and enterprises are committed to developing efficient reaction materials and reactor designs to improve hydrogen production efficiency and reduce costs. Some research teams abroad have made certain progress, such as using metal oxides or compounds as catalysts to improve the catalytic activity and stability of water decomposition reactions by constructing specific structures and interfaces. Guo *et al.* [50] proposed an efficient two-phase photocatalytic system consisting of integrated photothermal photocatalytic materials that convert liquid water into water vapor using charred wood substrates while decomposing hydrogen under light illumination without the need for additional energy. In China, some research institutions have also conducted relevant research to explore the potential of different reaction materials and reaction designs. Han *et al.* [51] constructed a surface carbonized wood/TiO_2 (C-wood/TiO_2) photothermal photocatalysis system, which can precipitate hydrogen gas from seawater (248.1 $\mu mol\ m^{-2}\ hour^{-1}$), demonstrating the strengths and hydrogen production potential of the photothermal photocatalytic biphasic system.

3.3.3 Biomass Hydrogen Production Technology

3.3.3.1 *Thermochemical Hydrogen Production Technology (THPT)*

The process of converting biomass into a hydrogen-rich combustible gas *via* thermochemistry is known as biomass THPT. The process can be created either directly from biomass raw materials or from dissolved biomass intermediate products (such as methanol and ethanol) to produce hydrogen. The approach can be further divided into steam gasification hydrogen technology, supercritical water gasification hydrogen technology, and biomass pyrolysis technology for hydrogen production in accordance with the variations of the particular hydrogen production process [52]. The production process of thermochemical H_2 can be seen in Figure 3.11.

Steam gasification hydrogen technology is classified based on the gasifying agent used for the process. Steam is used as the gasification medium during the process known as steam gasification. Pyrolysis usually takes place before gasification because of how much faster it happens. The removal of volatile materials by pyrolysis results in the formation of carbon (residue), which interacts with steam to create H_2, CO, and CO_2. It is

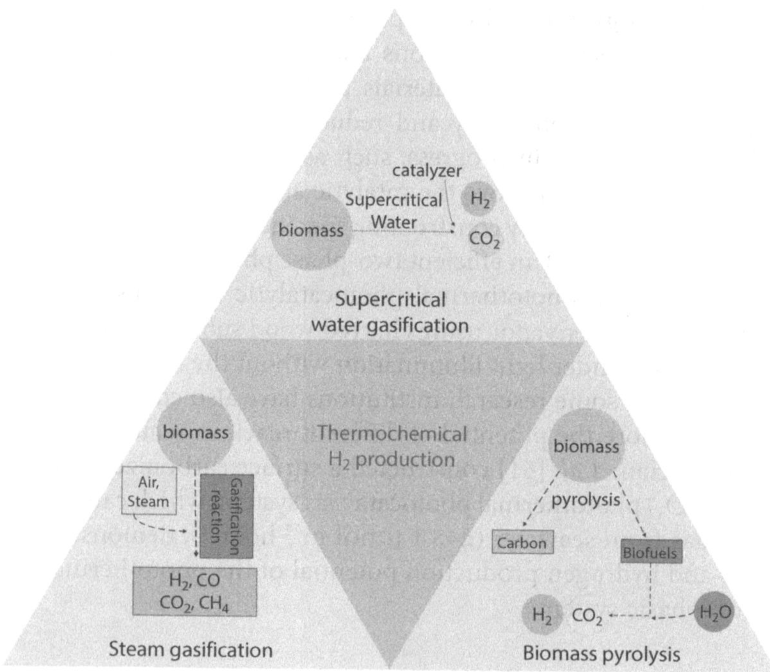

Figure 3.11 Thermochemical H_2 production.

thought that steam gasification, which produces high calorific value gas rich in hydrogen, is a promising technique. Reports state that the typical steam gasification product gas composition is 40% H_2, 25% CO, 8% CH_4, 25% CO_2, and 2% N_2. According to theory, the maximal biomass H_2 output is 17% (based on biomass weight). Rich hydrogen syngas (50% to 55%; on a dry basis) may be produced by this technique [53]. Supercritical water gasification hydrogen technology is an innovative hydrogen production technology that utilizes the properties of water in supercritical conditions to fully blend the water's gas and liquid conditions in 22.1 MPa and 374°C environments. In this state, the oxidizing properties of the water are activated, causing the carbon particles in carbon-containing substances, such as biomass, to be oxidized to CO, and then further converted to CO_2, while the H_2 molecules embedded in water and biomass are released to produce hydrogen. While this approach is considered to have enormous potential, there are also some shortcomings. The technology is more suitable for raw materials with a high water content, and the cost of hydrogen production is several times higher than traditional gas steam methane reconfiguration methods. Given that the technology is still in the development phase, considerable research will still be required in the future to validate it [54]. The heating and decomposition of biomass under the condition of isolating oxygen or introducing a small amount of air is called pyrolysis. Continuously high temperatures lead to the production of marsh, while marsh oil has sticky and unstable properties. At low temperatures, the marsh is difficult to gasify, while higher temperatures can cause piping clogging and reaction obstruction. Therefore, in order to improve the hydrogen output, it is necessary to re-integrate the manganese and bio-oil in the thermal solvent. Steam reconstitution discharges the thermally dissolved biomass residues from the system, and under the joint action of the catalyst and the water vapor, the heated solvent is second high-temperature treatment to dissolve the relatively larger molecular masses into hydrogen and methane, thereby increasing the content of hydrogen in the gas [55].

3.3.3.2 *Production of Hydrogen by Biological Method*

Biomass hydrogen technology converts water molecules and organic substrates in biomass into hydrogen through the catalytic activity of two key enzymes, hydrogenase, and nitrogenase. The common biological hydrogen technology mainly includes direct photolysis, indirect photolysis, light fermentation, dark fermentation, and light-dark coupling fermentation. The process of hydrogen by biological process is shown in Figure 3.12.

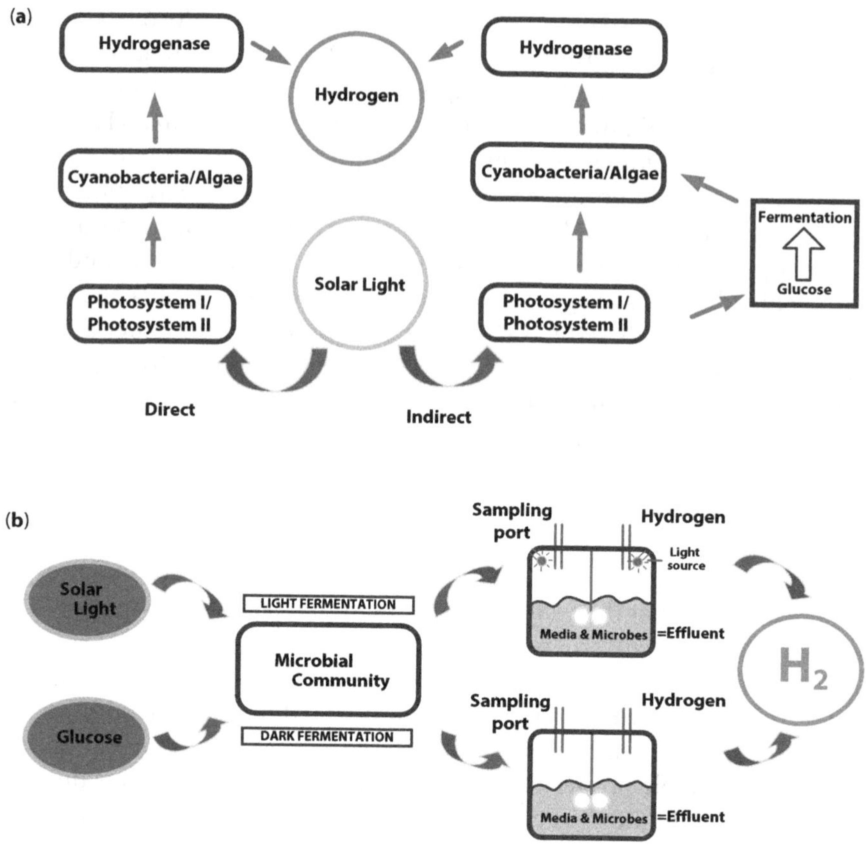

Figure 3.12 Biological process of hydrogen [56].

The direct photolysis hydrogen process occurs in algae or plant cells, where microorganisms split water molecules into hydrogen ions and oxygen through photosynthesis, and the resulting hydrogen ions are converted into hydrogen gas by hydrogenase [57, 58]. Indirect photolysis hydrogen technology uses other algae such as cyanobacteria or microalgae to produce hydrogen from starch or glucose; other processes are similar to direct photolysis. The efficient hydrogen-producing algae that can perform direct photolysis are Cyanobacteria, Chlorococcus, and Chlorella. Under anaerobic conditions, hydrogenases receive electrons from REDOX proteins, and the electron donor converts water into hydrogen. The photosystem participates in the CO_2 reduction reaction and the water decomposition reaction, splitting water molecules into hydrogen and oxygen, and at the end of the biomass photolysis, hydrogen is formed by hydrogenase or CO_2 reduction in the photosystem. Indirect photolysis hydrogen technology is the process of using cyanobacteria or microalgae to produce hydrogen from starch or glycogen, and after the formation of carbohydrates through photosynthesis, hydrogen is metabolized by plant cells under dark conditions. Electron transfer is always carried out under aerobic conditions, and after oxygen is exhausted, the cell changes to an anaerobic state, and the available electrons are transferred to the photosystem by inducing hydrogenase to produce hydrogen. The hydrogen produced by the nitrogenase in the algae can be reabsorbed by the hydrogenase while removing oxygen and indirectly increasing hydrogen production.

Using the respective advantages and complementary characteristics of light fermentation hydrogen–producing bacteria and dark fermentation hydrogen–producing bacteria, the method of combining them to improve hydrogen production capacity and substrate conversion efficiency is called the light-dark coupling fermentation method [59]. Dark fermentation hydrogen–producing bacteria can decompose large molecular organic matter into small molecular organic acids, obtain the energy needed to maintain their own growth, and produce hydrogen. The produced small molecular organic acids cannot be used by dark fermentation hydrogen–producing bacteria and accumulate in large quantities, which inhibits hydrogen production efficiency. Light fermentation hydrogen–producing bacteria can use small molecular organic acids produced by dark fermentation to eliminate the inhibition of organic acids on dark fermentation hydrogen production, and further release hydrogen. Therefore, coupling the two together can improve the efficiency of hydrogen production and expand the range of substrate utilization.

3.3.4 Nuclear Energy Hydrogen Production Technology

With the global attention to climate governance, various countries have launched carbon-neutral schedules, and the use of clean energy is the key. Both nuclear and hydrogen are clean and efficient sources of energy. Among them, nuclear hydrogen production is a combination of the nuclear reactor and advanced hydrogen production technology, which is a process of generating hydrogen from raw materials under high-temperature and high-pressure conditions by using nuclear reaction [60]. The principle is mainly based on the principle of nuclear energy reaction and electrolysis of water. Nuclear power reaction refers to the process of releasing large amounts of energy by means of nuclear fission or fusion. Nuclear fission refers to the process of splitting heavy nuclei (such as uranium and plutonium) into two or more light nuclei by collision or absorption of neutrons. Nuclear fusion is the process of fusing light nuclei (such as hydrogen and helium) into heavier nuclei at high temperatures and pressures.

Common reactors include light water reactor (1), heavy water reactor (2), supercritical water reactor (3), fast reactor (4), molten salt reactor (5), gas-cooled fast reactor (6), and high-temperature gas-cooled reactor (7). The respective exit temperature is shown in Figure 3.13. The molten salt reactor has the characteristics of high temperature, low pressure, and high heat capacity and can output high-temperature nuclear heat above 700°C.

At present, the main methods of nuclear hydrogen production are the thermochemical cycle for hydrogen production and high-temperature electrolysis for hydrogen production. The thermochemical cycle for hydrogen production [61] is a high-temperature thermochemical cycle process that uses the high-temperature heat provided by the reactor to make water vapor thermal cracking to prepare hydrogen with the participation of chemical substances such as I and S. The advantages of the thermochemical cycle hydrogen production are low cost, no need to use precious metal catalysts and economy of scale, and its all-fluid process is very easy to scale expansion and continuous operation. However, there are also high requirements for materials and equipment. At present, the direction of innovation is to optimize the technical route and solve the problem of reactor corrosion. High-temperature electrolysis for hydrogen production [62] uses the solid oxide electrolytic cell as the core reactor to achieve the decomposition of water vapor to produce hydrogen. High-temperature electrolysis

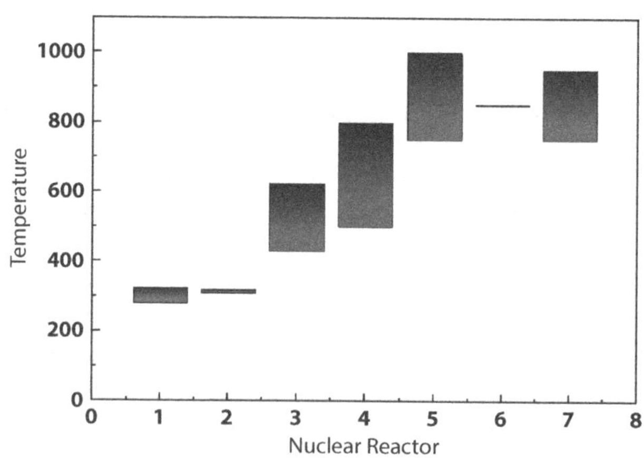

Figure 3.13 Nuclear reactors and outlet temperatures.

for hydrogen production mainly includes electrolyte, electrode material, electrolytic pile, and system. However, there are technical problems such as the high-temperature connection sealing of the electrolytic cell and the integration of large-scale hydrogen production systems, which are difficult to complete in the short term, so the future large-scale application prospects are still not very clear. In general, nuclear hydrogen production has the advantages of efficient energy conversion, potential renewable energy, low carbon emissions, and strong energy supply stability. But, there are also problems that cannot be ignored radioactive nuclear waste needs to be properly disposed of [63].

3.3.5 Hydrogen Production from Seawater

Using seawater as a raw material to produce hydrogen is an important pathway to achieving hydrogen economy and carbon neutrality. The principle of HPS mainly utilizes the decomposition of seawater to produce hydrogen and oxygen. Over the past decade, efforts have been devoted to utilizing high-purity freshwater resources for hydrogen production [64]. However, freshwater resources are currently scarce worldwide, while seawater reserves are abundant in comparison [65]. Therefore, this method will provide new approaches to address the shortage of freshwater resources, which carries significant practical implications for regions with limited freshwater resources but abundant seawater resources and renewable energy sources in coastal arid areas. Theoretical schematic diagram of HPS is shown in Figure 3.14.

Nevertheless, the complex characteristics of seawater present numerous challenges in practical applications. Seawater contains a substantial amount of chloride, which leads to an unavoidable competition between

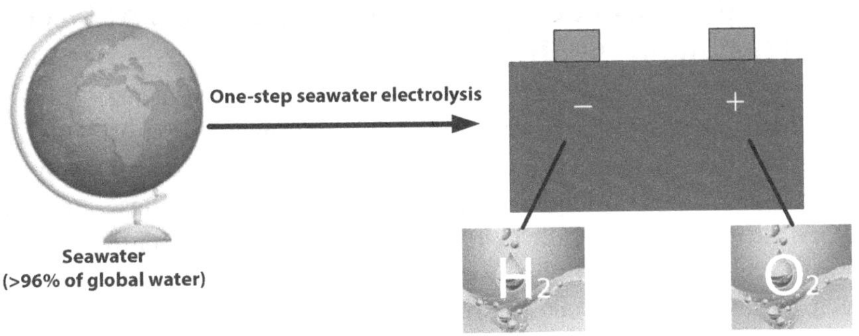

Figure 3.14 Theoretical schematic diagram of HPS.

chlorine evolution reaction and oxygen evolution reaction. Moreover, the high concentration of Cl^- also affects the catalytic activity and stability of transition metal catalysts. Consequently, the industrialization process of seawater electrolysis technology has been relatively slow. In the pursuit of commercializing seawater electrolysis, researchers have made continuous efforts to achieve HPS. Wang *et al.* attempted to simulate hydrogen generation from seawater using microwave plasma. The objective of this study was to construct a stable plasma reactor that enables direct HPS without the need for a catalyst. Results indicate that the application of this reactor has promising prospects for catalyst-free HPS [66]. Zadegan *et al.* developed a hybrid system for HPS and identified the optimal operating conditions. The research findings of this work present new perspectives and methods for further industrial applications of hydrogen production [67]. Additionally, developing efficient catalytic materials suitable for marine environments has become a hot research topic. Extensive studies have been conducted on noble metal catalysts due to their excellent performance in water electrolysis. However, their poor stability and high cost hinder their commercial applications. In recent years, transition metal phosphides have exhibited promising application prospects in seawater electrolysis as a novel type of electrolyte material. Feng *et al.* shed light on the current challenges and future development possibilities of HPS through their research on transition metal phosphides [68].

Although there are still many issues associated with the practical implementation of seawater electrolysis technology, with further research and optimization, significant advancements are expected to be made. In the foreseeable future, it will be possible to achieve industrial-scale applications of seawater electrolysis, thereby becoming an important direction in the development of hydrogen energy.

3.4 Challenges Facing the Industrialization of Green Hydrogen Production

The industrialization of green hydrogen production faces many challenges, such as energy cost, storage and transportation technology, supply and demand balance, technology maturity, and policy support. These challenges require the joint efforts of the government, industry, and scientific research institutions to solve in order to promote the development and commercial application of the green hydrogen production industry. Although the new green hydrogen production technology has some advantages in many aspects, due to the limitations of relevant technical conditions, there are bound to be many practical problems in the application process.

First of all, hydrogen production equipment and technology, although there are relevant enterprises to carry out the corresponding technology research and development, are in the small batch trial production stage and have not yet formed a mature hydrogen production process production line; the relevant core technology maturity is low; and system equipment localization degree is not high. Secondly, in terms of economy, high input cost is still the biggest factor limiting the development of some green hydrogen production technology, and most hydrogen production processes need to add subsequent hydrogen purification technology to obtain high purity H_2; how to effectively reduce the cost of hydrogen production is the main development direction in the future hydrogen production field.

The hydrogen production industry is in the early stage of development policy dependence. The hydrogen production industry is in the introduction period; in this stage, policy support is very important, through various forms of subsidies and government-specific administrative and regulatory means, in order to make the industry into the "scale–cost reduction–marketization" volume price cycle, forming a good situation of commercial competition development. However, in the past 2 years, due to the changes in the new coronavirus epidemic and the economic situation, as well as the policymakers' ideas in the development of hydrogen production are not completely unified, the "effective market" has not really emerged, many enterprises are facing the dual pressure of capital and the market, and even the situation of debt chain, some head enterprises are big risks, and many trailing enterprises are difficult to survive financing. This requires guiding enterprises to adapt to the situation and not rely on or even gamble on industrial policy, and enterprises that can seize opportunities should use industrial policy but not engage in industrial speculation.

Finally, in terms of the standard system, the current hydrogen production industry is single, the key technical indicators are lacking, and the mandatory national standards are less, which has made it difficult to meet the needs of the standardization of the hydrogen production industry. As a strategic industry, hydrogen production technology first needs scientific and technological innovation in industrial generic technology and product research and development, as well as innovation in the business model and policy environment. The huge pressure brought by the active or passive industrial structure adjustment brought by the Sino-US trade dispute and the new coronavirus epidemic, and the step-by-step decomposition of the goal of "carbon peak and carbon neutrality", has made many places choose hydrogen production technology as the local key cultivation and development of industrial direction. At present, the short-term focus of work all over the country is to support local existing enterprises and to attract foreign high-quality enterprises, there has been a distinctive slanting policy based on capital investment and resource transfer, which has also brought some policies that work in the short term but may not be beneficial in the long term, or some policies that are locally beneficial but affect the scientific layout of the industry. In fact, for enterprises that really want to do industry, it is more necessary for the government to create a healthy and sustainable development space and application scenarios, create a fair competitive market environment, and truly play the decisive role of the market in the allocation of scientific and technological and industrial innovation resources.

3.5 Conclusion

In conclusion, green hydrogen production stands as a revolutionary method that places environmental sustainability at its core in the quest for hydrogen generation. It starkly contrasts with traditional gray and blue hydrogen production methods, offering a significant reduction in carbon dioxide emissions and a substantial decrease in dependence on fossil fuels. By harnessing the power of renewable energy sources such as solar and wind, green hydrogen production utilizes electrolysis to split water molecules, generating hydrogen in an eco-friendly manner.

The applications of this green hydrogen are widespread, spanning across energy, transportation, and chemical industries. It emerges as a pivotal solution, not merely as an alternative but as a transformative force, actively contributing to the reduction of carbon emissions and mitigating the adverse effects of greenhouse gasses. This chapter delved into the intricate details of various green hydrogen production processes, elucidating their mechanisms and operational principles. Furthermore, a comprehensive comparison and analysis of advanced hydrogen production technologies were presented, highlighting their distinct advantages and limitations. Through this exploration, we uncovered existing bottlenecks and barriers concerning hydrogen production equipment, technology, and economics. These challenges, though formidable, underscore the need for continued research and innovation in the field. Looking ahead, the industrialization of green hydrogen production faces complex challenges. However, it is in overcoming these challenges that the true potential of green hydrogen can be realized. As technology continues to advance, green hydrogen production is poised to play an even more significant role. It is not just a chapter in the story of clean energy; it is a cornerstone, propelling the advancement and widespread application of sustainable, eco-friendly energy solutions.

In the face of global environmental concerns, the journey toward a greener, more sustainable future relies heavily on transformative technologies such as green hydrogen production. With steadfast dedication and collaborative efforts, we stand on the brink of a new era, where clean energy is not just a possibility but a reality. As we navigate the challenges ahead, the promise of a cleaner, greener world fuels our determination, guiding us toward a future where environmental harmony and energy efficiency coexist.

References

1. Ozcan, H. and El-Emam, R.S., Thermochemical looping technologies for clean hydrogen production – Current status and recent advances. *J. Clean. Prod.*, 382, 135295, 2023.
2. Gupta, A., Baron, G.V., Perreault, P., Hydrogen Clathrates: Next Generation Hydrogen Storage Materials. *Energy Storage Mater.*, 41, 69–107, 2021.
3. Abdin, Z., Zafaranloo, A., Rafiee, A., Hydrogen as an Energy Vector. *Renew. Sustain. Energy. Rev.*, 120, 109620, 2020.
4. Du, Q., Zhao, R., Chen, X., Synthesis of Ultrathin and Grid-Structural Carbon Nanosheets Coupled with Mo2C for Electrocatalytic Hydrogen Production. *Chem. Asian. J.*, 16, 15, 2107–2112, 2021.
5. Zou, X. and Zhang, Y., Noble Metal-Free Hydrogen Evolution Catalysts for Water Splitting. *Chem. Soc. Rev.*, 44, 15, 5148–5180, 2015.
6. Xu, Q., Wang, P., Wan, L., Superhydrophilic/Superaerophobic Hierarchical NiP2@MoO_2/Co(Ni)MoO_4 Core-Shell Array Electrocatalysts for Efficient Hydrogen Production at Large Current Densities. *ACS. Appl. Mater. Interfaces*, 14, 17, 19448–19458, 2022.
7. Ji, Y., Luo, W., Liu, Y., Bifunctional o-$CoSe_2$/c-$CoSe_2$/$MoSe_2$ Heterostructures for Enhanced Electrocatalytic and Photoelectrochemical Hydrogen Evolution Reaction. *Mater. Today Chem.*, 23, 100724, 2022.
8. Wang, P., Luo, Y., Zhang, G., Chen, Z., Ranganathan, H., Sun, S., Shi, Z., Interface Engineering of Nix Sy@MnOx Hy Nanorods to Efficiently Enhance Overall-Water-Splitting Activity and Stability. *Nano-Micro Lett.*, 14, 1, 120, 2022.
9. Li, R., Xie, M., Jin, H., Guo, L., Liu, F., Effect of swirl on gasification characteristics in an entrained-flow coal gasifier. *Int. J. Chem. React. Eng.*, 18, 20190203, 2020.
10. Kim, J., Choi, H., Lim, J., Rhim, Y., Chun, D., Kim, S., Lee, S., Yoo, J., Hydrogen production via steam gasification of ash free coals. *Int. J. Hydrogen Energy*, 38, 6014–6020, 2013.

11. Dai, F., Zhang, S., Luo, Y., Wang, K., Liu, Y., Ji, X., Recent Progress on Hydrogen-Rich Syngas Production from Coal Gasification. *Processes*, 11, 6, 1765, 2023.
12. He, H., Li, Q., Tang, J., Tang, J.S., Liu, P., Zheng, H., Zhao, F., Guan, W., Guo, E., Xi, C., Study of hydrogen generation from heavy oil gasification based on ramped temperature oxidation experiments. *Int. J. Hydrogen. Energy*, 48, 6, 2161–2170, 2023.
13. Okere, C.J. and Sheng, J.J., Review on clean hydrogen generation from petroleum reservoirs: Fundamentals, mechanisms, and field applications. *Int. J. Hydrogen Energy,* 48, 38188–38222, 2023.
14. Zolfaghari, A., Yun, J., Singh, V., Operational envelope prediction of fluoroelastomer seals for downhole operation in the oil and gas industry. *J. Appl. Polym. Sci.*, 139, 5, 51575, 2021.
15. Sanz, O., Velasco, I., Pérez-Miqueo, I., Poyato, R., Odriozola, J.A., Montes, M., Intensification of hydrogen production by methanol steam reforming. *Int. J. Hydrogen Energy*, 41, 10, 5250–5259, 2016.
16. Liao, P., Kang, J., Zhong, Y., Xiang, R., Wang, S., Li, S., Liu, X., Li, G., Recent advances of two-dimensional metal-organic frameworks in alkaline electrolysis water for hydrogen production. *Sci. China Chem.*, 66, 7, 1924–1939, 2023.
17. Huang, H., Xu, L., Woo, D.Y., Kim, S., Kim, S.M., Kim, Y.K., Byeon, J., Lee, J., Refining the surface properties of WO2.7 by vacancy engineering and transition metals doping for enhanced alkaline hydrogen evolution reaction. *Chem. Eng. J.,* 451, 138939, 2023.
18. Zhou, T., Huo, Y., Lei, N., Gong, Y., Modified cobalt oxalate heterostructure catalyst P–FeOOH@CoC_2O_4/NF for a highly efficient oxygen evolution reaction. *Int. J. Hydrogen Energy*, 48, 12, 4738–4749, 2023.
19. Tian, W., Zhang, X., Wang, Z., Cui, L., Li, M., Xu, Y., Li, X., Wang, L., Wang, H., Amorphization activated RhPb nanflowers for energy-saving hydrogen production by hydrazine-assisted water electrolysis. *Chem. Eng. J.*, 440, 135848, 2022.
20. Nnabuife, S.G., Ugbeh-Johnson, J., Okeke, N.E., Ogbonnaya, C., Present and Projected Developments in Hydrogen Production: A Technological Review*. *Carbon Capture Sci. Technol.*, 3, 100042, 2022.
21. Gopinath, M. and Marimuthu, R., A review on solar energy-based indirect water-splitting methods for hydrogen generation. *Int. J. Hydrogen Energy*, 47, 37742, 2022.
22. Lim, D., Lee, B., Lee, H., Byun, M., Lim, H., Projected cost analysis of hybrid methanol production from tri-reforming of methane integrated with various water electrolysis systems: Technical and economic assessment. *Renew. Sustain. Energy Rev.*, 155, 111876, 2022.
23. Lee, B., Lim, D., Lee, H., Lim, H., Which water electrolysis technology is appropriate: Critical insights of potential water electrolysis for green ammonia production. *Renew. Sustain. Energy Rev., 143,* 110963, 2021.

24. Mohammadi, A. and Mehrpooya, M., A comprehensive review on coupling different types of electrolyzer to renewable energy sources. *Energy*, 158, 632, 2018.
25. Lim, D., Lee, B., Lee, H., Byun, M., Cho, H.S., Cho, W., Kim, C.H., Brigljevic, B., Lim, H., Impact of voltage degradation in water electrolyzers on sustainability of synthetic natural gas production: Energy, economic, and environmental analysis. *Energy Convers. Manage.*, 245, 114516, 2021.
26. Shiva Kumar, S., Ramakrishna, S.U.B., Krishna, S.V., Srilatha, K., Devi, B.R., Himabindu, V., Synthesis of titanium (IV) oxide composite membrane for hydrogen production through alkaline water electrolysis. *S. Afr. J. Chem. Eng.*, 25, 54, 2018.
27. Schmidt, O., Gambhir, A., Staffell, I., Hawkes, A., Nelson, J., Few, S., Future cost and performance of water electrolysis: An expert elicitation study. *Int. J. Hydrogen Energy*, 42, 30470, 2017.
28. Weiss, A., Siebel, A., Bernt, M., Shen, T.H., Tileli, V., Gasteiger, H.A., Impact of Intermittent Operation on Lifetime and Performance of a PEM *Water Electrolyzer. J. Electrochem. Soc.*, 166, F487, 2019.
29. Lee, B., Heo, J., Kim, S., Sung, C., Moon, C., Moon, S., Lim, H., Economic feasibility studies of high pressure PEM water electrolysis for distributed H-2 refueling stations. *Energy Convers. Manage.*, 162, 139, 2018.
30. Chen, Y.H., Liu, C.F., Xu, J.C., Xia, C.F., Wang, P., Xia, B.Y., Yan, Y., Wang, X.Y., Key Components and Design Strategy for a Proton Exchange Membrane Water Electrolyzer. *Small Struct.*, 4, 2200130, 2023.
31. Buttler, A. and Spliethoff, H., Current status of water electrolysis for energy storage, grid balancing and sector coupling via power-to-gas and power-to-liquids: A review. *Renew. Sustain. Energy Rev.*, 82, 2440, 2018.
32. Shiva Kumar, S. and Lim, H., An overview of water electrolysis technologies for green hydrogen production. *Energy Rep.*, 8, 13793, 2022.
33. Dos Santos, K.G., Eckert, C.T., De Rossi, E., Bariccatti, R.A., Frigo, E.P., Lindino, C.A., Alves, H.J., Hydrogen production in the electrolysis of water in Brazil, a review. *Renew. Sustain. Energy Rev.*, 68, 563–571, 2017.
34. Guo, J., Zhang, Y., Zavabeti, A., Chen, K., Guo, Y., Hu, G., Fan, X., Li, G.K., Hydrogen production from the air. *Nat. Commun.*, 13, 5046, 2022.
35. Nasser, M., Megahed, T.F., Ookawara, S., Hassan, H., A review of water electrolysis-based systems for hydrogen production using hybrid/solar/wind energy systems. *Environ. Sci. Pollut. Res.*, 29, 86994–87018, 2022.
36. Wang, S., Zheng, S., Wang, Z., Cui, W., Zhang, H., Yang, L., Zhang, Y., Li, P., Superior lithium adsorption and required magnetic separation behavior of iron-doped lithium ion-sieves. *Chem. Eng. J.*, 332, 160–168, 2018.
37. Voloshchenko, G.N., Production of Hydrogen through High-Temperature Electrolysis of Water. *Nanotechnol. Russ.*, 15, 333–340, 2020.
38. Shen, R., Xie, J., Xiang, Q., Chen, X., Jiang, J., Li, X., Ni-based photocatalytic H2-production cocatalysts2. *Chin. J. Catal.*, 40, 240–288, 2019.

39. Kho, Y.K., Iwase, A., Teoh, W.Y., Mädler, L., Kudo, A., Amal, R., Photocatalytic H_2 evolution over TiO_2 nanoparticles. The synergistic effect of anatase and rutile. *J. Phys. Chem. C*, 114, 2821–2829, 2010.
40. Zhang, W., Wang, Y., Wang, Z., Zhong, Z., Xu, R., Highly efficient and noble metal-free NiS/CdS photocatalysts for H_2 evolution from lactic acid sacrificial solution under visible light. *Chem. Commun.*, 46, 40, 7631–7633, 2010.
41. Reza, M., Ahmad, N., Afroze, S., Taweekun, J., Sharifpur, M., Azad, A., Hydrogen Production from Water Splitting through Photocatalytic Activity of Carbon-Based Materials. *Chem. Eng. Technol.*, 46, 3, 420–434, 2022.
42. Wang, H., Zhang, L., Chen, Z., Hu, J., Li, S., Wang, Z., Wang, X., Semiconductor heterojunction photocatalysts: design, construction, and photocatalytic performances. *Chem. Soc Rev.*, 43, 15, 5234–5244, 2014.
43. Akach, J. and Ochieng, A., Monte Carlo simulation of the light distribution in an annular slurry bubble column photocatalytic reactor. *Chem. Eng. Res. Des.*, 129, 248–258, 2017.
44. Yat, L. and Jin, Z.Z., Hydrogen generation from photoelectrochemical water splitting based on nanomaterials. *Laser Photonics Rev.*, 4, 4, 517–528, 2010.
45. Yi, B.Z., Zhen, J.N., Ji, W.Z., Lan, X., Xian, Z.F., Jin, L.L., Recent Advancements in Photoelectrochemical Water Splitting for Hydrogen Production. *Electrochem. Energy Rev.*, 6, 14, 2023.
46. Chia, L.T., Chung, J.T., Jyh, C.C., Thermodynamic analysis of a photoelectrochemical hydrogen production system. *Int. J. Hydrogen Energy*, 35, 7, 2781–2785, 2010.
47. Lu, J., Shi, Y., Chen, Z. *et al.*, Photothermal effect of carbon dots for boosted photothermal-assisted photocatalytic water/seawater splitting into hydrogen. *Chem. Eng. J.*, 453, 139834, 2023.
48. Wang, Y., Huang, W., Guo, S. *et al.*, Sulfur-deficient $ZnIn_2S_4$/oxygen-deficient WO_3 hybrids with carbon layer bridges as a novel photothermal/photocatalytic integrated system for Z-scheme overall water splitting. *Adv. Energy Mater.*, 11, 46, 2102452, 2021.
49. Yan, K., Wen, C., Li, R. et al., Morphological optimized CeO_2 and Cu-doped CeO_2 nanocrystals for hydrogen production by solar photo-thermochemical water splitting based on surface photoinduced oxygen vacancies. *Appl. Surf. Sci.*, 636, 157779, 2023.
50. Guo, S., Li, X., Li, J. *et al.*, Boosting photocatalytic hydrogen production from water by photothermally induced biphase systems. *Nat. Commun.*, 12, 1, 1343, 2021.
51. Han, H., Huang, K., Yao, Y. *et al.*, Enhanced photocatalytic splitting of photothermally induced water vapor to evolve hydrogen. *Chem. Eng. J.*, 450, 138419, 2022.
52. Yin, Z., Fu, C., Han, K. *et al.*, Review on technologies of hydrogen production from biomass. *Thermal. Power Gener.*, 51, 37–48, 2022.

53. Parthasarathy, P. and Sheeba, K.N., Combined slow pyrolysis and steam gasification of biomass for hydrogen generation—a review. *Int. J. Energy Res.*, 39, 147–164, 2015.
54. Reddy, S.N., Nanda, S., Dalai, A.K. *et al.*, Supercritical water gasification of biomass for hydrogen production. *Int. J. Energy Res.*, 39, 6912–6926, 2014.
55. Lopez, G., Santamaria, L., Lemonidou, A. *et al.*, Hydrogen generation from biomass by pyrolysis. *N.R.M.P.*, 2, 20, 2022.
56. Yaashikaa, P.R., Devi, M.K., Kumar, P.S., Biohydrogen production: An outlook on methods, constraints, economic analysis and future prospect. *Int. J. Hydrogen Energy*, 47, 98, 41488–41506, 2022.
57. Frowijn, L.S. and van Sark, W.G., Analysis of photon-driven solar-to-hydrogen production methods in the Netherlands. *Sustain. Energy Technol.*, 48, 101631, 2021.
58. Zhang, Q., Zhu, S., Zhang, Z., Zhang, H., Xia, C., Enhancement strategies for photo-fermentative biohydrogen production: A review. *Bioresour. Technol.*, 340, 125601, 2021.
59. Cai, J., Zhao, Y., Fan, J., Li, F., Feng, C., Guan, Y., Tang, N., Photosynthetic bacteria improved hydrogen yield of combined dark-and photo-fermentation. *J. Biotechnol.*, 302, 18–25, 2019.
60. Yildiz, B. and Kazimi, M.S., Efficiency of hydrogen production systems using alternative nuclear energy technologie. *Int. J. Hydrogen Energy*, 31, 1, 77–92, 2006.
61. Gillis, R.J., Al-Ali, K., Green, W.H., Thermochemical production of hydrogen from hydrogen sulfide with iodine thermochemical cycles. *Int. J. Hydrogen Energy*, 43, 12939–12947, 2018.
62. Motazedi, K., Salkuyeh, Y.K., Laurenzi, I.J., *et al.* Economic and environmental competitiveness of high temperature electrolysis for hydrogen productio. *Int. J. Hydrogen Energy*, 46, 21274–21288, 2021.
63. Piera, M., Martínez-Val, J.M., Montes, M.J., Safety issues of nuclear production of hydrogen. *Energy Convers. Manage.*, 47, 2732–2739, 2006.
64. Feng, C., Chen, M., Yang, Z., Xie, Z., Li, X., Li, S., Abudula, A., Guan, G., Electrocatalytic seawater splitting for hydrogen production: Recent progress and future prospects. *J. Mater. Sci. Technol.*, 162, 203–226, 2023.
65. Gao, F., Yu, P., Gao, M., Seawater electrolysis technologies for green hydrogen production: challenges and opportunities. *Curr. Opin. Chem. Eng.*, 36, 100827, 2022.
66. Wang, Q., Wang, Y., Sun, J., Sun, S., Zhu, X., Sun, B., Hydrogen production from simulated seawater by microwave liquid discharge: A new way of green production. *Chem. Eng. J.*, 465, 142872, 2023.
67. Zadegan, M., Pedram, M., Sohani, A., Hoseinzadeh, S., A super-efficient method for hydrogen production from seawater. *Int. J. Hydrogen Energy*, 47, 26135–26155, 2022.

68. Feng, S., Yu, Y., Li, J., Luo, J., Deng, P., Jia, C., Shen, Y., Tian, X., Recent progress in seawater electrolysis for hydrogen evolution by transition metal phosphides. *Catal. Commun.*, 162, 106382, 2022.

Part 2

CARBON CAPTURING AND HYDROGEN

4

Biomass Waste-to-Hydrogen Conversion: Innovations in Methanol and Ammonia Production as Efficient and Sustainable Hydrogen Storage Pathways

Arif Darmawan[1], Abdul Hadi[1], Abdul Hamid Budiman[1], Eniya Listiani Dewi[1] and Muhammad Aziz[2]*

[1]Research Center for Energy Conversion and Conservation, The National Research and Innovation Agency, Tangerang Selatan, Indonesia
[2]Institute of Industrial Science, The University of Tokyo, Komaba, Meguro-Ku, Tokyo, Japan

Abstract

Hydrogen is garnering considerable attention as a fuel and energy storage (carrier) owing to its high energy density and environmental cleanliness. It offers advantages such as storage capability and minimal environmental impact during use. Nonetheless, the widespread adoption of hydrogen faces notable challenges, such as the high cost of production, storage, and transportation as well as limited social acceptance. On the other hand, the utilization of biomass wastes as energy sources is gaining increasing concern due to economic and environmental considerations. Biomass waste-to-energy technologies are important because they offer a sustainable solution for waste management, contribute to greenhouse gas emissions reduction, provide a reliable energy source, and support the transition to a circular economy. This chapter explores the potential for hydrogen carrier production from biomass wastes, including methanol and ammonia. If effectively managed, hydrogen holds promise as a potential substitute for conventional carbon-based fuels. The chapter delves into advanced ammonia and methanol production methods as hydrogen carriers, emphasizing total energy efficiency. Ensuring minimal exergy loss is crucial when designing the process, as it directly impacts the overall energy efficiency.

**Corresponding author*: maziz@iis.u-tokyo.ac.jp

Mehmet Sankir and Nurdan Demirci Sankir (eds.) Towards Green Hydrogen Generation, (123–148)

***Keywords*:** Hydrogen carrier, biomass waste, industrial waste, chemical looping, exergy efficiency, ammonia, methanol

4.1 Introduction

The United Nations (UN) member countries ratified the sustainable development goals (SDGs) in September 2015 as part of a UN General Assembly resolution focusing on the 2030 Sustainable Development Agenda [1]. One of these goals, SDG7, focuses on achieving affordable and clean energy through three main objectives: ensuring access to reliable and affordable modern energy, boosting the share of renewable energy within the global energy mix, and improving energy efficiency [2]. A study on SDG7 found that certain regions still lack clean and affordable energy, highlighting the need to tackle policy concerns [3]. Regulations should be drafted to benefit society's disadvantaged and marginalized segments. Additionally, it is important to evaluate how technology choices impact the interconnections among various SDGs. Initiatives promoting energy access and industrial development would contribute to the production and implementation of technologies [3].

The current energy system faces the pressing challenges of global climate changes and fossil fuel issues, demanding serious attention. Addressing these issues necessitates taking urgent measures. One potential solution is incorporating decentralized renewable energy sources such as biomass or waste (refused materials) within the energy system. This approach could enhance a sustainable energy system and alleviate the environmental impacts associated with these problems [4]. Biomass and its derivatives are regarded as renewable energy resources as they mainly originate from organic matter that can regenerate over time. Unlike finite fossil fuels, they can be continuously generated, making them a sustainable energy option. Biomass wastes consist of discarded materials from human activities and natural processes, as shown in Figure 4.1, typically requiring suitable treatment before conversion. Utilizing biomass waste for energy purposes offers several advantages. It is cost-effective while mitigating the negative environmental consequences of improper disposal methods like polluting open burning and disposal of waste in landfills [5].

Biomass waste conversion technologies provide efficient methods for waste management and energy generation [4, 7]. Converting them into energy reduces the volume of waste that would otherwise contribute to environmental problems. They present an opportunity to divert biomass wastes from conventional disposal methods, mitigating associated

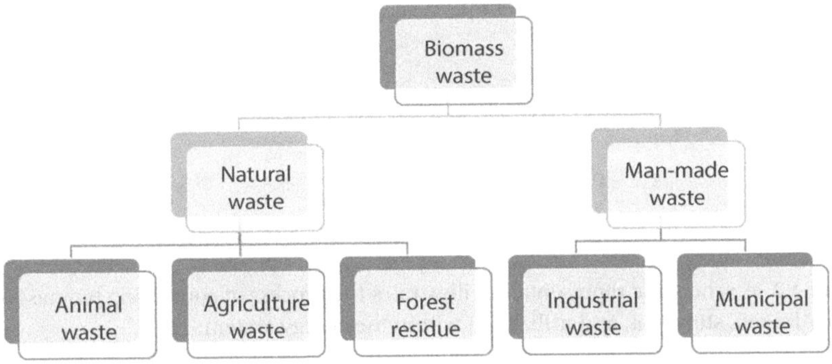

Figure 4.1 Simplified biomass waste classification [6].

environmental and health risks while significantly reducing greenhouse gas emissions. They offer an alternative to fossil fuels and contribute to diversifying the energy mix, reducing dependence on nonrenewable resources. Thus, the development of efficient biomass and waste conversion technologies is believed to be crucial from economic, environmental, and social perspectives. Moreover, converting these sources into alternative secondary energy sources like hydrogen can enhance economic value and bolster community energy resilience [8].

Hydrogen possesses advantages over electricity in terms of its ability to be stored for extended periods and serves as a raw material for various processes. The interchangeability of secondary energy sources involving power, hydrogen, and heat is vital for attaining heightened energy security and stability. To date, hydrogen has found extensive use in oil refineries (33%), ammonia, and methanol production (27% and 11%, respectively), as well as heat treatment of metal and food industries [9]. In addition, hydrogen can be used for power generation in stationary or mobile applications, such as employing dedicated combustion, co-combustion with other fuels (mixing), and fuel cells. This approach mostly emits water, contributing to an environmentally friendly process. Figure 4.2 illustrates the conceptual depiction of the conversion process of biomass waste into hydrogen, along with the subsequent storage and utilization. Biomass waste, such as agricultural residues, forestry residues, organic municipal waste, and dedicated energy crops, undergoes several pretreatment and conversion processes to produce hydrogen. Furthermore, it can be employed at the production location or stored using different methods, transported, and distributed to the final users. Before utilization at the final stage, the stored hydrogen may undergo decomposition or direct without requiring decomposition.

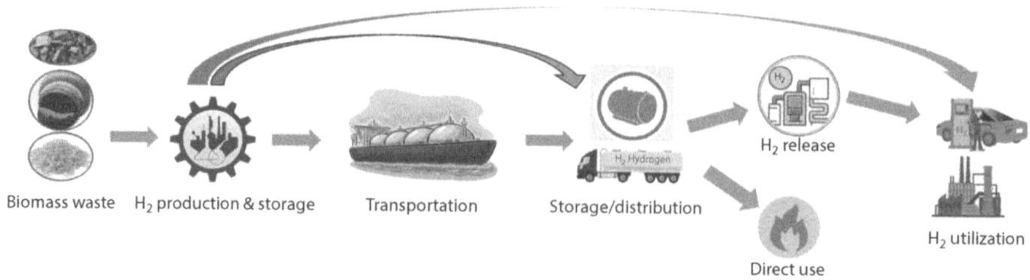

Figure 4.2 A schematic representation illustrates the process of converting biomass waste into hydrogen, storing it, and utilizing it within the energy system.

4.2 Hydrogen Production from Thermochemical Conversion and Methanol/Ammonia as Potential Hydrogen Carriers

With extensive global research in this field, biomass utilization for energy purposes has significantly expanded. Biomass waste as feedstock material can be converted into valuable thermal energy, power, or fuels. Conversion technology option depends on various factors, including biomass characteristics, the desired form of energy, environmental regulations, economic aspects, and project considerations [10]. Initially, prepared biomass undergoes specific methods to improve its quality, such as size reduction, density improvement, and drying [11]. The main pathways for hydrogen production from biomass can be categorized into thermochemical and biological methods, as depicted in Figure 4.3. Thermochemical conversion employs main technologies such as gasification, pyrolysis, and other thermochemical cycles.

Biological conversion, in contrast, utilizes microorganisms in decomposing and transforming biomass into biogas, consisting of methane (CH_4), carbon monoxide (CO), hydrogen (H_2), and carbon dioxide (CO_2). This approach encompasses various techniques such as fermentation, biophotolysis, enzymatic processes, and microbial electrolysis. Fermentation, a key biological conversion method, can be further categorized into dark and photo-fermentation. In dark fermentation, microorganisms break down biomass without light to produce gasses like hydrogen and methane. On the other hand, photo-fermentation relies on light energy, especially solar energy, in driving the conversion of biomass into gaseous fuels, predominantly hydrogen. Biophotolysis, another biological conversion process, encompasses both direct and indirect methods. Direct biophotolysis involves using photosynthetic microorganisms to convert biomass into

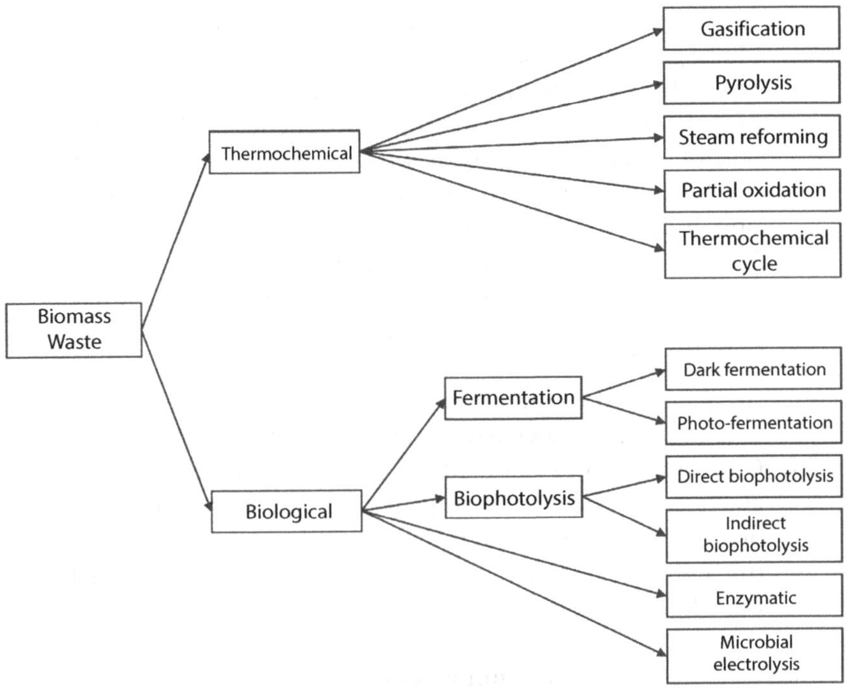

Figure 4.3 Various technologies for hydrogen production from biomass waste.

hydrogen directly in the presence of light. Indirect biophotolysis utilizes genetically modified microorganisms to produce hydrogen through multi-step biochemical pathways, often more efficiently than direct biophotolysis [4].

Thermochemical conversion is a preferable method for hydrogen production due to its numerous advantages:

- Enhanced efficiency: Thermochemical conversion processes exhibit high energy efficiency, ensuring a maximum feedstock conversion into hydrogen. This efficiency surpasses that of other methods.
- Versatility: Thermochemical conversion methods can utilize a diverse range of feedstocks, including water, hydrocarbons, and biomass. This flexibility allows the selection of the most suitable feedstock based on availability and cost.
- Utilization of high-temperature heat sources: Thermochemical conversion processes can effectively harness heat sources to supply the necessary energy to drive the chemical

reactions, improving overall efficiency and reducing greenhouse gas emissions.

- Scalability: Thermochemical conversion methods have the potential for large-scale hydrogen production. With efficient reactor designs and proper process integration, they can meet the increasing demand for hydrogen in the transportation, industry, and energy storage sectors.
- Carbon neutrality: Certain thermochemical conversion pathways, particularly those involving water splitting, have the potential to be carbon-neutral or low-carbon. Utilizing renewable heat sources or implementing carbon capture and storage (CCS) technologies can minimize or eliminate carbon emissions associated with hydrogen production.
- Co-production of value-added chemicals: Thermochemical conversion processes often generate value-added chemicals or byproducts alongside hydrogen. This co-production feature enhances the economic viability of the process by creating additional revenue streams and reducing waste.

Moreover, methanol and ammonia are potential carriers of hydrogen. Methanol, derived from CO_2 and hydrogen conversion, serves as an energy transporter to produce hydrogen *via* various methods, like steam reforming, autothermal reforming, partial oxidation, methanol and ammonia decomposition, and methanol-water electrolysis reactions. Additionally, methanol holds promise as a precursor for producing diverse chemical compounds like acetic acid, methyl tertiary butyl ether, dimethyl ether, and methylamine [12]. On the other hand, ammonia is a hydrogen-rich material. It can be used as an agent for hydrogen storage and transport and has been shown to have a good performance from an economic perspective for medium hydrogen production scale and short-distance transportation. Conventional ammonia production involves the Haber-Bosch process, the predominant method used globally. This process combines hydrogen (H_2) and nitrogen (N_2) to form ammonia (NH_3) through a catalytic reaction. The conventional ammonia production method has been widely employed due to its efficiency, scalability, and established infrastructure. However, there is ongoing research and development in alternative methods, such as green ammonia production using renewable energy sources, to reduce the environmental impact associated with the conventional process.

4.3 Enhancing Efficiency through Process Modeling and Exergy Optimization

Researchers are investing substantial efforts into exploring the optimum design, operation, and management of conversion systems to achieve high efficiency and effectiveness in converting biomass into usable energy sources. Because of their intricate nature, modeling and simulation techniques can be very important to understanding phenomena in each conversion process of biomass or the whole system itself. This approach can save time and money by reducing the reliance on large-scale and expensive experiments, especially at an industrial scale [13]. Process modeling is a method used to understand and illustrate the integrated system to convert biomass into various products, including energy. It focuses on enhancing the correlation among various material and energy flows between the current and future states of a process. Hence, process modeling involves both process integration and optimization. For example, a process modeling outcome might be a diagram depicting the integrated process of biomass conversion into hydrogen, including steps like gasification and steam reforming. These models serve as valuable tools for optimizing biomass energy conversions.

In integrated processes, two technical problems emerge. The first involves designing unit operations, while the second problem pertains to designing complete systems. This chapter primarily focuses on the latter issue, specifically addressing the design of the process flow sheet to enhance exergy efficiency. Figure 4.4, developed by Klemes *et al.* [14], presents the most

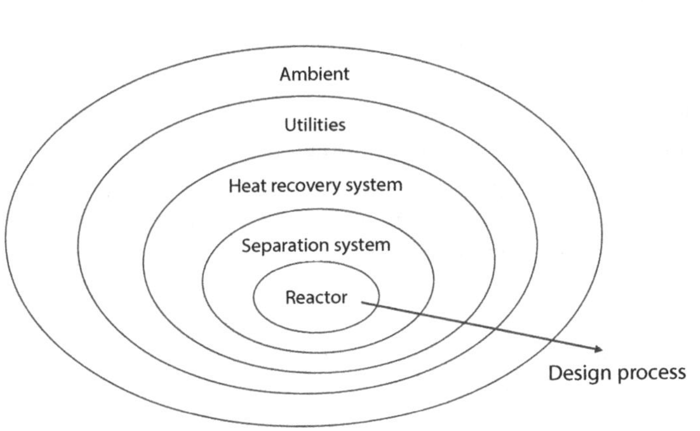

Figure 4.4 The conventional onion diagrams.

widely recognized representations for chemical process design, providing practical guidance for enhancing exergy recovery.

The heart of the entire process lies in the reaction layer, and its design significantly impacts the separation system found in the next layer. The design of both the reactor and separation system dictates the overall design and performance of the heat recovery system, which is the next focus of the design. Moreover, the energy utility system in the fourth layer is tailored to meet additional heating and cooling demands not fulfilled by the heat recovery system [15]. Finally, the outermost layer encompasses the ambient system, which manages various emissions and effluents generated by the systems before their eventual discharge into the surrounding environment. Various principles and guidelines for enhancing the exergy efficiency are detailed in Table 4.1.

Over 30 years, numerous onion diagrams have been suggested with the objective of a more optimal design process. We only discuss the second and third layers of the onion diagram and introduce an enhanced process integration approach, leading to a revised form of the previous onion diagram presented in Figure 4.4, now shown in Figure 4.5. This updated diagram incorporates the addition of the material-recovery system in the second layer and the inclusion of exergy elevation in the third layer. In process integration, the material-recovery system layer encompasses various strategies to recover the material, including direct reuse, recycling, and regeneration of the material. It emphasizes the importance of adopting a comprehensive strategy for process design, retrofitting, and operation, focusing on the integration of the process to achieve an optimal outcome.

A material recovery method is proposed, encompassing reuse, regeneration, and recycling, as illustrated in Figure 4.6. These strategies have gained wide acceptance and offer effective techniques for recovering materials. "Reuse" refers to diverting output materials from a process to another process rather than reintroducing them into the processes from which they originated. The recycling scheme allows effluents to reintroduce the processes in which they were generated. In regeneration schemes, output materials undergo partial treatment before being reused or recycled. Similar to pinch analysis, data extraction plays a vital role in adopting material-recovery strategies. These techniques, which encompass reuse, recycling, and regeneration, can be effectively implemented and integrated within a comprehensive plan that incorporates the principles of process integration. Employing these material-recovery strategies can achieve a more sustainable and efficient use of resources.

Table 4.1 Rules and guidelines for improving exergy efficiency [16].

	Components	**Specific guidelines**
Specific layer	Reactor	· Conduct exothermic processes at the maximum achievable temperature while carrying out endothermic processes at the minimum attainable temperature. · Execute exothermic and endothermic reactions simultaneously within a single reactor. · Reduce the utilization of diluents in both exothermic and endothermic processes.
	Separation system	· Optimize separation conditions to enhance separation efficiency. · Utilize multiple-effect systems and heat pumps to maximize heat reuse during evaporation and distillation. · Couple the separation system seamlessly into the chemical reactor for improved overall efficiency.
	Heat recovery system	· Ensure that streams with similar heat capacities are paired for efficient heat exchange. · Match streams for heat exchange using the pinch analysis approach, making sure that the end temperature of one stream nearly matches the beginning temperature of the other. · Minimize the exergy loss of heat from high temperatures to the environment or cooling water, maximizing its potential utilization. · Reduce or eliminate the adoption of intermediate heat transfer fluids to improve overall thermal efficiency.
	Utilities	· Position fans and compressors in locations with lower temperatures to maximize their efficiency. · Consider the utilization of expanders to recover energy waste · Avoid compressing steam that has already been compressed to optimize energy usage and minimize losses.

(Continued)

Table 4.1 Rules and guidelines for improving exergy efficiency [16]. (*Continued*)

	Components	Specific guidelines
	General rules	· Minimize excessive thermodynamic driving forces. · Preferentially implement counter-current instead of co-current processes. · Accept exergy losses only if there is a valid economic justification. · Avoid elongating thermodynamic process chains whenever possible to streamline the overall system. · Reduce the mixing of streams with varying temperatures, pressures, or chemical compositions to prevent unnecessary energy losses or complications.

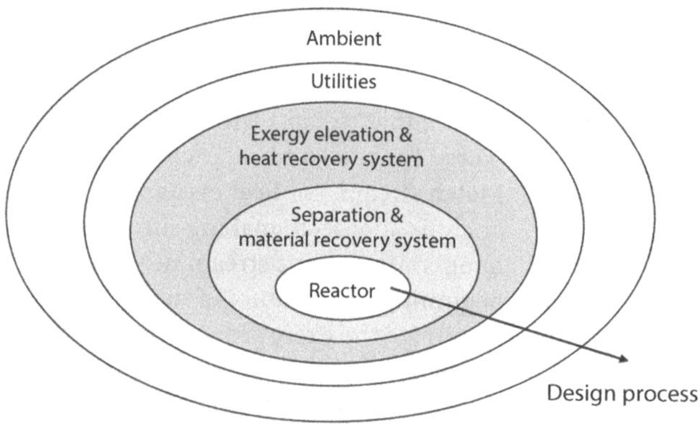

Figure 4.5 Developed onion diagram to improve efficiency.

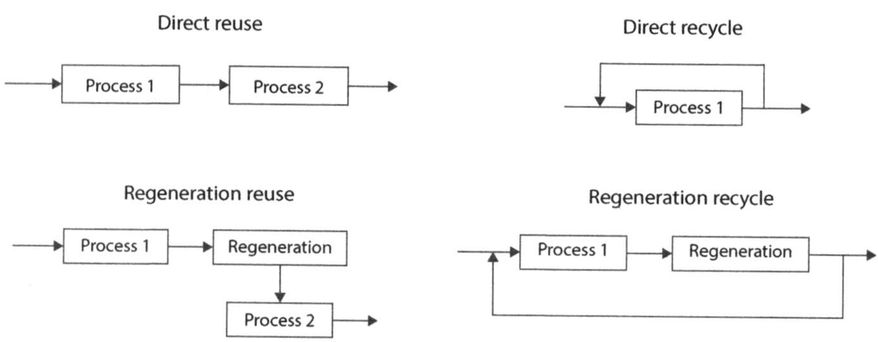

Figure 4.6 Recovery strategies for system efficiency improvement.

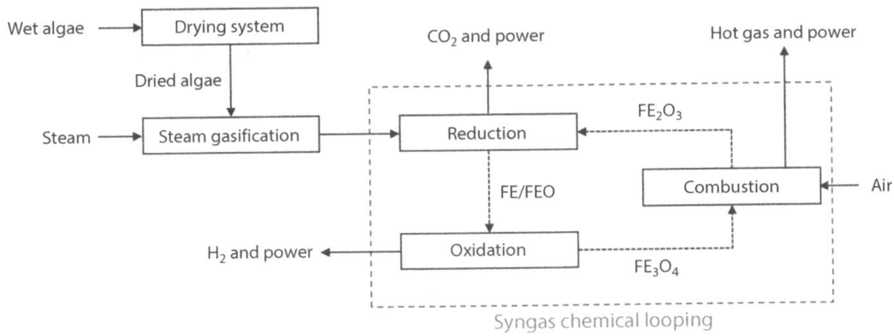

Figure 4.7 A conceptual system of algae utilization employing heat recovery and material recovery systems.

Figure 4.7 illustrates the schematic diagram of a previously published integrated cogeneration system from biomass algae [17]. The system incorporates advanced process integration technology to recover or recirculate optimally the heat, ultimately enhancing the overall efficiency of the energy conversion system, including hydrogen and power. Various components are integrated within the system, including pretreatment (drying), fuel conversion (steam gasification and chemical looping), and power generation. The process begins with fresh wet algae, which initially contains a moisture content of 85%. To enhance heat transfer during drying, the wet algae is uniformly sized by cutting. It is assumed that the mass flow rate of the wet algae is 100 tons per hour ($t\ h^{-1}$).

The integrated cogeneration system begins with wet brown macroalgae, which undergo a drying process to enhance their properties. In thermal drying systems, macroalgae dehydration relies on heating to evaporate moisture. However, this process is associated with significant exergy losses caused by large irreversibility, leading to high energy consumption. Imbalanced heat exchange between streams leads to exergy losses because of suboptimal heat coupling and the existence of unrecoverable heat, primarily stemming from limitations in exergy rate differences among the streams. Consequently, only a small fraction of the heat involved throughout the system can be recuperated, resulting in significant exergy losses. To tackle these challenges and enhance biomass drying efficiency, an innovative approach known as heat circulation technology has been developed, as depicted in Figure 4.8. This technology aims to decrease the required drying energy by employing heat circulation. Unlike traditional heat cascade utilization, which primarily focuses on energy recovery, heat circulation technology emphasizes exergy recovery. To achieve exergy recovery,

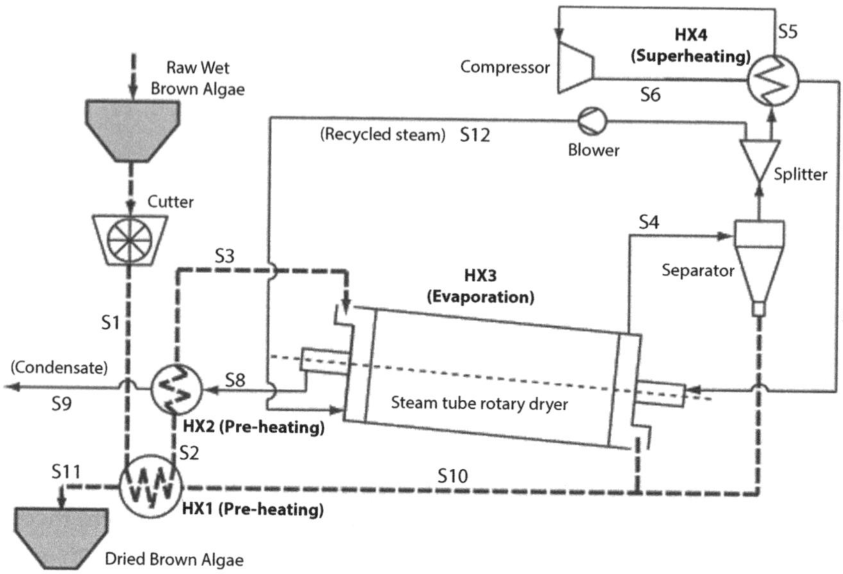

Figure 4.8 Schematic process flow diagram of the drying process adopting heat circulation technology.

the initial step involves elevating the exergy rate of the process stream by utilizing work, like *via* compression, to create a minimal temperature difference required for efficient heat exchange. The exergy losses can be reduced by combining the work and thermal energy having a lower exergy rate, to produce the thermal energy having a higher exergy rate.

The further step involves achieving balanced heat exchange by effectively pairing sensible and latent heat. This pairing is accomplished by identifying the optimal heat pairing for each stream based on temperature-enthalpy diagrams. This optimal heat pairing allows the drying energy required to be recuperated and used for the next drying process. By incorporating heat circulation technology and optimizing exergy recovery, the algae drying process's overall energy efficiency and effectiveness can be greatly enhanced. In the context of heat circulation technology and exergy elevation employed in a drying process, Figure 4.9 illustrates the underlying principle. The key concept involves the creation of a hot stream (process stream) through exergy elevation achieved by compressing the evaporated water to increase the exergy rate. By employing this method, self-heat exchange can be achieved, effectively minimizing exergy losses. Process integration is then implemented within a larger system, allowing for the effective combination of various modules and the efficient utilization of

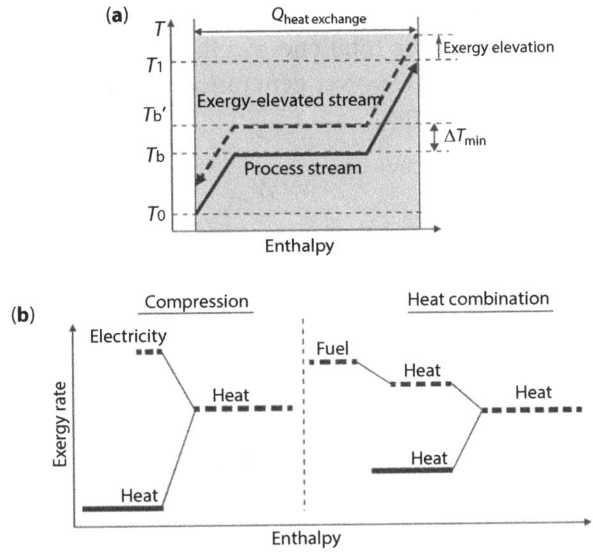

Figure 4.9 Principle of exergy recovery: (a) increasing the exergy level of a stream and heat exchange between the hot and cold (process) streams and (b) elevating the exergy through the combination of electricity, in the form of compression, and low-level thermal energy.

heat that cannot be recovered within individual modules. Optimization is achieved by integrating material and exergy recovery in the second and third layers of the system, enabling a larger amount of heat recovery. This integration of material and exergy recovery minimizes exergy destruction throughout the integrated system, leading to excellent total energy efficiency.

The dehydrated algae are then distributed into a gasification process to produce syngas employing a dual circulating fluidized-bed reactor. This technology presents benefits, including increased carbon conversion efficiency, conversion rate, mass and heat transfer, and overall efficiency. The system also incorporates a syngas chemical looping process with three reactors: a fuel reactor (reduction), a steam reactor (oxidation), and an air reactor (combustion). The syngas produced in the gasifier undergo reactions with carriers (OCs), resulting in emissions of CO_2 and water streams. In the steam reactor (oxidation), the previously reduced particles undergo a reaction with steam, resulting in the production of a hydrogen-rich gas stream exiting the system along with the excess/unreacted steam. The oxygen carriers are subsequently regenerated to their original state through a reaction with oxygen from the air (combustion). The heat produced within each reactor is recovered through expanders, converting it into electricity. For efficiency evaluation, the total energy efficiency (η_{total}) of the system is calculated by summing the power generation efficiency (η_{power}) and the H_2 production efficiency (η_{H2}).

$$\eta_{power} = \frac{W_{net}}{m_{dried\ algae} \times LHV_{algae}} \tag{4.1}$$

$$\eta_{H2} = \frac{m_{H2} \times LHV_{H2}}{m_{dried\ algae} \times LHV_{algae}} \tag{4.2}$$

$$\eta_{total} = \eta_{power} + \eta_{H2} \tag{4.3}$$

4.4 Municipal Solid Waste Conversion to Hydrogen and Methanol for Efficient Energy Storage

For many decades, the issue of municipal solid waste (MSW) has posed significant challenges, particularly in developing countries, causing numerous social and environmental problems. There is an urgent need for suitable treatment and conversion methods for this MSW. This subsection presents and investigates an efficient approach to convert MSW into hydrogen, which is subsequently stored in methanol, as shown in Figure 4.10. The developed system comprises multiple concurrent processes, including hydrothermal treatment, direct chemical looping, methanol synthesis, and a combined cycle for power generation. Hydrothermal treatment is utilized to improve the carbon content and calorific value of MSW and achieve material uniformity. The treated MSW is subsequently directed to the direct chemical looping process to convert it into hydrogen while simultaneously separating CO_2.

Chemical looping is employed as the primary method to convert the treated MSW and generate CO_2, hydrogen, and heat. This process is achieved by using an oxygen carrier (metal oxide). The system comprises three circulating reactors, namely, the reducer, oxidizer, and combustor as shown in Figure 4.11. These reactors are crucial in facilitating targeted reactions and heat transfer. The oxygen carriers, responsible for carrying

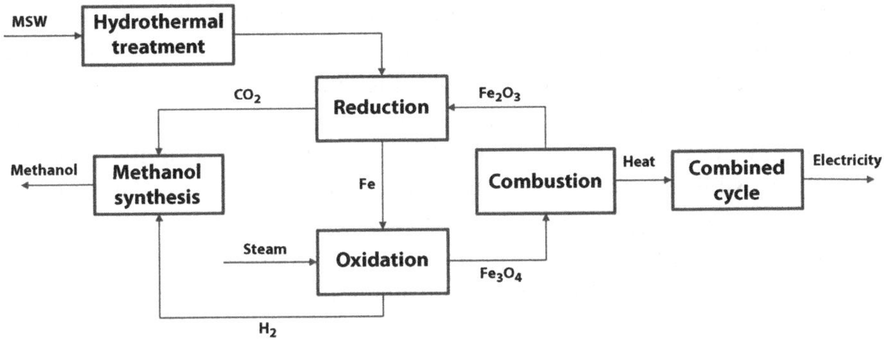

Figure 4.10 Conceptual design of the developed system that efficiently converts municipal solid waste into hydrogen [18].

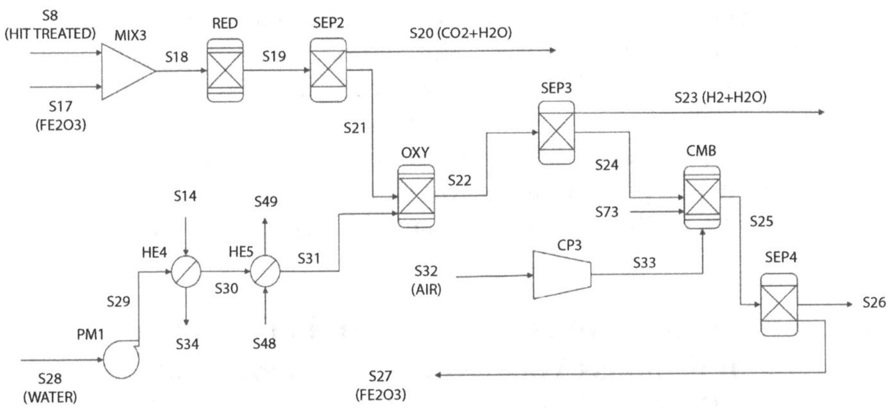

Figure 4.11 Process flow diagram of chemical looping module to generate H_2, CO_2, and heat.

out the chemical looping process, are circulated among these reactors. Iron-based materials are selected as the oxygen carriers in this system due to their numerous advantages. These materials possess a high oxygen-carrying capacity, exhibit good thermodynamic characteristics and excellent strength, and are cost-effective. Throughout the circulation process, the oxygen carrier particles transform hematite (Fe_2O_3) to wüstite (FeO) and to magnetite (Fe_3O_4) before ultimately being converted to hematite.

Both hydrogen and CO_2 generated in the chemical looping module are utilized as material inputs for methanol production. Ultimately, the system recovers heat through gas and steam turbines to generate electricity.

Various operating parameters, such as reactor pressures in the chemical looping module and temperature and pressure in the methanol synthesis module, are analyzed for total energy efficiency and the amount of produced methanol. The process design and integration of the developed system have been conducted meticulously, resulting in a highly promising and feasible approach for hydrogen-methanol production and power generation, focusing on achieving optimal total energy efficiency. The study's most notable outcome showcases the system's highest total energy efficiency, reaching 45.4%. This exceptional efficiency is achieved through specific operating conditions, including a reactor pressure of 4 MPa in the chemical looping module and temperature and pressure of 200°C and 5 MPa, respectively, in the methanol synthesis module. Using 100 kg h^{-1} of raw MSW as the feedstock, the system can produce approximately 41 kg h^{-1} of methanol.

Additionally, the system is self-sufficient in generating the required electricity throughout the process. This approach indicates a high level of competitiveness for the developed system. The attained energy efficiency exceeds twice the average energy efficiency typically observed in existing waste-to-energy systems. This result demonstrates the system's potential to outperform and provide an efficient and sustainable solution for MSW conversion into methanol while simultaneously generating power.

4.5 Innovations in Direct Ammonia Production from Biomass Waste through Carbonization and Thermochemical Cycles

A proposed integrated system comprises a two-step cyclic ammonia synthesis, biomass carbonization and power generation, to harvest the energy from empty fruit bunch (EFB). The waste generated from palm milling has been experiencing a significant increase due to the industry's substantial expansion, particularly in South-East Asia. According to production forecasts, there has been an annual growth rate of 7.2% between 2016 and 2021 [19]. Therefore, it is crucial to employ appropriate technologies to maximize the economic benefits and minimize the environmental impacts of high-potential biomass waste, including EFB, in this industry. Improper disposal practices like open burning can lead to numerous environmental issues.

Figures 4.12 and 4.13 illustrate the conceptual process design details for the proposed ammonia generation. Initially, the biomass feedstock

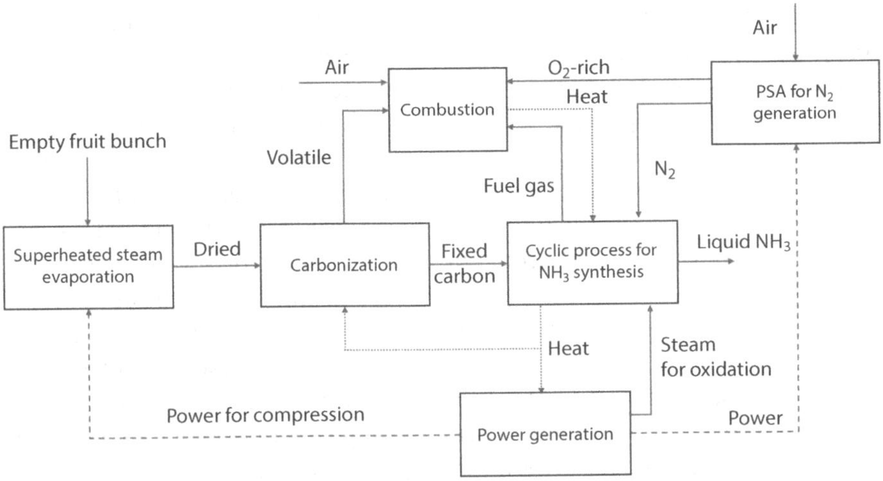

Figure 4.12 Integrated system and ammonia synthesis *via* cyclic process [20].

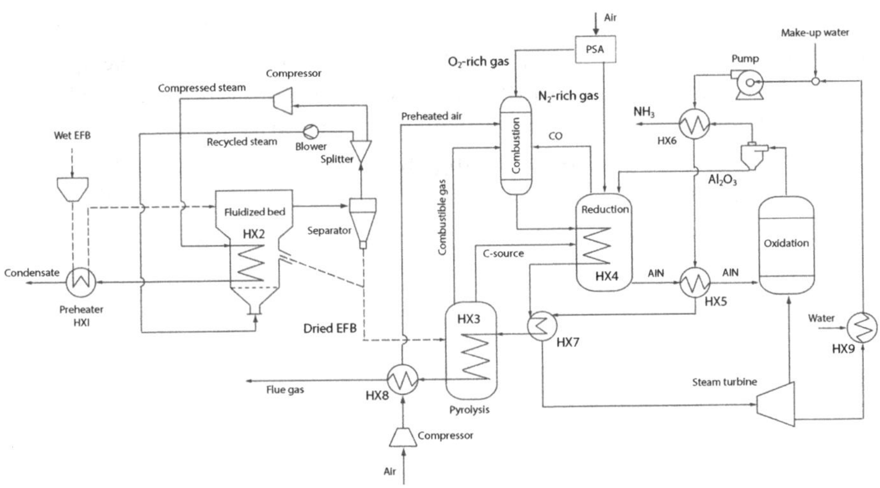

Figure 4.13 Detailed conceptual diagram of ammonia production.

undergoes evaporation and pyrolysis (carbonization) processes to yield a coal-like fuel. The evaporation process employs a fluidized bed system to ensure efficient particle mixing, uniform temperature distribution, and rapid heat and moisture transfer. Electricity is necessary to enable energy-efficient drying (evaporation), primarily consumed by the compressor (in the heat recovery process) and blower. Furthermore, biomass pyrolysis takes place at temperatures exceeding 300°C in an oxygen-deficient environment. This thermal cracking process increases the fixed carbon composition while removing moisture and volatile contents.

The gas generated from the pyrolysis process is combusted in the combustion chamber, supplying the necessary heat for ammonia synthesis. This part requires the highest energy, particularly during the reduction stage at temperatures exceeding 1500°C. The combustor also utilizes syngas, predominantly carbon monoxide (CO), generated during oxidation. The solid carbon and highly concentrated N_2 collected from the pressure swing adsorption unit are conveyed to the ammonia synthesis reactor. This process involves two separate reactors dedicated to the reduction and oxidation reactions, producing syngas and ammonia, respectively. In this cyclic process, alumina (Al_2O_3) and aluminum nitride (AlN) act as carriers for oxygen and nitrogen, correspondingly. Additionally, excess heat is recovered/recirculated, generating high-temperature steam that can be utilized for power generation.

Ensuring the reduction of exergy loss is crucial when designing the overall system, as it directly impacts energy efficiency. In the proposed system, the exergy destruction (loss) could be categorized into primary and secondary. The former represents the unavoidable loss of exergy that occurs during the progress of the process or reaction. On the other hand, secondary destruction refers to exergy losses that arise to support the primary reaction, such as losses during heat exchange, work, or heat loss. Heat exchange-related losses play a significant role in the overall exergy

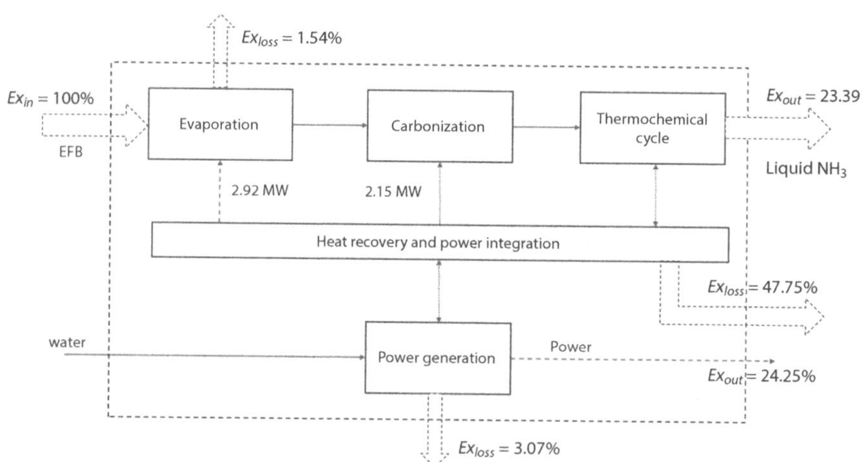

Figure 4.14 Comprehensive exergy balance analysis revealing key contributors to system efficiency loss.

destruction and are observed to be the most prominent contributor, as depicted in Figure 4.14. The unrecovered heat of the exhaust gas and the stream of ammonia and steam discharging into the environment cause the most significant exergy losses. These losses result in increased irreversibility due to heat exchange throughout the process, leading to decreased energy efficiency. Addressing these exergy losses and optimizing heat recovery mechanisms are crucial to improving the system's energy efficiency. The results demonstrate the high efficiency of the integrated system in simultaneously producing power and ammonia, with a remarkable total net efficiency of approximately 48%. By utilizing 100 t h^{-1} of EFB, the system can co-produce 8.95 t h^{-1} of ammonia and generate 46.35 MW of power.

4.6 Exergoeconomic Analysis, Combining Techno-Economic Analysis, and Exergy Analysis

The economic model is commonly developed by incorporating information obtained from analytical models, which provide insights into process temperature-pressure and dynamic simulations that account for the capacities of various components, consumption rates of utilities, and production rates of products [21]. Given the challenges associated with gathering all the necessary data simultaneously and in one location, certain assumptions can be made to facilitate the analysis. These assumptions are as follows:

- The capital costs (CAPEX) of biomass and main reactors are estimated using historical data from reliable sources.
- The CAPEX of equipment, such as heat exchangers, is determined using empirical equations that consider factors like temperature, pressure, and material properties. This approach enables an estimation of the capital costs based on the specific characteristics of the equipment.
- All costs considered in the economic model are adjusted to account for the effects of inflation. This adjustment ensures that the model captures the changing economic landscape over time.
- The operational costs and price of products are assumed to be flat throughout the system's operation. This simplifying assumption provides a baseline scenario for economic analysis but may not capture real-world fluctuations in operational costs or market prices.

These assumptions enable the construction of the economic model despite data collection challenges. They serve as reasonable approximations and allow for evaluating the economic viability and performance of the system under consideration.

Furthermore, exergoeconomic analysis is a comprehensive approach that integrates both thermodynamic exergy analysis and economic analysis. Examining the inefficiencies or exergy destruction of components, as well as the associated costs, provides a holistic assessment of energy conversion systems. This methodology has proven successful in various applications [22]. By combining thermodynamic and economic perspectives, exergoeconomic analysis offers valuable insights into system performance and economic viability. The analyses encompassed both exergy and economic aspects involving the following steps:

- Determining the largest contributors to exergy destruction: The analysis aimed to identify the components or processes within the system that cause the most significant exergy losses. This identification helps prioritize areas for improvement to enhance system efficiency.
- Examining the distribution of exergy destruction: The analysis involved understanding how exergy losses are distributed among different components or processes within the system. This distribution provides insights into the areas with the highest potential for reducing exergy destruction.
- Investigating the cost formation of internal streams and final products: This examination considered various factors, such as raw materials, energy consumption, and operational expenses.
- Balancing capital expenditure and exergy destruction costs: The aim was to find cost-effective solutions. This step involved identifying areas where investments could be made to minimize exergy destruction and improve overall system performance.
- Allocating the final production costs: The analysis allocated the total production costs, considering both the capital investment costs and the exergy destruction cost. This allocation provides a clear and thorough understanding of the total costs, as well as helps evaluate the system's economic feasibility.

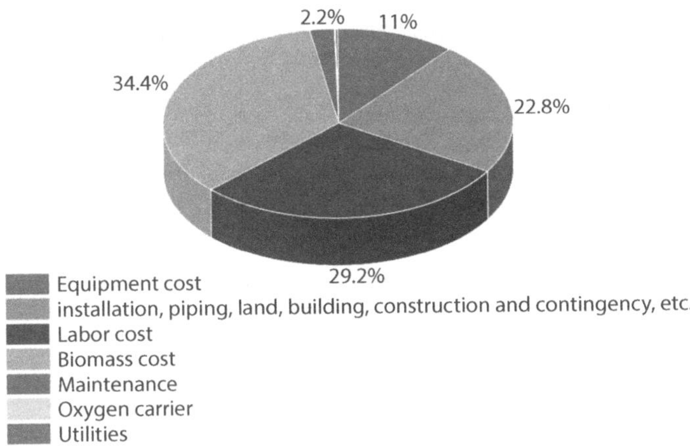

Figure 4.15 A typical various hourly cost for ammonia production [22].

Figure 4.15 illustrates the calculated hourly costs, including CAPEX, operational expenses (OPEX), and other expenses, as outlined in the study by Sun *et al.* [22]. Among these costs, the fuel cost (biomass) makes the largest contribution, accounting for 34.4% of the total, followed by labor costs at 29.2% and installation costs, including piping and construction expenses at 22.8%. It is important to note that equipment expenses represent about 11% of the total. Since the proposed system is able to provide electricity generation, the hourly utility costs are negligible, less than 1% of the total cost.

To improve the dissemination of renewable energy, Cui and Aziz proposed a range of potential infrastructure options for methanol and ammonia as alternative carriers [23]. Their study focuses on converting hydrogen into methanol or ammonia, not only for fuel transportation. The process's necessary carbon dioxide (CO_2) can be obtained from nearby power plants using CO_2 capture employing amine adsorption. Once methanol or ammonia is produced, it is transported to its respective destinations using four different transportation methods: trucks, railways, pipelines, or ships. Eight transportation combinations have been suggested and investigated, utilizing methanol and ammonia as carriers for hydrogen, as shown in Figure 4.16. These infrastructures outline different methods and approaches for the efficient distribution of hydrogen.

The levelized cost is determined by considering the CAPEX and OPEX of the processing equipment, as well as the transportation costs associated with different modes such as trucks, pipelines, ships, or trains. This computation snapshots and summarizes the cost-effectiveness of each infrastructure option and provides insights into the most economically

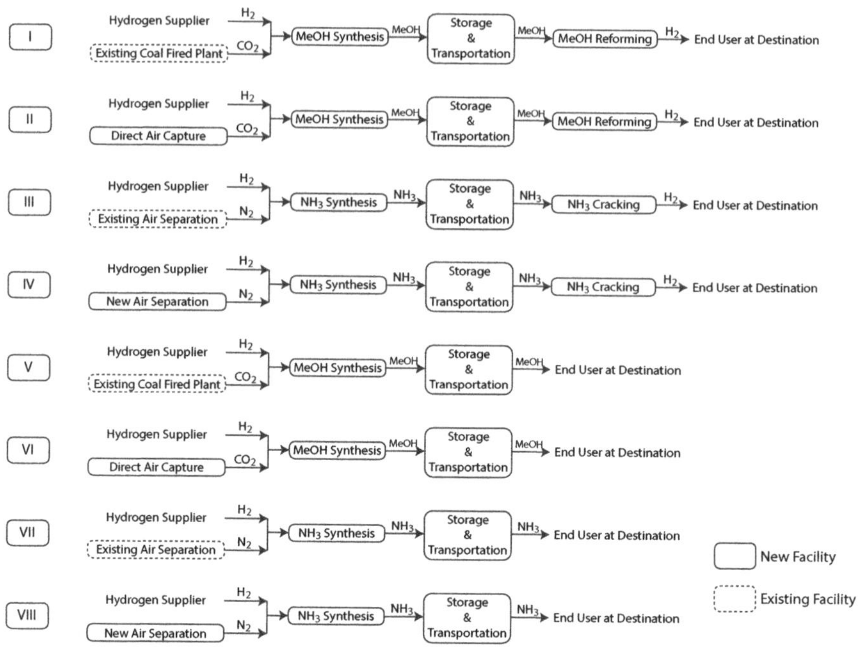

Figure 4.16 Various ammonia- and methanol-based hydrogen distribution [23].

viable methods for hydrogen transportation or energy transfer. Through a systematic analysis, it has been determined that the levelized costs to transport hydrogen in the forms of methanol and ammonia, under optimal conditions, are \$1879 per metric ton of hydrogen (t-H_2) and \$1479 per metric ton of hydrogen (t-H_2) [23], respectively.

4.7 Conclusion

Integrated systems have been developed to convert biomass into ammonia and methanol. The methanol production process comprises hydrothermal treatment, direct chemical looping, methanol synthesis, and a combined cycle. Similarly, ammonia production involves drying or evaporation for water elimination, carbonization of dried raw biomass, thermochemical cycle, and power generation. Various parameters have been investigated to understand the performance of integrated systems.

Process design development and its system integration have been executed effectively, indicating the potential and feasibility of the system for both methanol and ammonia production, particularly in terms of overall energy efficiency. Moreover, an exergoeconomic analysis approach, combining thermodynamic exergy analysis and techno-economic analysis, has been discussed. This analysis accounts for inefficiencies in system components (exergy destruction) and associated costs. The objective is to minimize costs by addressing the operating costs of related processes, capital costs, and internal/external exergy losses.

Overall, the integrated systems could be promising in terms of their performance and economic viability, and the application of exergoeconomic analysis provides valuable insights for optimizing system efficiency and reducing overall costs. This chapter also discussed a benchmark for the design of hydrogen transportation infrastructure. A thorough examination and evaluation set a standard against which future designs and implementations can be measured. The benchmark is a reference point for assessing the efficiency, feasibility, and economic performance of various H_2 transportation systems.

References

1. Sarre, A. D. and Davey, S. M., The Sustainable Development Goals, forests, and the role of Australian Forestry. *Aust. For.*, 84, 2, 41–49, 2021, https://doi.org/10.1080/00049158.2021.1920207.
2. Wu, J. and Wu, T., Ensure access to affordable, reliable, sustainable and modern energy for all, *UN Chron.*, 51, 17–8, 2015, https://doi.org/10.18356/24ef28d2-en.
3. Klemun, M.M., Ojanperä, S., Schweikert, A., Toward evaluating the effect of technology choices on linkages between sustainable development goals. *IScience*, 26, 105727, 2023, https://doi.org/10.1016/j.isci.2022.105727.
4. Aziz, M., Darmawan, A., Juangsa, F.B., Hydrogen production from biomasses and wastes: A technological review. *Int. J. Hydrogen Energy*, 46, 33756–81, 2021, https://doi.org/10.1016/J.IJHYDENE.2021.07.189.
5. Darmawan, A., Asyhari, T., Dunggio, I., Salmahaminati, Aziz, M., Energy harvesting from tropical biomasses in Wallacea region: scenarios, technologies, and perspectives. *Biomass Convers. Biorefin.*, 1, 3, 2023. https://doi.org/10.1007/s13399-023-04223-8.
6. Darmawan, A. and Aziz, M., An overview of biomass waste utilization, in: *Innovative Energy Conversion from Biomass Waste*, pp. 1–23, 2022, https://doi.org/10.1016/B978-0-323-85477-1.00007-5.
7. Abdallah, M., Hamdan, S., Shabib, A., Development of a dynamic optimization framework for waste management systems. *MethodsX*, 8, 101203, 2021, https://doi.org/10.1016/J.MEX.2020.101203.

8. Lui, J., Chen, W.H., Tsang, D.C.W., You, S., A critical review on the principles, applications, and challenges of waste-to-hydrogen technologies. *Renew. Sustain. Energy Rev.*, 134, 110365, 2020, https://doi.org/10.1016/J.RSER.2020.110365.
9. Lipman, T., An Overview of Hydrogen Production and Storage Systems with Renewable Hydrogen Case Studies, *Clean Energy States Alliance,* 2011, Vermont, United States, 2011.
10. Bajpai, P., Biomass conversion processes, in: *Biomass to Energy Conversion Technologies*, pp. 41–151, Elsevier, Amsterdam, Netherlands, 2020, https://doi.org/10.1016/b978-0-12-818400-4.00005-0.
11. Ptasinski, K.J., *Efficiency of Biomass Energy - An Exergy Approach to Biofuels, Power, and Biorefineries*, vol. 89, Wiley, New Jersey, United States, 2017, https://doi.org/10.1002/cite.201770067.
12. Dalena, F., Senatore, A., Basile, M., Knani, S., Basile, A., Iulianelli, A., Advances in Methanol Production and Utilization, with Particular Emphasis toward Hydrogen Generation via Membrane Reactor Technology. *Membranes (Basel)*, 8, 2018, https://www.mdpi.com/2077-0375/8/4/98.
13. Dahlquist, E. *et al.*, Modelling and Simulation of Biomass Conversion Processes in: *2013 8th EUROSIM Congress on Modelling and Simulation*, pp. 506–512, Cardiff, UK, 2013, doi: 10.1109/EUROSIM.2013.91.
14. Klemes, J., Friedler, F., Bulatov, I., Varbanov, P., *Sustainability in the Process Industry: Integration and Optimization*, McGraw-Hill Education, New York, United States, 2010.
15. Foo, D.C.Y. and Ng, D.K.S., Process Integration for Cleaner Process Design, in: *Handbook of Process Integration (PI): Minimisation of Energy and Water Use, Waste and Emissions*, pp. 443–60, Elsevier Inc., Amsterdam, The Netherlands, 2013, https://doi.org/10.1533/9780857097255.4.443.
16. Darmawan, A. and Aziz, M., Application of exergy analysis and enhanced process integration, in: *Innovative Energy Conversion from Biomass Waste*, pp. 61–105, 2022, https://doi.org/10.1016/B978-0-323-85477-1.00001-4.
17. Zaini, I.N., Nurdiawati, A., Aziz, M., Cogeneration of power and H2 by steam gasification and syngas chemical looping of macroalgae. *Appl. Energy*, 207, 134–145, 2017, https://doi.org/10.1016/j.apenergy.2017.06.071.
18. Hakandai, C., Sidik Pramono, H., Aziz, M., Conversion of municipal solid waste to hydrogen and its storage to methanol. *Sustain. Energy Technol. Assess.*, 51, 101968, 2022, https://doi.org/10.1016/J.SETA.2022.101968.
19. Zion Market Research, Palm Oil Market Analysis by Derivative (CPO, PKO, PKC, and Others), and for Edible Oil, Cosmetics, Bio-diesel, Lubricants, Surfactants and Other Applications - Global Industry Perspective, Comprehensive Analysis, and Forecast 2015-2021, Maharashtra, Florida, United States, 2016.
20. Darmawan, A., Lokahita, B., Tokimatsu, K., Aziz, M., Direct ammonia production via a combination of carbonization and thermochemical cycle from

empty fruit bunch. *Therm. Sci. Eng. Prog.*, 20, 100673, 2020, https://doi.org/10.1016/J.TSEP.2020.100673.

21. Wen, D. and Aziz, M., Techno-economic analyses of power-to-ammonia-to-power and biomass-to-ammonia-to-power pathways for carbon neutrality scenario. *Appl. Energy*, 319, 119272, 2022, https://doi.org/10.1016/J.APENERGY.2022.119272.
22. Sun, Z., Wang, S., Aziz, M., Highly integrated system for ammonia and electricity production from biomass employing direct chemical looping: Exergy and exergoeconomic analyses. *Energy Convers. Manage.*, 251, 115013, 2022, https://doi.org/10.1016/J.ENCONMAN.2021.115013.
23. Cui, J. and Aziz, M., Techno-economic analysis of hydrogen transportation infrastructure using ammonia and methanol. *Int. J. Hydrogen Energy*, 48, 15737–47, 2023, https://doi.org/10.1016/J.IJHYDENE.2023.01.096.

5

Advances and Challenges in Metal Oxide–Based Carbon Capture Technologies: An Overview

Berfu Kocabas[1], Olgu C. Cosar[1], Arpad Mihai Rostas[2], Ipek Deniz Yildirim[1], Ahmet Gungor[1] and Emre Erdem[1]*

[1]*Faculty of Engineering and Natural Sciences, Sabanci University, Istanbul, Turkey*
[2]*National Institute for Isotopic and Molecular Technologies, Donat, Cluj-Napoca, Romania*

Abstract
The global pursuit of sustainable industrial practices is confronted by the escalating challenge of controlling carbon dioxide emissions, primarily from key industrial processes such as power generation, cement production, and metal manufacturing. In response to this pressing issue, the deployment of carbon capture technologies has gained considerable traction as a viable strategy to mitigate the environmental impact of carbon dioxide emissions. This overview chapter critically examines the recent advancements and challenges in metal oxide-based carbon capture methodologies, focusing on their pivotal role in mitigating the adverse effects of climate change and fostering sustainable industrial operations.

Keywords: Carbon capture, carbon dioxide, sustainability, metal oxides

5.1 Introduction

The vital industrial processes contribute to societal prosperity and development conflict with the aims of controlling climate change and reducing carbon dioxide (CO_2) emissions [1]. The utilization of fossil fuels for generating electrical power and the methods used in cement production and the metal industry are significant sources of CO_2 emissions [2]. It is essential

**Corresponding author*: emre.erdem@sabanciuniv.edu

Mehmet Sankir and Nurdan Demirci Sankir (eds.) Towards Green Hydrogen Generation, (149–170)

to reduce emissions. Various alternatives across the globe are being evaluated to aid in achieving this objective of lessening the impact of climate change, with carbon capture among them [3]. Carbon capture, by definition, entails the extraction of CO_2 from raw materials either prior to or as a result of industrial procedures [4]. It is worth highlighting that this technology's capability to decrease these industries' carbon emissions is a significant advance toward sustainable operations.

Lately, there has been a growing focus on carbon capture to diminish the adverse impacts of climate change and the depletion of the ozone layer. Extensive research is centered on enhancing carbon capture technologies, aiming to decrease significantly CO_2 emissions while providing a cost-effective and easily implemented resolution [5]. The assessment of the carbon capture procedure usually involves three phases: post-combustion, pre-combustion, and fuel combustion [6]. Literature indicates that extensive research has been conducted to enhance and contrast these methods. Additionally, it involves material developments like adsorbents and membranes designed for employment in CO_2 capture technologies [7]. Another research field is that of the storage and utilization of captured CO_2. The location of CO_2 storage is a crucial and extensively studied factor [8]. Other critical considerations involve exploring industrial applications for stored CO_2 and the carbonization process [9]. Furthermore, significant engineering endeavors are required to refine carbon capture technologies and effectively incorporate them into industrial facilities emitting high levels of CO_2. While the reduction of CO_2 emissions is vital for sustainable environmental practices, it is equally crucial to mitigate potential detrimental environmental effects, like soil and water depletion, that may arise from the carbon capture process [10]. The current technologies used to capture CO_2, including amine washing and the utilization of substances such as $Ca(OH)_2$, $CaCO_3$, and $CaO + CO_2$, depend on the center of a thermal cycle that requires a high temperature to release the absorbed CO_2 [11, 12].

Recently, solid sorbents with a high surface area and pore structure, such as metal-organic frameworks (MOFs), have shown considerable potential for achieving high CO_2 capture [13]. Metal oxides (MOs) and MO-based systems have a rising popularity within the field of carbon capture technologies. The stability of the CO_2 molecule must be surmounted to facilitate its introduction into chemical reactions for the carbon capture process, which can be achieved through catalysts possessing certain specifications. MOs came up with good results as promising catalysts [14], mainly due to the remarkable capacity of MOs to capture carbon and their efficient ability to transform chemically CO_2. In recent years, there has been a growing interest in the utilization of porous MOs as viable candidates for the capture of CO_2 in several scenarios, including pre-combustion, post-combustion, and oxy-combustion. The interaction between exhaust gasses containing CO_2 and porous oxides results in the formation of carbonates. The enduring importance of these entities stems from their advantageous thermodynamic characteristics and widespread accessibility. However, these sorbents have been reported to exhibit low cyclic stability in humid environments [13]. Electrochemical carbon capture technologies, being an alternative CO_2 capture method, are typically preferred due to their efficiency and the possibility of harnessing renewable energy. However, they have encountered challenges such as low CO_2 capture capacity, sensitivity to O_2, and system complexity. Electrochemical carbon capture technology is based on redox-active carriers or pH swings [15, 16]. Quinones, which are redox-active carriers, bind CO_2 through redox reactions in the carbon capture design, providing high energy efficiency. However, in industrial applications, it exhibits low CO_2 capture capacity and is highly sensitive to O_2 gas [17, 18]. The pH swing method is an alternative electrochemical approach to carbon capture that involves the electrolysis of acidic or alkaline solutions to separate salts or water. As a result, the pH swing process is less susceptible to O_2 for CO_2 desorption following CO_2 adsorption and generates higher currents (100 mA cm^{-2}). Additionally, in some pH swing designs, the electrolysis of water and NaCl can produce byproducts such as H_2 and CO_2, further reducing the CO_2 capture capacity [19]. In a study conducted by Zhu *et al.* [20], an electrochemical carbon capture design was developed by integrating oxygen and water (O_2/H_2O) redox coupled with a solid electrolyte modular reactor. The design enables continuous carbon capture. The study showed that the CO_2 capture device performed reduction/oxidation reactions through oxygen reduction reaction (ORR) and oxygen evolution reaction (OER). The diluted CO_2 molecules were efficiently captured at the highly alkaline membrane–cathode interface, producing carbonate ions. The neutralization reactions were driven by an H_3O^+ anion flow, resulting in a high-purity CO_2 stream. Notably, no byproducts were detected during the carbon capture process. The solid electrolyte reactor designed for carbon capture purposes demonstrated exceptional carbon capture rates (440 mA cm^{-2}, 0.137 mmol CO_2 min^{-1} cm^{-2} or 86.7 kg CO_2 day^{-1} m^{-2}), high Faradaic efficiencies (above 90% on a carbonate basis), superior carbon removal efficiency in simulated flue gas (above 98%), and low energy consumption (starting from approximately 150 kJ mol^{-1} CO_2).

5.2 Metal Oxides as Carbon Capture Agents

The consumption of fossil fuels, increasing deforestation, and agricultural and industrial techniques all contribute to the increase of greenhouse gasses, particularly carbon dioxide and methane, into the atmosphere. These gasses absorb some of the heat given off by Earth after it has been warmed by sunlight, and higher levels of these gasses absorb more heat in the lower atmosphere, causing global warming [21]. The reduction of greenhouse gas emissions, especially CO_2, is important to promote sustainable living. Several studies have been performed to reduce the release of CO_2 from different industrial processes [22, 23]. The research carried out by Halliday and Hatton [24] evaluated the effectiveness of molten MOs as sorbents in the carbon capture process. The research aimed to minimize the cost of carbon capture and reduce CO_2 emissions. A model was developed for a coal-fired power station that uses molten MOs in the high-temperature CO_2 capture process while also eliminating other acidic gasses besides CO_2. The sorbent and desorber materials were introduced into the boiler during CO_2 capture at an elevated temperature as shown in Figure 5.1, enhancing the heat recovery efficiency. Based on the results of the simulations, the use of molten MO as sorbents for CO_2 capture could potentially improve the performance of currently used amine-based systems. Although

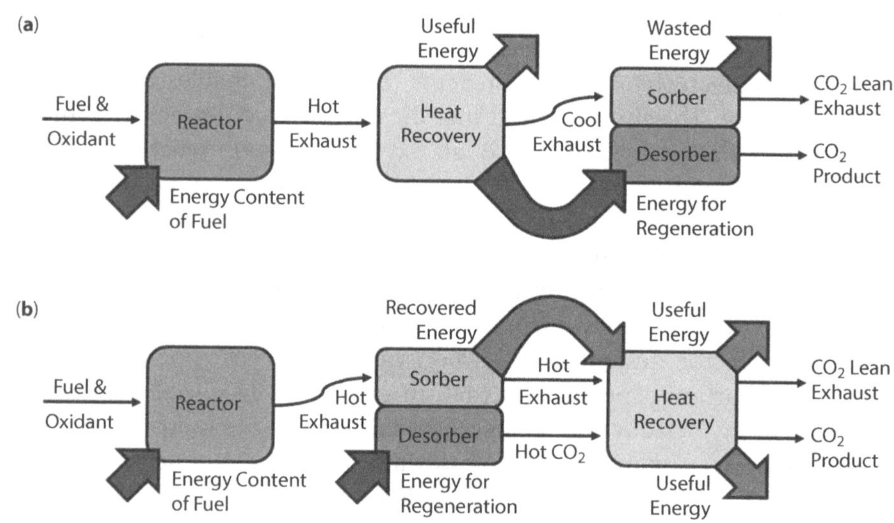

Figure 5.1 Block diagram showing carbon capture and associated energy flows with (a) separation at low temperatures and (b) separation at high temperatures.

amine-based CO_2 capture systems have an energy penalty of approximately 12%. The simulation studies demonstrate that using MO as a sorbent could reduce the thermodynamic energy penalty to approximately 6%. Furthermore, optimization studies have demonstrated an increase of about 40% in the electricity cost of the optimized design after CO_2 capture compared to a similar power plant without CO_2 capture. The cost of CO_2 avoided is approximately £26/tonne, around 38% lower than amine-based carbon capture systems commonly employed in the industry, making the social cost of CO_2 competitive. The simulation design results highlight the important role of corrosion and solids conditioning in reducing power plant energy costs. Hence, it is recommended to further investigate in future studies these aspects.

Carbon capture, utilization, and storage (CCUS) is crucial in overcoming the human-caused climate crisis and providing a sustainable environment and healthy living [25]. For facilities that use fossil fuels to generate energy, carbon capture is the most cost-effective approach to reducing the amount of CO_2 emitted into the atmosphere. Improving carbon capture processes and integrating them into power generation facilities allow for continued fossil fuel use [26]. Some recent studies have pointed out that bioenergy combined with carbon capture may lead to negative carbon emissions [27]. In industrial applications, aqueous amine sorbents are widely used for carbon capture. However, the poor efficiency of the adsorption/desorption cycle and the cost of the sorbents have stimulated the development of new "task-specific" alternatives to aqueous amine components [28]. Solid MO adsorption/desorption cycles provide numerous benefits for carbon capture compared to aqueous amine components, making them a highly investigated issue in the literature [29, 30]. Generally, calcium oxide (CaO) and magnesium oxide (MgO) are preferable choices for solid oxide adsorbents in the field of carbon capture [31, 32]. Sorbents used in the process are encountered in challenging environments, changing their physical form from solid to fluid, being exposed to high temperatures, and being regularly circulated in oxidation, reduction, or carbonation/calcination reactors. During the sintering process, there is a noticeable decline in the effectiveness of the carbon capture sorbents [33]. The fabrication of thermally stable sorbents is essential. Thermodynamically, only the stable (100) miller index corner/edge regions on the crystal phase surface of CaO and MgO can successfully perform carbon capture, decreasing carbon capture efficiency [34, 35]. However, the crystal phase surface with the (111) miller index contains regions that have a considerable concentration of low-coordinated areas, leading to a higher capacity for CO_2 capture [36]. In this research, Mutch *et al.* [34] employed MgO(111) nanolayers to capture

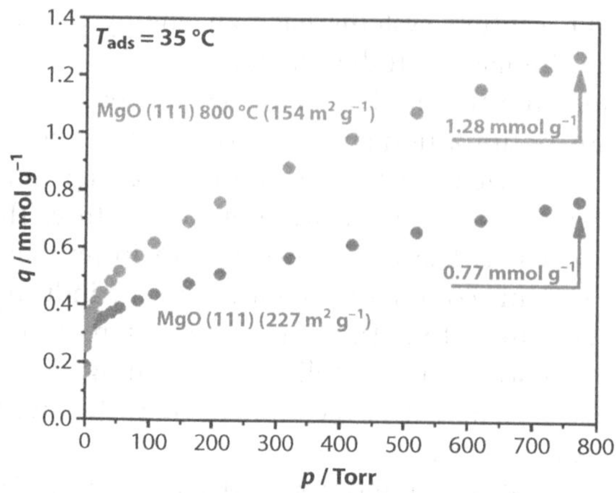

Figure 5.2 CO_2 adsorption isotherms (T_{ads} = 35°C) on MgO(111) and MgO(111) 800°C with specific surface areas inset. Lower surface area conversely provides higher CO_2 uptake.

CO_2. As shown in Figure 5.2, the study found that MgO(111) nanolayers have a high capacity for CO_2 adsorption, but a decrease of roughly 30% in surface area occurred after the sintering process. Despite the reduction in surface area, CO_2 adsorption capacity rose by around 65% (from 0.77 mmol g^{-1} at 227 m^2 g^{-1} to 1.28 mmol g^{-1} at 154 m^2 g^{-1}). Mutch and colleagues found that the utilization of MgO(111) nanolayers as the sorbent and the activation of MgO's polar (111) side at high temperatures were the primary reasons behind the increase in CO_2 adsorption capacity, despite the reduction in surface area. They also obtained a contradictory outcome, indicating that the sintering process did not always have an adverse impact on CO_2 adsorption capacity.

5.3 Enhancing Carbon Capture Efficiency

The selection of certain MOs significantly impacts various aspects, including their ability to adsorb CO_2, the rate of adsorption, the amount of heat needed for regeneration, the simplicity of obtaining them, their thermal stability, and their structural properties. Several types of porous MOs, including alkali MOs and alkaline-earth MOs such as MgO and CaO, can react with CO_2 and form carbonates that are thermodynamically stable. In recent times, there has been considerable interest in the potential of silicates, zirconates, and perovskites derived from lithium (Li), sodium (Na), and potassium (K) for their ability to enhance CO_2 capture methods [14]. For some porous MOs and metal carbonates (MCO_3), several examples of CO_2 capture cyclic processes are given below:

$$MgO(s) + CO_2(g) \leftrightarrow MgCO_3(g)$$

$$CaO(s) + CO_2(g) \leftrightarrow CaCO_3(g)$$

$$Fe_3O_4(s) + Fe(s) + 4CO_2(g) \rightarrow 4FeCO_3(s)$$

$$Li_5AlO_4 + 2CO_2 \rightarrow LiAlO_2 + 2Li_2CO_3$$

Like all the methods and technologies brought with science, MO-based systems also present unique and various challenges due to technological limitations and practical considerations. These challenges span a wide range, encompassing issues such as optimizing designs and developing cost-effective, scalable manufacturing processes. Moreover, there is a need to address the high-temperature operational requirements of certain MO systems, such as the hydrothermal synthesis method of core-shell materials [21]. Some methods like this require a high temperature and high pressure, which needs innovative solutions for managing heat effectively and improving the process to be more cost-efficient. Additionally, sustaining these materials' stability and durability, especially in the presence of corrosive flue gasses, is another challenge to deal with. Among all the technical and scientific issues, evaluating the economic viability of sourcing and utilizing transition metals for MO-based systems, including those rare and subject to supply chain fluctuations, creates more parameters to consider. Ongoing research endeavors are concentrated on innovative strategies to improve the efficiency of MO-based systems to enhance carbon capture technologies. The following subsections focus on these strategies and give insights into various methods developed for the abovementioned purposes.

5.3.1 Metal Silicates as Support Materials

Numerous sorbent materials have been employed to enhance the effectiveness of capturing and absorbing carbon dioxide, including substances like zeolite, activated carbon, and silica. Zeolite comprises crystalline aluminosilicate compounds known for their distinct chemical and physical properties, such as selectivity and strong thermal stability. However, zeolite requires higher temperatures for regeneration. MOs have emerged as a more cost-effective and less toxic alternative compared to zeolites and MOFs among these absorbents. When exposed to elevated temperatures, oxide-based sorbents exhibit remarkable selectivity in capturing carbon dioxide. Typically, alkaline earth MO sorbents such as magnesium oxide (MgO) and calcium oxide (CaO) are used in carbon dioxide absorption as support materials.

Nonetheless, these methods also have certain disadvantages [37]. Gamma-alumina (γ-Al_2O_3) is also a support material that is used in the preparation of potassium-based sorbents. The sorbents that are prepared with γ-Al_2O_3 show great carbon capture capacities. However, it is also observed that the capacity decreases rapidly as the cycle number increases. The reason behind this decrease is the formation of $KAl(CO_3)(OH)_2$ and similar byproducts. These byproducts are not fully transformed into active materials at temperatures below 473 K. Overcoming the mentioned challenges was made easier with the innovative approach to prepare potassium-based solid sorbents, which brings metal silicates into use for the impregnated material. Investigation below 473 K temperature was held, and potassium-based sorbents were compared in terms of their CO_2 sorption and regeneration, morphological, and structural properties [38].

Cho *et al.* prepared potassium-based sorbents using K_2CO_3 (30 wt%) and various MOs and metal silicates, including γ-Al_2O_3, CaO, SiO_2, $Al_6Si_2O_{13}$, $CaSiO_3$, and $ZrSiO_4$. The process starts with dissolving 4.5 g of K_2CO_3 in 50 ml of distilled water and adding 10.5 g of the chosen MO or metal silicate to the K_2CO_3 solution. This mixture was stirred at room temperature for 12 hours and then dried in a rotary evaporator under vacuum conditions. The dried mixture was then calcined in a furnace under a flow of nitrogen gas (100 ml min^{-1}) for 5 hours at a temperature of 600°C. The resulting sorbents were labeled as KAl, KCa, KSi, KAS, KCS, and KZS, where K represents K_2CO_3, and the other letters denote the specific MO or silicate used. Both MO-based and new metal silicate–used sorbents were tested with the help of their CO_2 sorption at 333.15 K and regeneration at 533.15 K under similar conditions.

KAl sorbent showed 95.9 mg CO_2/g sorbent capture capacity in the first cycle, as shown in Figure 5.3. Previous research supports that this capacity should be decreased dramatically after the initial cycle, and this experiment's expectation also came true [39]. Even for the KCa sorbent, which shows a capture capacity of 250 mg CO_2/g sorbent higher than the theoretical value at one cycle, it reduces to 43 mg CO_2/g sorbent at two cycles from 346.3 mg CO_2/g sorbent. While the KSi sorbent exhibits notably low capture capacity even after a single cycle, the KAS, KCS, and KZS sorbents

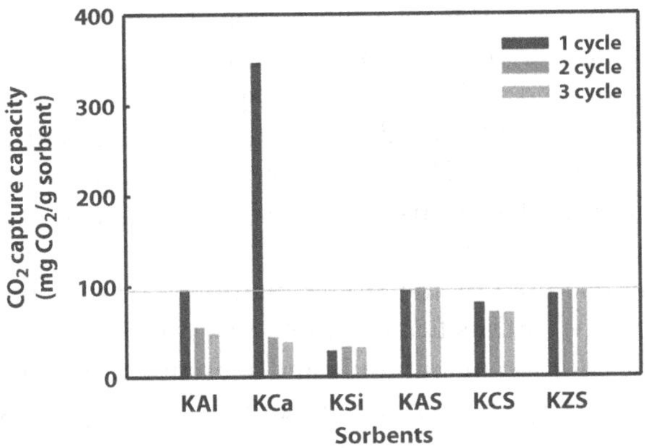

Figure 5.3 CO_2 capture capacities of KAl, KCa, KSi, KAS, KCS, and KZS sorbents during three cycles.

display significantly higher capacities, with approximately 95.4, 80.2, and 89.5 mg of CO_2 captured per gram of sorbent at 1 cycle. Additionally, when considering the regeneration of the sorbents, determined from their CO_2 capture capacities after 1 and 5 cycles, exceptional regeneration ratios of 100%, 86%, and 100% were observed for KAS, KCS, and KZS, respectively.

Structural properties and regeneration performances were also investigated with the help of various characterization methods such as XRD, TG, and FTIR. According to the characterization results, sorbents prepared with MOs have a poor regeneration performance compared to those prepared with metal silicates mainly because of the formation of byproducts in the case of MO-supported sorbents. For instance, the KCa sorbents' deactivation occurs due to the formation of byproducts such as $Ca(OH)_2$, $CaCO_3$, and $K_2Ca(CO_3)_2$ phases. In contrast, the KAS, KCS, and KZS sorbents' weight decreases at temperatures higher than 873.15 K, which are connected to structural changes brought on by thermal degradation above the calcination temperature. In contrast, weight losses in the KAl, KCa, and KSi sorbents after CO_2 absorption take place in the temperature range of 373.15 to 473.15 K, which corresponds to the release of CO_2 and H_2O brought on by the breakdown of $KHCO_3$ created during CO_2 absorption [40].

Copper silicate is another metal silicate that provides a strategy to enhance carbon dioxide capturing efficiency. The material finds a place in the synthesis of a novel amine-grafted mesoporous copper silicate named CSNS-TA [41]. Copper silicate nanospheres (CSNSs) have been modified with mono-, di-, and tri-aminosilanes. This modification process results in a notable increase in the adsorption capacity for CO_2 and exceptional cyclic regeneration under mild operating circumstances, indicating the potential of these solid sorbents for CO_2 capture.

The CSNSs were synthesized hydrothermally, wherein a mixture of $Cu(NO_3)_2 \cdot 3H_2O$ and $NH_3 \cdot H_2O$ was combined in deionized water. [41]. Diethylenetriamine (TA) is used as the grafting agent. Silicon dioxide (SiO_2) spheres are precursors of the copper silicate hollow spheres and they were manufactured by the Stöber method, a process that starts with dissolving tetraethyl orthosilicate (TEOS) in a water and ethanol mixture, followed by the addition of ammonia and subsequent stirring [42].

It is observed that types of amines, quantity of amine groups, and molecular chain steric hindrance affected CO_2 adsorption capacity and amine efficiency [41]. CSNS-TA adsorbent has a maximum CO_2 uptake of 47.88 mg g^{-1} and good stability. Additionally, CO_2 collection was performed under mild circumstances, such as 298 K adsorption, 373 K desorption, and ambient pressure. The straightforward approach for CSNS-TA adsorbent fabrication and mild CO_2 adsorption-desorption conditions suggested that this technology might be scaled up for commercial production and be useful for post-combustion CO_2 capture.

5.3.2 Mixed Metal Oxides

Direct air capture (DAC) is a carbon capture method slightly different from conventional CO_2 capture technologies. The main advantage of DAC is the method's location independence, which makes it suitable for removing CO_2 from point sources. The performance of DAC is also resistant to acidic gaseous contaminants and heavy metals [43]. Past research on these methods mentions that causticization is also a process of CO_2 extraction used in DAC systems [44]. Alkali earth hydroxides provide this process, but for thermal regeneration, the need for energy is high, also the amount of water and O_2 usage set a barrier in front of the development of the technology, and absorbent-based systems showed up to be better alternatives [45]. Solid-supported amines are great absorbents for DAC systems mainly due to their appropriate chemical properties. They display characteristics of chemisorption, enabling them to partake in chemical reactions with atmospheric CO_2, creating compounds like ammonium carbamate, ammonium carbonate, and ammonium bicarbonate [43]. They can be restored at modest temperatures below 393.15 K. Absorbents which involve impregnation of amines into mesostructured supports present an appealing equilibrium between cost-effectiveness and reasonably elevated uptake of CO_2 mainly where atmospheric CO_2 is highly diluted [46].

Moreover, silica materials facilitate improved contact between amines and CO_2 as solid supports owing to their high surface area and large pore volumes. However, increased CO_2 uptake cannot be prevented when the number of cycles increases. The stability of PEI/silica composites in kinetic and thermal aspects is difficult to sustain [47]. Wilfong mentions that 3-aminopropyltriethoxysilane and tetraethyl orthotitanate can improve the kinetic stability and CO_2 adsorption by increasing the strength of the bonds between PEI and silica surface [48]. On the other hand, a novel approach includes new adsorbents from mixed MOs (MMOs) derived from layered double hydroxide which PEI also loads. Layered double hydroxide consists of positively charged layers of metal cations coordinated with hydroxide ions. Zhu *et al.* synthesized MMO precursors using a combination of two methods: aqueous miscible organic (AMO) solvent treatment and co-precipitation [49]. Metal precursor solution containing 1 M ($Mg(NO_3)_2 \cdot 6H_2O$ and $Al(NO_3)_3 \cdot 9H_2O$) was slowly introduced to an equal volume of a 0.5 M Na_2CO_3 solution under vigorous stirring. They adjusted the pH of the mixture to 10.0 ± 0.1 using a NaOH solution. Following this, they allowed the mixture to age at a temperature of 298.15 K for 16 hours, and it was subsequently thoroughly rinsed with deionized water. An essential step in the AMO solvent method, executed by them, involves employing AMO solvents to cleanse and redisperse the moist sample cake before it is subjected to drying. They determined that this procedure was highly effective in eliminating the interlayered water. The AMO solvent, acetone in this particular study, was then removed through overnight vacuum drying to facilitate the exfoliation of the sample nanosheets. Finally, the MMOs were acquired *via* vacuum calcination performed at 723.15 K for a duration of 5 hours. The resulting MMO precursors and MMOs were designated as $Mg_xAl–CO_3$ and $Mg_xAl–O$ respectively, with x representing the Mg/Al molar ratio. In the process of preparing adsorbents for carbon capture, researchers opted for the wet impregnation technique. Firstly, the MMOs were subjected to an overnight vacuum drying process at 373.15 K to eliminate any previously adsorbed substances. Following this, they dispersed a specific amount of branched polyethyleneimine [50] in methanol and then introduced 0.5 g of the dried supporting material into this mixture. The stirring of this blend was conducted under nitrogen protection at 298.15 K for 3 hours. Subsequently, methanol was removed through rotatory evaporation, and the resulting powder was subjected to vacuum drying at 333.15 K for at least 12 hours. These samples were stored in vials under vacuum conditions at room temperature until further analysis. Additionally, commercially available mesoporous materials like SBA-15, γ-Al_2O_3, and light MgO were used as supporting materials for comparison purposes. The adsorbents prepared through this approach were labeled as PEIy/support, with 'y' denoting the weight ratio of PEI within the PEI/support composite. Through the AMO solvent treatment method and the utilization of low Mg/Al ratios in $Mg_xAl–CO_3$, MMO precursors were transformed into nanosheets with plate sizes ranging from 20 to 200 nm. Upon thermal treatment, these nanosheets are self-assembled into spherical particles. Effective dispersion of impregnated polyamines on the MMO nanosheet surface provided optimal accessibility for CO_2 molecules to interact with strong adsorption sites under ultra-dilute conditions. It is reported that the porous structure was preserved at high PEI loading ratios of up to 67 wt%, while polyamines began to coat the MMO frameworks, and their amine efficiency remained largely unchanged. PEI/MMO composites exhibited superior CO_2 uptake and kinetics compared to PEI/SBA-15 due to their well-suited morphologies and nanostructures [49]. Notably, the 67 wt% PEI-impregnated $Mg_{0.55}Al–O$ MMO achieved an impressive total CO_2 uptake of 2.27 mmol g^{-1} and a rapid adsorption rate of 1.1 mmol g^{-1} h^{-1} within 90 min at 298.15 K under 0.4 mbar CO_2. Furthermore, the defect-rich MMO layers exhibited strong electrostatic attraction toward the attached polyamines, ensuring the durability of the hybrid adsorbents over at least 20 cycles in both dry and humid conditions. These results underscore the potential of cost-effective, easily manufactured, amine-functionalized MMOs for scalable CO_2 capture applications under ultra-dilute conditions.

5.3.3 Sol-Gel and Core-Shell Methods

In recent decades, a significant amount of research has been dedicated to investigating solid oxide CO_2 sorbents, with particular emphasis on examining both the process and material-related factors. MO-based solid sorbents have garnered significant attention in CO_2 capture technologies, emerging as the subject of extensive research and investigation. Nevertheless, there is a lack of exploration about the requirement for synthetic sorbents with high performance, even if there have been reports indicating enhancements compared to natural sorbents such as limestone. The precise correlations among sorbent compositions, synthesis techniques, structural characteristics, and the efficiency of CO_2 collection have yet to be fully clarified.

Sorbents composed of calcium oxide (CaO) are frequently discussed, primarily due to their remarkable capacity for trapping CO_2 and high carbonation and regeneration rates. Furthermore, it is worth noting that these sorbents exhibit a high abundance inside the Earth's crust and are characterized by their comparatively low cost. Calcium oxide–based sorbents are extensively utilized in various applications owing to their notable characteristics, including a substantial capacity for CO_2 absorption, rapid rates of carbonation and regeneration, abundant availability in the Earth's crust, and cost-effectiveness. However, the effectiveness of unmodified CaO sorbents may be hindered throughout the process of uptake and regeneration due to sintering, which ultimately causes a decrease in both surface area and available pore volumes. Deactivated sorbents exhibit a state of inactivity, leading to a significant energy disadvantage when employed in continuous operations. The main challenge revolves around devising strategies to address the issue of insufficient cycle stability efficiently. Two separate strategies can be utilized to optimize the efficiency of calcium looping systems [51]. The initial approach entails the optimization of operational parameters, whereas the subsequent approach centers on developing synthetic sorbents with superior performance. Considerable investigation has examined the deactivation phenomenon reported in the context of calcium looping cycles [52].

Furthermore, the cyclic performance of CaO sorbents can be enhanced at a fundamental level by including inert refractory elements like Al_2O_3, ZrO_2, MgO, and their derivatives into the CaO matrix. This inclusion provides increased resistance to sintering and greater physical hardness. Several studies have documented the utilization of CaO-based sorbents supported by different quantities of refractory materials. The capacity for CO_2 uptake exhibited variability, which can be partly attributed to variations in the overall compositions [53].

The utilization of the sol-gel process has been investigated as an alternate approach for the synthesis of supported CaO sorbents. This methodology regulates the chemical interaction between inert supports and CaO, facilitating convenient particle size and surface area manipulation. Research findings indicate that the utilization of sorbents based on sol-gel technology has been associated with notable enhancements in specific surface areas and pore volumes. Consequently, these improvements have reduced particle size and the length of diffusion routes for carbon dioxide [54]. As an illustration, utilizing Zr-modified CaO with reduced particle sizes exhibited a notable increase in CO_2 adsorption and consistently maintained its CO_2 adsorption capacity during 10 successive cycles. Sol-gel methodologies yield sorbents with enhanced pore volumes, surface areas, CO_2 adsorption capacities, and cyclic stability.

Core-shell formations refer to nanoscale structures consisting of a central core enveloped by an outer shell. These formations often exhibit dimensions on the nanoscale. The mentioned sorbents are utilized in carbon capture devices to collect CO_2 efficiently. The central component consists of compact substances that effectively sequester carbon dioxide, while the external casing amplifies responsiveness, safeguards the central component, and facilitates the process of regeneration.

The sorbents with a core-shell structure are composed of CaO-based materials with a protective shell surrounding the core, thereby inhibiting fusion. Chen *et al.* mentioned that a sorbent consisting of a CaO core was synthesized successfully and encapsulated by an Al_2O_3 shell. The thickness of the Al_2O_3 shell varied between 4 and 8 nm. Their experimental efforts achieved a significant CO_2 uptake of 41 wt% after subjecting the sorbent to 20 cycles. The CO_2 capture performance can be influenced by the thickness of the shell [53]. As we explore the development of synthetic oxide sorbents for carbon dioxide capture, focusing on using group 1 and group 2 metals' alkalinity, we reach more comprehensive knowledge about various sorbents regarding their economic viability, availability, and efficacy. Synthetic methodologies, like incorporating active sorbent phases with thermal stabilizers and redox-active metals like iron, improve mass transfer and efficiency. Hybridization with membranes, polymers, and carbon-based materials can improve MO performance.

Recent research by Mat *et al.* gives insight into the new developments and ongoing challenges on MO-based core-shell nanoparticles (CSNs), which have enhanced functionality, stability, and characteristics of dispersion, and reduced consumption of material [55]. The MOs are selected based on their distinctive physicochemical characteristics, which result in decreased pore blockage and enhanced basicity of sites, hence enhancing the interaction and adsorption of CO_2. A wide range of synthesis techniques can be employed to produce CSNs, including microemulsion, reverse microemulsion, sol-gel, coating, gas-phase reaction, solvothermal treatment, post-treatment, hydrothermal, electrospun, space-confined strategy, washing and dehydrofluorination, one-pot, gas-bubbling–assisted membrane reduction precipitation (GBMR), photo-deposition, wet impregnation, and Stöber methods. Through adsorption techniques, MOs can potentially serve as catalysts for carbon dioxide (CO_2) capture. Enhanced porosity, surface area, and basicity of these MOs contribute to a greater capacity for CO_2 adsorption. The practicality of the physical adsorbent, the core-shell, is determined by the shape and composition of the materials it comprises, which directly impacts its flexibility. Carbon capture and storage networks (CSNs) have demonstrated their potential in various applications, including photoreduction for CO_2 utilization, methane reforming, Fisher-Tropsch synthesis, CO_2 photoconversion, CO_2 reduction through hydrogenation to CO, methane production through CO_2 hydrogenation, and CO_2 reduction. Nevertheless, producing efficient, cost-effective, and ecologically sustainable synthesis methods poses hurdles, as does the creation of uniformly shaped and scaled core-shell structures to avoid reaction errors.

Mesoporous titanium dioxide (TiO_2) is gaining interest due to its ability to adsorb hazardous elements, volatile organic molecules, and specific gasses. Its unsaturated surface atoms and restricted pore size distribution enable rapid and accurate adsorption. Recent materials science and engineering advancements suggest that TiO_2 can be enhanced by immobilizing TiO_2 particles onto nanostructured carbon materials like graphene oxide, a combination that has the potential to remove carbon dioxide from post-combustion flue gasses effectively. In 2015, Chowdhury *et al.* investigated the development of mesoporous titanium dioxide/graphene oxide (TiO_2/GO) nanocomposites as potential adsorbents for carbon dioxide capture [56]. The composites were created using a simple blending procedure and characterized using diffraction, spectroscopy, and microscopy methods. The CO_2 adsorption isotherms were determined at various temperatures and pressures, with the maximum pressure reaching 100 KPa. The results showed a robust agreement with the temperature-dependent Toth and Avrami models. As shown in Figure 5.4, the composite exhibited a notable adsorption capacity for carbon dioxide at room temperature, surpassing that of commonly used adsorbent materials. The material showed a significant reduction in the heat of adsorption and enhanced selectivity of carbon dioxide relative to nitrogen. The quantitative assessment of the adsorption equilibrium and kinetics was conducted using a Micromeritics ASAP 2020 adsorption machine. The research found a positive relationship between temperature reduction and adsorbate concentration rise, leading to augmented adsorption of pure component CO_2.

5.3.4 Photocatalysis-Electrocatalysis

Many other developing strategies are the target of research. Brief information on some of the methods is provided in the following.

Kim *et al.*, in their research, present a proposition for the electrochemical conversion of carbon dioxide into carbon-based compounds as a means of promoting. The aim is to achieve the selective manufacture of C2-C3 products at low overpotentials while also addressing the longstanding difficulty of retaining stability during the production of these specific products. Thus, a copper-based catalyst is prepared by the utilization of nanoparticles (NPs). Copper NPs were synthesized by reducing copper precursors at high temperatures using tetradecylphosphonic acid as surface ligands. NPs are subsequently immobilized onto a carbon-paper electrode, and

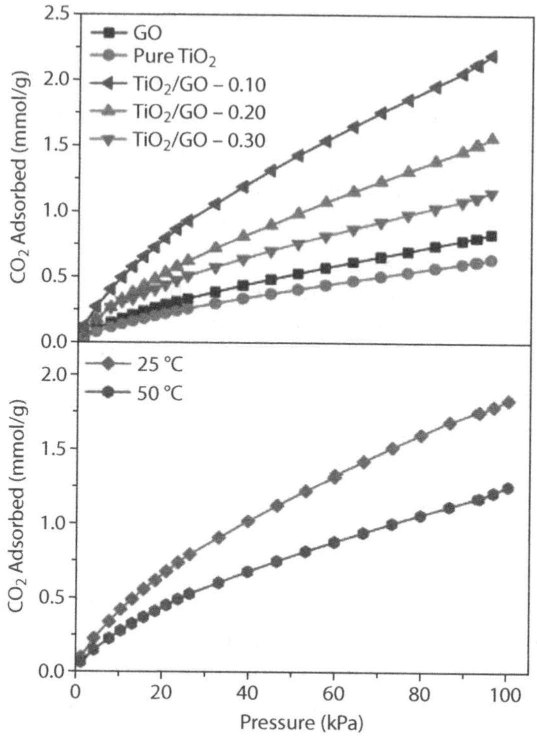

Figure 5.4 CO_2 adsorption isotherms for GO, TiO_2, TiO_2/GO-0.10, TiO_2/GO-0.20, and T TiO_2/GO-0.30 at 0°C (up) and CO_2 adsorption isotherms at 25°C and 50°C for TiO_2/GO-0.10 (below).

their efficacy in carbon dioxide reduction is evaluated under conditions of neutral pH. The products obtained are subjected to analysis by the utilization of gas chromatography and nuclear magnetic resonance spectroscopy techniques. Electron microscopy, X-ray photoelectron spectroscopy, and cyclic voltammetry are employed to analyze the NPs and their structures [50]. Results demonstrate that a cluster of copper NPs can efficiently convert carbon dioxide into C2-C3 hydrocarbons and oxygenates. The structural evolution of these nanostructures led to the formation of active nanostructures. The research underscores the need to further investigate NP and atom evolution under electrically biased and chemically relevant conditions.

A distinct study investigates the mechanism behind the plasma-photocatalytic conversion of carbon dioxide into carbon monoxide and oxygen. This procedure involves the usage of a dielectric barrier discharge (DBD) reactor and the deployment of photocatalysts, namely, $BaTiO_3$ and TiO_2. When the plasma is combined with these catalysts, it results in a modest elevation of the gas temperature within the plasma. The observed rise in temperature leads to a significant increase of 2.5 times in the conversion rate of CO_2 and a corresponding gain in energy efficiency. The presence of catalyst pellets within the plasma gap significantly influences plasma-related processes, such as the enhancement of the electric field and the generation of reactive species. However, the effect of catalyst pellets in the conversion of CO_2 is highly negligible. The study additionally illustrates a significant reduction in ultraviolet (UV) emissions generated within the CO_2 DBD when compared to external sources of UV, suggesting that the impact of UV emissions on the interaction between plasma and photocatalysis and its resulting synergistic effect is minimal or inconsequential. This study also investigates the role of plasma-generated highly energetic electrons as the principal catalyst in activating photocatalysts throughout the CO_2 conversion process. The synergistic effect arising from the integration of DBD and photocatalysis in low-temperature CO_2 conversion can be attributed to the simultaneous presence of a catalyst and the production of high-energy electrons during the CO_2 discharge mechanism [57].

Ola and Maroto-Valer employed a sol-gel methodology to immobilize MO-based TiO_2 (M-TiO_2) photocatalysts onto quartz plates [58]. Subsequently, the performance of these plates in reducing carbon dioxide under visible light exposure was evaluated. The research demonstrated that incorporating vanadium, chromium, and cobalt into the TiO_2 sol led to a reduction in the size of anatase crystallites and an improvement in optical absorption. The M-TiO_2 supports exhibited notable enhancement, particularly in the case of samples incorporating cobalt and titanium dioxide, which demonstrated the highest rates of photoconversion. The observed photocatalytic activity is hypothesized to be attributed to the influence of metal ion species. The supports based on M-TiO_2 generated from sol-gel processes exhibit promising potential for enhancing CO_2 applications.

5.4 Future Trends and Research Directions

In the field of carbon capture, researchers have increasingly focused on the potential of MOs due to their unique physicochemical attributes and large surface area, which make them effective in adsorbing CO_2. Current investigations in MO-based carbon capture emphasize fine-tuning MO structures and compositions to maximize their CO_2 capture capabilities. Scientists are actively exploring novel MO composites and nanomaterials to enhance both the selectivity and capacity of CO_2 adsorption. Moreover, efforts are dedicated to unraveling the fundamental mechanisms governing CO_2 adsorption on MOs to improve the overall efficiency of the capture process.

Anticipated breakthroughs in MO-based carbon capture are expected to arise from applying advanced synthesis techniques such as sol-gel and core-shell methods, enabling precise control over the morphology and surface characteristics of the materials. Additionally, the introduction of functionalized MO surfaces and customized dopants shows promise in improving the kinetics and selectivity of CO_2 adsorption, laying the groundwork for more efficient and economically viable carbon capture technologies. Emerging materials like mesoporous MOs and MOFs have also exhibited notable CO_2 uptake capacities and exceptional stability under harsh operational conditions, making them significant candidates for further exploration in the field.

The prospects for MO-based carbon capture hinge on exploring hybrid materials, including MO-polymer composites and MO-carbon nanotube hybrids, to leverage synergistic effects and enhance CO_2 adsorption performance. Furthermore, comprehensive assessments of the recyclability and durability of MO-based adsorbents are critical to ensure their long-term sustainability and cost-effectiveness in large-scale carbon capture applications. Extensive research into scalable and energy-efficient regeneration processes, combined with integrating renewable energy sources, can revolutionize the landscape of carbon capture technologies, offering a sustainable approach to mitigating CO_2 emissions and addressing global climate challenges.

5.5 Conclusion

This review investigates different strategies for improving the efficiency of carbon capture systems. MO-based systems being the main focus, other innovative approaches, and technologies are also mentioned briefly. It can be concluded that methods and materials suitable for CO_2 sorption are highly diverse. However, each strategy has its advantages and challenges. While MMO-based systems show great stability even at low regeneration temperatures, materials like metal-organic frameworks and zeolites emerge with high selectivity and adsorption capacities. Studies lead us to promising innovations and remark on the challenges that should be looked to overcome. Over the years, carbon capture technologies have been enhanced regarding energy efficiency, cost-effectiveness, environmental impacts, selectivity, and thermal and kinetic stability. The improvements came with different strategies that satisfy different needs and areas of use.

The future trends in carbon capture applications are likely to focus on improving efficiency and scalability and developing innovative solutions. In industrial applications, carbon capture in cement production and integration into processes like steel manufacturing is crucial. Biological carbon capture involves researching enhanced photosynthesis in plants and developing algae-based systems for carbon capture with applications in wastewater treatment and biofuel production. In transportation, technologies for direct CO_2 capture from vehicle exhaust and solutions for aviation sector challenges are emerging. CCUS innovations focus on using captured CO_2 as a feedstock for valuable products and enhancing storage techniques. Decentralized and modular systems aim at small-scale deployment in residential and commercial applications, driven by community initiatives. Policy and financial incentives, such as carbon pricing and government support, play a pivotal role. Technological integration involves AI and machine learning for process optimization and blockchain for transparent carbon credit tracking and trading.

References

1. Dale, A. *et al.*, Meeting the climate change challenge: local government climate action in British Columbia, Canada. *Clim. Policy*, 20, 7, 866–880, 2020.
2. Erdem, E., Why P2X must be the part of the energy solution? *Environ. Prog. Sustain. Energy*, 40, 3, e13545, 2021.
3. Cuéllar-Franca, R.M. and Azapagic, A., Carbon capture, storage and utilisation technologies: A critical analysis and comparison of their life cycle environmental impacts. *J. CO2 Util.*, 9, 82–102, 2015.
4. Hunt, A.J. *et al.*, Generation, Capture, and Utilization of Industrial Carbon Dioxide. *ChemSusChem*, 3, 3, 306–322, 2010.
5. Rubin, E.S. *et al.*, The outlook for improved carbon capture technology. *Prog. Energy Combust. Sci.*, 38, 5, 630–671, 2012.
6. Kheirinik, M., Ahmed, S., Rahmanian, N., Comparative Techno-Economic Analysis of Carbon Capture Processes: Pre-Combustion, Post-Combustion, and Oxy-Fuel Combustion Operations. *Sustainability*, 13, 24, 13567, 2021.
7. Al-Mamoori, A. *et al.*, Carbon Capture and Utilization Update. *Energy Technol.*, 5, 6, 834–849, 2017.
8. Liu, S.-Y. *et al.*, CO2 storage with enhanced gas recovery (CSEGR): A review of experimental and numerical studies. *Pet. Sci.*, 19, 2, 594–607, 2022.

9. Huijts, N.M.A., Molin, E.J.E., Steg, L., Psychological factors influencing sustainable energy technology acceptance: A review-based comprehensive framework. *Renew. Sustain. Energy Rev.*, 16, 1, 525–531, 2012.
10. Baena-Moreno, F.M. *et al.*, Carbon capture and utilization technologies: a literature review and recent advances. *Energy Sources Part A*, 41, 12, 1403–1433, 2019.
11. Trickett, C.A. *et al.*, The chemistry of metal–organic frameworks for CO2 capture, regeneration and conversion. *Nat. Rev. Mater.*, 2, 8, 17045, 2017.
12. Lyu, H. *et al.*, Covalent Organic Frameworks for Carbon Dioxide Capture from Air. *J. Am. Chem. Soc*, 144, 28, 12989–12995, 2022.
13. McDonald, T.M. *et al.*, Cooperative insertion of CO2 in diamine-appended metal-organic frameworks. *Nature*, 519, 7543, 303–308, 2015.
14. Reddy, M.S.B. *et al.*, Carbon dioxide adsorption based on porous materials. *RSC Adv.*, 11, 21, 12658–12681, 2021.
15. Datta, S. *et al.*, Electrochemical CO2 Capture Using Resin-Wafer Electrodeionization. *Ind. Eng. Chem. Res.*, 52, 43, 15177–15186, 2013.
16. Voskian, S. and Hatton, T.A., Faradaic electro-swing reactive adsorption for CO2 capture. *Energy Environ. Sci.*, 12, 12, 3530–3547, 2019.
17. Ranjan, R. *et al.*, Reversible Electrochemical Trapping of Carbon Dioxide Using 4,4′-Bipyridine That Does Not Require Thermal Activation. *J. Phys. Chem. Lett.*, 6, 24, 4943–4946, 2015.
18. Liu, Y. *et al.*, Electrochemically mediated carbon dioxide separation with quinone chemistry in salt-concentrated aqueous media. *Nat. Commun.*, 11, 1, 2278, 2020.
19. Eisaman, M.D. *et al.*, CO2 separation using bipolar membrane electrodialysis. *Energy Environ. Sci.*, 4, 4, 1319–1328, 2011.
20. Zhu, P. *et al.*, Continuous carbon capture in an electrochemical solid-electrolyte reactor. *Nature*, 618, 7967, 959–966, 2023.
21. Hussain, M. *et al.*, A comprehensive review of sectorial contribution towards greenhouse gas emissions and progress in carbon capture and storage in Pakistan. *Greenhouse Gases: Sci. Technol.*, 9, 4, 617–636, 2019.
22. Chu, S., Carbon capture and sequestration. *Science*, 325, 5948, 1599, 2009.
23. Zhang, Z., Borhani, T.N.G., El-Naas, M.H., Chapter 4.5 - Carbon Capture, in: *Exergetic, Energetic and Environmental Dimensions*, I. Dincer, C.O. Colpan, O. Kizilkan (Eds.), pp. 997–1016, Academic Press, London, United Kingdom, San Diego, CA, United States, 2018.
24. Halliday, C. and Hatton, T.A., The potential of molten metal oxide sorbents for carbon capture at high temperature: Conceptual design. *Appl. Energy*, 280, 116016, 2020.
25. Chen, S. *et al.*, A critical review on deployment planning and risk analysis of carbon capture, utilization, and storage (CCUS) toward carbon neutrality. *Renew. Sustain. Energy Rev.*, 167, 112537, 2022.
26. Zhao, X. *et al.*, Biomass-based chemical looping technologies: the good, the bad and the future. *Energy Environ. Sci.*, 10, 9, 1885–1910, 2017.

27. Bui, M., Fajardy, M., Mac Dowell, N., Bio-Energy with CCS (BECCS) performance evaluation: Efficiency enhancement and emissions reduction. *Appl. Energy*, 195, 289–302, 2017.
28. Kim, T.K. *et al.*, Nanoporous Metal Oxides with Tunable and Nanocrystalline Frameworks via Conversion of Metal–Organic Frameworks. *J. Am. Chem. Soc.*, 135, 24, 8940–8946, 2013.
29. Choi, S., Drese, J.H., Jones, C.W., Adsorbent Materials for Carbon Dioxide Capture from Large Anthropogenic Point Sources. *ChemSusChem*, 2, 9, 796–854, 2009.
30. Wang, Q. *et al.*, CO2 capture by solid adsorbents and their applications: current status and new trends. *Energy Environ. Sci.*, 4, 1, 42–55, 2011.
31. Li, L. *et al.*, Magnesia-Stabilized Calcium Oxide Absorbents with Improved Durability for High Temperature CO2 Capture. *Ind. Eng. Chem. Res.*, 48, 23, 10604–10613, 2009.
32. Alkadhem, A.M., Elgzoly, M.A.A., Onaizi, S.A., Novel Amine-Functionalized Magnesium Oxide Adsorbents for CO2 Capture at Ambient Conditions. *J. Environ. Chem. Eng.*, 8, 4, 103968, 2020.
33. Dean, C.C. *et al.*, The calcium looping cycle for CO2 capture from power generation, cement manufacture and hydrogen production. *Chem. Eng. Res. Des.*, 89, 6, 836–855, 2011.
34. Mutch, G.A. *et al.*, Carbon Capture by Metal Oxides: Unleashing the Potential of the (111) Facet. *J. Am. Chem. Soc.*, 140, 13, 4736–4742, 2018.
35. Noguera, C. and Goniakowski, J., Polarity in Oxide Nano-objects. *Chem. Rev.*, 113, 6, 4073–4105, 2013.
36. Geysermans, P. *et al.*, Combination of (100), (110) and (111) facets in MgO crystals shapes from dry to wet environment. *Phys. Chem. Chem. Phys.*, 11, 13, 2228–2233, 2009.
37. Khdary, N.H. *et al.*, Metal oxides as catalyst/supporter for CO2 capture and conversion, review. *Catalysts*, 12, 3, 300, 2022.
38. Cho, M.S. *et al.*, Characterization of new potassium-based solid sorbents prepared using metal silicates for post-combustion CO2 capture. *Process Saf. Environ. Prot.*, 117, 296–306, 2018.
39. Lee, S.C. *et al.*, CO2 absorption and regeneration of alkali metal-based solid sorbents. *Catal. Today*, 111, 3-4, 385–390, 2006.
40. Lee, S.C. *et al.*, Effects of alumina phases on CO 2 sorption and regeneration properties of potassium-based alumina sorbents. *Adsorption*, 20, 331–339, 2014.
41. Ren, Y. *et al.*, Amine-grafted mesoporous copper silicates as recyclable solid amine sorbents for post-combustion CO2 capture. *Appl. Energy*, 198, 250–260, 2017.
42. Wang, Y. *et al.*, One-pot synthesis of nanotube-based hierarchical copper silicate hollow spheres. *Chem. Commun.*, 48, 6555–6557, 2008.
43. Sanz-Pérez, E.S. *et al.*, Direct capture of CO2 from ambient air. *Chem. Rev.*, 116, 19, 11840–11876, 2016.

44. Stolaroff, J.K., Keith, D.W., Lowry, G.V., Carbon dioxide capture from atmospheric air using sodium hydroxide spray. *Environ. Sci. Technol.*, 42, 8, 2728–2735, 2008.
45. Yang, M. *et al.*, Recent advances in CO 2 adsorption from air: a review. *Curr. Pollut. Rep.*, 5, 272–293, 2019.
46. Chen, C.-H. *et al.*, The "missing" bicarbonate in CO2 chemisorption reactions on solid amine sorbents. *J. Am. Chem. Soc.*, 140, 28, 8648–8651, 2018.
47. Choi, S., Gray, M.L., Jones, C.W., Amine-tethered solid adsorbents coupling high adsorption capacity and regenerability for CO2 capture from ambient air. *ChemSusChem*, 4, 5, 628–635, 2011.
48. Wilfong, W.C. *et al.*, Spectroscopic investigation of the mechanisms responsible for the superior stability of hybrid class 1/class 2 CO2 sorbents: a new class 4 category. *ACS Appl. Mater. Interfaces*, 8, 20, 12780–12791, 2016.
49. Zhu, X. *et al.*, Efficient CO 2 capture from ambient air with amine-functionalized Mg–Al mixed metal oxides. *J. Mater. Chem. A*, 8, 32, 16421–16428, 2020.
50. Kim, D. *et al.*, Copper nanoparticle ensembles for selective electroreduction of CO2 to C2–C3 products. *Proc. Natl. Acad. Sci.*, 114, 40, 10560–10565, 2017.
51. de la Calle Martos, A. *et al.*, Effect of dolomite decomposition under CO 2 on its multicycle CO 2 capture behaviour under calcium looping conditions. *Phys. Chem. Chem. Phys.*, 18, 24, 16325–16336, 2016.
52. Hu, Y. *et al.*, Incorporation of CaO into inert supports for enhanced CO2 capture: A review. *Chem. Eng. J.*, 396, 125253, 2020.
53. Chen, H. *et al.*, A comparative study on energy efficient CO2 capture using amine grafted solid sorbent: Materials characterization, isotherms, kinetics and thermodynamics. *Energy*, 239, 122348, 2022.
54. Yoon, H.J. and Lee, K.B., Introduction of chemically bonded zirconium oxide in CaO-based high-temperature CO2 sorbents for enhanced cyclic sorption. *Chem. Eng. J.*, 355, 850–857, 2019.
55. Mat, N., Timmiati, S.N., Teh, L.P., Recent development in metal oxide-based core–shell material for CO2 capture and utilisation. *Appl. Nanosci.*, 13, 6, 3797–3817, 2023.
56. Chowdhury, S., Parshetti, G.K., Balasubramanian, R., Post-combustion CO2 capture using mesoporous TiO2/graphene oxide nanocomposites. *Chem. Eng. J.*, 263, 374–384, 2015.
57. Mei, D. *et al.*, Plasma-photocatalytic conversion of CO2 at low temperatures: Understanding the synergistic effect of plasma-catalysis. *Appl. Catal., B*, 182, 525–532, 2016.
58. Ola, O. and Maroto-Valer, M.M., Transition metal oxide based TiO2 nanoparticles for visible light induced CO2 photoreduction. *Appl. Catal., A*, 502, 114–121, 2015.

6

Coupling of Process Intensification to Life Cycle Assessment for Eco-Design of Biohydrogen Factory by Microalgae

Iván Ehecatl López-González[1], Pablo Antonio López-Pérez[2*] and Dulce Jazmín Hernández-Melchor[3]

[1]*Área Académica de Química, Instituto de Ciencias Básicas e Ingeniería, Universidad Autónoma del Estado de Hidalgo, Pachuca, Hidalgo, México*
[2]*Universidad Autónoma del Estado de Hidalgo, Escuela Superior de Apan, Carretera Apan-Calpulalpan Km.8, Col. Chimalpa Apan, Hidalgo, México*
[3]*Colegio de Postgraduados, Posgrado de Edafología, Microbiología de Suelos, Carretera México-Texcoco km 36.5, Montecillo, Estado de México, México*

Abstract

Currently, there is a need to conduct integrated management of all factors that can minimize the environmental impact in biohydrogen production processes from their origin. In this chapter, the environmental impact assessment was studied using life cycle impact assessment (LCIA) indicators under different scenarios of hydrogen production via fermentation, employing a photosynthetic microbial consortium without aeration. An important incentive of this work was to apply LCIA in combination with process design. The core of the methodology is based on the link between unit operations and process simulation tools, coupling flowcharts for both, energy production and processes with an LCIA module that generalizes and automates the assessment of environmental impacts. The interest in using such an approach is that different operating conditions and technologies can be systematically modeled and evaluated by a simulator such as the open-access CAPE-OPEN. Of course, experimental data reported in the literature at the laboratory level are considered necessary to identify the unit operations associated with the process per unit of hydrogen production under specific operating conditions.

Corresponding author: save1991@yahoo.com.mx; pablo_lopez@uaeh.edu.mx

Mehmet Sankir and Nurdan Demirci Sankir (eds.) Towards Green Hydrogen Generation, (171–204)

Keywords: CAPE-OPEN, multi-objective optimization, environmental impacts, photosynthetic microbial consortium, management strategies

6.1 Introduction

6.1.1 Biofuels

Nowadays, there are two major problems that the world faces in the energy sector, the decline in oil reserves and the pollution caused by burning fossil fuels; coupled with this, thermoelectric plants are responsible for about 80% of world electricity production, which through its process generates large amounts of CO_2, contributing to global warming [1]. For this, there arises a need to find new renewable sources that can generate environmentally friendly energy [2, 3]. No longer is organic matter simply burned to produce heat and light, as advanced equipment now converts raw biomass into more useful forms of energy that are easy to use and transport [4, 5]. Alternatively, hydrogen is a clean energy carrier, producing water as its only byproduct when it burns. The development of hydrogen biotechnology is progressing quickly as hydrogen becomes an increasingly important fuel source, in parallel to the demise of fossil fuels (coal, natural gas, oil, and nuclear) [6].

6.1.2 Microalgae

A promising option to generate electricity is using photosynthetic microorganisms that transform solar energy into chemical energy, which can be used to produce H_2 through its metabolism. In particular, microalgae are useful for the production of hydrogen (H_2) because they possess specific enzymes called hydrogenases, responsible for catalyzing the conversion of protons (H^+) and electrons (e^-) into gas [7].

Gaffron and Rubin first observed photosynthetic production of H_2 from green algae in 1942 [8]. They discovered that when *Chlamydomonas reinhardtii* was exposed to light, the microorganism could produce hydrogen gas as a metabolic byproduct, opening the door to the study of photobiological hydrogen production. From this discovery, researchers have detected that the presence of O_2 in the process can quickly deactivate hydrogenase functions because of the sensitivity of the active site of the enzyme to oxygen [9, 10]. Therefore, without further efforts to either inhibit PS II activity or separate O_2, sustainable H_2 production is challenging [11, 12].

Several authors have used inhibitors such as cysteine, sodium sulfite, ascorbic acid, and hydroquinone that reduce the concentration of O_2 in the medium for short periods, thus favoring hydrogen production [13]. From a technological point of view, microbial consortia have been frequently used for the generation of bioenergetics such as biogas [14, 15], biodiesel [16, 17], and biohydrogen [18–20] due to their higher efficiency of product synthesis and nutrient conversion at lower cost compared to pure cultures. In the case of hydrogen, maximum yields of 212.2 mL/(L·hour) have been reported from pure or mixed cultures and 1.968 mol H_2/mol glucose under continuous culture conditions, indicating stable production [21].

Given the high theoretical yields, several researchers have started looking into methods that allow for increasing its productivity, like mathematical modeling approaches that can contribute to large-scale H_2 production and generate potentially competitive operation strategies [22]. However, few published works have reported the mathematical modeling of hydrogen production [23–26].

In general, a wide range of mathematical models have been developed for natural systems, some include affinity models such as the Monod-type or Michaelis-Menten kinetics and others are based on product or substrate inhibition, such as those developed by Haldane, Andrew, Luong, Han-Levenspiel, Boulton, Moser, and Aiba [23–25, 27]. These types of models are complicated, though, for two reasons: i) the models are higher order, and some of them are described by partial differential equations; and ii) there are too many parameters that cannot be uniquely determined.

Therefore, in order to achieve an industrialized application, it is necessary to develop models with the following characteristics: a) few parameters and easy determination of its values, b) parameters with physical-biological meaning rather than empirical, and c) strong potential in on-line estimation and control of bioprocesses.

6.1.3 Process Intensification (PI)

The lack of scalability in biorefinery processes can be attributed to poor process integration and low techno-economic and environmental feasibility, which result in the development of complex conversion technologies and inadequate controls [28, 29]. In this sense, process intensification (PI) plays an indispensable role in technological, energetic, and sustainable development; maximizing the overall productivity of the system.

PI can be defined as any chemical engineering development that leads to the generation of cleaner, safer, and more energy-efficient technologies;

producing improvements in processes related to equipment size, waste production, reaction kinetics, capital costs, and other factors [30–32]. PI is applicable to each of the stages that comprise a bioprocess (upstream, midstream, and downstream), allowing the incorporation of new equipment, processing techniques, and development methods that offer substantial improvements [33–37].

It is characterized by the evaluation of five criteria: i) equipment size, ii) process performance, iii) inventory, iv) utilities and raw materials usage, and v) equipment efficiency [38]. These evaluation criteria are carried out at unit operation, task, and phenomena scales, applying common scientific areas of engineering and addressing problems related to synergy in processing units, reaction pathways, heat and electrical energy management, thermodynamic efficiency, and system operability [34, 39–42].

It is important to note that the classification and structure of PI are independent of a particular process or equipment and one of the most appreciated characteristics is its applicability at the molecular (microfluidic), meta (transport), macro (equipment), and mega (plants and companies) levels [38, 43, 44]. Figure 6.1 shows the elements and characteristics that makeup PI.

Generally, three basic areas are considered for understanding the PI framework: a) equipment (hardware), b) methods (software), and c)

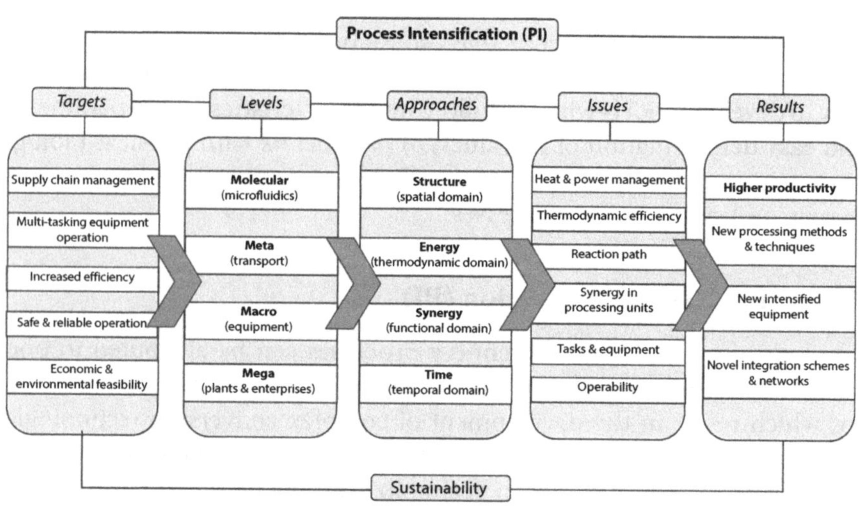

Figure 6.1 Elements of process intensification where there is a relationship between each group. Objectives are defined at different levels or scales that will lead to the solution of problems related to a particular domain, ultimately maximizing the overall productivity of the system. Own elaboration based on Tula *et al.* [38] and Rivas *et al.* [44].

plant design [30, 33, 34, 45]. Equipment includes new reactors and intensive mixing or heat and mass transfer devices. Methods involve all those techniques that use alternative energy sources, new reaction mechanisms, separation, heat exchange, or else new process control procedures. Plant design belongs to the mega level and refers to the modification of existing plant models, seeking to improve their techno-economic and environmental performance.

PI is aligned with the sustainable development goals established by the United Nations due to its approach related to energy efficiency, waste reduction, emissions, resource conservation, and circular economy [46]. In fact, new intensification techniques have been developed to reduce environmental impact, improve process efficiency, and enhance people's quality of life.

One such technique is computer-aided PI (CAPI), which involves the use of systematic model-based methods to determine the solution to a problem at one or multiple scales, satisfying a set of operational and process specifications [37, 38]. In particular, the environmental footprint of PI can be measured using life cycle assessment (LCA), which quantifies the environmental impacts of the process over its lifetime [47, 48].

6.1.4 Life Cycle Assessment (LCA)

LCA, also called cradle-to-grave analysis, is recognized as a useful framework for assessing the impacts of processes, products, and systems on the environment and human health [49]. LCA allows the evaluation of the overall scope of inputs, outputs, and environmental impacts of a system throughout its life cycle, *i.e.*, the set of stages that make up the useful life of a given product [50, 51]. It also provides a full range of quantitative analyses that facilitate long-term strategic decision-making.

LCA has been applied on multiple occasions for the environmental assessment of biofuels obtained from microalgae, estimating the impacts and identifying the dominant unit processes and operations [52–56]. Most of these and other studies base their LCA methodology on standards stipulated by the International Organization for Standardization (ISO), specifically ISO 14,040 and 14,044 [57]. The ISO 14,000 series specifies the steps and/or stages for performing the analysis, but not its scope and depth [58]. Thus, LCA comprises four main stages: i) definition of the objective and scope, ii) inventory analysis, iii) environmental impact assessment, and iv) life cycle interpretation (Figure 6.2).

The first stage of the LCA consists of determining the objectives and scope of the assessment, allowing to description of the boundaries,

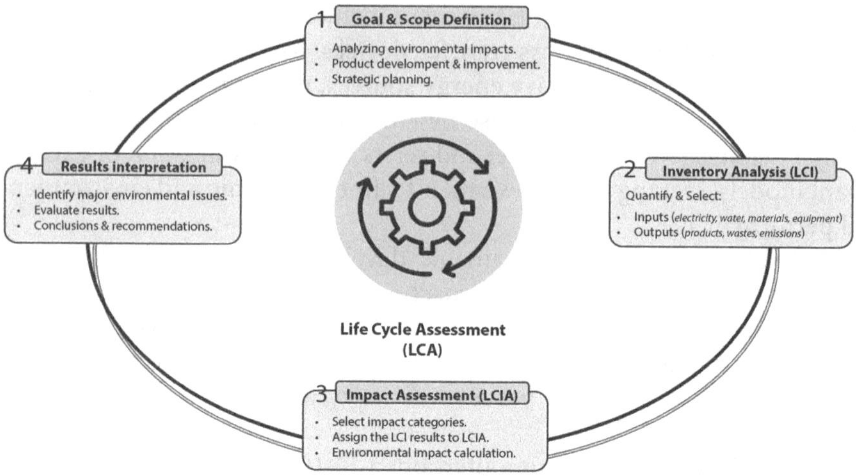

Figure 6.2 Stages of life cycle analysis. It comprises a process where each stage is related, from the definition of the objective to the interpretation of results. Own elaboration based on Mu *et al.* [58] and Osman *et al.* [49]. Process icon retrieved from Freepik (www.freepik.com).

collection methods, quality requirements, and the study functional units (FU) in charge of quantifying the environmental impacts of the system [58, 59]. Subsequently, the inventory analysis is carried out, where key data are collected and the methods/models for quantifying inputs and outputs of the unit processes are established [59].

The third stage is the most laborious and consists of environmental impact assessment (life cycle impact assessment, LCIA), which is quantified, distributed, and classified into different categories depending on the functional unit, system boundaries, and assessment objectives [49, 60]. Impact categories are usually classified into two groups: midpoint and endpoint. Midpoint indicators measure environmental problems caused by pollutant gas emissions or resource use at an intermediate level, while endpoint indicators comprise the total measurement of environmental impact for the estimation of damage to human health, ecosystem quality, and biodiversity or resource depletion [61, 62]. The difference between one approach and the other lies in their precision; midpoint indicators do not fully reflect the severity of the impact. In contrast, endpoint indicators show the results of environmental impacts in a comprehensive and consistent manner [63].

Finally, the life cycle interpretation is performed, involving the analysis of the results in terms of consistency, accuracy, and level of uncertainty

[50]. To achieve this, it is important to do a sensitivity and uncertainty analysis to identify which material, process, or life cycle stage has a significant impact on the environment, as well as to evaluate the precision and variability of the results [58]. As part of the LCA results report, it is required to add conclusions and recommendations that help reduce environmental impacts.

The selection of impact categories and methods to be used will depend on the objective and scope of the study. Nowadays, different software-assisted methods have been used to evaluate the environmental impact of a process or product by relating midpoint to endpoint indicators. Some methods reported in the literature are CML, TRACI, LIME2, IMPACT 2002+, MUIO-LCA, Eco-indicator 99, ReCiPe 2008, and ReCiPe 2016 [52, 60, 64–67].

The ReCiPe method offers an accurate analysis of environmental impacts using 18 midpoint and 3 endpoint indicators, standing out as one of the most widely used methodologies in the evaluation of biofuel processes. Broadly speaking, it allows reducing a large number of life cycle inventory results in a limited number of scores for each indicator [68]. As an example, the life cycle analysis in the production of biofuels from microalgae can have as a research objective to analyze the environmental impacts in each unitary process, from the growth and harvesting of the algae to the obtaining and distribution of the biofuel. In this case, the inventory would include as inputs electricity, water, nutrients, equipment, and materials for infrastructure, and as outputs the resulting products, co-products, wastes, and emissions.

Regarding the stages of impact assessment and interpretation of results, decision-makers would select environmental impact categories according to project requirements and formulate relevant conclusions and recommendations [58, 69]. In terms of biohydrogen production, LCA can be a useful tool for determining the most sustainable production system, identifying critical environmental impacts throughout the process and addressing challenges related to scaling, storage and transportation [70–72].

6.1.5 Eco-Design

The industrial and manufacturing sector is responsible for a significant part of environmental impacts, through overconsumption and discharge of waste and polluting emissions [73, 74]. Therefore, it is essential to improve the environmental performance of processes from the initial design stages, following a sustainability approach [75]. About 70% of the decisions related to the sustainability of a product are made in the design phase, of which, the initial stages represent the key period to establish the general conditions and destination of the final product(s), thus including the potential impacts to the environment [75–77].

Taking into account this and what the National Institute of Standards and Technology (NIST) regarding the need to integrate sustainability in the design stages has established, professionals have focused on developing useful, innovative, and environmentally friendly processes and products [78, 79].

The eco-design approach consists of integrating environmental aspects into process design, reducing negative impacts, and improving the environmental performance of products throughout their life cycle [80, 81]. Its importance lies not only in minimizing the environmental impacts of the product but also in improving the manufacturing process [82, 83]. In this sense, three aspects characterize eco-design: i) product design and development, ii) reduction of environmental impact, and iii) inclusion of the complete life cycle of the product(s) [84]. A general framework of the eco-design process is shown in Figure 6.3, highlighting the principles of its approach.

Most publications related to eco-design focus on the environmental dimension of sustainability, paying little attention to profitability (economic dimension) and product performance (social dimension) [84, 85]. However, sustainability has gained steady attention due to its ability to

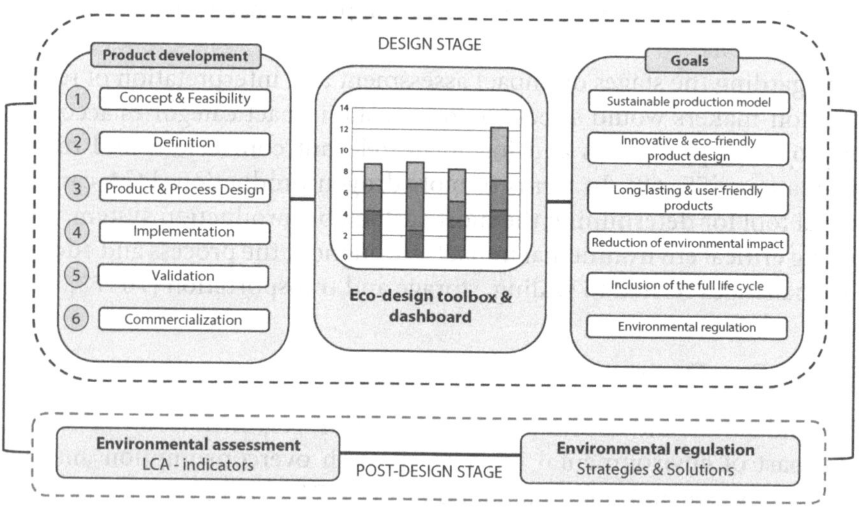

Figure 6.3 The general framework of eco-design; comprising the design and post-design stages, as well as product development, objectives and environmental assessment. Own elaboration based on Thai & Liu [82] and Schäfer & Lower [84].

evaluate and guide systems and products designed at multiple stages of the process, creating a new comprehensive approach to the three pillars [86].

It should be considered that although most eco-design activities originate in the engineering design and product development departments, other domains of the industry are affected [84]. For that reason, eco-design requires a mindset focused on developing a sustainable business model, making products more durable, with low environmental impact, and attached to consumer needs [87, 88].

The implementation of eco-design in the early stages of the process is essential since in addition to allowing the creation of different concepts and scenarios, it provides engineers with a wealth of data, numbers, and facts that facilitate decision-making [77, 89].

According to Vallet *et al.* [90], there are two stages of post-design that should be emphasized: a) environmental assessment (LCA) and b) environmental improvement. Environmental assessment aims to evaluate the environmental impacts of a product or service through the use of indicators, while environmental improvement focuses on the search for eco-friendly solutions [49, 90]. The post-design stage represents a tedious process for engineers and product designers, as generating alternative eco-designs and evaluating the environmental hotspots of various scenarios is time-consuming [91].

The evolution of eco-design in different regions of the world has allowed the generation of new approaches to ensure an environmentally responsible product. One of these is parametric LCA (PLCA), which involves modeling using parameterization techniques, multi-objective optimization, LCA, and sensitivity analysis by tools such as MATLAB, SimaPro, or CAD/CAM (computer-aided design and manufacturing) [77]. Its main function is to support designers to assess the sustainability of multiple scenarios during the eco-design process and thereby identify, control, monitor, and optimize the environmental performance of the entire life cycle considering product requirements and quality [77, 89, 91, 92].

6.1.6 Biohydrogen Factory

Biohydrogen is a clean fuel considered a promising source of energy and an alternative to the unsustainable use of fossil fuels. It offers multiple advantages over other types of biofuels that make it stand out, such as its high-energy content, lower pollutant production, carbon neutrality, and applicability [93–95].

The market of biohydrogen is broad and encompasses its use as a fuel, energy storage, heat and electricity generator, and as a means of production of other compounds of industrial interest [96]. In fact, about 55 million tons per year of hydrogen is currently produced globally and a market value of more than $191 billion is expected by 2024 [97, 98].

Biohydrogen is produced from the conversion of organic matter by chemical, thermochemical, and biological methods [99]. In particular, biological hydrogen production involves the use of microorganisms such as bacteria, algae, or archaea by photo-fermentation, direct and indirect photolysis, dark fermentation, or hybrid systems using organic waste streams [100–102].

In recent years, biohydrogen production with microalgae has been distinguished as a promising and sustainable method due to its photosynthetic efficiency, low environmental impact, and ability to produce significant amounts of gas (high yields) [103, 104]. This process involves a series of unit operations that perform specific functions in the general system:

i. Microalgae cultivation: Microalgae are grown in large-scale photobioreactors or open ponds by adjusting necessary conditions for their growth, such as control of pH, temperature, nutrient availability, and light exposure [105–107].
ii. Biomass harvesting: Once the cultured microalgae reach a suitable density, they are harvested by filtration, centrifugation, or flocculation methods [107].
iii. Pretreatment: The microalgae are subjected to a physical/mechanical treatment consisting of pressing and/or drying of the biomass to reduce water content.
iv. Hydrogen production: Pretreated microalgae are subjected to specific conditions to induce hydrogen production, generally through fermentation or biophotolysis processes [104]. On the one hand, fermentation consists of an anaerobic process of decomposition of organic substrates to produce hydrogen in the absence (dark fermentation) or the presence of light (photo-fermentation). On the other hand, biophotolysis uses light energy to split water molecules into hydrogen and oxygen, either directly or indirectly. As a result, the choice of production method will depend on the specific objectives of the experiment, the available raw material, and the environmental conditions.
v. Control and monitoring: Use of sensors, control systems, and data acquisition tools to optimize and control parameters during the production process [108–111].

vi. Capture and separation: Hydrogen generated during the process is captured and separated by absorption techniques in liquid solvents or through hydrogen-selective membranes [112].
vii. Purification: The collected hydrogen is subjected to a purification process by intermittent pressure adsorption (PSA) or with the use of high-performance selective membranes to remove impurities and traces of other gasses [112, 113].
viii. Storage: The purified hydrogen is stored in containers or tanks suitable for later use.
ix. Distribution and use: Hydrogen is distributed and used as a source of energy, fuel for transportation, or industrial processes.

In this context, the concept of a "biohydrogen factory" refers to the design of a hydrogen production system that is efficient, economically viable, and environmentally sustainable. For this purpose, it is necessary to use modeling and simulation tools that not only enable the design of the plant but also determine the feasibility of scaling it up [114]. Therefore, its objective is to represent and simulate the ideal operating conditions for the bioprocess, improving overall productivity in terms of production rates, substrate sources, location, costs, investment, and concentrations [115].

The design of a biohydrogen plant can be done using modeling platforms such as Aspen Plus, Aspen Hysys, SuperPro Designer, EMSO simulator, or COCO simulator and should include microalgae cultivation equipment (photobioreactors), fermentative or photosynthetic systems, a controlled light source, separators, filters, storage tanks, among others. COCO simulator, in particular, has been used regularly due to its interface designed to facilitate the integration, flexibility, and efficiency of process modeling and simulation, thus allowing the use of a wide range of software components compatible with CAPE-OPEN (Computer Aided Process Engineering – Open) [116].

6.2 Methodology

The present work comprises the integration of an experimental system, modeling, simulation, and environmental impact assessment of biohydrogen production processes through fermentation carried out in

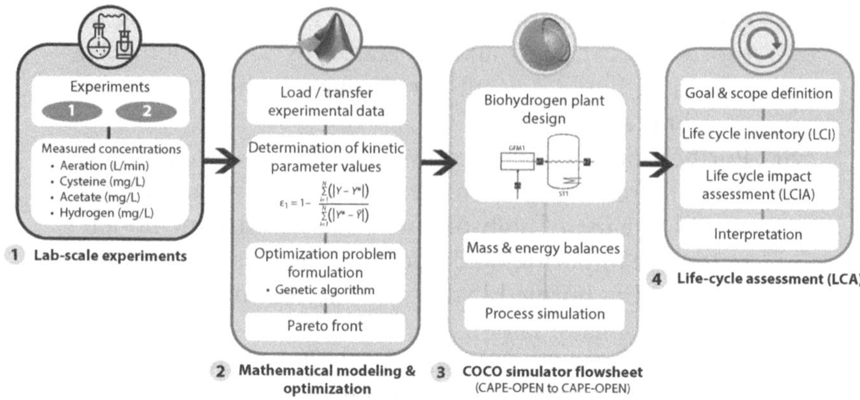

Figure 6.4 Conceptual framework developed in this study. Source: Own elaboration with icons obtained from Flaticon (www.flaticon.com) and USDA (data.nal.usda.gov/lca-icon).

photobioreactors using a photosynthetic microbial consortium, considering different concentrations of cysteine, acetate, hydrogen, and reaction times (Figure 6.4).

Based on Figure 6.4, the framework operates in the following way:

(1) The starting point of this study includes the design and performance of a series of experiments at a laboratory scale, where the kinetics of biomass growth were performed in batch culture photobioreactors.

(2) The second stage covers the mathematical modeling and optimization of the hydrogen production process using a multi-objective method assisted by the MATLAB genetic algorithm.

(3) The third stage consists of the design and modeling of the biohydrogen plant in COCO simulator (CAPE-OPEN to CAPE-OPEN), where the basic unit operations of the process were considered and the role of the bubble column photobioreactors in the microalgae cultivation was highlighted.

(4) The last stage involves the environmental impact assessment of the hydrogen production process from micro algae by means of OpenLCA software, using the ReCiPe method with information obtained from the literature and the Ecoinvent database.

6.2.1 Experimental Set-Up

Biomass growth kinetics were performed in 250 mL bubble column photobioreactors in batch culture with 200 mL of BG_{11} medium, 10% inoculum (v/v), and a reaction time (RT) of 5 days. The operating batch conditions were fermentation time of 120 hours without air injection, 80 (light) μmol/m^2/s photon flux density at room temperature with a constant photoperiod (12:12 light:dark cycle), and shaken manually twice a day. Cysteine (400 to 500, optimal of 498 mg/L) was used to reduce the concentration of dissolved oxygen in the medium, and acetate (200 mg/L) as a carbon source [13, 117]. Two controls were placed to compare microalgal growth under the aforementioned light and temperature conditions. Biomass production was evaluated using spectrophotometry at 750 nm (PerkinElmer UV-vis), the acetate concentration was measured using HPLC equipment (Shimadzu LC10Ai) connected to a UV detector (k = 210 nm) and the produced hydrogen was determined by gas chromatography in a Gow-Mac chromatograph model 350. Experiments were performed in duplicate.

6.2.2 Multi-Objective Optimization

To explain the kinetics of hydrogen production in a batch operation, a mathematical model was created. The model utilizes an unstructured kinetic method and takes into account two substrates as well as the effect of the starting cysteine concentration. The equations used are shown below:

- Acetate mass balance (Ac):

$$\frac{dAc}{dt} = -\frac{\delta_1 Ac}{Y} \tag{6.1}$$

- Biomass balance (Xb):

$$\frac{\mathrm{dXb}}{\mathrm{dt}} = \delta_1 \mathrm{Ac} - \mathrm{k_d Xb} \tag{6.2}$$

- Nitrogen mass balance (N_2):

$$\frac{\mathrm{dN_2}}{\mathrm{dt}} = -\frac{\delta_1 \mathrm{N_2 H_2}}{\mathrm{Y}} \tag{6.3}$$

- Hydrogen mass balance (H_2):

$$\frac{dH_2}{dt} = \delta_3\left[-\frac{\delta_2 H_2}{Y_p} - k_{H_{2(cys)}} H_2\right] \tag{6.4}$$

$$\delta_3 : conversion\ factor\left(\frac{mg}{L}\right) to(ml)$$

- Kinetic term:

$$\delta_1 = \delta_{max}\left(1 - \frac{Xb}{K_{X_1}}\right)^{\rho} \tag{6.5}$$

- Hydrogen production kinetics:

$$\delta_2 = \delta_{maxH_2}\left(1 - \frac{Xb}{K_{X_1}}\right)^{\tau} N_2Ac \tag{6.6}$$

The kinetic parameters of the proposed model were determined *via* nonlinear regression based on the Levenberg-Marquardt least squares minimization algorithm. For the PI study, a well-established non-dominated sorting genetic algorithm II (NSGA II) was utilized. The application was run under gamultiobj and Pareto search by means of finding the Pareto front of multiple fitness functions using a genetic algorithm in MATLAB [118, 119].

6.2.3 System Modeling

The design and modeling of the biohydrogen plant were carried out with the free access software COCO simulator. For this, it was necessary to carry out a systematic review of the literature to obtain experimental data at the laboratory level and identify the unit operations associated with the process per unit of hydrogen production under specific operating conditions. In addition, a block diagram was developed to facilitate the identification of inputs and outputs of the overall process (Figure 6.5).

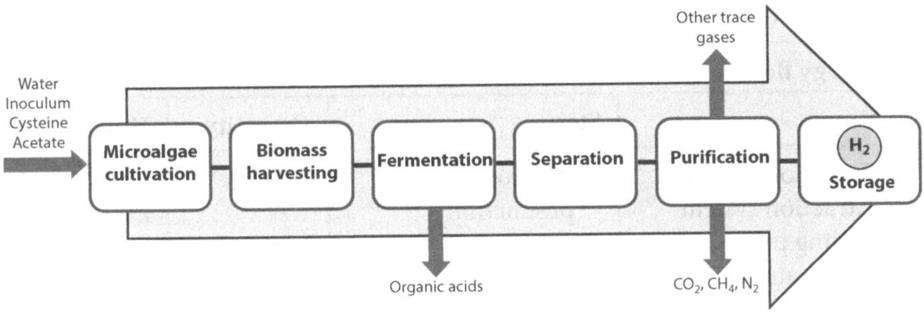

Figure 6.5 Block diagram of the hydrogen production process from microalgae. Inputs include water, inoculum, cysteine, and acetate; outputs are organic acids, CO_2, trace gasses, and the main product (biohydrogen). Source: Own elaboration.

6.2.4 Life Cycle Assessment (LCA)

6.2.4.1 Goal and Scope Definition

In this case, the LCA's goal is to evaluate the environmental impacts of the biohydrogen production process from a photosynthetic microbial consortium. The boundary of the proposed system is defined from cradle-to-gate. This approach was considered appropriate because one of the main objectives of the chapter is to analyze the potential impacts of hydrogen production *via* fermentation under different scenarios (cysteine/acetate ratio), from microalgae cultivation to hydrogen storage. It should be noted that the harvesting, pretreatment, use, and distribution of hydrogen (transport) steps were not involved.

6.2.4.2 Inventory

The inventory contemplated in the analysis was carried out by obtaining and calculating input and output data, corresponding to the energy flows and material used. This information was obtained from the Ecoinvent 3.6 database, literature, and the results of our previous study based on laboratory-scale hydrogen production [119]. The required calculations were confined to the proposed functional unit, assuming 100% total conversion in the fermentation stage and 100% energy efficiency in the equipment used. The life cycle inventory (LCI) data for the biohydrogen production scenarios are presented in Table 6.1.

Table 6.1 General inventory flow for hydrogen production from microalgae.

Energy flow			
Name	**Step**	**Quantity**	**Unit**
Photobioreactor Extraction system Feeding pump Temperature control	Biohydrogen production	1.5 1.2 1.78×10^3 3.46×10^3	kWh/m^3 kWh/m^3 kWh kWh
Membrane separator	Separation	0.5	kWh/m^3
Gas purification system	Purification	1.3	kWh/m^3
Compressed hydrogen tank	Storage	2.0	kWh/m^3
Mass flow			
Name	**Step**	**Quantity**	**Unit**
Inoculum	Microalgae cultivation and biohydrogen production	0.2248*	kg
Water		4.5000*	L
Cysteine		4.98×10^5	g/L
Acetate		2.0×10^5	g/L
Biomass		2.0×10^5	g/m^2d
CO_2	Separation and purification	0.00435*	kg
N_2		0.0764*	kg

Source: Own elaboration based on Acién *et al.* [120], Toledo-Cervantes *et al.* [121], Ferreira *et al.* [122], Akhbari *et al.* [123], Barghash *et al.* [124], and Ecoinvent database. The energy flow quantity corresponds to the energy consumption of the equipment. *The average data for the inoculum, water, CO_2, and N_2 correspond to the basic unit (1L H_2/1 kg H_2).

6.2.4.3 *Life Cycle Impact Assessment (LCIA)*

The environmental impact assessment for each hydrogen production scenario was carried out with OpenLCA 1.11.0 software, following the ReCiPe method, which considers the calculation of 18 midpoint and three endpoint indicators. The selected version used in this study was the hierarchist midpoint and only five categories related to ecosystem, human impact, and

Table 6.2 Impact categories considered in the LCIA.

Impact category	Reference unit	Description
Climate change (GWP100)	kgCO_2-eq	-Global warming potential of greenhouse gas emissions.
Terrestrial ecotoxicity (TETPinf)	kg 1.4-DCB-eq	-Potential toxic effects on terrestrial ecosystems.
Human toxicity (HTPinf)	kg 1.4-DCB-eq	-Potential carcinogenic toxicity to humans from exposure to hazardous substances.
Ionizing radiation (IRP_HE)	kBq Co-60-eq	-Potential impacts associated with the release of ionizing radiation from various sources.
Water depletion (WDP)	m^3	-Amount of water that is consumed or withdrawn from local water bodies and not returned.

Source: Own elaboration based on the Ecoinvent database.

resource depletion were analyzed. Table 6.2 shows the impact categories contemplated in the analysis, along with their descriptions.

6.3 Numerical Experiments

This section presents the results of the multi-objective optimization, design, and modeling of the biohydrogen plant, as well as the environmental impact assessment of the proposed case studies.

6.3.1 Process Intensification

This chapter aimed to calculate new optimal values for hydrogen production based on experimental data from Hernández-Melchor *et al.* [119]. According to the regression analysis, a polynomial fit model was proposed for the optimal conditions and predictions of Equation 6.7, Table 6.3, and Figure 6.6. The normalized data are shown in Figure 6.7 and Table 6.4.

Formal definition and classification were established according to the multiple regression of hydrogen production. It is well known that the

Table 6.3 Multiple regression (R-square = 0.99571).

Parameter	Value	Error	t-Value	Prob>|t|
α	−585.98403	166.52135	−3.51897	0.00176
β_1	227.36588	60.92113	3.73213	0.00103
β_1	−34.68291	8.75633	−3.9609	5.81597E^{-4}
β_1	2.58634	0.61846	4.18188	3.3244E^{-4}
β_1	−0.0936	0.021	−4.35843	2.12309E^{-4}
β_1	0.00132	2.93781E−4	4.48734	1.52957E^{-4}

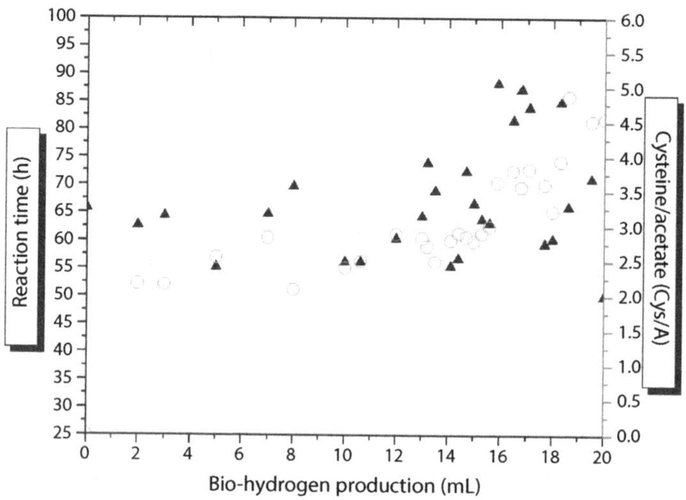

Figure 6.6 Pareto frontier of the hydrogen production process where the optimization values are shown.

response of a system can be characterized by polynomial n-order (in the case of fifth order) as follows:

$$¥ = \alpha + \beta_1\lambda + \beta_2\lambda^2 + \beta_3\lambda^3 + \beta_4\lambda^4 + \beta_5\lambda^5 \tag{6.7}$$

where $(mlH_2)¥ \in \Re^n$ represents the state variable of the system.

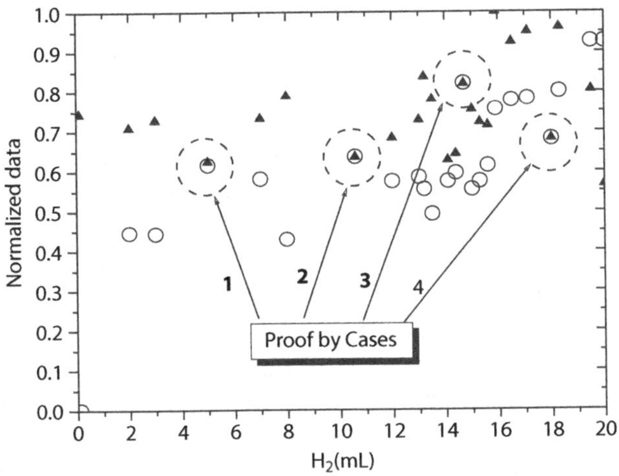

Figure 6.7 Normalized data of the multi-objective optimization.

Table 6.4 Values derived from data optimization and normalization.

Case	Cysteine/acetate ratio	Retention time (hour)	Optimal data H_2 (mL)
1	55.17	30	5
2	56.17	31	10.6
3	72.8	40	14.7
4	60	32	18

All the above quandary formulations provide useful information about the system, facilitating the choice of various process variables [125]. In general, results showed that three independent variables (aeration, acetate, and cysteine) along with their interaction exerted significant influences on the hydrogen production potential. The results indicated that a proper concentration of cysteine (<700 mg/L) and acetate (<200 mg/L) could promote hydrogen production, while excessive cysteine and acetate could decrease it [126].

Over the past few years, research on microalgae-based hydrogen production has been effective with a production rate of 0.6 to 40 ml/L/hour in batch, continuous, and semicontinuous cultivation [13, 127, 128]. According to Xie *et al.* [126], the maximal production rate increased as the concentration of cysteine grew from 0.0 to 1000 mg/L, but it dropped as the concentration of cysteine climbed further, between 1000 and 1500 mg/L. Furthermore, a hydrogen generation rate of 32.85 mL/L/hour was attained at a cysteine concentration of 1000 mg/L. The highest cysteine concentration in this experiment was 700 mg/L and the total production rates for cases 1, 2, and 3 were 5.85, 3.40, and 8.74 ml/L/hour, respectively.

It is possible to take into consideration the predictions and the suggested model under various initial conditions, depending on a comparison of simulated outcomes and experimental data. In fact, the BG_{11} medium with 200 mg/L of acetate, 210 mg/L of N_2, and 498 mg/L of cysteine (without aeration) represented the optimal conditions for hydrogen fermentation. Under these favorable conditions, the highest amount of experimental H_2 gas was 11 mg/L. The minimum biomass optimized the hydrogen concentration to 11.4 mg/L, the simultaneous optimization for the cysteine and RT minimization led to the maximum hydrogen concentration of 28.19 mg/L and the in-parallel optimization for the minimum ratio of cysteine-acetate produced a hydrogen concentration of 15 mg/L from Hernández-Melchor *et al.* [119].

6.3.2 Biohydrogen Plant

A biohydrogen plant designed for *in situ* fermentation in bubble column photobioreactors was modeled. Followed by a membrane separator, a gas purification system, and a storage tank (Figure 6.8).

The process begins in bubble column photobioreactors, where the microalgae are cultivated under controlled conditions and the inoculum, water, and nutrient solutions such as the BG_{11} medium containing cysteine and acetate are introduced. The photobioreactors are equipped with a light source (panels) to provide the necessary illumination to carry out photosynthesis and thermostat control to maintain the optimal range of temperature for microbial growth. Likewise, a feed pump is used to circulate the culture medium within the equipment, ensuring that microorganisms receive enough nutrients.

After the fermentation process, the culture medium and the hydrogen are directed to a membrane separator responsible for selectively capturing and separating gas from the liquid culture. Hydrogen is collected through an extraction system consisting of a series of ducts connecting the top of the photobioreactor to a gas collection device. Then, the separated hydrogen, which contains impurities such as CO_2, CH_4, N_2, and trace gasses, is

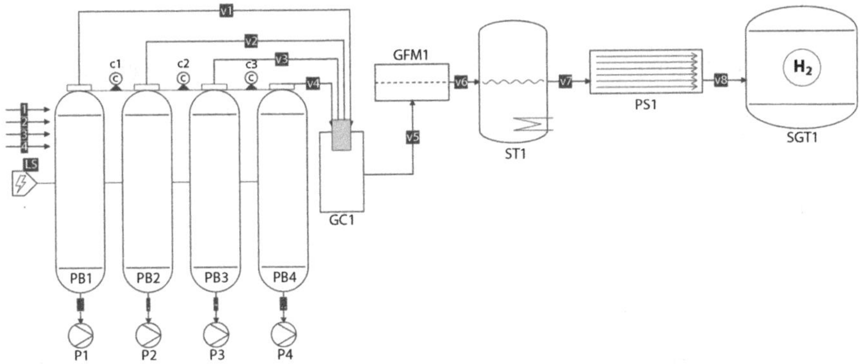

Figure 6.8 Biohydrogen plant designed in COCO simulator. LS (light source), PB (photobioreactor), P (pump), C (controller), GC1 (gas collector), GMF1 (gas flow meter), ST1 (separator tank), PS1 (purification system), SGT1 (storage tank), V (valves), and 1-4 (inputs). Source: Own elaboration.

directed to a gas purification system. The purified product is finally stored in a dedicated storage tank until it is ready for use and distribution.

6.3.3 Environmental Impact Assessment

As discussed in section two, an evaluation of the environmental impacts of the biohydrogen production process from microalgae was done, which considered the industrial scaling of four scenarios corresponding to different concentrations of hydrogen, cysteine/acetate, and reaction times (optimization and normalization data). The results obtained in the analysis are shown in Tables 6.5 and 6.6.

The categories responsible for the major environmental impacts were climate change and terrestrial ecotoxicity while ionizing radiation, human toxicity, and water depletion had minor impacts. Climate change is an impact category based on long-term variations in the Earth's weather patterns, which may include changes in precipitation, temperature, and extreme weather events [129]. In particular, global warming is a direct consequence of this phenomenon and is mainly driven by the accumulation of greenhouse gasses (GHGs) in the atmosphere [129, 130]. On the other hand, terrestrial ecotoxicity refers to the ability of a chemical, pollutant, or toxic agent to cause adverse effects on terrestrial organisms and ecosystems, being mainly affected by the presence of heavy metals and industrial chemicals [131].

Table 6.5 Average contribution per equivalent unit.

Impact category	Reference unit	Average impact
Climate change (GWP100)	$kgCO_2$-eq	2.99123×10^5
T. ecotoxicity* (TETPinf)	kg 1.4-DCB-eq	1.23109×10^6
Human toxicity (HTPinf)	kg 1.4-DCB-eq	6.58690×10^2
Ionizing radiation (IRP_HE)	kBq Co-60-eq	2.94367×10^3
Water depletion (WDP)	m^3	6.27232×10^2

Note: The reported values correspond to the impacts according to their industrial scaling. *Terrestrial ecotoxicity.

Table 6.6 Contribution by equivalent units per case (LCIA).

Impact category	Reference unit	Case 1	Case 2	Case 3	Case 4
Climate change (GWP100)	$kgCO_2$-eq	1.2386×10^5	2.6258×10^5	3.6415×10^5	4.4589×10^5
T. ecotoxicity* (TETPinf)	kg 1.4-DCB-eq	5.0979×10^5	1.0807×10^6	1.4987×10^6	1.8352×10^6
Human toxicity (HTPinf)	kg 1.4-DCB-eq	2.7275×10^2	5.7823×10^2	8.0188×10^2	9.8190×10^2
Ionizing radiation (IRP_HE)	kBq Co-60-eq	1.2189×10^3	2.5841×10^3	3.5836×10^3	4.3881×10^3
Water depletion (WDP)	m^3	2.5972×10^2	5.5061×10^2	7.6359×10^2	$9.3501 x10^2$

Note: The reported values correspond to the impacts according to their industrial scaling. *Terrestrial ecotoxicity.

First, it can be deduced that the energy source used in the biohydrogen production process could contribute significantly to the estimated environmental impact. Akhbari *et al.* [123] and Barghash *et al.* [124] reported in their studies that the main contributor to the indicator is electricity; hence, the high energy consumption in the plant led to an increase in GHG emissions. Regarding the concentrations of cysteine and acetate used in the cultivation of microalgae, it should be considered that although they are not toxic substances for the soil, the release of large amounts could alter its chemical composition and have negative effects on the ecological balance.

In this context, it is necessary to develop strategies to improve the energy efficiency of the process and appropriately manage substances; including their handling, storage, and disposal. As part of the LCA, the following recommendations are proposed to enhance the environmental performance and overall productivity of the system:

- Enhance the plant's energy efficiency by implementing technologies that minimize energy losses and maximize energy generation.
- Develop an effective waste management plan for safe control and disposal [132–140].
- Ensure continuous monitoring and evaluation of both input and output levels throughout the process.
- Optimize hydrogen separation, purification, and storage routes.
- Perform a complete LCA that includes all stages of the process, from microbial consortium isolation to the utilization and transport of the produced hydrogen.

6.4 Conclusion

In this chapter, we analyzed the interaction between PI, modeling, and LCIA. A mathematical model based on an unstructured kinetic strategy and the Plackett-Burman experimental design matrix was proposed to predict hydrogen production considering the effect of acetate, cysteine, and non-aeration on the kinetic growth of the consortium.

The optimal process conditions for hydrogen production with biomass, reaction time (RT), and cysteine/acetate ratio were calculated using a multi-objective optimization algorithm. Accordingly, four optimal and normalized scenarios were considered for plant design and environmental impact assessment with a cradle-to-gate approach.

The results demonstrate that the suggested mathematical model can be useful for maximizing hydrogen production through microalgae inoculation. Moreover, process modeling and simulation played an indispensable role in the life cycle analysis execution, as it allowed addressing the complexity of systems and processes, reducing time, and effectively evaluating scenarios for more efficient management and facilitating decision-making.

LCIA revealed a higher environmental impact in the climate change and terrestrial ecotoxicity categories, while showing lower proportions in ionizing radiation, human toxicity, and water depletion. In order to improve the environmental performance of the system, it is necessary to adjust the process by implementing more efficient production, purification, and storage methods, as well as adopting technologies that reduce waste, energy consumption and emissions. The complete environmental assessment of biohydrogen production systems using LCA can be integrated into *in situ* microalgae fermentation schemes to move forward in the upcoming years with more potentially sustainable processes.

Disclaimer

None.

Acknowledgments

Conahcyt awarded the doctoral scholarship to I.E.L.G. (CVU #1234673).

References

1. Kalair, A., Abas, N., Saleem, M.S., Kalair, A.R., Khan, N., Role of energy storage systems in energy transition from fossil fuels to renewables. *Energy Storage*, *3*, 1, e135, 2021.
2. Megía, P.J., Vizcaíno, A.J., Calles, J.A., Carrero, A., Hydrogen production technologies: From fossil fuels toward renewable sources. A mini review. *Energy Fuels*, *35*, 20, 16403–16415, 2021.
3. Deshmukh, M.K.G., Sameeroddin, M., Abdul, D., Sattar, M.A., Renewable energy in the 21st century: A review. *Mater. Today: Proc.*, *80*, 1756–1759, 2023.
4. Cherubini, F., The biorefinery concept: Using biomass instead of oil for producing energy and chemicals. *Energy Convers. Manage.*, *51*, 7, 1412–1421, 2010.
5. Bonatto, C., Camargo, A.F., Scapini, T., Stefanski, F.S., Alves, S.L., Müller, C., Treichel, H., Biomass to bioenergy research: current and future trends for biofuels, in: *Recent developments in bioenergy research*, pp. 1–17, 2020.
6. Sahrin, N.T., Khoo, K.S., Lim, J.W., Shamsuddin, R., Ardo, F.M., Rawindran, H., Cheng, C.K., Current perspectives, future challenges and key technologies of biohydrogen production for building a carbon-neutral future: A review. *Bioresour. Technol.*, 364, 128088, 2022.
7. Wang, Y., Yang, H., Zhang, X., Han, F., Tu, W., Yang, W., Microalgal hydrogen production. *Small Methods*, 4, 3, 1900514, 2020.

8. Gaffron, H. and Rubin, J., Fermentative and photochemical production of hydrogen in algae. *J. Gen. Physiol.*, 26, 2, 19–240, 1942.
9. Ghirardi, M.L., Togasaki, R.K., Seibert, M., Oxygen sensitivity of algal H 2-production. *Appl. Biochem. Biotechnol.*, *63*, 141–151, 1997.
10. Lu, Y. and Koo, J., O2 sensitivity and H2 production activity of hydrogenases—A review. *Biotechnol. Bioeng.*, *116*, 11, 3124–3135, 2019.
11. Das, D. and Veziroglu, T.N., Advances in biological hydrogen production processes. *Int. J. Hydrogen Energy*, *33*, 21, 6046–6057, 2008.
12. Ishaq, H., Dincer, I., Crawford, C., A review on hydrogen production and utilization: Challenges and opportunities. *Int. J. Hydrogen Energy*, *47*, 62, 26238–26264, 2022.
13. Márquez-Reyes, L.A., del Pilar Sánchez-Saavedra, M., Valdez-Vazquez, I., Improvement of hydrogen production by reduction of the photosynthetic oxygen in microalgae cultures of Chlamydomonas gloeopara and Scenedesmus obliquus. *Int. J. Hydrogen Energy*, *40*, 23, 7291–7300, 2015.
14. Vasseur, C., Bougaran, G., Garnier, M., Hamelin, J., Leboulanger, C., Le Chevanton, M., Fouilland, E., Carbon conversion efficiency and population dynamics of a marine algae–bacteria consortium growing on simplified synthetic digestate: First step in a bioprocess coupling algal production and anaerobic digestion. *Bioresour. Technol.*, *119*, 79–87, 2012.
15. Sahil, S., Karvembu, P., Kaur, R., Katyal, P., Phutela, U.G., Enhanced biogas production from rice straw through pretreatment with cellulase producing microbial consortium. *Energy Nexus*, 12, 100246, 2023.
16. Hena, S., Fatimah, S., Tabassum, S., Cultivation of algae consortium in a dairy farm wastewater for biodiesel production. *Water Resour. Ind.*, *10*, 1–14, 2015.
17. Anand, U., Dey, S., Parial, D., Federici, S., Ducoli, S., Bolan, N.S., Bontempi, E., Algae and bacteria consortia for wastewater decontamination and transformation into biodiesel, bioethanol, biohydrogen, biofertilizers and animal feed: a review. *Environ. Chem. Lett.*, 21, 1–25, 2023.
18. Shobana, S., Kumar, G., Bakonyi, P., Saratale, G.D., Nemestóthy, N., Bélafi-Bakó, K., Chang, J.S., A review on the biomass pretreatment and inhibitor removal methods as key-steps towards efficient macroalgae-based biohydrogen production. *Bioresour. Technol.*, *244*, 1341–1348, 2017.
19. Kumar, S., Herrmann, M., Blohm, A., Hilke, I., Frosch, T., Trumbore, S.E., Küsel, K., Thiosulfate-and hydrogen-driven autotrophic denitrification by a microbial consortium enriched from groundwater of an oligotrophic limestone aquifer. *FEMS Microbiol. Ecol.*, *94*, 10, fiy141, 2018.
20. Ergal, I., Bochmann, G., Fuchs, W., Simon, K., Rittmann, M.R., Design and engineering of artificial microbial consortia for biohydrogen production. *Curr. Opin. Biotechnol.*, *73*, 74–80, 2022.
21. Qian, C.X., Chen, L.Y., Hui, R.O.N.G., Yuan, X.M., Hydrogen production by mixed culture of several facultative bacteria and anaerobic bacteria. *Prog. Nat. Sci.: Mater. Int.*, *21*, 6, 506–511, 2011.

22. López-Pérez, P.A., Neria-González, M., II, López, R.A., Increasing the bio-hydrogen production in a continuous bioreactor via nonlinear feedback controller. *Int. J. Hydrogen Energy*, *40*, 48, 17224–17230, 2015.
23. Nasr, N., Hafez, H., El Naggar, M.H., Nakhla, G., Application of artificial neural networks for modeling of biohydrogen production. *Int. J. Hydrogen Energy*, *38*, 8, 3189–3195, 2013.
24. Gadhe, A., Sonawane, S.S., Varma, M.N., Ultrasonic pretreatment for an enhancement of biohydrogen production from complex food waste. *Int. J. Hydrogen Energy*, *39*, 15, 7721–7729, 2014.
25. Zhang, D., Xiao, N., Mahbubani, K.T., del Rio-Chanona, E.A., Slater, N.K.H., Vassiliadis, V.S., Bioprocess modelling of biohydrogen production by Rhodopseudomonas palustris: model development and effects of operating conditions on hydrogen yield and glycerol conversion efficiency. *Chem. Eng. Sci.*, *130*, 68–78, 2015.
26. Badur, J., Stajnke, M., Ziółkowski, P., Jóźwik, P., Bojar, Z., Ziółkowski, P.J., Mathematical modeling of hydrogen production performance in thermocatalytic reactor based on the intermetallic phase of Ni3Al. *Arch. Thermodyn.*, 40, 3–26, 2019.
27. Gayathiri, E., Prakash, P., Kumaravel, P., Jayaprakash, J., Ragunathan, M.G., Sankar, S., Govindasamy, R., Computational approaches for modeling and structural design of biological systems: A comprehensive review. *Prog. Biophys. Mol. Biol.*, 185, 17–32, 2023.
28. López-Pérez, P.A., López, R.A., Femat, R., State Estimation and Observers. Part II: Observability and Control Concepts, in: *Control in Bioprocessing. Control in Bioprocessing: Modeling, Estimation and the Use of Soft Sensors*, John Wiley & Sons Ltd, Chichester, UK, 2020, https://doi.org/10.1002/9781119296317.ch3.
29. Usmani, Z., Sharma, M., Awasthi, A.K., Lukk, T., Tuohy, M.G., Gong, L., Gupta, V.K., Lignocellulosic biorefineries: the current state of challenges and strategies for efficient commercialization. *Renew. Sustain. Energy Rev.*, *148*, 111258, 2021.
30. Stankiewicz, A., II and Moulijn, J.A., Process intensification: transforming chemical engineering. *Chem. Eng. Prog.*, *96*, 1, 22–34, 2000.
31. Reay, D., Ramshaw, C., Harvey, A., *Process Intensification: Engineering for efficiency, sustainability and flexibility*, Butterworth-Heinemann, Oxford, UK, 2013.
32. Baldea, M., From process integration to process intensification. *Comput. Chem. Eng.*, *81*, 104–114, 2015.
33. Huther, A., Geisselmann, A., Hahn, H., Process intensification-A strategic option for the chemical industry. *Chem. Ing. Tech.*, *77*, 11, 1829–1837, 2005.
34. Keil, F.J., Process intensification. *Rev. Chem. Eng.*, *34*, 2, 135–200, 2018.
35. López-Pérez, P.A., López, R.A., Femat, R., Introduction. Part I: Overview of the Control and Monitoring of Bioprocesses and Mathematical Preliminaries, in: *Control in Bioprocessing. Control in Bioprocessing: Modeling, Estimation*

and the Use of Soft Sensors, John Wiley & Sons Ltd, Chichester, UK, 2020, https://doi.org/10.1002/9781119296317.ch1.

36. Boodhoo, K.V.K., Flickinger, M.C., Woodley, J.M., Emanuelsson, E.A.C., Bioprocess intensification: A route to efficient and sustainable biocatalytic transformations for the future. *Chem. Eng. Process. - Process Intensif.*, *172*, 108793, 2022.
37. López-Pérez, P.A., López, R.A., Femat, R., Dynamical Behavior of a 3-Dimensional Continuous Bioreactor, in: *Control in Bioprocessing. Part III Software Sensors and Observer-Based Control Schemes for Bioprocess. In Control in Bioprocessing. Control in Bioprocessing: Modeling, Estimation and the Use of Soft Sensors*, John Wiley & Sons Ltd, Chichester, UK, 2020, https://doi.org/10.1002/9781119296317.ch5.
38. Tula, A.K., Eden, M.R., Gani, R., Computer–aided process intensification: Challenges, trends and opportunities. *AIChE J.*, *66*, 1, 2019.
39. Van Gerven, T. and Stankiewicz, A., Structure, energy, synergy, time.The fundamentals of process intensification. *Ind. Eng. Chem. Res.*, *48*, 5, 2465–2474, 2009.
40. Lutze, P. and Gorak, A., Reactive and membrane-assisted distillation: Recent developments and perspective. *Chem. Eng. Res. Des.*, *91*, 10, 1978–1997, 2013.
41. Siirola, J. and Gani, R., Process Intensification Preface. *Comput. Chem. Eng.*, *105*, 1–1, 2017.
42. Alvarado-Santos, E., Mata-Machuca, J.L., López-Pérez, P.A., Garrido-Moctezuma, R.A., Pérez-Guevara, F., Aguilar-López, R., Comparative Analysis of a Family of Sliding Mode Observers under Real-Time Conditions for the Monitoring in the Bioethanol Production. *Fermentation*, *8*, 9, 446, 2022.
43. Moulijn, J.A., Stankiewicz, A., Grievink, J., Górak, A., Process intensification and process systems engineering: a friendly symbiosis. *Comput. Chem. Eng.*, *32*, 1-2, 3–11, 2008.
44. Rivas, D.F., Boffito, D.C., Faria-Albanese, J., Glassey, J., Afraz, N., Akse, H., Weber, R.S., Process intensification education contributes to sustainable development goals. Part 1. *Educ. Chem. Eng.*, *32*, 1–14, 2020.
45. Abiev, R.S., Process intensification in chemical engineering: General trends and Russian contribution. *Rev. Chem. Eng.*, *37*, 1, 69–97, 2021.
46. Segovia-Hernández, J.G., Hernández, S., Cossío-Vargas, E., Sánchez-Ramírez, E., Challenges and Opportunities in Process Intensification to achieve the UN's 2030 Agenda: Goals 6, 7, 9, 12 and 13. *Chem. Eng. Process. - Process Intensif.*, 12, 109507, 2023.
47. Kalakul, S., Malakul, P., Siemanond, K., Gani, R., Integration of life cycle assessment software with tools for economic and sustainability analyses and process simulation for sustainable process design. *J. Cleaner Prod.*, *71*, 98–109, 2014.

48. Demirel, S.E., Li, J., Hasan, M.F., Systematic process intensification. *Curr. Opin. Chem. Eng.*, *25*, 108–113, 2019.
49. Osman, A., II, Mehta, N., Elgarahy, A.M., Al-Hinai, A., Al-Muhtaseb, A.A.H., Rooney, D.W., Conversion of biomass to biofuels and life cycle assessment: a review. *Environ. Chem. Lett.*, *19*, 4075–4118, 2021.
50. ISO, *Environmental Management: Life Cycle Assessment - Principles and Framework*, p. 14040, ISO, Geneva, Switzerland, 2006.
51. Cucurachi, S., Scherer, L., Guinée, J., Tukker, Life cycle assessment of food systems. *One Earth*, *1*, 3, 292–297, 2019.
52. Wu, W., Lei, Y.C., Chang, J.S., Life cycle assessment of upgraded microalgae-to-biofuel chains. *Bioresour. Technol.*, *288*, 121492, 2019.
53. Banu, J.R., Kavitha, S., Gunasekaran, M., Kumar, G., Microalgae based biorefinery promoting circular bioeconomy-techno economic and life-cycle analysis. *Bioresour. Technol.*, *302*, 122822, 2020.
54. Chen, P.H. and Quinn, J.C., Microalgae to biofuels through hydrothermal liquefaction: Open-source techno-economic analysis and life cycle assessment. *Appl. Energy*, *289*, 116613, 2021.
55. Li, S., Chang, H., Zhang, S., Ho, S.H., Production of sustainable biofuels from microalgae with CO2 bio-sequestration and life cycle assessment. *Environ. Res.*, 227, 115730, 2023.
56. Nasir, A.S.M., Mohamed, B., Li, L.Y., Comparative life cycle assessment of co-pyrolysing sewage sludge and wastewater-grown microalgae for biofuel production. *Resour. Conserv. Recycl.*, *190*, 106780, 2023.
57. Lewandowski, I., Clifton-Brown, J.C., Scurlock, J.M.O., Huisman, W., Miscanthus: European experience with a novel energy crop. *Biomass Bioenergy*, *19*, 4, 209–227, 2000.
58. Mu, D., Xin, C., Zhou, W., Life cycle assessment and techno-economic analysis of algal biofuel production, in: *Microalgae Cultivation for Biofuels Production*, 281-2922020.
59. Hauschild, M.Z., Rosenbaum, R.K., Olsen, S.I., *Life cycle assessment*, Springer International Publishing, Cham, 2018, https://doi. org/10.1007/978-3-319-56475-3 Book.
60. Hosseinzadeh, A., Zhou, J.L., Li, X., Afsari, M., Altaee, A., Techno-economic and environmental impact assessment of hydrogen production processes using bio-waste as renewable energy resource. *Renew. Sustain. Energy Rev.*, 156, 111991, 2022.
61. Shonnard, D.R., Klemetsrud, B., Sacramento-Rivero, J., Navarro-Pineda, F., Hilbert, J., Handler, R., Donovan, R.P., A review of environmental life cycle assessments of liquid transportation biofuels in the Pan American region. *Environ. Manage.*, 56, 1356–1376, 2015.
62. Smurthwaite, M., Jiang, L., Williams, K.S., A review of the LCA literature investigating the methods by which distinct impact categories are compared. *Environ. Dev. Sustain.*, 1, 1–17, 2023.

63. Dong, Y.H. and Ng, S.T., Comparing the midpoint and endpoint approaches based on ReCiPe—a study of commercial buildings in Hong Kong. *Int. J. Life Cycle Assess.*, *19*, 1409–1423, 2014.
64. Landis, A.E. and Theis, T.L., Comparison of life cycle impact assessment tools in the case of biofuels, in: *2008 IEEE international symposium on electronics and the environment*, IEEE, pp. 1–7, 2008.
65. Cavalcanti, E.J., Carvalho, M., Ochoa, A., Exergoeconomic and exergoenvironmental comparison of diesel-biodiesel blends in a direct injection engine at variable loads. *Energy Convers. Manage.*, *183*, 450–461, 2019.
66. Sills, D.L., Van Doren, L.G., Beal, C., Raynor, E., The effect of functional unit and co-product handling methods on life cycle assessment of an algal biorefinery. *Algal Res.*, *46*, 101770, 2020.
67. Zapata-Boada, S., Gonzalez-Miquel, M., Jobson, M., Cuéllar-Franca, R.M., Life Cycle Environmental Evaluation of Alternative Solvents Used in Lipid Extraction—The Case of Algae Biodiesel. *ACS Sustain. Chem. Eng.*, *11*, 32, 11934–11946, 2023.
68. Huijbregts, M.A.J., Steinmann, Z.J.N., Elshout, P.M.F.M., Stam, G., Verones, F., Vieira, M.D.M., Zijp, M., van Zelm, R., ReCiPe 2016 - A harmonized life cycle impact assessment method at midpoint and endpoint level. Report I: Characterization, The International Journal of Life Cycle Assessment, 22, 2016.
69. Castro, J.S., Ferreira, J., Magalhães, I.B., Junior, M.J., Marangon, B.B., Pereira, A.S.A.P., Calijuri, M.L., Life cycle assessment and techno-economic analysis for biofuel and biofertilizer recovery as by-products from microalgae. *Renew. Sustain. Energy Rev.*, *187*, 113781, 2023.
70. Rodríguez, M.R., De Ruyck, J., Diaz, P.R., Verma, V.K., Bram, S., An LCA based indicator for evaluation of alternative energy routes. *Appl. Energy*, *88*, 3, 630–635, 2011.
71. Rathore, D. and Singh, A., Biohydrogen production from microalgae, in: *Biofuel Technologies: Recent Developments*, pp. 317–333, Springer, Berlin, Heidelberg, 2013.
72. Singh, A., Sevda, S., Abu Reesh, I.M., Vanbroekhoven, K., Rathore, D., Pant, D., Biohydrogen production from lignocellulosic biomass: technology and sustainability. *Energies*, *8*, 11, 13062–13080, 2015.
73. Cai, W., Lai, K.H., Liu, C., Wei, F., Ma, M., Jia, S., Lv, L., Promoting sustainability of manufacturing industry through the lean energy-saving and emission-reduction strategy. *Sci. Total Environ.*, *665*, 23–32, 2019.
74. Siddique, H.M.A. and Kiani, A.K., Industrial pollution and human health: evidence from middle-income countries. *Environ. Sci. Pollut. Res.*, *27*, 11, 12439–12448, 2020.
75. Belucio, M., Rodrigues, C., Antunes, C.H., Freire, F., Dias, L.C., Eco-efficiency in early design decisions: A multimethodology approach. *J. Cleaner Prod.*, *283*, 124630, 2021.

76. Basic, S., Hollberg, A., Galimshina, A., Habert, G., A design integrated parametric tool for real-time Life Cycle Assessment–Bombyx project. *IOP Conf. Ser.: Earth Environ. Sci.*, 323, 1, 012112, IOP Publishing, 2019.
77. Kamalakkannan, S. and Kulatunga, A.K., Optimization of eco-design decisions using a parametric life cycle assessment. *Sustain. Prod. Consumption, 27*, 1297–1316, 2021.
78. Eddy, D.C., Krishnamurty, S., Grosse, I.R., Wileden, J.C., Lewis, K.E., A normative decision analysis method for the sustainability-based design of products. *J. Eng. Des., 24*, 5, 342–362, 2013.
79. Shahbazi, S. and Jönbrink, A.K., Design guidelines to develop circular products: Action research on nordic industry. *Sustainability, 12*, 9, 3679, 2020.
80. Muys, B., Wouters, G., Spirinckx, C., Cleaner production: a guide to information sources, Final report EEA contract. *Denmark, Trans. Ecol. Environ., 1*, 1–112, 1997.
81. Lee, K.M., Ecodesign: Best practice of ISO/TR 14062, Eco-product Research Institute, Ajou University. *Ministry of Commerce, Industry and Energy Republic of Korea, 1*, 1–66, 2005.
82. Thai, R. and Liu, G., Eco-design pilot project in China-Monsoon offer 2 upgrade. *IOP Conf. Ser.: Earth Environ. Sci.*, 94, 1, 012170, IOP Publishing, 2017.
83. Corsi, I., Venditti, I., Trotta, F., Punta, C., Environmental safety of nanotechnologies: The eco-design of manufactured nanomaterials for environmental remediation. *Sci. Total Environ., 864*, 161181, 2023.
84. Schäfer, M. and Löwer, M., Ecodesign—A review of reviews. *Sustainability, 13*, 1, 315, 2020.
85. Thomé, A.M.T., Scavarda, A., Ceryno, P.S., Remmen, A., Sustainable new product development: a longitudinal review. *Clean Technol. Environ. Policy, 18*, 2195–2208, 2016.
86. Kim, H., Cluzel, F., Leroy, Y., Yannou, B., Yannou-Le Bris, G., Research perspectives in ecodesign. *Des. Sci., 6*, e7, 2020.
87. Van Weenen, J.C., Towards sustainable product development. *J. Cleaner Prod., 3*, 1-2, 95–100, 1995.
88. Bangsa, A.B. and Schlegelmilch, B.B., Linking sustainable product attributes and consumer decision-making: Insights from a systematic review. *J. Cleaner Prod., 245*, 118902, 2020.
89. Ostad-Ahmad-Ghorabi, H. and Collado-Ruiz, D., Tool for the environmental assessment of cranes based on parameterization. *Int. J. Life Cycle Assess., 16*, 392–400, 2011.
90. Vallet, F., Eynard, B., Millet, D., Mahut, S.G., Tyl, B., Bertoluci, G., Using eco-design tools: An overview of experts' practices. *Des. Stud., 34*, 3, 345–377, 2013.
91. Vandervaeren, C., Galle, W., De Temmerman, N., Parametric life cycle assessment of a reusable brick veneer. *IOP Conf. Ser.: Earth Environ. Sci.*, 323, 1, 012137, IOP Publishing, 2019.

92. Lolli, N., Fufa, S.M., Inman, M., A parametric tool for the assessment of operational energy use, embodied energy and embodied material emissions in building. *Energy Proc.*, *111*, 21–30, 2017.
93. Gassanova, L.G., Netrusov, A., II, Teplyakov, V.V., Modigell, M., Fuel gases from organic wastes using membrane bioreactors. *Desalination*, *198*, 1-3, 56–66, 2006.
94. Hallenbeck, P.C. and Ghosh, D., Advances in fermentative biohydrogen production: the way forward? *Trends Biotechnol.*, *27*, 5, 287–297, 2009.
95. Nikolaidis, P. and Poullikkas, A., A comparative overview of hydrogen production processes. *Renew. Sustain. Energy Rev.*, *67*, 597–611, 2017.
96. Goria, K., Kothari, R., Singh, H.M., Singh, A., Tyagi, V.V., Biohydrogen: Potential applications, approaches, and hurdles to overcome, in: *Handbook of Biofuels*, pp. 399–418, Academic Press, Cambridge, Massachusetts, US, 2022.
97. Wang, J. and Yin, Y., Fermentative hydrogen production using pretreated microalgal biomass as feedstock. *Microb. Cell Fact.*, *17*, 1, 1–16, 2018.
98. Bolatkhan, K., Kossalbayev, B.D., Zayadan, B.K., Tomo, T., Veziroglu, T.N., Allakhverdiev, S., II, Hydrogen production from phototrophic microorganisms: reality and perspectives. *Int. J. Hydrogen Energy*, *44*, 12, 5799–5811, 2019.
99. Demirbas, A., *Biohydrogen*, pp. 163–219, Springer, London, 2009.
100. Dinesh, G.H., Nguyen, D.D., Ravindran, B., Chang, S.W., Vo, D.V.N., Bach, Q.V., Arun, A., Simultaneous biohydrogen (H2) and bioplastic (poly-β-hydroxybutyrate-PHB) productions under dark, photo, and subsequent dark and photo fermentation utilizing various wastes. *Int. J. Hydrogen Energy*, *45*, 10, 5840–5853, 2020.
101. Ramu, S.M., Thulasinathan, B., Hari, D.G., Bora, A., Jayabalan, T., Mohammed, S.N., Alagarsamy, A., Fermentative hydrogen production and bioelectricity generation from food based industrial waste: An integrative approach. *Bioresour. Technol.*, *310*, 123447, 2020.
102. Ananthi, V., Ramesh, U., Balaji, P., Kumar, P., Govarthanan, M., Arun, A., A review on the impact of various factors on biohydrogen production. *Int. J. Hydrogen Energy*, 52, 33–45, 2022.
103. Gilmour, D.J., Microalgae for biofuel production. *Adv. Appl. Microbiol.*, *109*, 1–30, 2019.
104. Limongi, A.R., Viviano, E., De Luca, M., Radice, R.P., Bianco, G., Martelli, G., Biohydrogen from microalgae: production and applications. *Appl. Sci.*, *11*, 4, 1616, 2021.
105. López-Pérez, P.A., Peña, C., Ruiz, C., Aguilar, L., Increasing of lipid productivity in microalgae cultures via dynamic analysis and closed-loop operation. *Eur. Chem. Bull.*, *6*, 4, 145–150, 2017.
106. Dębowski, M., Zieliński, M., Kazimierowicz, J., Kujawska, N., Talbierz, S., Microalgae cultivation technologies as an opportunity for bioenergetic system development—advantages and limitations. *Sustainability*, *12*, 23, 9980, 2020.

107. Udayan, A., Sirohi, R., Sreekumar, N., Sang, B., II, Sim, S.J., Mass cultivation and harvesting of microalgal biomass: Current trends and future perspectives. *Bioresour. Technol.*, *344*, 126406, 2022.
108. Amabilis-Sosa, L., Rodríguez-Mata, A., Baray-Arana, R., Robles-Vega, I., Gonzalez-Huitrón, V., López-Pérez, P., Robust fractional control based on high gain observers design (RNFC) for a Spirulina maxima culture interfaced with an advanced oxidation process. *Open Chem.*, *21*, 1, 20220214, 2023.
109. Aguilar-López, R., Ruiz Camacho, B., Neria-González, M.I., Rangel, E., Santos, O., López Pérez, P.A., State estimation based on nonlinear observer for hydrogen production in a photocatalytic anaerobic bioreactor. *Int. J. Chem. Reactor Eng.*, *15*, 5, 20170004, 2017.
110. López-Pérez, P.A., Lopez, R.A., Femat, R., *Control in bioprocessing: Modeling, estimation and the use of soft sensors*, John Wiley & Sons, Hoboken, New Jersey, US, 2020.
111. López-Pérez, P.A., Rodriguez-Mata, A.E., Hernández-González, O., Amabilis-Sosa, L.E., Baray-Arana, R., Leon-Borges, J., Design of a Robust sliding mode controller for bioreactor cultures in overflow metabolism via an interdisciplinary approach. *Open Chem.*, *20*, 1, 120–129, 2022.
112. Lee, H.S., Lee, S.Y., Yoo, K., Kim, H.W., Lee, E., Im, N.G., Biohydrogen production and purification: Focusing on bioelectrochemical systems. *Bioresour. Technol.*, *363*, 127956, 2022.
113. Dawood, F., Anda, M., Shafiullah, G.M., Hydrogen production for energy: An overview. *Int. J. Hydrogen Energy*, *45*, 7, 3847–3869, 2020.
114. Rodríguez-Mata, A.E., Gómez-Vidal, E., Lucho-Constantino, C.A., Medrano-Hermosillo, J.A., Baray-Arana, R., López-Pérez, P.A., State Estimation in a Biodigester via Nonlinear Logistic Observer: Theoretical and Simulation Approach. *Processes*, *11*, 4, 1234, 2023.
115. Sahinoglu, M., Modeling and simulation in engineering. *Wiley Interdiscip. Rev.: Comput. Stat.*, *5*, 3, 239–266, 2013.
116. Van Baten, J.M. and Szczepanski, R., A thermodynamic equilibrium reactor model as a CAPE-OPEN unit operation. *Comput. Chem. Eng.*, 35, 1251–1256, 2011.
117. Barbosa, M.J., Rocha, J.M., Tramper, J., Wijffels, R.H., Acetate as a carbon source for hydrogen production by photosynthetic bacteria. *J. Biotechnol.*, *85*, 1, 25–33, 2001.
118. Weuster-Botz, D., Puskeiler, R., Kusterer, A., Kaufmann, K., John, G.T., Arnold, M., Methods and milliliter scale devices for high-throughput bioprocess design. *Bioprocess. Biosyst. Eng.*, *28*, 109–119, 2005.
119. Hernández-Melchor, D.J., Camacho-Pérez, B., Ríos-Leal, E., Alarcón-Bonilla, J., López-Pérez, P.A., Modelling and multi-objective optimization for simulation of hydrogen production using a photosynthetic consortium. *Int. J. Chem. Reactor Eng.*, *18*, 7, 20200019, 2020.

120. Acién, F.G., Fernández, J.M., Magán, J.J., Molina, E., Production cost of a real microalgae production plant and strategies to reduce it. *Biotechnol. Adv.*, *30*, 1344–1353, 2012.
121. Toledo-Cervantes, A., Estrada, J.M., Lebrero, R., Muñoz, R., A comparative analysis of biogas upgrading technologies: Photosynthetic vs physical/chemical processes. *Algal Res.*, *25*, 237–243, 2017.
122. Ferreira, A.F., Toledo-Cervantes, A., de Godos, I., Gouveia, L., Munoz, R., Life cycle assessment of pilot and real scale photosynthetic biogas upgrading units. *Algal Res.*, *44*, 101668, 2019.
123. Akhbari, A., Onn, C.C., Ibrahim, S., Analysis of biohydrogen production from palm oil mill effluent using a pilot-scale up-flow anaerobic sludge blanket fixed-film reactor in life cycle perspective. *Int. J. Hydrogen Energy*, *46*, 68, 34059–34072, 2021.
124. Barghash, H., AlRashdi, Z., Okedu, K.E., Desmond, P., Life-Cycle Assessment Study for Bio-Hydrogen Gas Production from Sewage Treatment Plants Using Solar PVs. *Energies*, *15*, 21, 8056, 2022.
125. Link, H. and Weuster-Botz, D., Genetic algorithm for multi-objective experimental optimization. *Bioprocess. Biosyst. Eng.*, *29*, 385–390, 2006.
126. Xie, G.J., Liu, B.F., Xing, D.F., Nan, J., Ding, J., Ren, N.Q., Photo-fermentative bacteria aggregation triggered by L-cysteine during hydrogen production. *Biotechnol. Biofuels*, *6*, 1, 1–14, 2013.
127. Oncel, S. and Kose, A., Comparison of tubular and panel type photobioreactors for biohydrogen production utilizing Chlamydomonas reinhardtii considering mixing time and light intensity. *Bioresour. Technol.*, *151*, 265–270, 2014.
128. Khetkorn, W., Rastogi, R.P., Incharoensakdi, A., Lindblad, P., Madamwar, D., Pandey, A., Larroche, C., Microalgal hydrogen production–A review. *Bioresour. Technol.*, *243*, 1194–1206, 2017.
129. Houghton, J.T., Callander, B.A., Varney, S.K. (Eds.), *Climate change*, Cambridge University Press, Cambridge, UK, 1992.
130. Al-Ghussain, L., Global warming: review on driving forces and mitigation. *Environ. Prog. Sustain. Energy*, *38*, 1, 13–21, 2019.
131. Backes, J.G. and Traverso, M., Life cycle sustainability assessment as a metrics towards SDGs agenda 2030. *Curr. Opin. Green Sustain. Chem.*, 100683, 2236–2452, 2022.
132. López-Pérez, P., López-López, M., Núñez-Colín, C., Mukhtar, H., Aguilar-López, R., Peña-Caballero, V., A novel nonlinear sliding mode observer to estimate biomass for lactic acid production. *Chem. Prod. Process Model.*, 18, 4, 565–580, 2023.
133. Vallejo, E. and López–Pérez, P.A., Strong chemical adsorption of CO2 and N2 on a five-vacancy graphene surface. *Solid State Commun.*, 356, 114934, 2022.
134. Tamayo-Román, E.D., Ordaz-Oliver, J.P., López-Pérez, P.A., Gonzalez-Viveros, I., Hernández-Hernández, J.L., Observador robusto de orden

reducido aplicado a un biodigestor. *Pädi Boletín Científico De Ciencias Básicas E Ingenierías Del ICBI*, 10, 48–56, 2022.
135. Alvarado-Santos, E., Aguilar-López, R., Neria-González, M.I., Romero-Cortés, T., Robles-Olvera, V.J., López-Pérez, P.A., A novel kinetic model for a cocoa waste fermentation to ethanol reaction and its experimental validation. *Prep. Biochem. Biotechnol.*, 53, 167–182, 2023.
136. Flores, G., Rodriguez-Mata, A., Amabilis-Sosa, L., Gonzalez-Huitron, V., Hernández-González, O., Lopéz-Peréz, P., A turbidity sensor development based on NL-PI observers: Experimental application to the control of a Sinaloa's River Spirulina maxima cultivation. *Open Chem.*, 18, 1, 1349–1361, 2020.
137. Rodriguez-Mata, A.E., Bustos-Terrones, Y., Gonzalez-Huitrón, V., Lopéz-Peréz, P.A., Hernández-González, O., Amabilis-Sosa, L.E.A., Fractional High-Gain Nonlinear Observer Design—Application for Rivers Environmental Monitoring Model. *Math. Comput. Appl.*, 25, 44, 2020.
138. Hernández-Melchor, D.J., López-Pérez, P.A., Carrillo-Vargas, S. *et al.*, Experimental and kinetic study for lead removal via photosynthetic consortia using genetic algorithms to parameter estimation. *Environ. Sci. Pollut. Res.*, 25, 21286–21295, 2018.
139. López Pérez, P., Neria González, M., Aguilar López, R., Cadmium concentration stabilization in a continuous sulfate reducing bioreactor via sulfide concentration control. *Chem. Pap.*, *67*, 3, 326–335, 2013.
140. López-Pérez, P.A., Alvarado-Santos, E., Hernández Melchor, D.J., Martınez-Farıas, F.J., Jiménez Muñoz, E., Chapter 4. A Generic Mathematical Model for Pest Control with Time Delay Dynamics Based on Silico Approaches, in: *Biology, Behavior and Ecology*, pp. 95–128, Nova, Science and Technology, USA, 2023.

7

In Situ Monitoring for Biohydrogen Production Using a Low-Cost Sensor

Pablo Antonio López Pérez[1*], Patricia Meneses Martínez[2], Emmanuel Vallejo Castañeda[1] and Ricardo Aguilar López[3]

[1]*Universidad Autónoma del Estado de Hidalgo, Escuela Superior de Apan, Carretera Apan-Calpulalpan Km.8, Col. Chimalpa, Apan, Hidalgo, México*
[2]*Centro de Investigación en Tecnologías de Información y Sistemas, Universidad Autónoma del Estado de Hidalgo, Hidalgo, México*
[3]*Department of Biotechnology and Bioengineering, CINVESTAV-IPN, San Pedro Zacatenco, Mexico City, México*

Abstract

The main obstacle in the control industry process is to measure the input variable and the output variable in order to drive this output into the set point of the bioprocess. Monitoring systems for biohydrogen production processes is an issue that needs precise solutions for its control and automation which is fundamental for wide-scale industrial credibility. Several instruments have been designed the biohydrogen production with a high budget of prices. The chapter mainly implemented and validated an instrument capable of measuring hydrogen using a low-cost sensor. The learning model and sensor were experimentally *in situ* using the LabVIEW and MATLAB platforms. The experimental data of hydrogen production were determined in photobioreactors in batch culture in anaerobic conditions using a photosynthetic consortium. In addition, a methodology of learning model coupled to a low-cost sensor is presented through iterative analysis of data-based detection results. It is worth mentioning that the sensor allows the detection of hydrogen concentrations in a range of 5 to 300 (ppm) with 85% accuracy. Finally, it will allow for the generation of a more robust monitoring system by reducing the data load and at the same time improving detection capabilities.

Keywords: Online, photosynthetic consortium, LabVIEW

**Corresponding author*: save1991@yahoo.com.mx; pablo_lopez@uaeh.edu.mx

Mehmet Sankir and Nurdan Demirci Sankir (eds.) Towards Green Hydrogen Generation, (205–236)

7.1 Introduction

In the bioprocess industry, correct management of control variables is essential to achieve the detection and diagnosis of system failures and adequate performance of control loops in general. Because these types of processes are nonlinear and present high variations over time, microbiological kinetics are very complex and limit the real-time measurement of process variables. This disadvantage can occur either due to a lack of reliable and precise devices, high acquisition and maintenance costs, or the type of operation of the systems themselves. It is possible that sometimes the sensors have not been designed or their availability is low in the market.

Another impediment to the measurement of some variables is that they are affected by the nature of the system; they can become corrosive or have extreme temperatures, making direct measurement of the variable within the production line impossible. For a bioreactor, complexity increases, since any slight change in a variable can cause cell destruction and, therefore, degradation of the product [1].

Because most biological processes have intrinsic transient dynamics, bioreactors operate nonstationarily and the effects of system inputs depend on the history of the system. To model this type of dynamic system, mathematical expressions are used to describe the kinetics of the reaction, which are then used to form mass balance equations to describe the temporal behavior of all the biochemical species that interact in the bioreactor. Currently, several nonlinear control techniques are used to model bioprocesses and in most cases, it is assumed that each variable in the model is available for online measurement. This fact is not always possible in reality; therefore, to solve this problem, virtual sensors, commonly called state estimators, are used [2].

7.1.1 Automation of Industrial Bioprocesses

Bioprocesses are a fundamental part of several food, pharmaceutical, and chemical industries. They are based on using microbial, animal, or plant cells and their components, such as enzymes, to generate new products and destroy harmful waste. As in any type of manufacturing industry, there is a requirement for knowledge, accurate information, and automated feedback control to ensure robustness and reliability in the process. The complex nature of a biological system makes the process even more difficult and, when it comes to metabolites or byproducts of biological reactions, imposes extremely high demands on quality issues [3].

The global control system of a plant is structured by levels of hierarchy, thus forming the so-called automation pyramid. The position of each level within the pyramid reflects the treatment that will be given to the information. At the base of the pyramid, a faster response is required (in milliseconds) since greater precision is needed, while as you go up, response times increase given the complexity of the calculations to be performed and the large volume of information that requires computer processing for better use of the data [4].

The supervision stage is part of the third link of the hierarchy pyramid of automation. In this, a large amount of data is handled through human-machine interfaces, supervisory control and data acquisition (SCADA), specific devices for supervision, or control computer desks. All these data are received from the plant controller, which constitutes the brain of the process. In the control stage (second level) are the programmable controllers and processors such as programmable logic controller (PLC), programmable automation control (PAC), distributed control system (DCS), and field-programmable gate array (FPGA), among others. These controllers have input to all field devices as actuators (valves, motors, and frequency converters) and sensors (flow meters, pressure, level, and temperature.) and each action to be performed is managed through programming, imposing references and admission thresholds to subsequently monitor from the computer [4].

Once the information is in the supervision and monitoring equipment, not only is interaction with the user achieved through graphics, alarm signals, report generation, and global analysis of the plant but also it can directly influence the dynamic behavior of the system variables through programming of observers, filters, virtual sensors, among other components that are incorporated into the plant. Supervision achieves the reconfiguration of system parameters through the controller, executing a set of actions to bring the process to its normal operation using model-based self-tuning methods or without mathematical knowledge of the process, among others. This constitutes a crucial difference in relation to monitoring that only covers detection and diagnosis and sometimes those that undertake only surveillance tasks have been mistakenly called supervision systems [5].

Finally, the last two levels of the pyramid, manufacturing execution system (MES) planning and enterprise resource planning (ERP), are manufacturing execution systems that organize the resources necessary to execute the plant's production plan that covers raw materials, order of priorities, change of production instructions, the controllers, and measurement interval of the sensors, among others. Already in enterprise resource management (fifth level), a centralized database of the entire company and user interfaces are presented to execute the activities of the entity such as sales, human resources, finances, suppliers, and others [5].

The case study of this research has an approach based on the monitoring of a biohydrogen production bioprocess, which includes the connection to the embedded system of the instrumentation relevant to a bioreactor to the graphical interaction of the user through HMI screens and computers.

In the bioprocess industry, due to their complexity, measurement instruments are used that usually have high costs and high-end features; however, there are alternatives to these instruments with features not comparable to instruments on the market but with acceptable precision and an affordable cost (Figure 7.1) [6].

7.1.2 Bioprocess Monitoring

One of the great challenges of the bioprocess industry is to achieve a favorable environment for the development and growth of microorganisms in bioreactors (biomass) and, in this way, consume enough substrate to obtain the desired final product with high performance and productivity (Figure 7.2). Biotechnological processes are very dynamic and involve continuous changes in the physicochemical conditions of the environment in which they develop. For this, real-time monitoring becomes a crucial

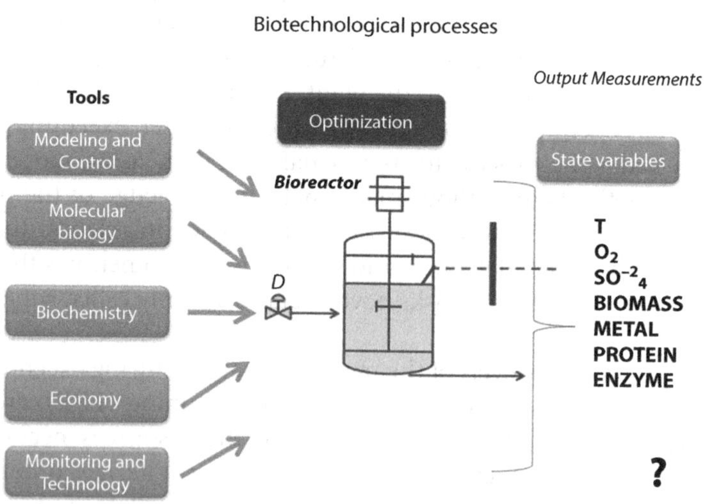

Figure 7.1 Bioprocess and bioreactors to improve bioprocessing with digital sensors.

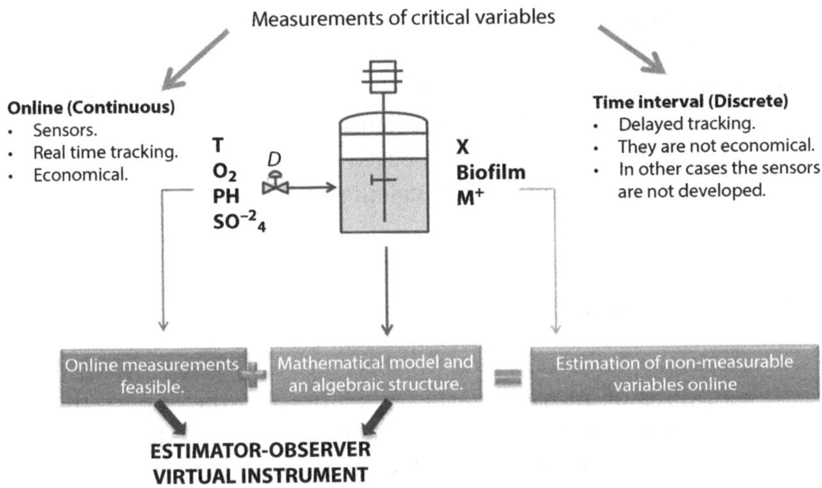

Figure 7.2 Sensor systems for bioprocess monitoring.

task to select the most favorable conditions, and therefore, it is essential to strengthen the tools that guarantee this work [7].

Monitoring is based on running continuous real-time tasks to determine the conditions of a physical system. The recording of information allows for recognizing and indicating anomalies in behavior. As mentioned, it is impossible to monitor a process if a fault detection and diagnosis (FDD) system is not carried out, which determines the faults present in a system and the detection time (fault detection). Then, its type and location are established, to later determine the size and temporal behavior of the failure (fault diagnosis). Below, some essential concepts are defined to introduce this topic, which was compiled from the IFAC SAFEPROCESS technical committee [8].

- **Fault:** Unexpected or impermissible deviation of one or more system parameters from their acceptable or standard condition.
- **Damage:** Permanent interruption of the system's ability to perform a required function under specified operating conditions.
- **Malfunction:** Intermittent irregularity in the fulfillment of the desired function.
- **Disturbance:** Input acting on a system, which results in a temporary deviation from its current state.

- **Symptom:** It is the change in the magnitude of a variable from its normal behavior.
- **Error:** Deviation between the measured value of an output variable and the actual value (specified value).
- **Residual:** Failure indicator, based on the deviation between measurements and values calculated based on the system model.

7.1.2.1 Failure Classification [9]

- **Actuator failures:** Defective bearings and gears, aging or jamming of moving parts. Actuators are the inputs that allow the process to be controlled mechanically. If they are damaged, they will not be able to generate the desired control inputs.
- **Sensor failures:** Scale errors, drifts, short circuits, and false contacts may occur. The sensors provide the necessary measurements to the controller to generate the control inputs; therefore, they must be perfectly calibrated and meet the basic requirements for their use.
- **Abnormal deviations in process parameters:** Changes in system components. When making these variations, the original model is altered and the designed controller may not be able to detect these variations and lead the process to the expected performance.
- **Abrupt failures (step type):** They are associated with unforeseen damage to some element of the system. They are the most dangerous since they can lead the system to instability.
- **Incipient or evolutionary failures (drift type):** They are generally associated with the aging of the system components. Due to their gradual nature, they are more difficult to detect.
- **Intermittent failures:** In reference to their name, they are those that appear and disappear intermittently, this is due to partially damaged elements or interruptions in communication.

According to the representation within a control loop, failures can be additive or multiplicative.

- **Additive faults:** These are caused by unknown inputs or disturbances that act on the system and can cause changes in the plant outputs independently of the inputs [10].
- **Multiplicative failures:** These are sudden or gradual failures of the plant parameters. They represent deterioration of equipment, dirt, and loss of power, among other factors. They also model component failures [10].

Next (Figure 7.3), there is a summary of the different types of failures that can occur in a control loop.

7.1.2.2 Classification of Fault Detection and Diagnosis Methods

Model-Free Methods

Physical redundancy: Also known as hardware redundancy, it employs multiple sensors, actuators, and control components for a particular variable. The disadvantage of this method is the cost of additional equipment and its maintenance, in addition to the space required within the production line.

Special sensors: Special sensors dedicated to the detection and diagnosis of failures such as vibration, and elongation, among others, are directly installed.

Analysis of signs or symptoms: Failures are indicated when there are deviations in the measured variables and they exceed certain thresholds pre-established by the programmer.

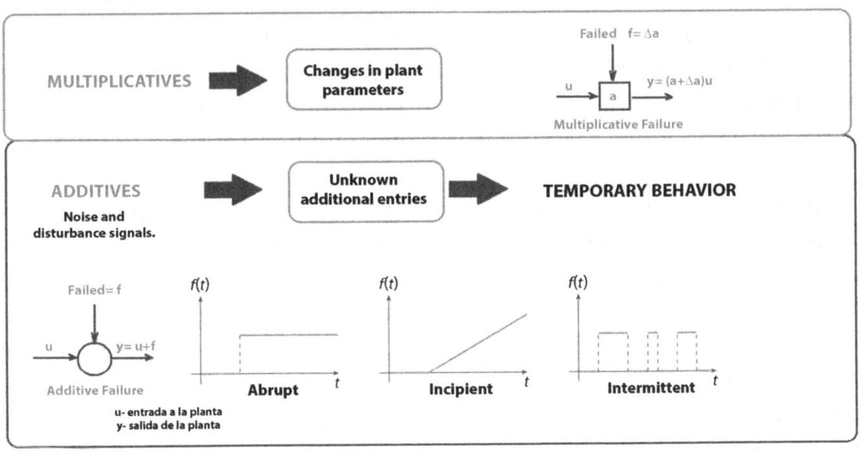

Figure 7.3 Types of process failures.

Spectral analysis: Certain types of faults exhibit a characteristic spectral representation, its analysis facilitating the detection of anomalies in the system.

Statistical analysis: Taking into account the distribution of the study system, the mean value and standard deviation under normal conditions are known. The anomaly is detected when these elements go out of the normal operating range, causing a failure in the system.

Expert systems: They address problems by having a domain of knowledge based on experience. They are capable of acquiring knowledge, choosing its representation, encoding it in a knowledge base, and developing inference procedures for the diagnosis of the failure [11].

Model-Based Methods

Parity Equations: A linear dynamic transformation is applied to the input-output equations of the model. The transformed waste is used for fault detection and isolation.

Parameter Estimation: It is the most intuitive way to detect and isolate multiplicative faults. A reference model is obtained by identifying the fault-free plant. Subsequently, online parameters are identified continuously and a comparison with the reference model is performed to isolate and detect faults. This method, despite dispensing with greater computing capacity and input excitation requirements, is considered the most reliable among analytical redundancy methods.

Diagnostic Observers: An observer is used to estimate the outputs of the current system. Subsequently, residuals are constructed from the process measurements and their estimation. The design of the observer can control the dynamics of the response to faults, within its limits by its gains.

Kalman Filters: In principle, the application of the Kalman filter is similar to that of state observers, but it is especially suitable in processes with relatively large variations of the state variables and noise at the output [12].

Much research has been carried out for FDD, through the application of mathematical algorithms and signal models. The first investigations were carried out at the beginning of the 70s, based on the design of observers and the use of filters. Later, the principle of analytical redundancy was developed. Subsequently, for the first time, a specialized monitoring system was implemented for the detection and diagnosis of failures in a nuclear plant. In recent years, promising techniques have been investigated and applied, such as parity space, fuzzy logic, and neural networks, among others.

Next, we will briefly address how a robust computer system influences the FDD processes within a bioprocess plant.

A real-time computer system is one that must comply with specific time restrictions both at the beginning and at the end of activities, which implies associating time-dependent values with the latter. One of the main requirements of real time is to respect the individual temporal requirements of each task or process, in addition to ensuring adequate processing results. This tool is very useful for carrying out online monitoring of process variables and relieving the user of the tiring activity of manually detecting failures since bioprocesses are influenced by a large number of parameters given by complex dynamics at the cellular level. It is necessary to emphasize that although the FDD system is responsible for monitoring process variables, graphing them, and triggering alarm signals, the human being will always be left free to choose in decision-making [13].

As a summary, Figure 7.4 presents a supervision system (logic to be followed by a computer system in real time). It includes the detection and diagnosis of faults as well as the reconfiguration of parameters to correct faults presented online and avoid stopping the process.

7.1.3 Embedded Systems in Bioprocess Control

Integrated intelligence is key to improving system performance in terms of functionality, security, energy efficiency, environmental impacts, and costs. The use of virtual sensors in the industry manages to optimize the operational performance, safety, functionality, and reliability of the bioprocess. This is because they can estimate a quantity that is difficult to measure either due to costs or the nonavailability of measuring instruments, using mathematical models along with lower-cost physical sensors.

Some might argue that all the data needed to provide optimal monitoring, diagnosis, and control could be provided more reliably and robustly

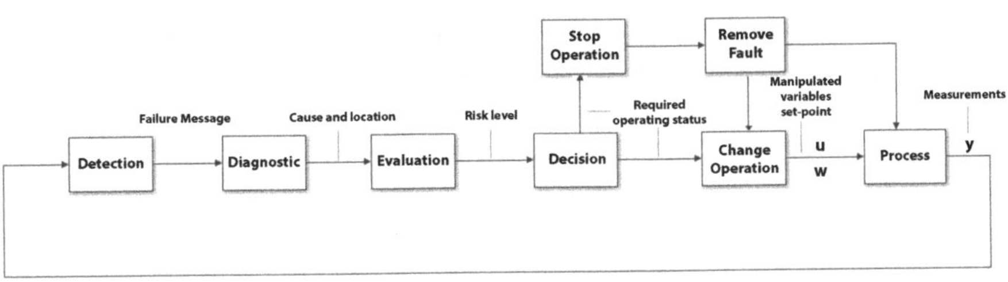

Figure 7.4 Supervision system in the event of failures.

using physical sensors. However, virtual sensors have several advantages in addition to lower cost. For example, virtual sensors could be more easily added as retrofits in a number of important applications, such as biomass or metabolite measurement [14].

An embedded system is any device that is made up of a programmable computer (microprocessor or microcontroller); however, its main function is not computational. The word "embedded" means that it is part of a larger whole, hidden from view within the general system. It has three important characteristics:

- Computer hardware: Its physical components differ from those that make up a general-purpose system such as a computer; it is related to the software and the specific application in question.
- Embedded software: It makes up the main application and carries out one or more particular tasks.
- Operating system: Allows applications to be monitored and provides mechanisms for the execution of processes; in many cases, it is required to operate in real time.

For the use of any embedded system, it is necessary to analyze its advantages and disadvantages in a general way and how these affect the applications where they are applied. This analysis will then be carried out.

Advantages:

- Low production cost.
- Easy maintenance as it is not a compact system. In case of breakdowns, the affected component is easily replaced without completely replacing the entire system.
- They have a small size, so they easily adapt to any industrial application without requiring a large workspace.
- It is adaptable. Any module that is necessary can be integrated. In addition, it adapts to any desired peripheral as it is modular.
- They consume a minimum of energy, which causes the life of the batteries to be extended.
- They do not require the additional implementation of cooling elements.
- They have high performance in processing data at high speed and in real time. Many of these react to eternal stimuli

or environmental changes, performing calculations quickly since fixed-time results must be obtained in the event of any anomaly or alarm condition [14]

Disadvantages:

- They are specific operating systems. It generally executes a specific program repetitively. The opposite of a desktop system that runs multiple tasks. However, there may be exceptions where the embedded system program presents updates.
- The software of an embedded system presents some restrictions such as small amounts of memory (generally in the order of KB), limited processing capabilities (generally the speed of the processors does not exceed the order of MHz), and limits the consumption of instant energy whether in execution state or not.
- They may present cybersecurity risks. Because they feature weak encryption, data shared between two devices can be easily intercepted and decrypted. Because most of these do not present certificates recognized by large sectors of the industry, manufacturers can falsify them to deceive users [14].

Data acquisition systems play an important role in the bioprocess industry. In [15], a data acquisition system applied to a bioreactor to achieve alcoholic fermentation is characterized. This system monitors parameters such as O_2, CO_2, temperature, flow, humidity, and pH based on the sensors that will collect data and the controllers to correspond to the set points of the variables. The measured data have the potential to practice another improved software tool for the estimation of biomass growth in solid-state fermentation and the determination of some kinetic parameters from process data.

Why Control and Monitor?
The measurement of variables in processes is a necessary task:

1. Know internal behaviors.
2. Fault diagnosis.
3. Monitoring.
4. Visualization of critical variables.
5. Biological processes subject to disturbances.

6. Nonlinear systems.
7. Unstable.
8. Maintain productivity.
9. Quality standards.

In addition:

Arduous, expensive, and difficult:

1. No trusted devices.
2. Time delays.
3. Errors in the measurement system.
4. High device costs.
5. Measurement conditions.
6. There are no sensors.

Using:

- Nonlinear, adaptive, predictive controllers, etc.
- Model-based state observers.

7.1.4 Virtual Sensors and State Estimators

There are excellent articles and sources in the literature on the state of the art and procedures for performing condition estimation techniques. Respectively, these have their advantages and disadvantages depending on their ability to handle model imbalances and measurement perturbations. According to their classification, state estimators can be as follows: Luenberger observers, finite-dimensional observers, Bayesian estimators, fault and disturbance detection observers, artificial intelligence–based observers, and hybrid observers [16]. Currently, although there is great information about state estimators and virtual sensors. Very little has been investigated in the implementation of these techniques in physical plants, specifically in those that make up bioprocesses [17].

One of the most important components in FDD systems is the correct measurement of process variables in real time [17]. This is an essential aspect since work is highly expensive and difficult to perform due to several factors such as the low availability of reliable equipment, time delays, errors in the measurement system, high costs of measurement devices, or that these are found in hostile environments within the plant. One of the alternatives used to develop the previous objectives is virtual sensors or soft sensors [18].

A virtual sensor is a mechanism that associates a physical or hardware sensor with an estimation algorithm that is integrated into monitoring, optimization, and control systems since it allows obtaining magnitudes in real time, such as data on the specific growth rate, the concentration of biomass, product, among other variables whose measurement may present economic or technical restrictions. Virtual sensors are classified in two ways: model-based or data-based [19]. The first group describes the physical and chemical generalities of the processes through mathematical models, which are based on the ideal steady states of the process, this being a difficulty in basing soft sensors on this type. On the other hand, the second group of soft sensors bases its measurement on the measured process data, describing the real conditions within the plant. This becomes a major advantage of these sensors compared to model-based ones [20].

The correct estimation of state variables requires two fundamental elements: a mathematical model of the process and some hardware measurements of some other signals, such as the concentrations of substrate or biomass existing in the bioprocess.

Virtualizing technology has seen rapid development over the past decade in several different domains, such as avionics, autonomous robots, telemedicine, traffic, automotive, nature, building monitoring, and control. In particular, virtual sensing has found widespread applications in bioprocess monitoring. Virtual sensors have been developed in other fields to obtain measurements indirectly in a cost-effective, noninvasive, and/or practical way. In the context of this article, virtual sensing is considered to include an indirect method of determining a quantity that uses results from other physical sensors along with models that describe the dynamics of the process [21].

7.1.5 Biohydrogen Production

The use of bioelectrochemical systems allows the development of systems for the production of renewable energy and simultaneously the treatment of wastewater and carbon fixation. In different investigations, microbial fuel cells have been worked on using pure cultures and consortia of microalgae for the production of electrical energy. Microalgae are unicellular photosynthetic organisms that can come from both marine and freshwater environments. In addition, they have the ability to hydrolyze water through oxidation-reduction through hydrolase enzymes. Green algae have a photosynthetic system similar to that of plants and have the ability to produce hydrogen using carbon dioxide, sunlight as an energy source, and water as an electron donor.

Currently, there is a wide variety of methods to obtain clean energy. These turn out to be highly expensive both in their implementation and in ensuring their correct functioning. A particular case is alcoholic fermentation through anaerobic digestion (AD) or natural fermentation with the absence of oxygen, where the biodegradable material (substrate) is transformed by microorganisms (biomass) into a mixture of gasses that constitute biogas. Thanks to the wide versatility of installation and the feasibility of obtaining organic waste, a material that is abundant in nature, the implementation of these systems is generally not expensive [22].

In recent years, studies have been carried out with different microorganisms that have the ability to produce H_2 from organic matter or from the oxidation of water catalyzed by enzymes such as nitrogenase and hydrogenase. Because the hydrogenase enzyme found in *Chlamydomonas reinhardtii* is sensitive to O_2, removal or separation of O_2 is critical to sustaining H_2 production. Methods have included the use of oxygen scavengers, reducers, and inert gas purges. Due to added costs and possible cell death, these measures may not be economically viable for H_2 production [23].

Hydrogen is generated through biochemical reactions, mainly in anaerobic fermentation processes such as biophotolysis of water using green algae and blue-green algae (cyanobacteria), direct biophotolysis and indirect biophotolysis, and photo-fermentation and dark fermentation. There are also hybrid systems composed of dark-fermentative and photo-fermentative bioreactors (semiconductor photocatalysts). The latter use bioelectrochemically assisted microbial bioreactors; therefore, hydrogen yields by pure or mixed cultures have been reported to be the average maximum hydrogen production [212.2 ml/(L·hour)], but it also had a stable hydrogen production under continuous culture conditions, which was 1.968 mol H_2/mol glucose [24].

Considering the theoretical high yields, several researchers have begun to explore approaches to increase hydrogen production. Mathematical modeling approaches could contribute to large-scale H_2 production, as several microorganisms can produce H_2 under moderate conditions from readily available renewable substrates, making operational strategies potentially competitive. The constant search for renewable energies entails the study of the generation, regulation, control, and distribution of energy. The information presented contributes to the control and generation of hydrogen as a clean fuel [25].

7.2 Methodology

7.2.1 Process Parameter Monitoring Methods

According to the literature, there are several methods to monitor critical quality attributes and process parameters: online, *in situ*, at-line, and offline (Figure 7.5) [26].

The online method is when a sample that was captured by the sensors is diverted from the bioreactor and then returned to it, giving continuity to the production process. It must be taken into account that the reactor is specifically designed to achieve this action. On the other hand, *in situ* monitoring is when the sensor captures data directly from inside the bioreactor, without intervening with the actual execution of the process. On the other hand, in at-line monitoring, samples are removed from the system and analyzed close to the production process. It is very similar to *in situ* monitoring, where correct sterilization of the sample measurement tools must be achieved. Unlike online monitoring, samples taken are not returned to the production line.

Finally, offline monitoring is when the samples extracted from the system are analyzed in a laboratory remote from the production line and pre-treatment methods are required for their subsequent analysis. Disadvantages of this method include manual intervention and time delay in obtaining analysis results.

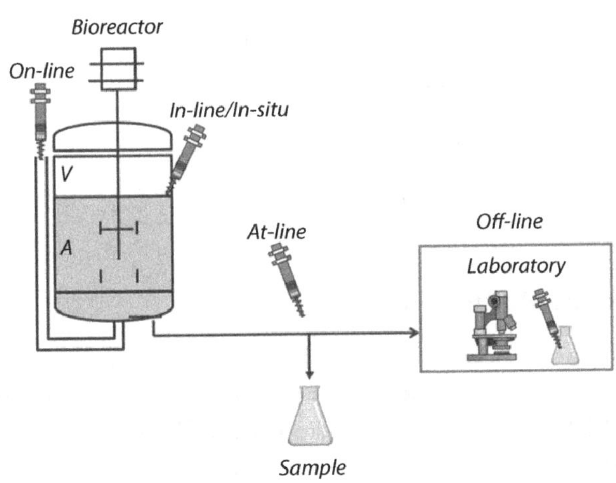

Figure 7.5 Data monitoring methods related to each other and to a laboratory pipeline.

Table 7.1 Regression values.

Hydrogen	Rs/Ro
200	9
300	5
400	3.5
500	2.8
600	2
700	1.8
800	1.5
900	1.3
1000	1
1500	0.6
2000	0.4
3000	0.2
4000	0.15
5000	0.1
6000	0.07
7000	0.05
8000	0.04
9000	0.03
10000	0.02

7.2.2 General Methodology

In general, the process can be defined in terms of three steps described in the following: (i) data collection and pre-processing, (ii) experimental bio-process, and (iii) sensor implementation and validation. In the first stage of data collection and preprocessing, the components and devices that guarantee the appropriate hardware for the application are taken into account. In this case, a data acquisition card and a hydrogen sensor are chosen as the measuring instrument for the process variable. Then, the physical connection of these elements is made and the voltage signal sent by the sensor is acquired when it detects the hydrogen present in the medium. The first stage concludes with the pre-processing of the signal, which is when it is conditioned either through filtering or amplification, among other techniques, to achieve better manipulation [27].

The sensor has a sensitivity that can be configured to measure different concentrations of gasses, including air, alcohol CO, CH_4, LPG, and H_2. The calibration of the sensor is according to the type of gas to be detected, it is necessary to say that the potential of the sensor (see Table 7.1). The sample shows how air is the element with better sensitivity than other gasses, the measurement is in ppm (parts per million) particles of the element. Based on this, it was decided to use an average calibration value for RL of 10 kΩ recommended by the manufacturer [28].

This graph (Figure 7.6) is used to calculate through exponential regression $y = ax^b$; using the built-in Microsoft Excel Table 7.1 analysis tool, we obtain the following regression expression and graph

The voltage signal delivered by the sensor is acquired (Figure 7.7). For this, the **NI USB 6008 OEM** data acquisition card from the manufacturer *National Instruments* is used. This device provides eight single-ended analog input (AI) channels, two analog output (AO) channels, 12 DIO channels, and a 32-bit counter with a full-speed USB interface. For this case, was used a USB cable from the manufacturer with a Manufacturer Part Number 184125-02 in order to connect the computer to the device properly. It is necessary to mention that in addition to the LabVIEW software, the *NI MyRio* module must be installed.

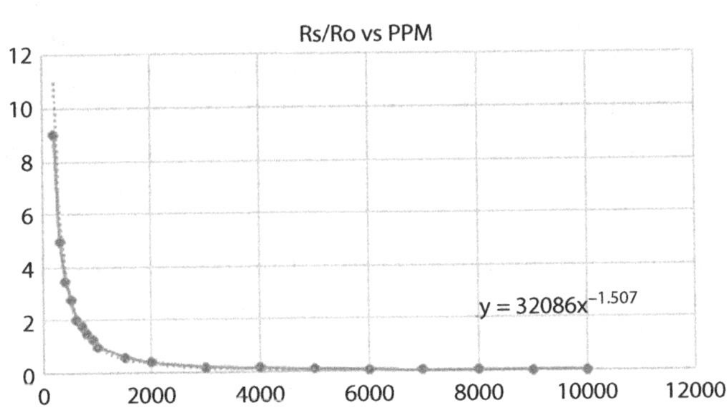

Figure 7.6 Regression expression and graph.

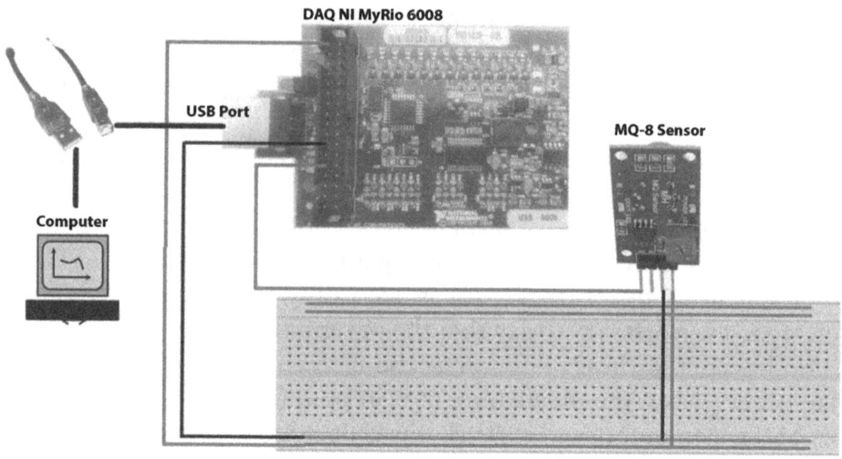

Figure 7.7 Connection diagram of the DAQ with the MQ-8 sensor and the computer.

Subsequently, using the LabVIEW software from the same manufacturer, the DAQ channel that will acquire the signal is configured according to the sensor parameters (Figure 7.8).

7.2.3 Sensor MQ-8

The MQ-8 analog gas sensor (Figure 7.9) is used in the detection of gas leaks from equipment in consumer and industrial markets, this sensor is appropriate for the detection of LP gas, butane, propane, methane, alcohol, and hydrogen. It has a high sensitivity, a fast response time, and this sensitivity can be adjusted by a potentiometer.

This small analog sensor detects the presence of combustible gas and smoke at concentrations of 10 to 1000 ppm. It incorporates a simple analog voltage interface that only requires one analog input pin of the microcontroller. With the 5-V connection on the power pin, the sensor remains hot enough for proper operation, delivering an output voltage depending on the hydrogen concentration it captures. The sensitivity of the detector is adjusted with a resistive load between the output and ground pins.

The sensor is composed of a ceramic micro tube (Al_2O_3), a sensitive layer of tin dioxide (SnO_2), and the measuring electrode, and the heater is fixed in a shell made of plastic and stainless steel net. The heater runs the required working conditions for the work of sensitive components.

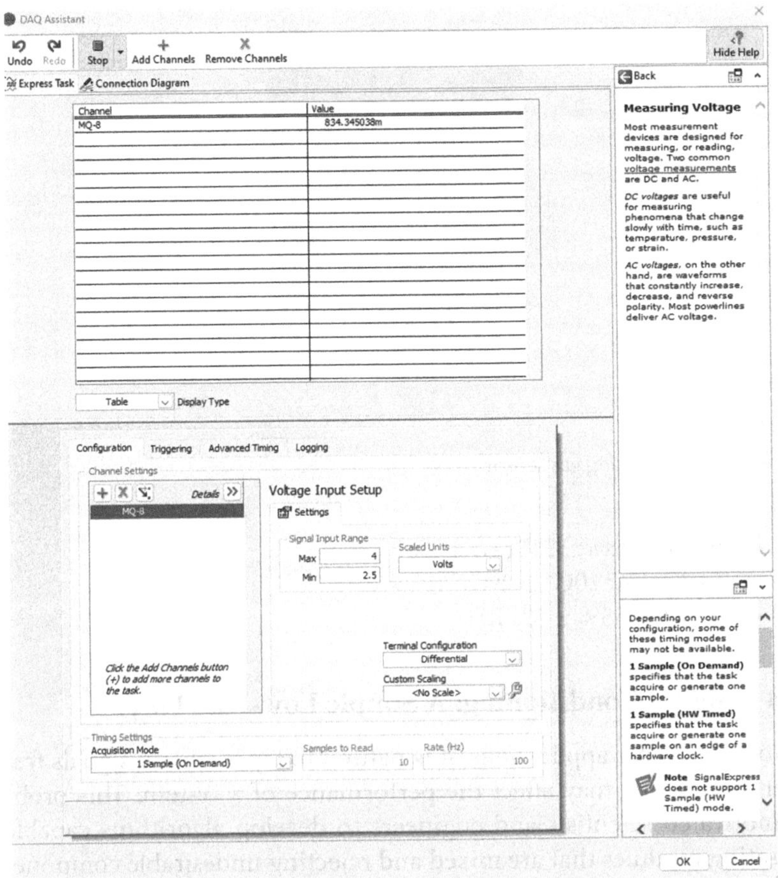

Figure 7.8 DAQ AI 1 channel configuration for signal acquisition.

The MQ-8 housing has six pins, four of them are used to collect signals, and another two are used to provide a heating current.

MQ-8 Technical Data:

- *Sensor type*: Semiconductor
- *Detection range:* 100~1000 ppm (H_2 gas)
- *Circuit voltage:* V_c = 5 V± 0.1 V
- *Load resistance* R_L: Adjustable
- *Output voltage:* Vs = 2.5~4 V

Figure 7.9 MQ-8 analog gas sensor.

- *Heater resistance:* 33Ω (+/−) 5%
- *Temperature:* 20C° ± 2°C
- *Humidity:* 55% ± 5% RH

7.2.4 Signal Conditioning: A Simple Low-Pass Filter

In most electronic applications, it is natural to try to obtain signals free of uncertainties that may affect the performance of a system. This problem has motivated scientists and engineers to develop algorithms capable of separating modules that are mixed and rejecting undesirable components.

Figure 7.10 shows the connection diagram that describes the programming of the process using the graphical language of the NI LabVIEW

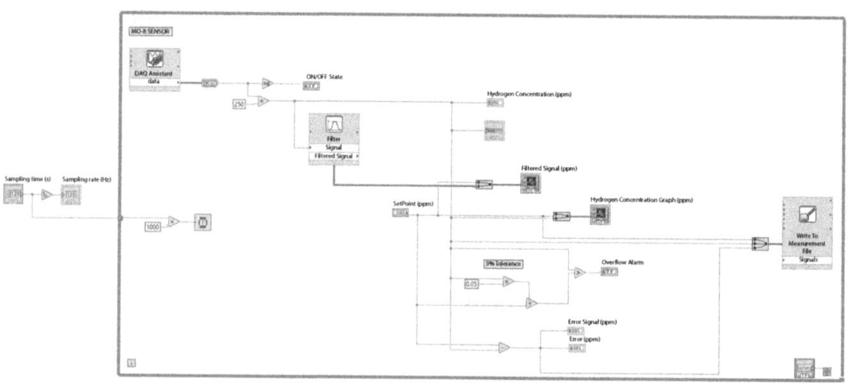

Figure 7.10 Implementation of the low-pass filter in real time.

platform. The communication of the controls, indicators, and graphics of the process with the signal acquired through the DAQ through analog input channel 1 (AI 1) is observed. Finally, to filter the output signal of the MQ-8 hydrogen sensor in real time, a first-order low-pass filter is used. In this type of filter, the desired cutoff frequency is set at 0.1 Hz; allowing the passage of signals with a lower cutoff frequency and attenuating signals with frequencies higher than this.

7.3 Results

The calibration is the process of comparing the values obtained by a determining instrument with the corresponding measurement of a reference standard (or standard). According to the International Bureau of Weights and Measures, calibration is an operation under specific conditions that presents two stages. In the first, a relationship is established between the values and measurement uncertainties provided by standards and corresponding indications with the associated measurement uncertainties. As a second step, use this information to establish a relationship and obtain a measurement result from a prompt.

From this definition, it can be deduced that to calibrate an instrument or a standard it is required to have one of greater precision (standard) that provides the conventionally verifiable value. This value will be used to compare it with the indication of the instrument being calibrated. This is done through an uninterrupted and fully documented chain of comparisons until reaching the primary standard, which constitutes what is known as traceability. The objective of calibration is to maintain and verify the proper functioning of the equipment, respond to the requirements established in the quality standards, and guarantee the reliability and traceability of the measurements.

However, the sensor to be calibrated is an experimental sensor based on trial and error. Therefore, the verification will be carried out through two methods, the first based on the controlled chemical reaction. We can observe that the bioreactor has the greatest presence of hydrogen production after 24 hours of rest. Calibration of the sensor is proposed using biomass bioreactors (Figure 7.11), which produce controlled amounts of hydrogen. It is possible to predict the quantities of hydrogen produced for quantities and hours of product concentration. As a second step, the sensor signal is acquired using the data acquisition card. Once the data are acquired, the processing stage is carried out using a low-pass filter to reduce the noise present at high frequencies. Finally, the process is monitored through the

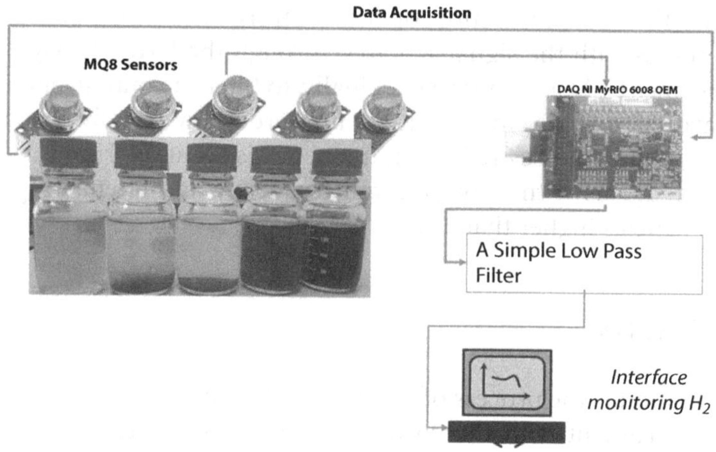

Figure 7.11 Study of the experimental validation of the sensor in the bioprocess.

computer, using alarms, graphs that describe the trend of the variable over time, and a user-friendly HMI interface.

7.3.1 Acquisition and Filtering of Signal from the MQ-8 Sensor Under Ambient Conditions

A human-machine interface (Figure 7.12) was designed in the LabVIEW platform to show a user-friendly visualization of the industrial process and thus facilitate real-time monitoring of the process. The designed interface has three fundamental parts. The first is to fill in process data, where the hydrogen concentration value in the bioreactor and its respective error with respect to the set point are shown. In addition, the sampling period in which the system will operate must also be indicated, and based on this, the sampling frequency of the process is defined. On the other hand, the process is monitored using an alarm to indicate signal overshoot with respect to the reference, having a tolerance of 5% and an LED that indicates whether the sensor is delivering voltage to its output. In the second part, there is the emergency stop button and a small display of the hydrogen concentration.

Finally, the process graphs are observed (Figure 7.13) that show the representation of the hydrogen concentration with the reference, the output of the low-pass filter with the signal already filtered in real time, and the error

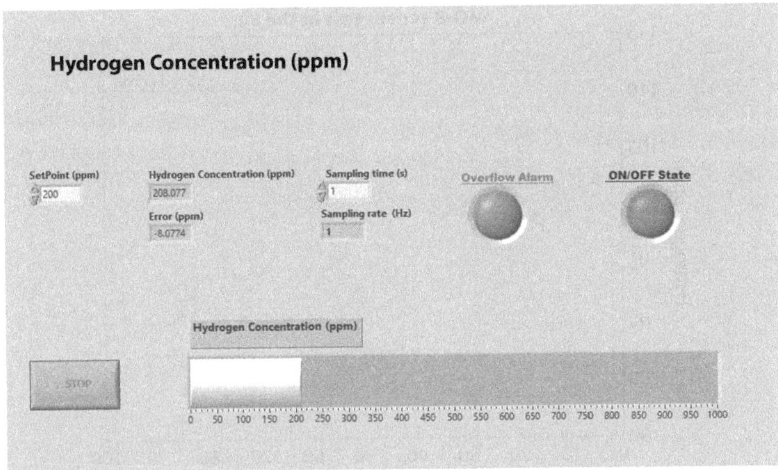

Figure 7.12 Human-machine interface.

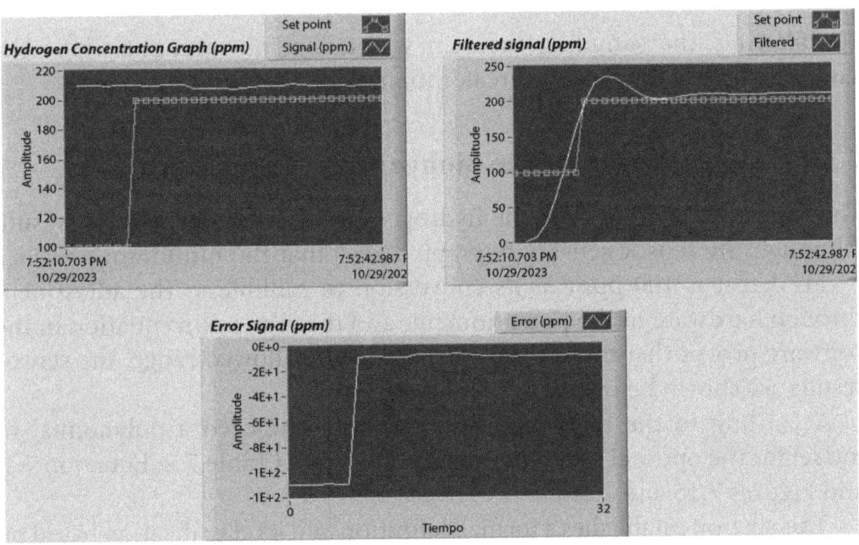

Figure 7.13 Process graphics.

signal between the real value measured by the sensor and the set point imposed by the user.

When testing the MQ-8 sensor in the air (ambient conditions) and after processing and filtering the signal with a low-pass filter with a cutoff frequency of 0.1 Hz, the response is observed as shown in Figure 7.14.

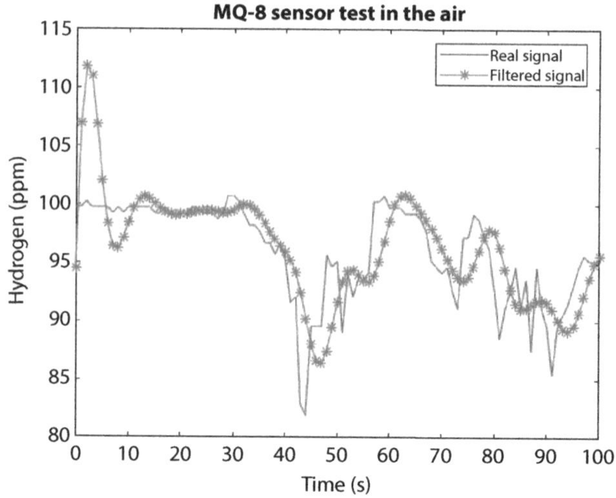

Figure 7.14 Real and filtered hydrogen signal (ppm) under ambient conditions.

For filtering, the sampling theorem was taken into account where the sampling frequency must be greater than twice the cutoff frequency.

7.3.2 MQ-8 Sensor for the Biohydrogen Validation

The algorithm used to read the hydrogen produced had a positive result. Although the sensor sensitivity graph shows that the minimum presence it can detect is 100 ppm or its conversion to milliliters. The adjustment through hardware in the potentiometer and the regression equation in the software proves that the measurement is in the allowed range, the sensor results are shown below:

According to the regression analysis, was proposed a polynomial fit model for the optimal conditions and predictions (Table 7.2, Equation 7.1, and Figures 7.15 and 7.16).

This section establishes a formal definition and its classification based on the multiple regression of hydrogen production. R-Square: 0.97791, Prob> |t| <0.0001. It is well known that the response of a system can be characterized by polynomial n-order (in the case of fourth order) as follows:

$$¥ = \alpha + \beta_1\lambda + \beta_2\lambda^2 + \beta_3\lambda^3 + \beta_4\lambda^4 \quad (7.1)$$

Table 7.2 Multiple regression R-square = 0.99571.

Parameter	Value	Error
α	1.15663	0.13686
β_1	1.27604	0.08667
β_2	−0.0424	0.00359
β_3	4.8248E−4	4.86599E−5
-	−1.86001E−6	2.14599E−7

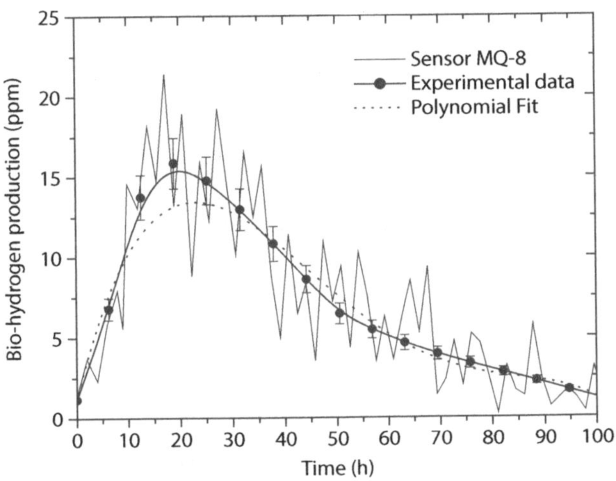

Figure 7.15 MQ-8 sensor output, experimental data, and polynomial fit.

where
$¥ \in \Re^n : (H_2 ppm)$, represents the state variable of the system.

Next, the signal is filtered using an averaging filter that takes three samples for each filtering iteration (Figure 7.17). This process is done offline using MATLAB software. The data collected from the experiment is processed and saved in real time in an Excel file from the LabVIEW platform.

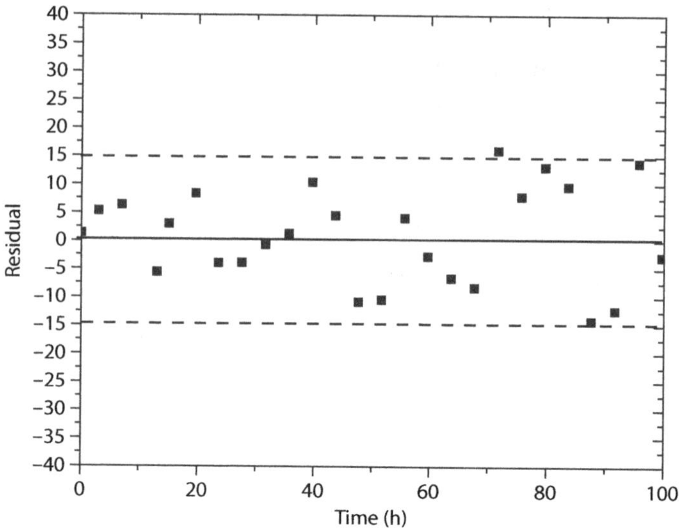

Figure 7.16 Residual values from experimental data and signal.

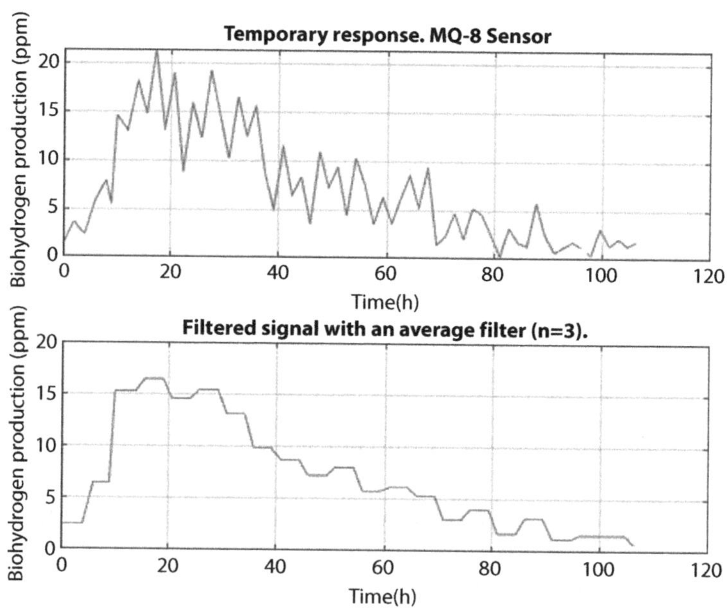

Figure 7.17 Average filter. Hydrogen signal obtained from MQ-8 (ppm).

A generalized sensor for measurements of hydrogen has previously been used to model pairwise interactions considering key parameters in allowing periodic behaviors. The existence of chaos in dynamic behavior and the chaotic parameter space can be identified using various optimization techniques to be controlled at the industrial level. In some bioreactors, it has been observed that the oscillating mode of operation may be due to the manipulation of a parameter, such as the dilution rate, temperature, inhibitions, and potential of hydrogen; controllers that stabilize self-oscillating bioreactors have been designed by manipulating the input of these parameters.

However, the design of operational policies began from open-loop approaches performed by expert plant operators, followed by closed-loop operations that were improved by applying the linear controllers; this kind of control law has been one of the most successful advances of the control theory. Therefore, these policies force industrial equipment to operate in highly demanding conditions.

The lack of real-time data is another important issue, and the new observer-based controller designs, where the separation principle for the satisfactory tuning of nonlinear observers coupled with nonlinear controllers, is an important research topic. The above are examples of the complexity of the real-time implementation of nonlinear techniques in practical applications. These contributions in the control area will only be possible if there are suitable sensors that can be fed to the state estimators in real time to close the control loop in real time with a view to optimizing *n* sub-loops integrated into a control automation approach [29–38].

7.4 Conclusions

This chapter presented a way to obtain, with a degree of error, the amount of hydrogen produced by bioreactors. As a form of calibration, the theoretical production ranges of parts per million of hydrogen particles generated, in contrast to the measurements obtained experimentally, prove that although the sensor is not designed for very small quantities, it is possible to carry out the application at an industrial level. The calibration-validation methodology and implementation are presented and the (85%) accuracy and precision are evaluated. The commercial online sensor proposal may be a suitable solution for online biohydrogen concentration measurement. Future mechanisms will consider the use of the MQ-8 sensor in a lab-scale process to an industrial scale [39, 40].

References

1. Pérez, P.A.L., López, R.A., Femat, R., Introduction. Part I: Overview of the Control and Monitoring of Bioprocesses and Mathematical Preliminaries, in: *Control in Bioprocessing, Control in Bioprocessing: Modeling, Estimation and the Use of Soft Sensors*, P.A. Pérez, R.A. López, R. Femat (Eds.), John Wiley & Sons Ltd., Chichester, UK, 2020.
2. Busse, C., Biechele, P., de Vries, I., Reardon, K., Solle, D., Scheper, T., Sensors for disposable bioreactors. *Eng. Life Sci.*, 17, 940–952, 2017.
3. Barragán, A., Silva, M., Olmos, S., Polanco, M., Biotechnology and Bioprocesses: Their Contribution to Sustainability. *Processes*, 8, 436, 2020.
4. Diaz, M., Chaccara, V., Aquije, G., Atoche, R., Villanueva, V., Gamarra, C., Samanamud, O., Supervision, control, and data acquisition system of a heat exchanger. *Indones. J. Electr. Eng. Comput. Sci.*, 28, 155, 2022.
5. Daniyan, I.A., Daniyan, L., Adeodu, A., Uchegbu, I., Levels of Automation, in: *Principles of Automation and Control*, I.A. Daniyan (Ed.), pp. 50–59, Bentham Science Publishers Pte. Ltd., Singapore, 2023.
6. Rösner, L.S., Walter, F., Ude, C., Beutel, S., Sensors and Techniques for On-Line Determination of Cell Viability in Bioprocess Monitoring. *Bioengineering*, 9, 12, 762, 2022.
7. Rahul, S., Chitra, R., Kumar, V., Abhishek, P., Reddy, B., Virtual Instrumentation Based Graphical User Interface for Fermentation Bioprocess Monitoring Using LabVIEW, in: *Innovations in Mechanical Engineering*, G.S.V.L. Narasimham, A.V. Babu, S.S. Reddy, R. Dhanasekaran, (Eds.), Springer, Singapore, 2022.
8. Reyes, S.J., Durocher, Y., Pham, P.L., Henry, O., Modern Sensor Tools and Techniques for Monitoring, Controlling, and Improving Cell Culture Processes. *Processes, 10*, 189, 2022.
9. Chimmiri, V. and Karri, R., Optimal state and parameter estimation for fault detection and diagnosis of a nonlinear batch beer fermentation process, in: *Optimal State Estimation for Process Monitoring, Fault Diagnosis and Control*, G.S.V.L. Narasimham, A.V. Babu, S.S. Reddy, R. Dhanasekaran (Eds.), Springer, Singapore, 2022.
10. Yang, T., Hongbo, S., Bing, S., Shuai, T., Parallel supervised additive and multiplicative faults detection for nonlinear process. *J. Franklin Inst.*, 356, 11716–11740, 2019.
11. Luo, L., Lovelett, R.J., Ogunnaike, B.A., Hierarchical monitoring of industrial processes for fault detection, fault grade evaluation, and fault diagnosis. *AIChE J.*, 63, 2781–2795, 2017.
12. Wang, B., Sun, Z., Jiang, X., Zeng, J., Liu, R., Kalman Filter and Its Application in Data Assimilation. *Atmosphere*, 14, 8, 1319, 2023.
13. Akande1, S., Adeyemi, W., Aremu, F., Automated Processes and Systems, in: *Principles of Automation and Control*, I.A. Daniyan (Ed.), pp. 20–49, Bentham Science Publishers Pte. Ltd., Singapore, 2023.

14. Agrawal, R. and Badhoutiya, A., Learning of Embedded System for Incorporating: Organization, in: *Advances in Industrial and Production Engineering*, R.K. Phanden, R. Kumar, P.M. Pandey, A. Chakraborty (Eds.), Springer, Singapore, 2023.
15. Soccol, C.R., Spier, M.R., Porto de Souza, L., Pedroni, A.B., Junior, L.A., Sturm, W., Data Acquisition Systems in Bioprocesses, in: *Data Acquisition Applications*, Z. Karakehayov (Ed.), pp. 1–28, InTech, Brazil, 2012.
16. Pérez, P.A., López, R.A., Femat, R., Development of Virtual Sensor Based on the Just-In-Time Model for Monitoring of Biological Control Systems. Part III Software Sensors and Observer-Based Control Schemes for Bioprocess, in: *Control in Bioprocessing: Modeling, Estimation and the Use of Soft Sensors*, P.A. Pérez, R.A. López, R. Femat (Eds.), John Wiley & Sons Ltd., Chichester, UK, 2020.
17. López, R.A., Camacho, B.R., Neria, R., Santos, O., López, P.A., State Estimation Based on Nonlinear Observer for Hydrogen Production in a Photocatalytic Anaerobic Bioreactor. *Int. J. Chem. Reactor Eng.*, 15, 5, 20170004, 2017.
18. Hernández, A., Vallejo, E., Martínez, F., Pelayo, J.J., Hernández, L.A., Pescador, J.A., Tamayo, L., Morales, A., López, P.A., Cortes, E.R., Changes to the Dissociation Barrier of H_2 Due to Buckling Induced by a Chemisorbed Hydrogen on a Doped Graphene Surface. *J. Mol. Model.*, 24, 244, 2018.
19. Pérez, P.A.L., López, R.A., Femat, R., Regulation of a Continuously Stirred Bioreactor via Modeling Error Compensation. Part III Software Sensors and Observer-Based Control Schemes for Bioprocess, in: *Control in Bioprocessing: Modeling, Estimation and the Use of Soft Sensors*, P.A. Pérez, R.A. López, R. Femat (Eds.), John Wiley & Sons Ltd., Chichester, UK, 2020.
20. Pérez, P.A.L., López, R.A., Femat, R., Virtual Sensor Design for State Estimation in a Photocatalytic Bioreactor for Hydrogen Production. Part III Software Sensors and Observer-Based Control Schemes for Bioprocess, in: *Control in Bioprocessing: Modeling, Estimation and the Use of Soft Sensors*, P.A. Pérez, R.A. López, R. Femat (Eds.), John Wiley & Sons Ltd., Chichester, UK, 2020.
21. Pérez, P.A.L., López, R.A., Femat, R., Production System Myco-Diesel for Implementation of "Quality" of the Observability. Part III Software Sensors and Observer-Based Control Schemes for Bioprocess, in: *Control in Bioprocessing. Control in Bioprocessing: Modeling, Estimation and the Use of Soft Sensors*, P.A. Pérez, R.A. López, R. Femat (Eds.), John Wiley & Sons Ltd., Chichester, UK, 2020.
22. Havlik, I., Beutel, S., Scheper, T., Reardon, K.F., On-Line Monitoring of Biological Parameters in Microalgal Bioprocesses Using Optical Methods. *Energies*, *15*, 875, 2022.
23. López, P.A.L., Cuervo, J.A., Robles, V.J., Rodríguez, G.D., España, V.H., Romero, T., Development of a Novel Kinetic Model for Cocoa Fermentation Applying the Evolutionary Optimization Approach. *Int. J. Food Eng.*, 14, 5–6, 2018.

24. López, P.A.L., Neria, R., Aguilar, L., Increasing the Bio-Hydrogen Production in a Continuous Bioreactor via Nonlinear Feedback Controller. *Int. J. Hydrogen Energy*, 40, 48, 17224–30, 2015.
25. Yardley, J.E., Kell, D.B., Barrett, J., Davey, C.L., On-line, real-time measurements of cellular biomass using dielectric spectroscopy. *Biotechnol. Genet. Eng. Rev.*, 17, 3–35, 2000.
26. Biopharma PAT, Quality Attributes, Critical Process Parameters & Key Performance Indicators at the Bioreactor, 2018, https://processanalytics.hamiltoncompany.com/hubfs/00%20-%20Process%20Analytics/Documents%20(PA)/PDFs/Hamilton_WhitePaper_Biopharm_PAT.pdf.
27. Guadarrama, O., Hernández, J., García, L., Gutiérrez, T., Estrada, E.B., Simultaneous Bio-Electricity and Bio-Hydrogen Production in a Continuous Flow Single Microbial Electrochemical Reactor. *Environ. Prog. Sustain. Energy*, Wiley Online Library, 1–5, 2018.
28. Pérez, P.A.L., López, R.A., Femat, R., State Estimation and Observers. Part II: Observability and Control Concepts, in: *Control in Bioprocessing: Modeling, Estimation and the Use of Soft Sensors*, P.A. Pérez, R.A. López, R. Femat (Eds.), John Wiley & Sons Ltd, Chichester, UK, 2020.
29. López-Pérez, P., López-López, M., Núñez-Colín, C., Mukhtar, H., Aguilar-López, R., Peña-Caballero, V., A novel nonlinear sliding mode observer to estimate biomass for lactic acid production. *Chem. Prod. Process Model.*, 18, 4, 565–580, 2023.
30. Vallejo, E. and López–Pérez, P.A., Strong chemical adsorption of CO2 and N2 on a five-vacancy graphene surface. *Solid State Commun.*, 356, 114934, 2022.
31. Flores, G., Rodriguez-Mata, A., Amabilis-Sosa, L., Gonzalez-Huitron, V., Hernández-González, O., Lopéz-Peréz, P., A turbidity sensor development based on NL-PI observers: Experimental application to the control of a Sinaloa's River Spirulina maxima cultivation. *Open Chem.*, 18, 1, 1349–1361, 2020.
32. Rodriguez-Mata, A.E., Bustos-Terrones, Y., Gonzalez-Huitrón, V., Lopéz-Peréz, P.A., Hernández-González, O., Amabilis-Sosa, L.E., A Fractional High-Gain Nonlinear Observer Design—Application for Rivers Environmental Monitoring Model. *Math. Comput. Appl.*, 25, 44, 2020.
33. López Pérez, P., Neria González, M., Aguilar López, R., Cadmium concentration stabilization in a continuous sulfate reducing bioreactor via sulfide concentration control. *Chem. Pap.*, *67*, 3, 326–335, 2013.
34. López-Pérez, P.A., Alvarado-Santos, E., Hernández Melchor, D.J., Martınez-Farıas, F.J., Jiménez Muñoz, E., Chapter 4. A Generic Mathematical Model for Pest Control with Time Delay Dynamics Based on Silico Approaches, in: *Biology, Behavior and Ecology*, pp. 95-128, Nova, Science and Technology, USA, 2023.
35. Amabilis-Sosa, L., Rodríguez-Mata, A., Baray-Arana, R., Robles-Vega, I., Gonzalez-Huitrón, V., López-Pérez, P., Robust fractional control based on

high gain observers design (RNFC) for a Spirulina maxima culture interfaced with an advanced oxidation process. *Open Chem.*, 21, 1, 20220214, 2023.

36. Hernández-Melchor, D.J., Camacho-Pérez, B., Ríos-Leal, E., Alarcón-Bonilla, J., López-Pérez, P.A., Modelling and multi-objective optimization for simulation of hydrogen production using a photosynthetic consortium. *Int. J. Chem. Reactor Eng.*, *18*, 7, 20200019, 2020.
37. Rodríguez-Mata, A.E., Gómez-Vidal, E., Lucho-Constantino, C.A., Medrano-Hermosillo, J.A., Baray-Arana, R., López-Pérez, P.A., State Estimation in a Biodigester via Nonlinear Logistic Observer: Theoretical and Simulation Approach. *Processes*, *11*, 4, 1234, 2023.
38. Alvarado-Santos, E., Mata-Machuca, J.L., López-Pérez, P.A., Garrido-Moctezuma, R.A., Pérez-Guevara, F., Aguilar-López, R., Comparative Analysis of a Family of Sliding Mode Observers under Real-Time Conditions for the Monitoring in the Bioethanol Production. *Fermentation*, 8, 9, 446, 2022.
39. Hernández Melchor, D.J., López, P.A., Carrillo, S., Alberto, A., González, E., Camacho, B., Experimental and Kinetic Study for Lead Removal via Photosynthetic Consortia Using Genetic Algorithms to Parameter Estimation. *Environ. Sci. Pollut. Res.*, 25, 21286–95, 2018.
40. Pérez, P.A.L., López, R.A., Femat, R., Observability Analysis Applied to 2D and 3D Bioreactors with Inhibitory and Non-inhibitory Kinetics Models. Part III Software Sensors and Observer-Based Control Schemes for Bioprocess, in: *Control in Bioprocessing. Control in Bioprocessing: Modeling, Estimation and the Use of Soft Sensors*, P.A. Pérez, R.A. López, R. Femat (Eds.), John Wiley & Sons Ltd., Chichester, UK, 2020.

Part 3

GREEN HYDROGEN GENERATION

8

Green Hydrogen Production by PEM Water Electrolysis

Sergey A. Grigoriev[1,2,3,4]

[1]National Research University "Moscow Power Engineering Institute", Moscow, Russia
[2]National Research Center "Kurchatov Institute", Moscow, Russia
[3]HySA Infrastructure Center of Competence, Faculty of Engineering, North-West University, Potchefstroom, South Africa
[4]A.N. Nesmeyanov Institute of Organoelement Compounds, Russian Academy of Sciences, Moscow, Russia

Abstract

Proton exchange membrane (PEM) water electrolysis is a key mature technology of green hydrogen production, which has been used in various applications over many decades. PEM electrolysis systems demonstrate high efficiency and durability, fast start-up (including at subzero temperatures) and shutdown, are able to operate in a wide load range (5% to 100%), at increased pressure (up to several hundred bar), and are successfully integrated with power plants based on renewable energy sources with power consumptions up to several megawatts. Green hydrogen produced using PEM electrolyzers has been successfully implemented in a number of industrial-scale processes: food, glass, steel, and aluminum manufacturing; agricultural chemical production; grid stabilization; and so on. PEM water electrolysis systems fully meet the paradigms of decarbonization, digitalization, and decentralization of energy.

***Keywords*:** Green hydrogen production, water electrolysis, proton exchange membrane, polymer electrolyte membrane

Email: sergey.grigoriev@outlook.com

Mehmet Sankir and Nurdan Demirci Sankir (eds.) Towards Green Hydrogen Generation, (239–266)

Introduction

According to the existing terminology, green hydrogen is produced from water using renewable energy. It may be produced through the different methods being developed on the lab scale and having a low technology readiness level (TRL), but only a few technologies are commercialized [1, 2]. The key mature technology for the production of green hydrogen is proton exchange membrane (PEM) water electrolysis, which is already implemented in the industry with power consumptions up to several MW (in particular, within the framework of power-to-gas solution).

Due to several advantages, PEM water electrolysis stands out among other hydrogen production technologies [1]. PEM electrolysis systems are characterized by a minimum start-up (including at subzero temperatures) and shutdown time, the ability to operate in a wide load range (5% to 100%), the possibility of operation at increased pressure (up to several hundred bar), and are successfully integrated with power plants based on renewable energy sources. When considering the advantages of PEM water electrolyzers over other types of electrolyzers, it is necessary to mention the following facts. The first fact is the possibility of PEM water electrolysis cell operation at voltages below the thermoneutral voltage (≈1.48 V) [3], which, in addition to electricity, also requires heat supply. Although the current densities under this endothermic regime are low, it can be applied within the so-called trigeneration concept to produce cold directly within the electrolysis cell. The second significant fact is the achievement of operating current densities up to 18 to 20 A/cm^2 [4], which is a kind of record in the field of water electrolysis. These facts demonstrate the possibility of PEM electrolyzer operation in a wide range of loads, including endothermic mode. The third fact to be mentioned is the achievement of a working pressure of up to 875 bar in a PEM electrochemical single cell in a single stage. Such pressures, in particular, were implemented in PEM cells by the HyET Hydrogen company [5]. Therefore, current electrochemical technologies allow the refueling of high-pressure hydrogen (and oxygen) storage systems without the use of additional compressors. For example, it may potentially be possible to use PEM water electrolyzers at hydrogen refueling stations for charging zero-emission vehicles with both 350- and 700-bar storage tanks. The direct charging of conventional gas cylinders (pressure, 150 or 200 bar) from the PEM electrolysis system (without additional compressors) has been demonstrated already.

As for the creation of large-scale PEM electrolysis systems, the most sizable Silyzer 300 systems with a hydrogen capacity of up to 2000 kg/h (22,270 m^3/h) have been created by Siemens AG for the so-called Power-to-X technologies (X = gas, heat, mobility, etc.), which converts electrical energy derived from renewable sources into chemical energy of liquid or gaseous fuels, including hydrogen, without carbon footprint. Finally, due to its advantages, PEM-based hydrogen electrochemical systems and, in particular, water electrolyzers fit well into the so-called Internet of Energy concept [6].

Historical Milestones and State-of-the-Art

The electrolysis of water is a mature technology. The principles of water electrolysis were discovered more than two centuries ago. The history of electrolysis dates back to 1789 when it was first reported by two Dutchmen, namely, Adriaan Paets van Troostwijk and Johan Rudolph Deiman [7, 8]. Later, electrolytic splitting of water using woolen cloth soaked in alkali was observed by William Nicholson and Anthony Carlisle in England in 1800 [7]. These first electrolyzers were of the alkaline type.

The first PEM water electrolyzers were developed in the 1960s by General Electric Co., as part of the U.S. space program. It was associated with the invention of a sulfonated polystyrene ion-exchange electrolyte membrane by W. Thomas Grubb in the early 1950s [9]. Today, PEM water electrolyzers continue to be used in space. In particular, since 2006, they have been using the U.S. orbital segment of the International Space Station. This oxygen generation system employs a 28-cell operating at a current density of *ca.* 0.22 A/cm^2 to produce oxygen at ambient pressure with 80% efficiency [10].

The most widely used perfluorosulfonic acid and commercially available membranes are sold under the brand name of Nafion®, some alternative membrane materials are available on the market as well (Figure 8.1). Today, the global manufacturer of Nafion® membranes is Chemours, which spun off (sold) from DuPont (previously produced these membranes). The process of manufacturing polymer electrolyte membranes is quite complex and consists of about 30 steps. In particular, the process chain includes the synthesis of starting monomers (tetrafluoroethylene and alkyl vinyl ether). Usually, the synthesis results in membranes in protonated (SO_3–H^+) or sodium form (SO_3–Na^+), the proton in the Nafion® membrane is quite easily replaced by other cations (Li^+, K^+, Mg^{2+}, etc.). The type of membrane formation may be extrusion or casting.

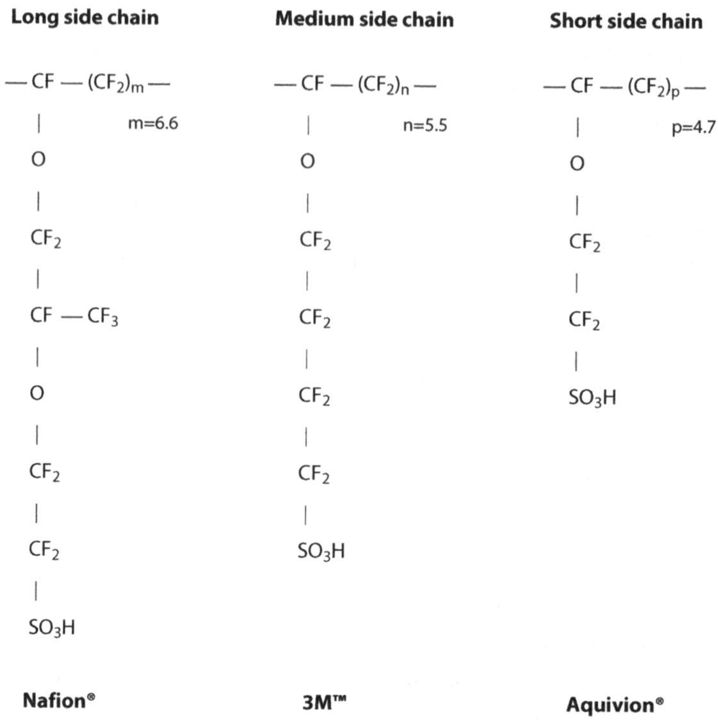

Figure 8.1 Chemical structures of poly(perfluorosulfonic acid) membranes.

Despite the extensive use of the main types of membrane materials shown in Figure 8.1, alternative membranes are being developed. Several strategies such as cross-linking, hybridization, doping by nanoparticles, and reinforcement could be employed in order to additionally improve membrane stability, proton conductivity, and mechanical strength. For example, incorporating metal-organic frameworks (MOFs) with protic functional groups into PEM can effectively improve its proton conductivity [11].

Today, commercially available PEM water electrolysis stacks and systems are offered by Elogen/ArevaH2Gen and McPhy Energy (France); Siemens, Hoeller Electrolyzer, and H-TEC Systems (Germany); ITM Power (UK); Cummins/Hydrogenics (USA/Canada); Nel Hydrogen/Proton Onsite (Norway/USA); Giner, Plug Power, and Electric Hydrogen (USA); Ohmium International (USA/India); Toshiba and Kobelco Eco-Solutions (Japan); Elchemtech (Korea); PERIC Hydrogen Technologies/718th Research Institute of CSIC, Suzhou Jingli, and Hele Titanium Technology (China); and other suppliers. The most powerful megawatt class Silyzer

Figure 8.2 PEM-based Silyzer 300 electrolysis system (Siemens Energy) with hydrogen productivity of 2000 kg/h in the minimum configuration [12]. Reprinted with permission from Siemens.

300 PEM electrolyzers of modular design (Figure 8.2) are produced by Siemens Energy (Germany). It is expected to scale up these systems to reach a capacity of 3 GW by 2025.

Fundamentals of PEM Water Electrolysis

The main components of PEM water electrolysis cells are the membrane, anode and cathode electrocatalytic layers, and porous transport layers/gas diffusion electrodes (Figure 8.3), which form the so-called membrane-electrode assembly (MEA). To achieve the required power/productivity level, individual MEAs are assembled into a filter-press structure, commonly referred to as a stack (Figure 8.4). Cells in the stack are electrically connected in series and gas-disconnected by so-called bipolar plates having distribution channels/ribs for the supply/removal of the reactant/reaction products, electric current, and heat from/to the MEA. The term "bipolar" is used due to the fact that one part of the plate is in the anode compartment (+) of one element and the other in the cathode compartment (–) of the adjacent one.

Along with the planar design of PEM electrolysis cells and stacks, a tubular cell design (Figure 8.5) could be employed. The tubular type of cell has the potential to reduce production costs by using innovative approaches such as co-extrusion, additive manufacturing, and atomic layer deposition [13].

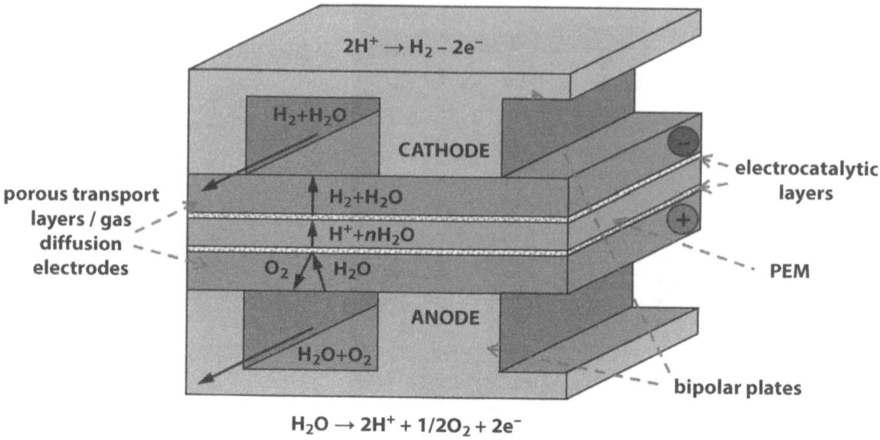

Figure 8.3 Schematic diagram of PEM water electrolysis cell of planar type.

Figure 8.4 PEM water electrolysis stack assembling.

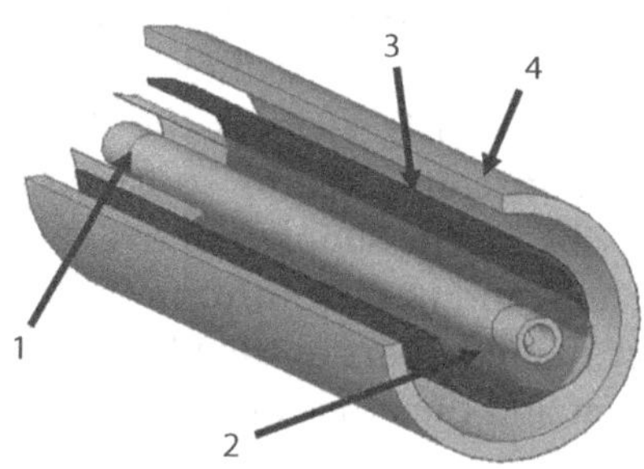

Figure 8.5 Schematic representation of the tubular design of PEM water electrolysis cell [13]. (1) Anode porous transport layer, (2) extruded tubular membrane, (3) cathode porous transport layer, and (4) outer cathode current collector.

Despite this, the tubular design of the PEM electrolysis cell still needs considerable improvements to compete with the planar cell design.

In the course of PEM water electrolyzer operation, the external DC supply system provides the movement of negatively charged electrons. Electrons participate in half-reactions flowing on the electrodes allowing the conversion of electrical energy into chemical energy.

Water is decomposed with oxygen evolution on the positive electrode (anode):

$$H_2O \rightarrow 1/2O_2 + 2\bar{e} + 2H^+$$

The positively charged hydrogen ions (protons) are transported across the PEM, and hydrogen is generated at the negative electrode (cathode):

$H^+ + 2\bar{e} \rightarrow H_2$

The overall reaction is

$$H_2O \rightarrow H_2 + 1/2O_2 \tag{8.1}$$

Electron current is compensated by proton current, and, together with each proton, water molecules are carried out through the membrane from the anode to the cathode zone (Figure 8.3).

According to thermodynamics, at reversible conditions and P,T=const,

$$\Delta G = \Delta H - T\Delta S,$$

where ΔG is the change of Gibbs free energy (work on transfer of electric charge, i.e., the amount of electric energy that must be supplied to the electrolysis cell to decompose water); ΔH is the enthalpy change [the total amount of energy (electricity + heat) that must be supplied to the electrolysis cell to decompose water]; T is the absolute temperature; ΔS is the entropy change; $T\Delta S$ is the amount of heat.

The "useful" electrical work required for the electrolysis of one mole of water is

$$\Delta G° = nFE°,$$

where n is the number of exchanged electrons in the water electrolysis reaction, F is the Faraday constant in C/mol, and $E°$ in V is the standard thermodynamic voltage (for liquid water $E° = 1.229$ V at 25°C).

In standard conditions (25°C and 1 atm), $\Delta G^\circ = 237.23$ kJ/mol, $\Delta H^\circ = 285.83$ kJ/mol, and $\Delta S^\circ = 163.09$ J/(mol*K).

An important thermodynamic parameter is the thermoneutral voltage, i.e., the voltage of electrolysis without heat exchange with the surroundings

$$E_{tn} = \Delta H/nF.$$

At 298 K and 1 bar, the thermoneutral voltage E_{tn} is ≈ 1.48 V and there is a slight decrease of E_{tn} with temperature rise. However, at 100°C the step-wise decay of E_{tn} takes place (Figure 8.6). According to thermodynamics, for electrolytic water splitting into hydrogen and oxygen, it is necessary to supply the electric energy, and at a voltage less than the thermoneutral voltage (E_{tn}), also the thermal energy (Figure 8.6). On the right vertical axis, the power consumption corresponds to the electric energy consumption. In case $U < E^\circ$, electrolysis is not possible.

Specific electric energy consumption W_e (expressed in kW*h/m³ or kW*h/kg) for electrolytic gas generation is commonly used for the comparison of electrolyzers' efficiencies and could be estimated using Faraday's law

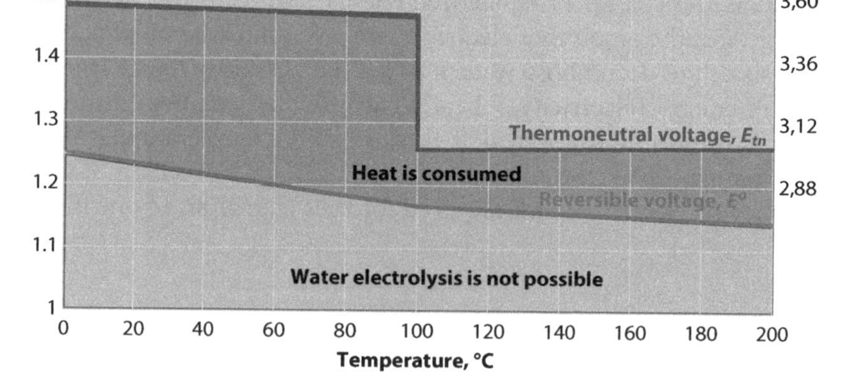

Figure 8.6 The reversible and thermoneutral voltages as functions of temperature. Specific electric energy demand of electrolysis cell (right vertical axis) as a function of electrolysis cell voltage (left vertical axis).

$$V = ke\, I\, t\, /\, \rho,$$

where V in m^3 is the theoretical volume of gas evolved, k_e is the electrochemical equivalent of gas, kg/C = kg/(A·s), I in A is the total current, t in s is the time period, and ρ in kg/m^3 is the gas density.

For hydrogen generation,

$$W_e = I\, U\, t\, /\, V = 0.001 * 0.0899\, /\, (0.0104 * 3600 * 0.000001) \approx 2.40 * U\ [kW*h/m^3],$$

where U is the electrolysis cell voltage, V.

Specific heat consumption is

$$W_h \approx 2.40 * (E_{tn} - U)\ [kW*h/m^3].$$

It should be noted that the above-stated W_e and W_h are theoretical values. However, in practice, at voltages below E_{tn}, due to ohmic and polarization resistance (as will be shown below), heat is released, the value of which at low temperatures usually exceeds the W_h value, and the heat should be removed from the electrolyzer.

The polarization (U-i) curve of the PEM water electrolysis cell (Figure 8.7) can be approximated using the equation

$$U = E^{\circ} + \eta_a + \eta_c + iR, \tag{8.2}$$

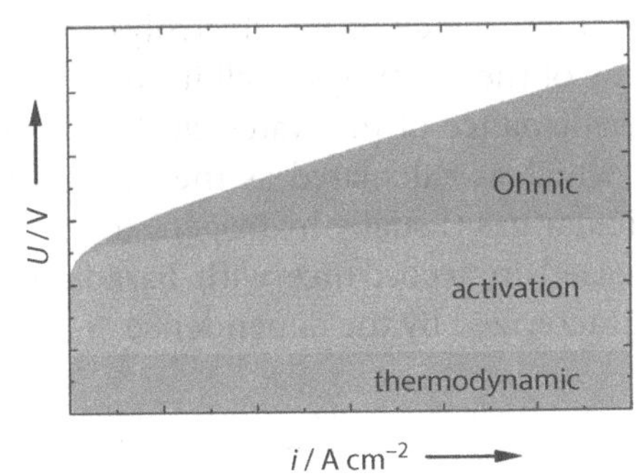

Figure 8.7 Components of polarization (U-i) curve of PEM water electrolysis cell according to equation $U = E^{\circ} + \eta_c + \eta_a + iR$. On this plot, "thermodynamic" + "activation" = $E^{\circ} + \eta_c + \eta_a$.

where $E°$ is mentioned above the standard value of the equilibrium cell voltage (referred also as the counter-EMF value of the cell at the given temperature and pressure), η_a is the anode overvoltage, η_c is the cathode overvoltage, and iR are the ohmic losses (primarily due to losses on the membrane).

As can be seen from Figure 8.7, the efficiency of the electrolyzer is dependent on the operation mode: Electrolyzers operate more efficiently at lower loads. More precise approaches may be applied to simulate the *U-i* curve of the PEM water electrolysis cell with higher accuracy [14]. The authors of the paper [15] have formulated a novel model parameter estimation problem with eight unknown parameters and solved it using a modified Honey Badger algorithm.

For practical applications, the widely used formula for estimation of the water electrolysis cell efficiency is

$$\eta = E_{tn} / U.$$

However, it should be noted that various approaches could be used to estimate the efficiency of the electrolysis cell [16]. In particular, sometimes, efficiency is estimated as $E°/U$. A common practice in industry is to estimate efficiency through specific energy consumption for hydrogen production (in kW*h/kg or kW*h/m^3). In an ideal situation, at voltages near E_{tn} and low current densities, about 3.55 kW*h of electricity or even less is required to generate 1 m^3 of H_2. However, in fact, due to the need to operate at higher (practically significant) current densities, certain polarization and ohmic losses always occur (as can be seen in Figure 8.7), and the demand for electricity is higher. In addition, the subsystems and auxiliary equipment included in the electrolysis plant (Figure 8.8) require additional energy, and the overall efficiency of the electrolysis system is lower compared to the efficiency of the electrolysis cell itself.

An important performance of the water electrolysis cell is the current (faradic) efficiency, which is calculated as the ratio of the actual volume produced by a particular gas (usually hydrogen) to its theoretical amount, which could be released in accordance with Faraday's laws. The current efficiency (η_F) is characterized by the dependence

$$\eta_F = \frac{V_{actual}}{V_{theor.}}$$

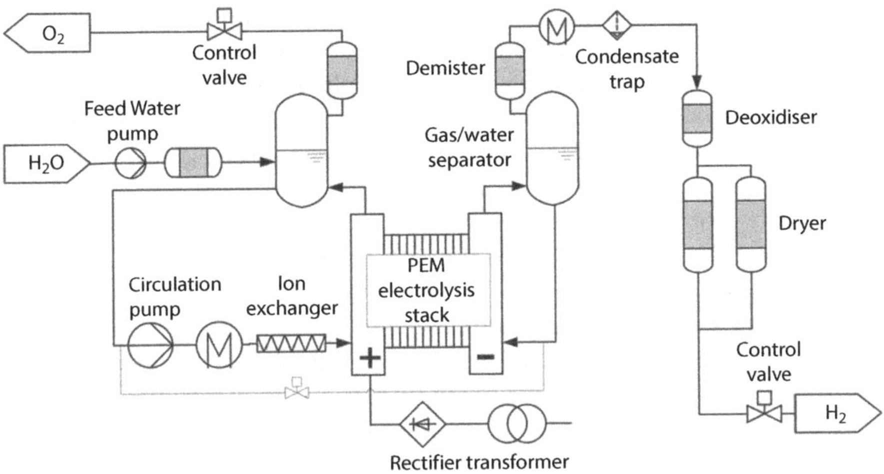

Figure 8.8 Typical layout of a PEM water electrolyzer for pressure-balanced operation with anode water feeding [17]. Reprinted with permission 5716390117518.

Usually, for the PEM electrolysis process, $\eta_F \approx 1$, but lower values are sometimes observed.

Electrocatalysis

Platinum group metal (MPG) nanoparticles are used as electrocatalysts in PEM water electrolyzers. With respect to the anode oxygen evolution reaction (OER), by activity, the PGM can be ranked in the following order:

$$\text{Ru} > \text{Ir} > \text{Rh} > \text{Pd} > \text{Pt}.$$

Unfortunately, Ru, which has maximum catalytic activity in the OER, is not stable under anode conditions (potentials higher than 1.23 V in an acidic environment), and currently, the most widely used anode catalyst is Ir or its oxide. Platinum black is sometimes used as an anode catalyst, but in this case, the voltage of the electrolysis cell is 100 to 200 mV higher than when using iridium black or iridium oxide. Mixed oxides based on Ru, Ir, and non-PMG components are intensively studied.

Other recent approaches applied by electrocatalyst developers and suppliers such as TKK and NEChemCat (Tokyo, Japan), Johnson Matthey (London, UK), ETEK/BASF (Ludwigshafen, Germany), and Umicore (Antwerpen, Belgium) are the use of catalysts with a core-shell structure, the electrocatalytic layer with a concentration gradient of the components.

Carbon-supported platinum is mainly used as cathode electrocatalysts for hydrogen evolution reactions (HERs). In the past years, several non-MPG catalysts have been developed and applied on the cathode of PEM water electrolyzers instead of platinum. In particular, the performances of iron- and cobalt-based clathrochelates are encouraging and comparable with that of the most promising non-platinum HER electrocatalysts [18]. Single-atom catalysts (the 100% PMG or non-PMG atom utilization) at least for HER are under development. In particular, it was shown that single-platinum atoms supported on nitrogen-doped graphene nanosheets fabricated by atomic layer deposition method demonstrated HER activity higher than that of commercial Pt/C catalyst [19]. An interfacial engineering strategy to obtain a low-Pt loading catalyst with atomically Pt-doped molybdenum carbide quantum dots decorated on conductive porous carbon (Pt-MoC_x@C) for high-rate and stable HER in PEM water electrolyzer has been applied in [20]. Fe_2P/C electrocatalysts were proposed in [21]. The ability of an organic electrode, more specifically poly(3,4-ethylenedioxythiophene) (PEDOT), to be an electrocatalyst for HER in PEM water electrolysis cells has been demonstrated [22].

Key Performance Indicators of PEM Water Electrolyzers

A number of key performance indicators (KPIs) have been developed to evaluate and compare the performances of PEM water electrolyzers, some of them are listed in Table 8.1. KPIs can be used as performance assessment criteria and allow comparison of commercial products on the market, in particular, for grid stabilization services, etc. It should be noted that Table 8.1 demonstrates only a few basic KPIs, other of them could be found in [1].

Table 8.1 List of main KPIs used for water electrolysis [4].

KPI 1	Efficiency
KPI 2	Flexibility
KPI 3	Reactivity
KPI 4	Durability
KPI 5	Safety
KPI 6	Capex

One of the KPIs is efficiency, it usually expressed as specific electric energy consumption (kW*h/kg or kW*h/m^3 of H_2) and could refer to the electrolyzer itself, or to the entire water electrolysis plant (including the energy consumption of ancillary equipment). The typical electricity consumption of modern PEM electrolyzers is 50 kW*h/kg or 4.5 kW*h/Nm^3 at a current density of 1.0 A/cm^2. Besides electricity, around 1.3 L of water per cubic meter of hydrogen is required. In practical conditions, however, there is always a loss of energy in the course of the electrolysis process, and the electricity consumption will be higher. Larger units work more efficiently and so the energy consumption decreases with volume [23].

Regarding KPI 2, PEM water electrolysis stacks/plants demonstrate the required flexibility and reactivity to enter the primary and secondary energy storage markets. In particular, according to the Silyzer 300 datasheet, the load changing interval is 5 to 100% from the nominal load, typical load change rate is 10%/s. The minimum load is mainly limited by gas crossover across the PEM. The warm start takes usually <1 min.

An important KPI is the durability of the electrolyzer. The long-term voltage increase of only 16 μV/h over 20,000 h of operation demonstrates that the PEM water electrolysis cell can be operated over long periods [4]. The service life of a commercial electrolysis plant reported by some suppliers is up to 80,000 h of continuous operation. One of the factors limiting PEM water electrolyzer lifetime is cation contamination of the MEA. The presence of contaminants (such as Fe, Ti, Na, Mg, K, Cl, S, F, O, C, and Ca) in the feed water increases the cell voltage due to poisoning catalyst sites resulting in kinetic losses at the anode and cathode, increase in the ohmic resistance of the membrane and decrease in its protonic conductivity, inhibiting oxygen transport, etc. [24, 25]. To reduce MEA contamination, the use of filters and/or regular water changes in the electrolysis system may be recommended.

One of the KPIs of electrolysis plants is gas purity, sometimes expressed as dew point, i.e., the condensation temperature of water contained in electrolysis hydrogen. The dew point depends on the partial pressure of the water vapor contained in the hydrogen and is the temperature to which the hydrogen must be cooled so that the water vapor contained therein reaches saturation and begins to condense into the dew. The dew point also depends on the gas pressure. The lower the dew point, the lower the moisture content of hydrogen, i.e., the higher the hydrogen purity.

In the course of water electrolyzer operation, hydrogen is released in a wet environment, hence hydrogen is saturated with water vapors at the temperature of the electrolysis stack. As a rule, the dew point of electrolysis hydrogen exiting directly from the electrolysis module is not higher than minus 15°C. If the consumer needs hydrogen with a lower moisture content, the electrolysis unit is equipped with hydrogen drying systems that allow reaching dew point values, for example, –72°C, which corresponds to a hydrogen purity of 99.9998%, or –73°C, which corresponds to a hydrogen moisture content of less than 1.5 ppm. Drying of hydrogen can be carried out by cooling or using sorbents, for example, silica gel. Regeneration of silica gel can be carried out with the same gas at 200°C to 250°C. The balance of the plant may include a pressure swing adsorption (PSA) unit, where hydrogen is additionally dried and purified from the traces of oxygen that penetrate the hydrogen side of the cell.

KPI 5 evaluates the safety level. The primary risk associated with PEM electrolysis systems is the potential for the formation of flammable or explosive mixtures. To avoid the accumulation of such mixtures in the internal chambers and lines, the following measures are usually taken:

- control of pressure and the pressure difference between hydrogen and oxygen lines;
- controlling the hydrogen concentration (<0.4 vol% H_2) [26];
- minimization of free volumes in the electrolysis system (for example, between the stack and high-pressure case housing) where hydrogen/oxygen mixture could potentially be located.

Several levels of risk reduction measures are usually considered for PEM water electrolysis systems. In particular, Explosion Protection and Prevention guidelines recommend "primary", "secondary", and "tertiary" safety measures. Siemens decided to go beyond these requirements and enlarged these definitions by implementing the extended primary, secondary, and tertiary safety measures, making them applicable to its complete PEM system and herewith adapting its design from the start toward a systematic risk limitation [27].

Regarding the Capex of PEM water electrolyzers (KPI 6), they are still more expensive (≈1000 \$/kW) than alkaline electrolyzers, but significant progress is expected in the coming years to reduce the cost down to 500 \$/kW. Further development and cost reduction of PEM fuel cells (having similar element bases) may have an additional impact on the cost reduction of PEM water electrolyzers.

Current Trends and Perspectives

Current trends in the development of MEAs and electrochemical stacks are compaction, reducing material consumption, increasing specific characteristics, automation of design, and production [28]. In particular, 3M has developed the nanostructured thin-film (NSTF) catalyst-coated membranes (CCMs) for water electrolyzers with an increased specific surface area of contact between the membrane and electrocatalytic layers (Figure 8.9). Such CCMs are characterized by low mass transport losses, high proton and electron conductivity, good heat transfer, and high hydrophilic properties, their use allows for the improvement of the performance of PEM electrolysis cells.

An example of the fabrication of such patterned CCMs is provided in Figure 8.10. By increasing the membrane surface area and improving catalytic performance, these patterned CCMs exhibit higher current densities and achieve higher hydrogen production rates.

In order to meet the challenges of the future, it is necessary to develop a new generation of PEM water electrolyzers. The main requirements for advanced PEM electrolysis systems are compactness; improved gas, liquid, and temperature management; high load flexibility, durability, and maintainability; cost reduction, improved stack design using components suitable for mass production and their unification; highly efficient power electronics; and improved diagnostics and control, including by using artificial neural network (ANN) and machine learning.

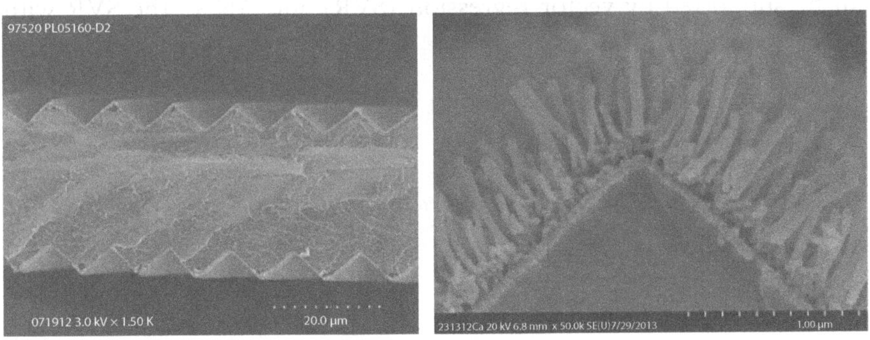

Figure 8.9 Cross-sectional micrographs of NSTF CCM developed by 3M [29]. Copyright (Insert © Year) From (Insert Title) by (Insert Author/Editor Name). Reproduced by permission of Taylor and Francis Group, LLC, a division of Informa plc. This permission does not cover any third-party copyrighted work which may appear in the material requested. The user is responsible for obtaining permission for such material separately from this grant.

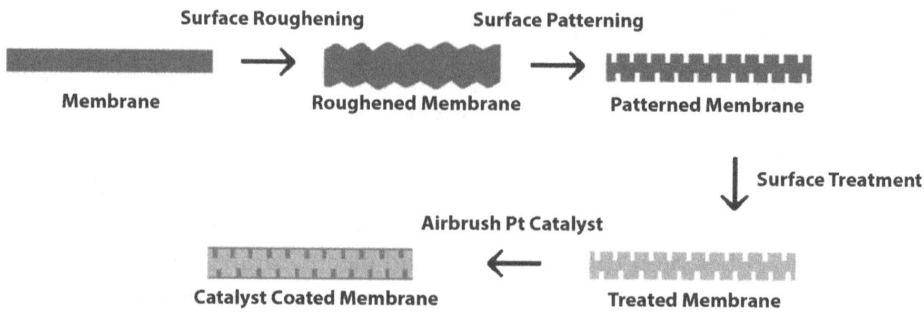

Figure 8.10 Schematic diagram of the surface modification and catalyst deposition processes for fabrication of patterned CCM [30]. Reprinted with permission 5716380728132.

Implementation of Artificial Intelligence and Machine Learning

Today, there is an active use of artificial intelligence and, in particular, machine learning for the design, production, and optimization of PEM electrochemical systems [31, 32]. Thus, the integration of ANN and machine learning, with a particular focus on their ability to gather and analyze data, is able to predict the relationship between the operation parameters of the PEM water electrolyzer and its output characteristics (Figure 8.11), to select/optimize electrocatalysts and other components, to estimate the dynamic characteristics and stability of electrochemical system [33].

Machine learning algorithms can be used to study the effects of various factors including operating temperature, current density, PTL/electrocatalytic layer interface, and catalyst loading. Machine learning methods can be supported by vector regression (SVR) and ANN. The SVR with

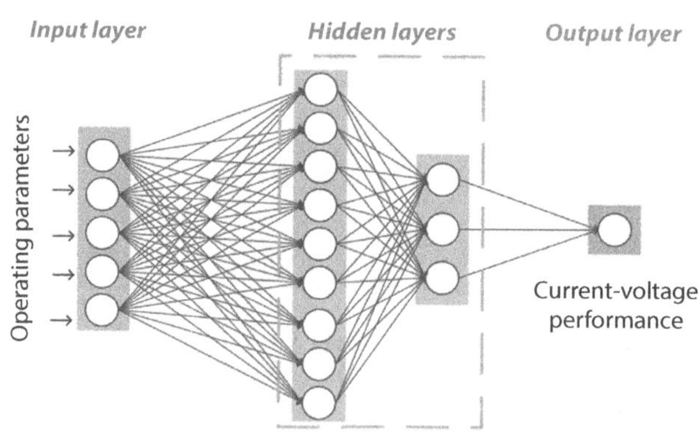

Figure 8.11 The structure of ANN linking the operating parameters of the cell and its output performance.

hyper-parameters fine-tuned using a genetic algorithm can achieve prediction accuracy comparable to an ANN with a fully connected architecture consisting of two hidden layers, each with 30 nodes. The algorithms' predictions allow us to recommend PEM water electrolyzer operation modes enhancing its performance and lifetime, subsequently leading to cost reduction at large-scale hydrogen production [34].

Digital Replicas (Digital Twin Technology) of PEM Water Electrolysis Systems

In recent years, great attention from researchers and industry has been paid to the development of digital replicas of water electrolysis systems. A digital replica (also known as a digital twin) is a software-based virtual copy of the complete physical system. MATLAB/Simulink provides a high degree of maturity and versatility to implement digital replicas of PEM electrolyzers [35]. In particular, equivalent circuit models could describe the behavior of the individual PEM water electrolysis cells [36].

Developed digital twins become an invaluable tool to analyze various "what if" scenarios, and various BoP improvements. A digital twin can also validate the optimized control and safety schemes, prove the compliance of the electrolysis facility with regulatory requirements, and validate its integration within industrial plants, microgrids, etc [37].

Operation at Increased Pressure and Temperature

Usually, in practice, there are three different ways to deliver pressurized hydrogen to the customers (e.g., filling stations, storage systems, and industry). The first way is to use a water electrolyzer operating at atmospheric pressure with further compression of hydrogen to the pressure of interest using a mechanical compressor. The second approach is to use a balanced high-pressure water electrolyzer that delivers hydrogen and oxygen at similar pressures. The third pathway is to use an unbalanced high-pressure water electrolyzer that delivers hydrogen under increased pressure, but oxygen at atmospheric pressure [4].

A comparison of the compression energies required by three different approaches (Figure 8.12) shows that up to a pressure of *ca.* 40 bar, less energy is required for unbalanced electrolysis of water than for the other two approaches. However, for pressures >40 bar, the energy consumption for the atmospheric electrolyzer and mechanical compressor becomes less than the energy consumption of electrolyzers with high-pressure water electrolyzers (for example, up to 16% at 100 bar compared to the low-pressure electrolyzer). However, differences are small and arguments other than energetic ones (e.g., overcost associated with the compression vessel

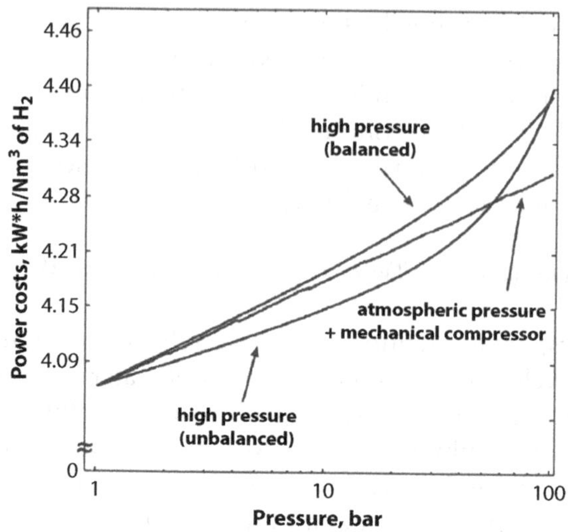

Figure 8.12 Specific power costs required for three different ways of high-pressure hydrogen generation: (1) atmospheric PEM water electrolysis with subsequent hydrogen mechanical compression, (2) balanced high-pressure PEM water electrolysis, and (3) unbalanced high-pressure PEM water electrolysis. Reproduced from [4]. Reprinted with permission 5716380274458.

and gas purity management) might play a role in deciding which technology is the most appropriate one for a given application [4].

As mentioned above, one of the approaches is to carry out the high-pressure electrolysis with asymmetry of pressure between the hydrogen and oxygen circuits (so-called unbalanced pressure scheme, i.e., with a pressure drop between the anode and cathode chambers). The use of an asymmetric (unbalanced pressure) electrolysis system allows us to significantly simplify the technological scheme of the electrolysis plant due to a decrease in pressure in the anode circuit, and a reduction in the energy consumption for the production of high-pressure hydrogen by 7 to 15% compared to the conventional (balanced pressure) scheme. For instance, the high-pressure electrolysis stacks with unbalanced pressure (Figure 8.13) have been developed by the National Research Centre "Kurchatov Institute". These self-pressurized PEM electrolysis stacks are capable of producing electrolytic gasses at pressures up to 200 bar directly in electrolysis stacks and are intended for charging standard gas cylinders, excluding additional compression, with a decrease in capital and operating costs [38]. Hydrogen Innovation GmbH in collaboration with HIAT (Germany) has achieved

Figure 8.13 Unbalanced-type high-pressure PEM electrolysis stack with hydrogen productivity up to 1 m^3/h operating at up to 200 bar (National Research Centre "Kurchatov Institute").

differential pressures of up to 100 bar (1450 psi) by using innovative construction and manufacturing techniques.

It is noteworthy that the operation of PEM electrolysis stacks at a pressure of a few tens of bar allows one to reduce energy costs for electrolysis. This eliminates the most energy-demanding first compression stage (from 0 to *ca.* 30 to 50 bar) of multistage mechanical hydrogen compressors. Compared with piston-type compressors, an economy of the electric power of 600 kW*h/t of hydrogen is obtained, by increasing the operating pressure from 20 to 60 bar [4].

The upper temperature level of PEM water electrolysis is mainly limited by the thermal stability of membrane materials. For perfluorosulfonic acid Nafion® membranes, the operating temperature is limited by 100°C at ambient pressure due to dehydration of the membrane at higher temperatures. However, by using short-side-chain perfluorosulfonic acid membranes (for example, Aquivion™; Figure 8.1), the upper temperature limit of water electrolyzer operation can be raised up to 130°C to 150°C.

Aquivion™ is another type of PFSA polymer that differs from Nafion® in its equivalent weight and length of sulfonic acid–terminated side chains (Figure 8.1). Compared to conventional Nafion® membranes, this type of PFSA membrane has significantly higher softening temperatures, which should facilitate electrolysis at elevated temperatures. Its high elastic modulus has also been identified as the sole reason for its superior performance in PEM fuel cells [39].

The alternative solution is to apply a membrane based on polybenzimidazoles (PBI) doped by phosphoric acid (PA), capable of operation up to 200°C. Compared with the electrolysis of liquid water at low temperature, medium temperature water vapor electrolysis technology, first of all, exhibits faster reaction kinetics, primarily the OER, and thus reduces the overpotentials that are connected to kinetic limitations. Recyclable residual heat exhibits higher energy conversion efficiency.

Second, the total thermodynamic energy requirement for water splitting (ΔH) decreases from *ca.* 283 kJ mol^{-1} (liquid water, 100°C) to *ca.* 243 kJ mol^{-1} for steam electrolysis at 180°C. In addition, the reversible voltage calculated from the change in Gibbs free energy (ΔG) decreases slightly as the temperature increases from 80°C (1.18 V, liquid water) to 130°C (1.16 V, steam). The decrease in electrical energy supplied must be compensated for by an increase in heat added to the system, but it is assumed that this heat will be taken from the waste heat generated by ohmic losses in the system. This heat could also be used for preheating and evaporation of the feed water [37]. Finally, the application of the PBI-PA membrane allows us to significantly simplify the water treatment system since PBI-based membranes are less sensitive to the impurities in the water (in contrast to perfluorinated membranes Nafion® and its analogs). It allows us not to use the bi-distilled deionized water, but to use steam with lower quality (obtained, for example, using evaporator-type steam generators characterized by a minimum transition of impurities from water to steam).

Cold Start

Currently, significant attention is paid to the stability of the performances of electrochemical systems and aspects of their cold start. Thus, cold start strategies and corresponding technical solutions (water removal, heating, the use of special materials, antifreeze, etc.) have been developed. Their use allows for eliminating or minimizing degradation processes caused by water freezing and ensuring stable and effective operation of electrolysis systems at subzero temperatures.

Storage, start-up, and operation of PEM water electrolysis systems at subzero temperatures can lead to degradation processes in electrocatalytic layers, gas diffusion electrodes/porous transport layers, membranes, and MEAs as a whole as a result of freezing of water in these components. Repeated cold starts (from a state where the temperature of the electrochemical system is equal to the ambient temperature and is below 0°C) may result in irreversible damage to the above components and loss of performance of the electrochemical system. However, there are cold start strategies and corresponding technical solutions, the use of which eliminates or minimizes degradation processes and ensures stable and effective operation of the electrolysis system at subzero temperatures.

For instance, purging or drying of the system when it is turned off, heating, the use of antifreeze, and so forth can be used. A comprehensive set of measures to ensure the functioning of water electrolyzers at negative temperatures should affect both the start-up and shutdown of the electrochemical system. In addition, the design of the electrochemical system, in general, and the MEAs, in particular, must be adapted to start operation at temperatures below zero. The integrated approach allows us to work out the issues of storage, transportation, and cold start of PEM-based water electrolyzers at deeply negative temperatures.

Strategies and technical solutions similar to those used in PEM fuel cells and corresponding power plants can be used to prevent damage to the elements of the electrolysis stack and its piping during water freezing [40]. In particular, freezing of the electrolysis system should be preceded by purging the electrode chambers and water-containing piping with gas (e.g., nitrogen).

Start-up of the water electrolyzer at subzero temperatures can be carried out using water-alcohol solutions [41]. In particular, instead of water, an aqueous solution of ethanol, for example, at a concentration of 40 vol%, having a freezing point of about −28°C, may be supplied to the cell. The overall reaction of the electrochemical oxidation of ethanol is as follows:

$$C_2H_5OH + 3H_2O \rightarrow 2CO_2 + 12e + 12H^+, \tag{8.3}$$

its standard potential is 0.085 V [vs. standard hydrogen electrode (SHE)].

After heating the electrolysis cell to positive temperatures, the supply of pure water to the anode begins, and instead of reaction (3), reaction (1) takes place.

In addition to ensuring the start-up of electrolysis systems at subzero temperatures, operation on alcohol solutions reduces energy costs for the production of hydrogen by reducing the potential of the anode reaction. As follows from the data presented in Figure 8.14, at low current densities (up to 0.3 A/cm^2) in the presence of ethanol solution, a significant decrease in cell voltage is observed (curves 5 to 8), up to 50% in the case of anode catalysts Pt and $Pt_{0.5}Ru_{0.5}$ (curves 5 and 6) [42].

It can be seen from Figure 8.14 that aqueous ethanol solutions can be used to reduce energy consumption for the electrochemical production of hydrogen, as well as to provide a cold start of electrolysis systems at

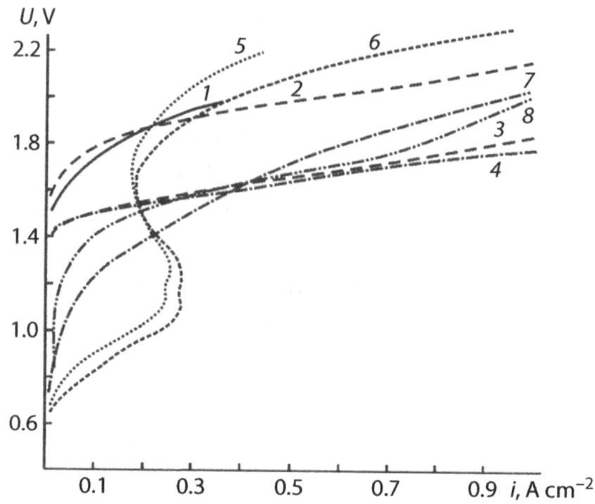

Figure 8.14 Current-voltage characteristics of PEM electrochemical cell for decomposition of water (1 to 4) and aqueous ethanol solution (5 to 8). (*U*) Voltage and (*i*) current density. Electrocatalysts used on the anode: $Pt_{0.5}Ru_{0.5}$ (1, 5), Pt (2, 6), $Ir_{0.3}Ru_{0.3}Sn_{0.4}O_2$ (3, 7), and Ir (4, 8) [42]. Reprinted with permission 5716390445349.

subzero temperatures. Another possibility is the electrochemical oxidation of 1-propanol aqueous solutions [43].

Power Grid Regulation and Energy Storage Systems

PEM water electrolyzers can be efficiently used for grid balancing services and energy storage systems [44]. During off-peak night hours or other periods of excess energy, low-cost electricity from a renewable (or nuclear) power plant can be used to electrolyze water, with further accumulation of electrolytic gasses in a storage system. At peak electrical demand hours, the stored gasses can be used to generate electricity (using fuel cells or gas turbines) to make up for electricity shortages.

Standardization, Testing, and Certification

Standardization and certification are important components of the transition to hydrogen energy, as well as essential tools for the functionality of the hydrogen economy. The PEM electrolysis system (according to ISO 22734:2019 – A Hydrogen Generator Using Electrolysis of Water) is CE marked, indicating its conformance with the Pressure Equipment Directive 97/23/EC, the Low Voltage Directive 2006/95/EC, the ATEX Directive 94/9/EC, the Electromagnetic Compatibility Directive 2004/108/EC, and the Machinery Directive 2006/42/EC. The development and actualization of technical regulations and test protocols [45] allow the further development and implementation of water electrolysis technologies. In order to increase the level of interchangeability of components and maintainability of electrolysis stacks, improve their quality, and protect consumers, it is necessary to unify and harmonize the element base produced by different companies.

Effective assessment of the durability of PEM water electrolysis stack components over time is possible by using the accelerated stress test (AST) protocols. According to recent reports [46], the main degradation mechanisms are associated with anode catalyst dissolution, membrane thermochemical decomposition, and the formation of semiconducting oxides on the metal components. However, there are still no widely accepted AST protocols in the community, so its development is of high priority.

Regulations, codes, and standards used to form the legal framework for water electrolysis technologies should become more flexible. At the same time, safety-related issues of PEM electrolysis systems should be carefully addressed to avoid technical incidents and skepticism of public opinion.

Conclusions

PEM water electrolysis is the optimal method of green hydrogen generation due to rapid dynamic response time (fast start-up, including at subzero temperatures, and shut-down, excellent load dynamics), wide operational current density range (maneuverability), high efficiency and durability, and possibility to operate at high pressure and increased temperature, coupling with renewable energy sources and grid stabilization. Over the past decades, green hydrogen produced using PEM electrolyzers has been successfully implemented in a number of industrial-scale processes: food, glass, steel aluminum manufacturing, agricultural chemical production, and so on. PEM water electrolysis systems fully comply with the paradigms of decarbonization, digitalization, and decentralization of energy.

The KPIs for PEM water electrolyzers are efficiency, durability, and cost. For water electrolysis stacks, the target values of efficiency and service life are 63 to 68% and 90,000 h (or 10 years). Regarding the Capex, a reduction to the target value of 500 $/kW is expected. The important issue is the unification and harmonization of the element base produced by various suppliers.

Acknowledgment

The work was supported by the Ministry of Science and Higher Education of the Russian Federation under the project FSWF-2023-0014.

References

1. Chatenet, M. *et al.*, Water electrolysis: from textbook knowledge to the latest scientific strategies and industrial developments. *Chem. Soc. Rev.*, 51, 4583–4762, 2022, https://doi.org/10.1039/d0cs01079k.
2. Zhou, J., Ji, W., Cao, X. *et al.*, A Current Perspective on the Renewable Energy Hydrogen Production Process. *J. Therm. Sci.*, 32, 542–596, 2023, https://doi.org/10.1007/s11630-023-1749-3.
3. Sone, Y., Hernandez-Mendoza, O.S., Shima, A. *et al.*, Water electrolysis by the direct water supply to the solid polymer electrolyte through the interdigitated structure of the electrode. *Electrochemistry*, 89, 138–140, 2021.
4. Bessarabov, D. and Millet., P., *PEM Water Electrolysis*, vol. 2, B.G. Pollet (Ed.), Elsevier, Cambridge, United States, 2018, ISBN: 978-0-08-102830-8, https://doi.org/10.1016/B978-0-08-102830-8.00002-3, DOI: 2018 Academic Press.
5. High yield Energy Technologies, The Netherlands, https://hyethydrogen.com/hcs100/.
6. Grigoriev, S.A., Fateev, V.N., Bessarabov, D.G., Millet, P., Current status, research trends, and challenges in water electrolysis science and technology. *Int. J. Hydrogen Energy*, 45, 49, 26036–26058, 2020.
7. De Levie, R., The electrolysis of water. *J. Electroanal. Chem.*, 476, 1, 92–93, 1999.
8. Trasatti, S., Water electrolysis: who first? *J. Electroanal. Chem.*, 476, 90–1, 1999.
9. Davenport, R.J. and Schubert, F.H., Space water electrolysis: space station through advanced missions. *J. Power Sources*, 36, 235–250, 1991.
10. Roy, R., Making Space Safer with Electrolysis, The American Society of Mechanical Engineers, New York, United States, Aug 26, 2011, https://www.asme.org/topics-resources/content/making-space-safer-with-electrolysis.
11. Wang, Q., Shen, D., Tu, Z., Li, S., Improved performance of lanthanide-doped UIO-66/Nafion hybrid proton exchange membrane for water electrolyzer. *Int. J. Hydrogen Energy*, 56, 1249–1256, 2024, https://doi.org/10.1016/j.ijhydene.2023.12.288.
12. https://assets.new.siemens.com/siemens/assets/api/uuid:abae9c1e48d-6d239c06d88e565a25040ed2078dc/version:1524040818/ct-ree-18-047-db-silyzer-300-db-de-en-rz.pdf.

13. Laube, A., Sanchez Batalla, B., Weidlich, C., Hofer, A., Bachmann, J., Zallmann, S., Körner, C., Fischer, S., Chica, A., Struckmann, T., Tubular PEM electrolysis cells with a 3D-printed oxygen electrode and ALD catalyst coating. *Int. J. Hydrogen Energy*, 47, 94, 39719–30, 2023, 10.1016/j.ijhydene.2023.08.084.
14. Grigor'ev, S.A., Khaliullin, M.M., Kuleshov, N.V., Fateev, V.N., Electrolysis of Water in a System with a Solid Polymer Electrolyte at Elevated Pressure. *Russ. J. Electrochem.*, 37, 8, 819–822, 2001, https://doi.org/10.1023/A:1016735003101.
15. Khajuria, R., Yelisetti, S., Lamba, R., Kumar, R., Optimal model parameter estimation and performance analysis of PEM electrolyzer using modified honey badger algorithm. *Int. J. Hydrogen Energy*, 49, Part C, 238–259, 2024, https://doi.org/10.1016/j.ijhydene.2023.07.172.
16. Lamy, C. and Millet, P., A critical review on the definitions used to calculate the energy efficiency coefficients of water electrolysis cells working under near ambient temperature conditions. *J. Power Sources*, 447, 227350, 2020.
17. Smolinka, T., Ojong, E.T., Garche, J., Chapter 8 - Hydrogen Production from Renewable Energies—Electrolyzer Technologies, in: *Electrochemical Energy Storage for Renewable Sources and Grid Balancing*, P.T. Moseley and J. Garche (Eds.), pp. 103–128, Elsevier, Amsterdam, Netherlands, 2015, ISBN 9780444626165, https://doi.org/10.1016/B978-0-444-62616-5.00008-5.
18. Grigoriev, S.A., Pushkarev, A.S., Pushkareva, I.V., Millet, P., Belov, A.S., Novikov, V.V., Belaya, I.G., Voloshin, Y.Z., Hydrogen production by proton exchange membrane water electrolysis using cobalt and iron hexachloroclathrochelates as efficient hydrogen-evolving electrocatalysts. *Int. J. Hydrogen Energy*, 42, 46, 27845–27850, 16 November 2017, DOI: 10.1016/j.ijhydene.2017.05.048.
19. Cheng, N., Stambula, S., Wang, D., Banis, M.N., Liu, J., Riese, A., Xiao, B., Li, R., Sham, T.-K., Liu, L.-M., Botton, G.A., Sun, X., Platinum single-atom and cluster catalysis of the hydrogen evolution reaction. *Nat. Commun.*, 7, 13638, 2016.
20. Chen, L., Huang, Y., Ding, Y. *et al.*, Interfacial engineering of atomic platinum-doped molybdenum carbide quantum dots for high-rate and stable hydrogen evolution reaction in proton exchange membrane water electrolysis. *Nano Res.*, 16, 12186–12195, 2023, https://doi.org/10.1007/s12274-023-5666-2.
21. Prieto, N., da Silva, E.L., Castiglioni, J.R., Cuña, A., Synthesis and characterization of non-noble metal cathode electrocatalysts for PEM water electrolysis. *Electrochim. Acta*, 473, 143474, 2024, https://doi.org/10.1016/j.electacta.2023.143474.
22. Valiollahi, R., Vagin, M., Gueskine, V., Singh, A., Grigoriev, S.A., Pushkarev, A.S., Pushkareva, I.V., Fahlman, M., Liu, X., Khan, Z., Berggren, M., Zozoulenko, I., Crispin, X., Electrochemical hydrogen production on a metal-free polymer. *Sustain. Energy Fuels*, 3, 12, 3387–3398, 2019.

23. Smolinka, T., Günther, M., Garche, J., Stand und Entwicklungspotenzial der Wasserelektrolyse zur Herstellung von Wasserstoff aus regenerativen Energien, Fraunhofer-Institut für Solare Energiesysteme (ISE), Freiburg, Germany, 2011.
24. Sin, P.P.S., Wai, S.H., Ota, Y., Nishioka, K., Suzuki, Y., Performance recovery of proton exchange membrane electrolyzer degraded by metal cations contamination. *Int. J. Hydrogen Energy*, 53, 86–92, 2024, https://doi.org/10.1016/j.ijhydene.2023.12.049.
25. Grigoriev, S.A., Bessarabov, D.G., Fateev, V.N., Degradation Mechanisms of MEA Characteristics during Water Electrolysis in Solid Polymer Electrolyte Cells. *Russ. J. Electrochem.*, 53, 3, 318–323, 2017, https://doi.org/10.1134/S1023193517030065.
26. https://seshydrogen.com/en/h2-safety-2-electrolyzer/.
27. https://h2tools.org/sites/default/files/2019-08/paper_193.pdf.
28. Baranov, I.E., Nikolaev, I.I., Soloviev, M.A., Grigoriev, S.A., Automation and Control of the Electrocatalytic Layers Formation Using a Two-Dimensional Coordinate Spraying Machine. *Mekhatronika Avtom. Upr.*, 23, 5, 246–255, 2022, https://doi.org/10.17587/mau.23.246-255.
29. Lewinski, K., 3M NSTF for PEM Water Electrolysis: Principles and Applications, in: *PEM Electrolysis for Hydrogen Production*, pp. 87–117, 2015, https://doi.org/10.1201/b19096.
30. Hindson, W.A. and James, S., Effects of surface modification on a proton exchange membrane for improvements in green hydrogen production. *Int. J. Hydrogen Energy*, 49, Part C, 1040–1047, 2024, https://doi.org/10.1016/j.ijhydene.2023.10.322.
31. Günay, M.E. and Tapan, N.A., Analysis of PEM and AEM electrolysis by neural network pattern recognition, association rule mining and LIME. *Energy AI*, 13, 100254, 2023, ISSN 2666-5468, https://doi.org/10.1016/j.egyai.2023.100254.
32. Arjmandi, M., Fattahi, M., Motevassel, M. *et al.*, Evaluating algorithms of decision tree, support vector machine and regression for anode side catalyst data in proton exchange membrane water electrolysis. *Sci. Rep.*, 13, 20309, 2023, https://doi.org/10.1038/s41598-023-47174-w.
33. Günay, M.E. and Tapan, N.A., Evaluation of polymer electrolyte membrane electrolysis by explainable machine learning, optimum classification model, and active learning. *J. Appl. Electrochem.*, 53, 415–433, 2023, https://doi.org/10.1007/s10800-022-01786-8.
34. Hayatzadeh, A., Fattahi, M., Rezaveisi, A., Machine learning algorithms for operating parameters predictions in proton exchange membrane water electrolyzers: Anode side catalyst. *Int. J. Hydrogen Energy*, 56, 302–314, 2024, https://doi.org/10.1016/j.ijhydene.2023.12.149.

35. Gaspar, F., Godoy, A., Pérez, I., Godoy, M., Calero, J., Martín, D., Design of a Simulation Platform to Test the Suitability of Different PEM Electrolyzer Models to Implement Digital Replicas, in: *Proceedings of the 11th International Conference on Simulation and Modeling Methodologies, Technologies and Applications - Volume 1: SIMULTECH*, pp. 430–437, 2021, ISBN 978-989-758-528-9, DOI: 10.5220/0010582104300437.
36. Folgado, F.J., González, I., Calderón, A.J., PEM Electrolyzer Digital Replica based on internal resistance determination applied to hydrogen energy storage. *J. Energy Storage*, 75, 109694, 2024, https://doi.org/10.1016/j.est.2023.109694.
37. https://www.emerson.com/documents/automation/article-digital-twin-technology-transforms-hydrogen-production-mimic-en-7425216.pdf.
38. Millet, P., Grigoriev, S.A., Porembskiy, V.I., Development and characterisation of a pressurized PEM bi-stack electrolyser. *Int. J. Energy Res.*, 37, 5, 449–456, April 2013, DOI: 10.1002/er.2916.
39. Hansen, M.K., PEM Water Electrolysis at Elevated Temperatures. Ph.D. Thesis. Department of Energy Conversion and Storage, Technical University of Denmark, Technical University of Denmark, Kongens Lyngby, Denmark, 2012.
40. Ivanova, N.A., Spasov, D.D., Mensharapov, R.M., Kukueva, E.V., Zasypkina, A.A., Fateev, V.N., Grigoriev, S.A., Efficient and stable subzero operation of a PEM fuel cell with a composite anode using hydrogen-methanol composition during freeze/thaw cycles. *Int. J. Hydrogen Energy*, 48, 30, 11410–11420, 2023, https://doi.org/10.1016/j.ijhydene.2022.05.298.
41. Rodríguez-Gómez, A., Dorado, F., de Lucas-Consuegra, A., Sánchez, P., de la Osa, A.R., Ethanol electrolysis in a PEM cell: Dependence of the PtRu surface composition on the EOR activity under dynamic potential application. *Catal. Today*, 427, 114411, 2024, https://doi.org/10.1016/j.cattod.2023.114411.
42. Pushkareva, I.V., Pushkarev, A.S., Grigoriev, S.A., Lyutikova, E.K., Akel'kina, S.V., Osina, M.A., Slavcheva, E.P., Fateev, V.N., Electrochemical conversion of aqueous ethanol solution in an electrolyzer with a solid polymer electrolyte. *Russ. J. Appl. Chem.*, 89, 12, 2109–2111, December 2016, DOI: 10.1134/S1070427216120260.
43. Abdelnasser, S., Hakamata, T., Ogihara, H., Kurokawa, H., Electrochemical oxidation of 1-propanol through proton exchange membrane electrolysis. *J. Electroanal. Chem.*, 928, 117009, 2023, https://doi.org/10.1016/j.jelechem.2022.117009.
44. Allidières, L., Brisse, A., Millet, P., Valentin, S., Zeller, M., On the ability of PEM water electrolysers to provide power grid services. *Int. J. Hydrogen Energy*, 44, 20, 9690–9700, 2019, https://doi.org/10.1016/j.ijhydene.2018.11.186.

45. Tsotridis, G. and Pilenga, A., EU harmonised protocols for testing of low temperature water electrolysers. EUR 30752 EN, Publications Office of the European Union, Luxembourg, Publications Office of the European Union, Luxembourg, 2021, doi: 10.2760/58880, JRC122565. ISBN 978-92-76-39266-8.
46. Aßmann, P., Gago, A.S., Gazdzicki, P., Friedrich, K.A., Wark, M., Toward Developing Accelerated Stress Tests for Proton Exchange Membrane Electrolyzers. *Curr. Opin. Electrochem.*, 21, 225–233, 2020.

9

Wind Driven Hydrogen Production in Eastern Morocco: Suitability Atlas Development and Techno-Economic Analysis

Salaheddine Amrani[1], Samir Touili[1,2], Abdellatif Azzaoui[1], Ahmed Alami Merrouni[1]* and Hassane Dekhissi[1]

[1]Materials Science, New Energies and Applications Research Group, LPTPME Laboratory, Department of Physics, Faculty of Sciences, Mohammed First University, Oujda, Morocco
[2]Centre de Recherche de l'École des Hautes Études d'Ingénierie, Oujda, Morocco

Abstract

The objective of this chapter is to evaluate the potential suitability of the Oriental region of Morocco for hosting wind-generated hydrogen production units (Wind-H_2). In pursuit of this goal, we developed a high-resolution geographic information system (GIS) database, incorporating precise maps that integrate information on wind resources, infrastructure, terrain characteristics, ecological factors, and economic considerations. Subsequently, we utilized GIS analysis to identify areas unsuitable for accommodating the investigated technology, resulting in the creation of an exclusion mask. Using the GIS analysis results, we generated a refined land suitability map designed to pinpoint optimal locations in the Oriental region of Morocco capable of supporting large-scale Wind-H_2 units.

Once suitability for Wind-H_2 units was established, techno-economic analysis was done to evaluate the capacity of hydrogen production and the costs—based on the levelized cost of hydrogen ($LCOH_2$) metric—for eastern Morocco. The results indicate that the study area is exceptionally well-suited for hosting Wind-H_2 units, as unsuitable locations account for only 23% of the overall area. Furthermore, a 100-MWe wind plant in the region can produce 3538 tons of hydrogen annually, with an $LCOH_2$ of 14.6 \$/kg.

**Corresponding author*: alami.univ.oujda@gmail.com

Mehmet Sankir and Nurdan Demirci Sankir (eds.) Towards Green Hydrogen Generation, (267–292)

***Keywords*:** Hydrogen, GIS, wind, exclusion mask, Eastern Morocco, $LCOH_2$

9.1 Introduction

Wind energy represents one of the rapidly growing green energy sources worldwide [1]. It has seen exponential growth in recent years due to decreasing costs and increased awareness of environmental protection [2].

Compared to other renewable sources, wind energy is considered one of the most competitive energy sources in terms of cost—in comparison to geothermal and solar [3]. However, there are still challenges to be overcome for wind energy to be more widely adopted, including the variability of wind energy production due to its intermittence.

In Morocco, the context of wind energy is very favorable [4]. The country benefits from a climate favorable to wind energy production and supportive public policies to encourage the development of the integration and installation of wind farms. In recent years, Morocco has significantly increased its wind energy capacity by installing new wind turbines across the country. The country plans to have 5 GW of installed renewable wind energy installed by 2030 [5].

Regarding hydrogen production, there are several technologies that can be used to produce hydrogen from wind energy [6]. The least expensive technology for hydrogen production from wind is water electrolysis. This technology uses electricity to separate water into hydrogen and oxygen [7]. However, there are also other technologies under development that could be more competitive in the future, such as alkaline photolysis.

The objective of this chapter is to determine the capacity of eastern Morocco for hydrogen production from wind energy using a GIS database and excluding non-favorable zones involves several steps [8]:

- Data collection: It is necessary to collect data on weather conditions and wind patterns for the eastern Morocco region. In our case, historical wind data, from real-time measurements, have been collected for several locations in eastern Morocco. Also, a high-resolution GIS database includes maps of the infrastructure, the accessibility, and the roads.
- Spatial analysis: Once the data were collected, a spatial analysis using GIS software was conducted to create a wind atlas.
- Exclusion of non-favorable zones: Zones considered non-favorable for wind energy production, such as those with

land restrictions and unstable weather conditions (low wind speed locations), were excluded from the analysis.

- Site suitability analysis Atlas: A final map presenting the suitable locations to host large-scale production of hydrogen units from wind power was developed by combining the hydrogen production Atlas and the restriction map.

Techno-economic analysis in which simulations of the production of electricity and hydrogen from the Best-selected locations from Table 9.1 (Bouarfa and Ich, where we have average wind speeds of 6.32 and 6.3 m/s). Also, the production costs have been evaluated based on the $LCOH_2$. The findings show that eastern Morocco has great potential to host wind-based hydrogen units as the favorable locations account for 77% of the total land area. Furthermore, an analysis of the levelized cost of hydrogen ($LCOH_2$) has been performed, demonstrating that the manufacturing expenditures exhibit competitiveness and practicality within the designated geographic zones. This underscores the potential of wind-based hydrogen production as a feasible alternative for advancing renewable energy within this specific locale.

This study aims to contribute to the existing knowledge on renewable energy planning and support policymakers, energy companies, and other stakeholders involved in the deployment of wind-powered hydrogen production facilities. The results of this study will provide valuable insights into the spatial distribution of feasible and exclusion areas for hydrogen production in eastern Morocco [20], facilitating informed decision-making and sustainable energy development practices.

9.2 Materials and Methods

In this study, we focus on utilizing GIS to analyze the feasibility and exclusion areas for hydrogen production from wind power in eastern Morocco. Hydrogen production from wind power offers significant potential for clean energy generation and can contribute to the region's sustainable development goals. This section aims to introduce the datasets and spatial analyses to identify areas with high wind power potential and assess their suitability for hydrogen production. Nevertheless, before that, a brief introduction to our field of study (eastern Morocco) is provided.

Eastern Morocco, one of Morocco's twelve administrative divisions, is positioned in the northeastern part of the country with a total land area of 90,127 km^2. The region is not very populated—in comparison to other regions in Morocco—where the population is 2,505,730 residents [9]. Geographically, Oriental is located in the northeast of Morocco, with its northern border along the Mediterranean Sea. It shares its eastern borders with Algeria. Additionally, the region shares a border with Spain, which makes it a strategic region to host green energy platforms and hydrogen production units. Also, this is the main reason to select this region to conduct this study.

9.2.1 Methodology

A geographic information system (GIS) can be used to identify favorable and exclusion areas for wind hydrogen [10] GIS is a powerful simulation tool enabling the analysis and the conception of geospatial data, making it an ideal tool for assessing the potential for the development of wind energy.

The first step in the analysis is the creation of a GIS database of relevant geospatial data, such as wind speeds and patterns, land use, population density, wildlife activity, and cultural and historical sites. In our case, some of these data were collected as shape files from government organizations, and others were prepared by us.

The subsequent phase involves the application of GIS tools for data analysis, with the objective of discerning regions conducive to wind energy development. This process entails generating maps illustrating wind speed distributions and patterns, along with the identification of suitable locations for installing wind turbines. Additionally, GIS would be used to identify exclusion areas, such as areas with high population density, areas with high wildlife activity, and areas with high cultural or historical significance. Finally, the GIS would be used to conduct environmental impact assessments, by creating maps that show the potential impacts of wind energy development on the local environment [11], such as noise and visual impacts on nearby communities. The methodology used is presented on Figure 9.1.

9.2.2 Wind Resources

It is widely recognized that the main factor to be considered while assessing the wind energy potential at a specific site is the wind speed. Wind speeds

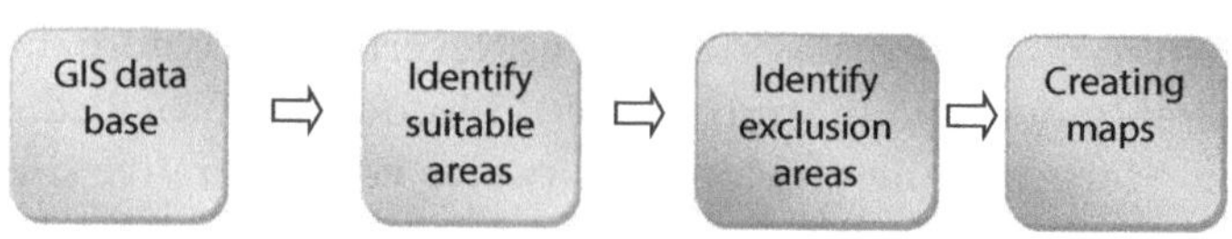

Figure 9.1 The steps followed for producing maps.

are generally higher in areas that are elevated and have fewer obstructions such as trees, buildings, and other structures. Besides, wind energy is typically most abundant near coastlines and in mountainous areas due to the influence of the ambient temperature and pressure differences.

In this study aimed at assessing the wind resource potential and establishing a wind speed atlas [12], the following procedure can be pursued:

- Determine the location: Identify the specific location where you want to assess the wind resource potential (eastern Morocco in this article); see Table 9.1.
- Gather wind data: Use measurements from global atlas wind data where we have wind speed and direction for the selected location.

Before using the wind data to create the maps, these data were carefully analyzed and their quality check and control were conducted. Figure 9.2 presents the wind rose and Figure 9.3 presents the monthly wind speed values of the selected locations.

As can be noticed from Figures 9.2 and 9.3, the wind in Oujda predominantly comes from the west totaling 13.8%. As for the wind speed, the diagram illustrates heightened wind velocities from December to April and

Table 9.1 Location of the meteorological stations [13].

Station	Latitude	Longitude	Elevation (m)	1.1 Average wind speed (m/s) (altitude, 10 m)
Oujda	34.65	−1.9	617	2.49
Nador	34.98	−3.01	178	3.25
Taourirt	34.39	−2.89	415	3.87
Bouarfa	32.52	−1.92	1202	6.32
Ich	32.51	−1	1278	6.3
Jerada	34.2	−2.09	1015	3.64
Mechraa Hammadi	34.73	−2.79	180	4.44
Mengoube	32.27	−2.35	995	4.46

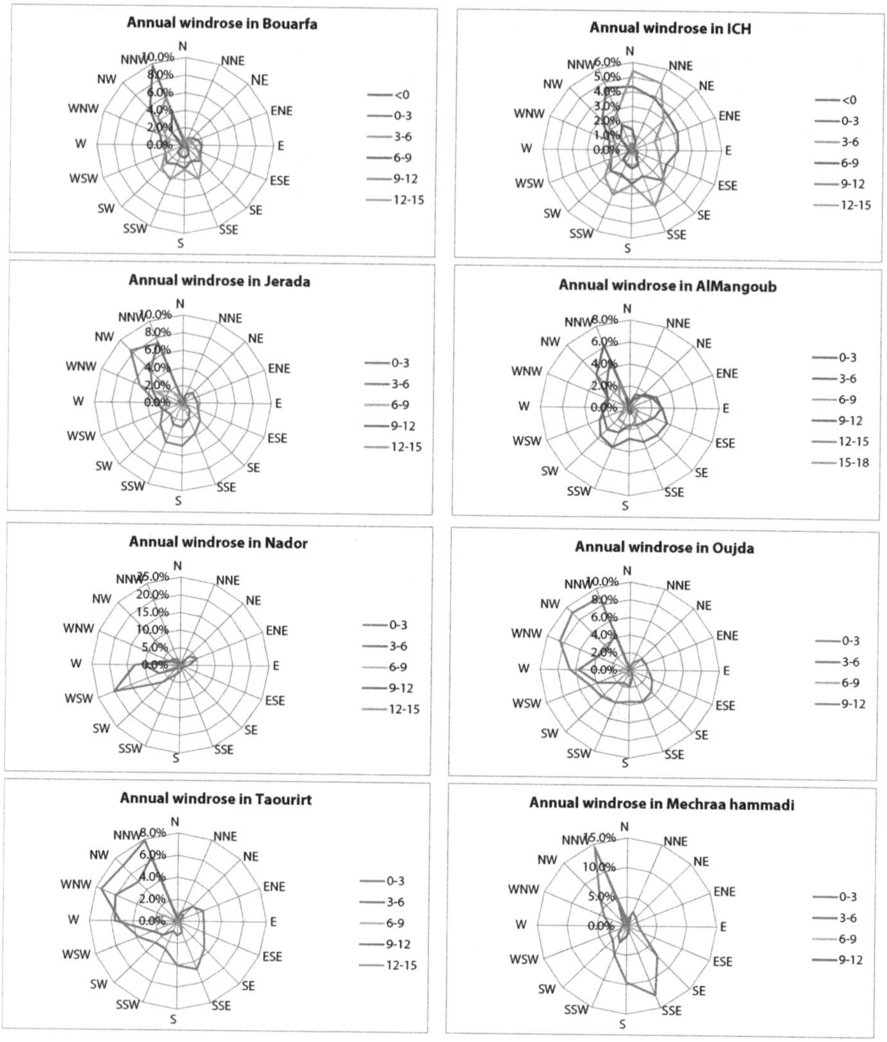

Figure 9.2 Wind frequency rose in cities in eastern Morocco.

calm winds from May to November. The minimum recorded wind speed is 2 m/s, while the maximum occurs at 4 m/s in January and February.

Likewise, in Almangoub, the prevailing wind direction is predominantly from the north-northwest, persisting for 14.2% annually. Elevated winds are observed from February to May, while subdued winds prevail from June to August. The minimum recorded wind speed, at 3 m/s, is noted in June and August, while the maximum occurs at 7 m/s in May.

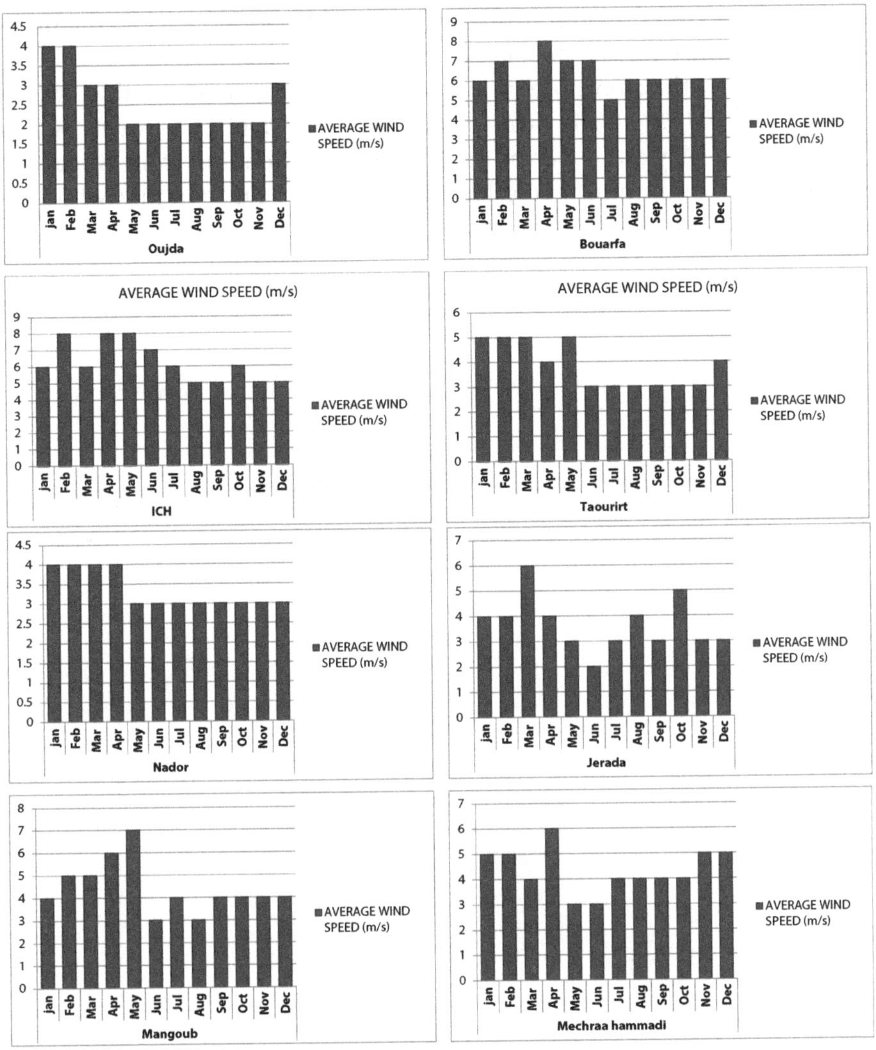

Figure 9.3 The monthly wind speed values of the selected locations.

In Bouarfa, the wind rose reveals that the wind blows from the north-northwest for 20.45%. Notably, heightened winds are experienced from February to June, while tranquil winds prevail from July to December. The recorded minimum wind speed, at 5 m/s, is consistently observed in July, while the maximum, reaching 8 m/s, occurs in April.

Similarly, in Mechraa Hammadi, the predominant wind direction is also from the north-northwest, totaling 17.8%. The recorded minimum wind speed, at 3 m/s, is consistently observed from May to June, while the maximum, reaching 6 m/s, occurs in April.

In Taourirt, the prevailing winds come from the west-northwest at a percentage of 15.1%. The wind speed diagram depicts elevated winds from April to May and a period of calm winds extending from August to December. The minimum recorded wind speed is 1 km/h in July, while the maximum reaches 21 km/h in April.

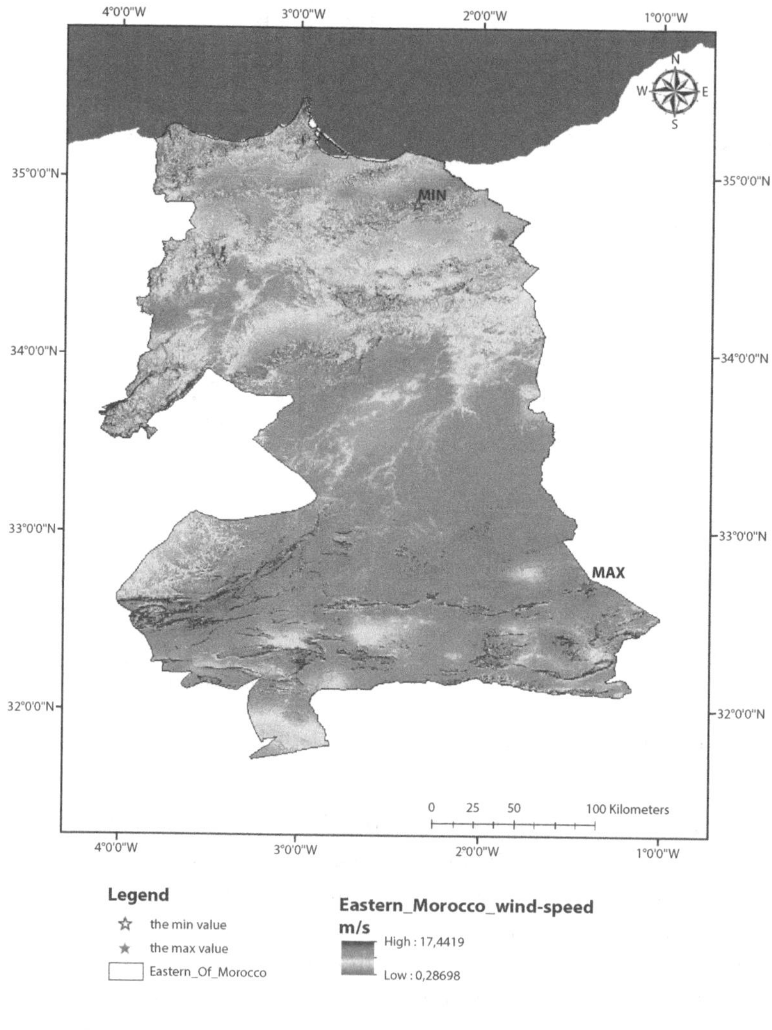

Figure 9.4 The wind map of eastern Morocco.

Moving on to Jerada, the wind predominantly blows from the north-northwest for 15.51%. The Jerada wind speed diagram illustrates heightened winds from Mars to April, with calm winds prevailing from June to July. The minimum recorded wind speed is 2 m/s in June, and the maximum is 6 m/s in Mars.

In Nador, winds predominantly come from the west-southwest for 27.3%. The diagram highlights heightened winds from January to April and calm winds from May to December. The minimum recorded wind speed is 3 m/s, while the maximum reaches 4 m/s.

Ich experiences wind from the north for 11.7%. The wind speed diagram indicates elevated winds from February to May and calm winds from August to December. The minimum recorded wind speed is 5 m/s, while the maximum is 8 m/s in February, April, and May [14]. To better evaluate the wind resources of our field of study, the wind speed atlas of Eastern Morocco is presented on Figure 9.4.

9.2.3 Land Restriction Maps

A restriction mask is a cartographic representation delineating the criteria to be excluded during the spatial analysis for the identification of areas unsuitable for hydrogen production from wind energy. In the literature, the development of a restriction map relies on the following steps:

- A land use analysis to determine the non-suitable areas for wind farm implementation. This analysis includes the determination of sensitive areas, such as nature reserves, national parks, and wetlands.
- An examination and exclusion of areas with high population density, significant wildlife activity, and notable cultural or historical importance assessing the environmental aspects having ecological consequences by wind farm implementation. This assessment would include an evaluation of impacts on wildlife, habitats, and landscapes, as well as an assessment of potential noise and visual impacts on neighboring communities.
- Technical limitations analysis by excluding all areas having low wind speed values as they are not suitable to host wind farms and, consequently, hydrogen production units.

To better present the exclusion and restriction parameters, Table 9.2 presents a literature review of the most used restriction criteria/parameters

Table 9.2 A description of the restriction criteria in the literature.

	Siyal *et al.* [15]	**Nguyen *et al.* [16]**	**Baban *et al.* [17]**	**Hofer *et al.* [18]**	**Villacreses *et al.* [19]**
Restrictions	Buffer	Buffer	Buffer	Buffer	Buffer
Main roads	200 m	100 m	10,000 m	V	—
Railways	200 m	100 m	—	100 m	—
Electricity network	200 m	—	10,000 m	100 m	—
Airports	2500 m	2500 m	—	—	2500 m
Military zones	V	—	—	—	—
Lakes, water sources, and shorelines	100 m	400 m	400 m	50 m	4000 m
Urban areas/large settlement	1000 m	2000 m	2000 m	550 m	3000 m
Single residential houses	500 m	—	500 m	—	—
Protected areas	V	500 m	—	300 m	250 m
Nature, culture/ woodland	V	500 m	500 m	300 m	—
Wind speed less than 2 m/s	—	—	—	—	—
Slope less than 10%	—	—	V	—	—
Slope less than 30%	—	—	—	V	—
Wind speed greater than 5m/s	—	—	V	—	—
Historic sites	—	—	1000 m	V	V
Natural trust property/ areal ecological value	—	—	1000 m	V	—
Wind speed greater than 6m/s	—	—	—	V	—
Mixed use area	—	—	—	400 m	—
Flood area	—	—	—	—	V
Volcanic hazard	—	—	—	—	V

in the case of wind farm implementation. As for our study, the restriction maps and parameters are grouped in Table 9.3.

In this study, for a technical aspect, we excluded all areas having wind speeds less than 2 m/s or greater than 5 m/s. Also, all areas with a slope higher than 10% have been excluded.

Similarly, we excluded (with a buffer of 200 m) all the lakes and water resources as a hydrogen production unit cannot be installed in such locations. Urban areas and large settlements were also excluded with a buffer zone of 1500 m.

Also, in this study, we exclude all locations of interest: nature, culture, and environment, and protected areas have a buffer zone of 500 m.

In the same vein, this study excluded the national roads, railroads, and highways with a buffer of 200 m. Similarly, the airports of the region were excluded with a buffer zone of 2500 m. The maps of the excluded area used in this study are summarized in Figure 9.5.

Table 9.3 A description of the restriction criteria in this study.

Restrictions	**Buffer zone**
Airports	2500 m
Locations of interest: nature, culture, and environment	500 m
Electricity grid	200 m
Lakes, water sources, and shorelines	200 m
National roads	200 m
Protected areas	500 m
Railroads	200 m
Slope less than 10%	10%
Urban areas/large settlement	1500 m
Wind speed greater than 5m/s	5 m/s
Wind speed less than 2m/s	2 m/s

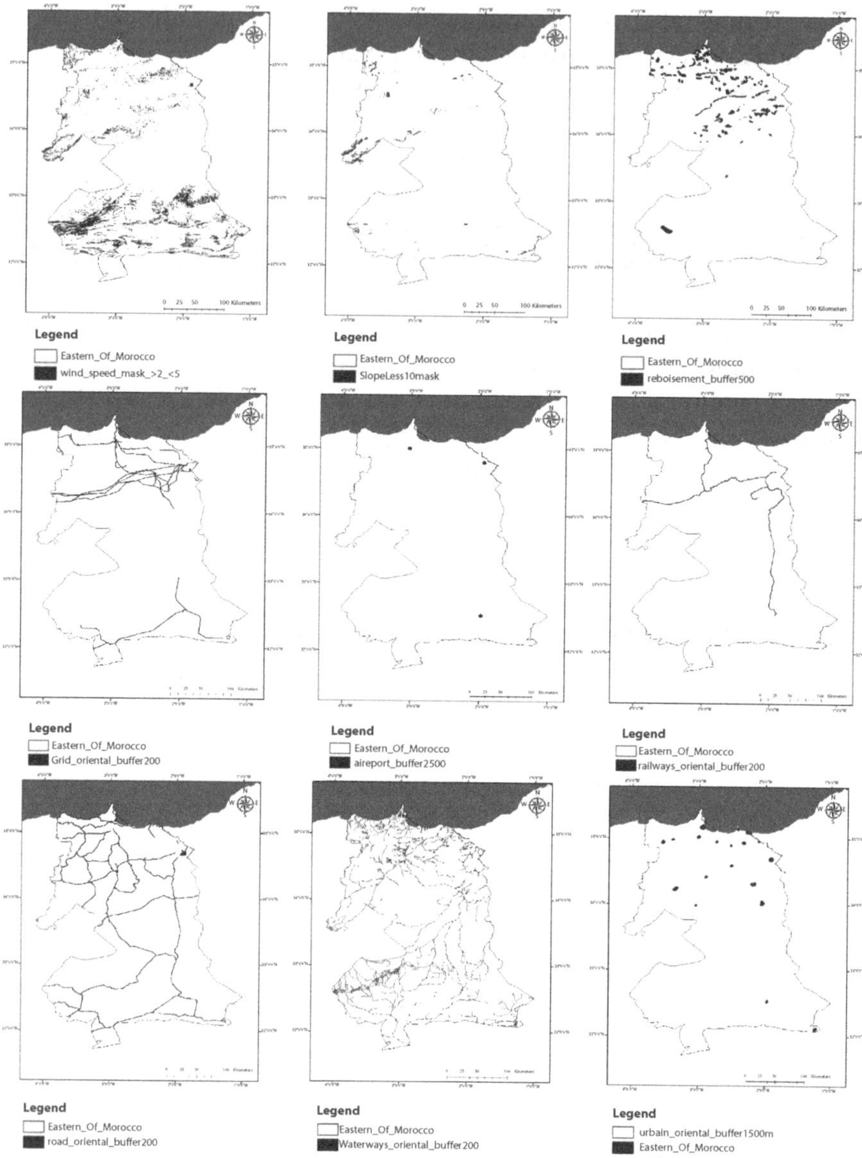

Figure 9.5 Zone of restrictions area in eastern Morocco.

9.2.4 Hydrogen Production and Economic Analysis

9.2.4.1 *Technical and Economic Analysis*

After determining eastern Morocco's capacity for hosting green hydrogen production plants, this section will delve into a techno-economic evaluation of the performance of a 100-MWe wind energy plant. This plant is designed for the dual production of electricity and hydrogen, situated in two selected sites, Bouarfa and Ich, known for their abundant wind resources. The analysis will extend to calculating and discussing the production costs associated with both electricity and hydrogen.

For technical simulations, we employ the Green Energy System Analysis Tool (Greenius), recognized as one of the most adept simulation software tools for analyzing renewable energy systems [19]. Following is a concise depiction of the model utilized by Greenius for simulating wind turbines:

The expression for the kinetic energy in the wind, denoted as power (in kilowatts, kW), is as follows [20]:

$$P_w = \frac{1}{2} \times \rho \times \mu \times v^3 \tag{9.1}$$

where ρ is the air density (Kg/m^3), μ. is the surface swept by the wind turbine (m^2), and v is the wind speed (m/s).

The energy produced by the wind turbine is noted P_{el}. and expressed by Equation 9.2 [21]:

$$P_{el} = \frac{1}{2} \times \rho \times C_e \times \mu \times v^3 \tag{9.2}$$

Where C_e is the efficiency factor, which relates to the system design, cversion unit, and wind speed.

$$C_e = C_P \times \eta_g \times \eta_m \tag{9.3}$$

Where C_P is the coefficient of performance of the wind turbine, η_g is the efficiency of the generator, and η_m is the efficiency of the gearbox [21].

The energy produced by the wind turbine is given by

$$E_{el} = P_{el} \times \Delta t \tag{9.4}$$

Table 9.4 The technical data of the wind energy plant.

Wind turbine	Gamesa G128
Number of turbines	20
Turbine height	95 m
Rotor diameter	128 m
Cut-in speed	2 m/s
Rated speed	13 m/s
Cutoff speed	27 m/s
Turbine rated power	5 MW

Greenius requires the hourly meteorological data of an entire year to run the simulation which includes wind speed, direction, ambient temperature, and pressure. In addition, Greenius requires the technical data of the simulated plant; the simulation inputs used in this study are gathered in Table 9.4.

9.2.4.2 *Hydrogen Production*

The electrical energy produced by the wind is used to produce hydrogen through an electrolyzer. In this study, we used the PEM electrolyzer (proton exchange membrane) which has high efficiency and long lifetime; in addition, it can work well with electricity produced from renewable energies, which may be unstable at times [23].

The masse of hydrogen produced (in kg) is calculated using the following expression [21, 23, 24]:

$$M_{H2} = \frac{E_{el} \cdot \eta_{ele}}{HHV_{H_2}} \tag{9.5}$$

where E_{el} (kWh) is the quantity of electricity produced by the wind turbine and η_{ele} is the efficiency of the electrolyzer (75%). The chemical energy content of a fuel is expressed by HHV (higher heating value), for the case of hydrogen, the $\mathrm{HHV_{H_2}}$ taken to be 39.4 kWh/kg [5]. In addition, a converter unit must be added to convert the AC current produced by the wind energy plant into a DC current required for the electrolysis process. The efficiency of such a converter is typically 92% [25].

9.2.4.3 *Economic Analysis*

In order to conduct an economic analysis of a renewable energy project, levelized costs of electricity (LCOE) and hydrogen $\left(LCO_{H_2}\right)$ were used. Indeed, LCOE and LCO_{H_2} are respectively the prices by which electricity and hydrogen are sold on the market.

LCOE ($/kWh) is calculated using the following formula [23, 24]:

$$LCOE = \frac{\sum_{i=0}^{n} (C_e(1+T)^{-i})}{\sum_{i=0}^{n} (E_{eli}\left(1+T\right)^{-i})} \tag{9.6}$$

where E_{el} represents the electricity generated by the wind turbine system, n is the project lifetime, T is the discount rate, and C_e ($) is the investment cost of the wind turbine system which can be calculated using Equation 9.6.

$$C_e = C_w + C_i + C_{OM} \tag{9.7}$$

where C_w (1150 $/k$W_p$) represents the cost of a wind turbine, C_i and C_{OM} are the installation and the maintenance and operation costs, where they represent respectively 20% and 25% of the wind turbine cost [23, 26, 27]. In addition, the converter unit cost in the case of hydrogen production is 130$/kW with a lifetime of 10 years [25].

The LCO_{H_2} ($/kg) is defined as follows:

$$LCO_{H_2} = \frac{\sum_{i=0}^{n} ((C_e + C_{ele})(1+T)^{-i})}{\sum_{i=0}^{n} (M_{H_2i}(1+T)^{-i})} \tag{9.8}$$

M_{H_2i} is the annual mass of hydrogen produced, and *i*, *n*, and *T* are the considered year, the lifespan of the project, and the discount rate, respectively. C_{ele} is the investment cost of the electrolyzer (expressed in $) which is expressed as a function of the electrolyzer cost, the replacement C_{rep}. cost, and the maintenance and operation cost C_{OM}. by the following equation:

$$C_{ele} = C_{el} + C_{rep} + C_{OM}. \quad (9.9)$$

electrolyzer capital cost is given by the following equation [28, 29]:

$$C_{el} = C_{el,u} \times \frac{M_{H_2} K_{el,th}}{8760\eta_{el}}. \quad (9.10)$$

Where $C_{el,u}$. (368 \$/kW_e) and $K_{el,th}$., (52.5 kWh/kg) are the electrolyzer unit price and the theoretical energy required, respectively, and η_{el} is the electrolyzer efficiency.

9.3 Results and Discussion

9.3.1 Wind Atlas of Eastern Morocco and Exclusion Mask

Wind Speed Variation: The average wind speeds in eastern Morocco show significant variation. Stations like Bouarfa and Ich, located in areas with higher elevations, experience much higher average wind speeds compared to stations at lower elevations like Nador. This elevation-related variation is typical, and it is an important factor to consider when assessing wind resources for potential energy production.

Figure 9.6 presents the exclusion mask of eastern Morocco. This mask gathers all the restriction criteria presented in the section above (Table 9.3). As it can be noticed the share of the non-suitable areas for hosting wind-based hydrogen units is 23%. This share is acceptable and makes our field of study a good location to host (see Figure 9.7) such technology.

9.3.2 Sites Suitability Analysis

Figure 9.8 presents the suitable locations to host wind farms and, consequently, wind-based hydrogen production units. As calculated, the share of the suitable areas represents 77% of the total land share of eastern Morocco. With this value, we can say that our field of study is a good spot for installing wind-hydrogen production units. Also, the best locations are mainly located in the southern part of the region and in the mid-west. The next section will discuss the techno-economic analysis of hydrogen unit production and costs.

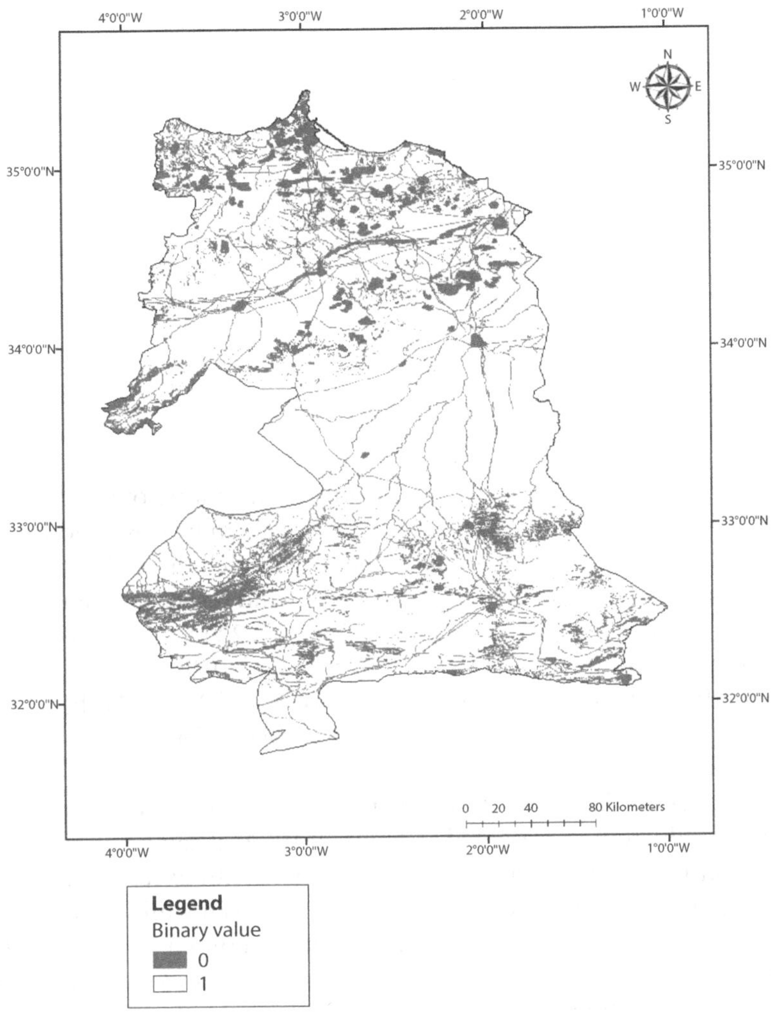

Figure 9.6 The exclusion mask of eastern Morocco for hydrogen production from wind energy.

9.3.3 Techno-Economic Analysis of Wind-Hydrogen Potential

9.3.3.1 Simulation Results

9.3.3.1.1 Electricity Production

The daily variation of electricity generation from the simulated wind energy plant during the year in the two selected sites is presented in Figure 9.9. It can be noticed that the production differs from one day to another

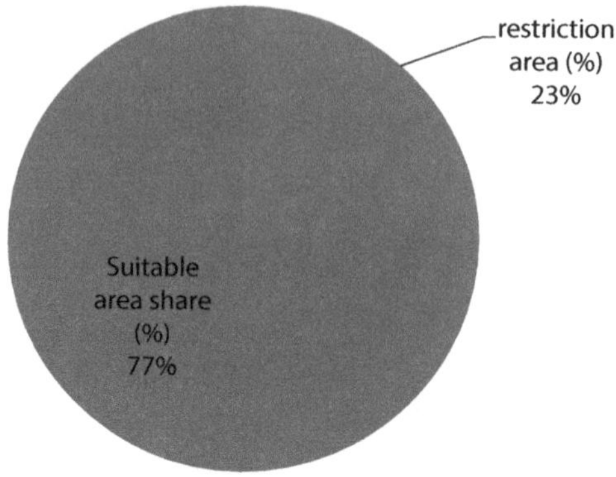

Figure 9.7 Statistic of restriction and feasible area.

with no apparent pattern; this is mainly due to the nature of wind velocity which is very fluctuant in space in time. Also, it can be observed that the electricity production in Bouarfa is higher than in Ich throughout the year which was expected since Bouarfa has higher wind resources than Ich with an annual mean speed at hub height of 5.55 and 4.22 m/s, respectively. The daily electricity production varies between 11 and 1926 MWh in Bouarfa, while for Ich the variation is in the range of 8 and 1491 MWh, whereas the daily average production is 553 and 315 MWh for Bouarfa and Ich, respectively.

For a better display of the electricity production throughout the year, the monthly production is presented in Figure 9.10. As it can be observed, the production in Bouarfa surpasses Ich in every month except in February, in fact, the monthly average in Bouarfa is 16.83 GWh while for Ich is 9.58 GWh. As for the monthly variation, it ranges between 10.13 and 26.42 GWh with the lowest production recorded in December and the highest in May, whereas in Ich the maximal production occurs in February with 15.2 GWh and the lowest in October with 6.68 GWh.

9.3.3.1.2 Hydrogen Production

Similarly to the previous section, the daily and monthly hydrogen productions are plotted in Figures 9.11 and 9.12, respectively. The first observation is that hydrogen production has a similar pattern as electricity production since the amount of hydrogen generated depends directly on the available electricity from the wind energy plant.

The daily hydrogen production (Figure 9.11) shows also a great fluctuation due to the high fluctuation in wind speed in both sites as it varies

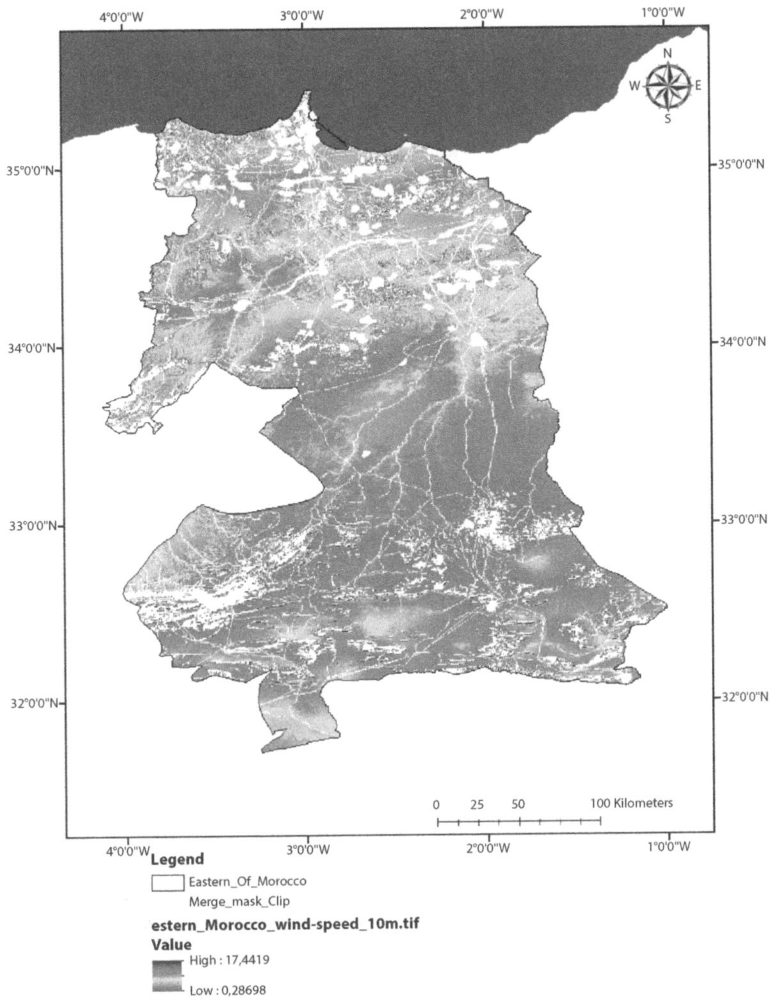

Figure 9.8 Suitable sites for the production of hydrogen from wind in eastern Morocco.

between 192.63 kg and 33.72 tons in Bouarfa and between 140.1 kg and 26.11 tons in Ich with a daily average of 9.68 and 5.51 tons in Bouarfa and Ich, respectively.

With respect to the monthly hydrogen production, the results are depicted in Figure 9.12, and it shows that the highest production is recorded in May for Bouarfa with 462.84 tons, and in February for Ich with 266.35 tons, whereas the lowest amounts are observed in December for Bouarfa with 177.49 tons and in October in ich with 117.12 tons while the monthly average is 249.8 and 167.84 tons in Bouarfa and Ich, respectively.

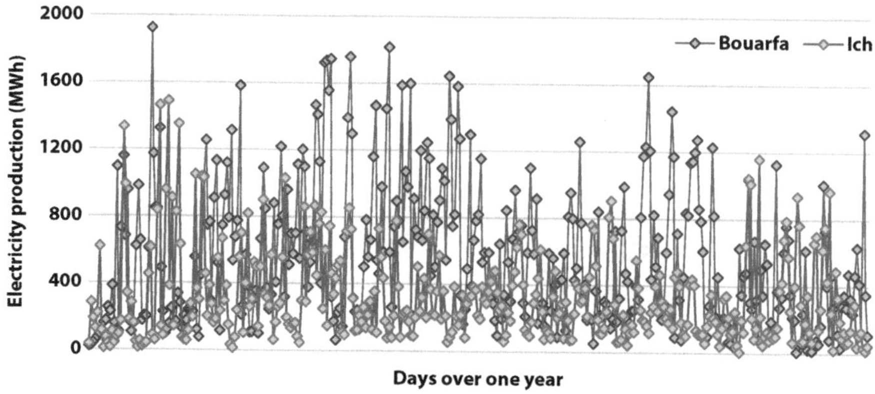

Figure 9.9 The daily electricity production in Bouarfa and Ich.

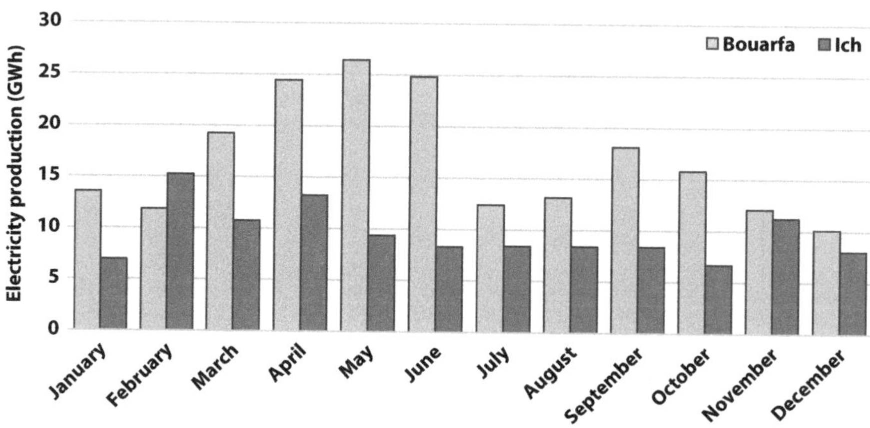

Figure 9.10 The monthly electricity production in Bouarfa and Ich.

The annual electricity production and its cost (LCOE) are presented in Figure 9.13, and it shows that in the site of Bouarfa, the generation is greater than Bouarfa as it reaches 202 GWh, whereas in Ich is 115 GWh. In addition, it can be noticed that the LCOE in Bouarfa is lower than Ich which is predictable since the LCOE depends on the amount of electricity generated, and it decreases with the increase of the electricity production. Accordingly, the LCOE in Bouarfa is 0.198 \$/kWh and 0.348 \$/kWh in Ich.

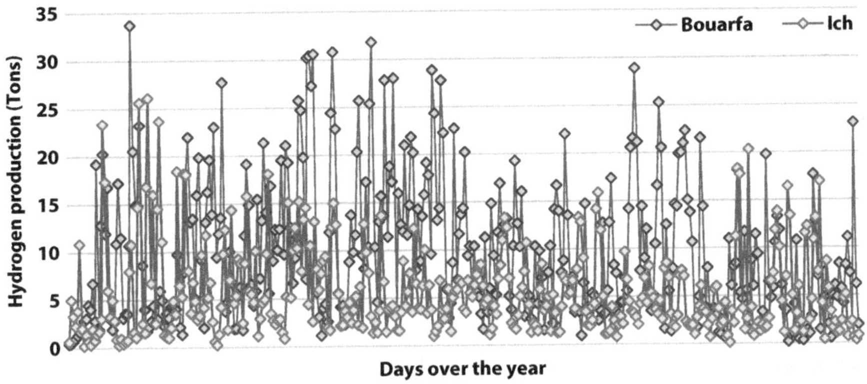

Figure 9.11 The daily hydrogen production in Bouarfa and Ich.

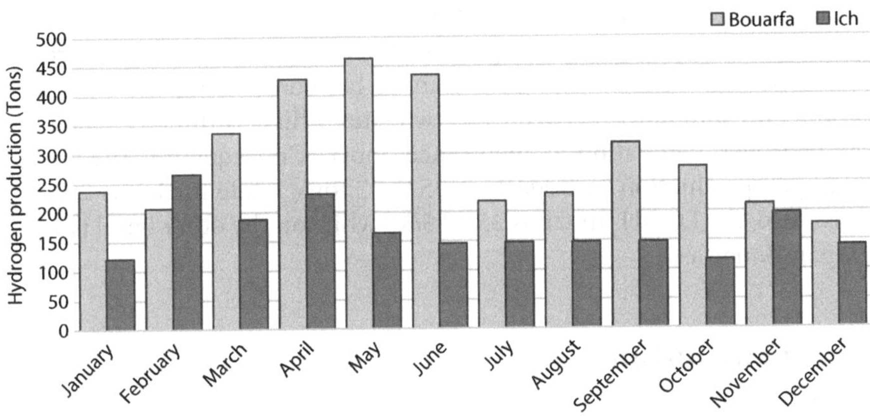

Figure 9.12 The monthly hydrogen production in Bouarfa and Ich.

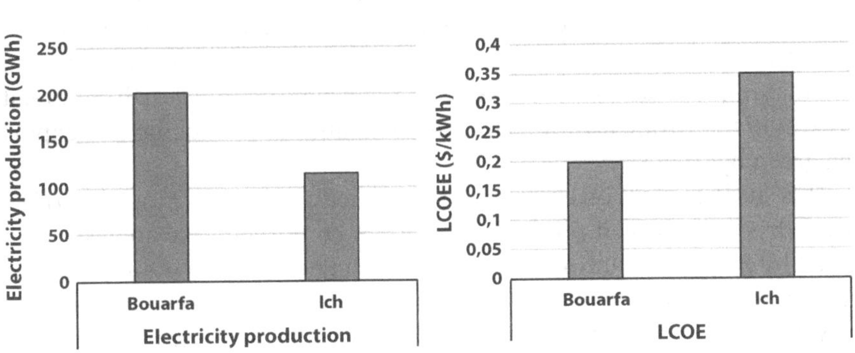

Figure 9.13 Annual electricity production and LCOE in Bouarfa and Ich.

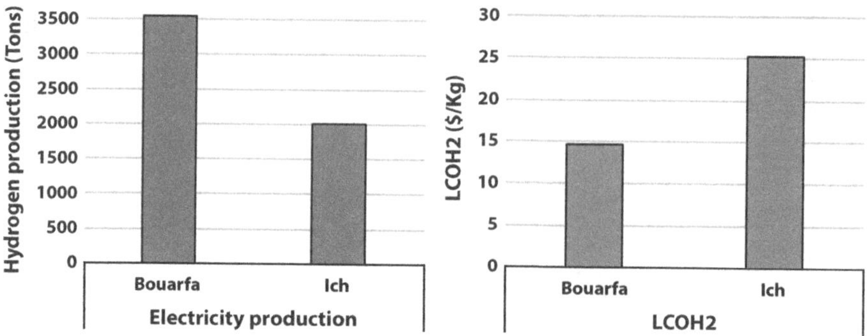

Figure 9.14 Annual hydrogen production and $LCOH_2$ in Bouarfa and Ich.

9.3.3.2 Techno-Economic Comparison Between the Selected Sites

With respect to hydrogen production, the yearly productions as well as the $LCOH_2$ are presented in Figure 9.14. Also in this case, the hydrogen production in Bouarfa is higher than in Ich considering the electricity production difference between the two sites which is directly related to the hydrogen generation as mentioned above. Consequently, the annual hydrogen production in Bouarfa is 3537.77 tons, while for Ich is 2015.01 tons. Also, the $LCOH_2$ in Ich is 25.29 $/kg while on the other hand is 14.60 $/kg in Bouarfa.

9.4 Conclusion

Given the escalating significance of green hydrogen amid the global shift toward sustainable energy, our assessment of the Oriental region of Morocco for deploying wind-based hydrogen production units (Wind-H_2) holds particular relevance. This is especially pertinent for Morocco, a nation heavily reliant on importing more than 90% of its energy needs and making substantial investments in renewable energy. Furthermore, its proximity to Europe positions it as an ideal hub for producing and exporting green hydrogen.

This study focuses on evaluating the capacity of eastern Morocco to host wind-driven hydrogen production units. A comprehensive analysis, facilitated by a high-resolution GIS database, ranging from wind resources to economic considerations, highlights the pivotal role green hydrogen could play in Morocco's sustainable energy landscape. By identifying optimal locations for large-scale Wind-H_2 units, our study not only contributes to the broader discourse on renewable energy practices but also strategically positions eastern Morocco as a promising region for advancing Wind-H_2 technology. With only 23% of the region deemed unsuitable, the majority of the landscape appears favorable for Wind-H_2 units, reinforcing the potential for green hydrogen to become a crucial component of Morocco's renewable energy portfolio.

After analyzing exclusion zones and areas favorable for hydrogen production from wind energy, coupled with a review of wind speed data from various stations in the Oriental region of Morocco, we selected two cities with consistently high average wind speeds (BOUARFA and ICH). These cities serve as focal points for simulating the operation of a hydrogen production plant powered by wind energy.

In conclusion, our findings reveal that in Bouarfa, more electricity is produced (202 GWh) compared to Ich (115 GWh), and the cost of electricity in Bouarfa is lower ($0.198/kWh) than in Ich ($0.348/kWh). This correlation aligns with the expectation that the cost of electricity tends to decrease as production increases.

Shifting the focus to hydrogen production, simulation results project that a 100-MWe wind plant in Bouarfa can produce more hydrogen (3537.77 tons) than Ich (2015.01 tons) annually. The cost of producing hydrogen (LCOH_2) is also more economical in Bouarfa ($14.60/kg) compared to Ich ($25.29/kg). In simpler terms, Bouarfa emerges as the preferred choice for establishing wind-based hydrogen production units in the eastern region of Morocco due to its higher electricity and hydrogen production and lower production costs.

References

1. Khan, F.H., Pal, T., Kundu, B., Roy, R., Wind Energy: A Practical Power Analysis Approach, in: *2021 Innovations in Energy Management and Renewable Resources(52042)*, Kolkata, India, pp. 1–6, 2021, doi: 10.1109/IEMRE52042.2021.9386754.
2. de Simón-Martín, M. *et al.*, Wind energy planning for a sustainable transition to a decarbonized generation scenario based on the opportunity cost of the wind energy: Spanish Iberian Peninsula as case study. *Energy Proc.*, 157, 1144–1163, 2019, https://doi.org/10.1016/j.egypro.2018.11.282.
3. Ochoa, G.V., Alvarez, J.N., Acevedo, C., Research Evolution on Renewable Energies Resources from 2007 to 2017: A Comparative Study on Solar, Geothermal, Wind and Biomass Energy. *Int. J. Energy Econ. Policy*, 9, 6, 242–253, Oct. 2019, https://doi.org/10.32479/ijeep.8051.

4. Eisner, P., Continental-scale assessment of the African offshore wind energy potential: Spatial analysis of an under-appreciated renewable energy resource. 405, 394–407, 2019, https://doi.org/10.1016/j.rser.2019.01.034.
5. Hutchinson, M. and Zhao, F., Global Wind Energy Council, 2023, https://gwec.net/wp-content/uploads/2023/04/GWEC-2023_interactive.pdf.
6. Ogden, J.M., Hydrogen: The Fuel of the Future? 71, 69–75, 2002, https://doi.org/10.1063/1.1480785.
7. Alavi, O. *et al.*, Influence of noise of wind speed data on a wind-hydrogen system. 22751, 22751–22759, 2016, https://doi.org/10.1016/j.ijhydene.2016.10.032.
8. Siyal, S. *et al.*, Mapping key economic indicators of onshore wind energy in Sweden by using a geospatial methodology. 224, 211–226, 2016, https://doi.org/10.1016/j.enconman.2016.09.055.
9. Haut Commissariat au Plan, Le Maroc en chiffres, 2023, (version arabe & française).pdf.
10. Griffiths, J.C. and Dushenko, W.T., Effectiveness of GIS suitability mapping in predicting ecological impacts of proposed wind farm development on Aristazabal Island, BC. *Environ. Dev. Sustain.*, 13, 957–991, 2011, https://doi.org/10.1007/s10668-011-9300-1.
11. Copping, A.E., Gorton, A.M., May, R., Bennet, F., DeGeorge, E., Repas Goncalves, M., Rumes, B., Enabling Renewable Energy While Protecting Wildlife: An Ecological Risk-Based Approach to Wind Energy Development Using Ecosystem-Based Management Values. *Sustainability, 12*, 9352, 2020, https://doi.org/10.3390/su12229352.
12. Badger, J. *et al.*, 2023, https://globalwindatlas.info/en.
13. Buck, A., 2023, https://www.meteoblue.com/en/weather/historyclimate/climatemodelled.
14. Pogumirskis, M., Sīle, T., Seņņikovs, J., Bethers, U., PCA analysis of wind direction climate in the baltic states. *Tellus A: Dyn. Meteorol. Oceanogr.*, 73, 1, 116, 2021, doi: 10.1080/16000870.2021.1962490.
15. Siyal, S.H. *et al.*, Wind energy assessment considering geographic ans environmental restrictions in Sweden a GIS based approach. 450, 447–461, 2015, https://doi.org/10.1016/j.energy.2015.02.044.
16. Nguyen, K.Q. *et al.*, Wind energy in Vietnam Resource assessment development status and future implication. 1408, 1405–1413, 2007, https://doi.org/10.1016/j.enpol.2006.04.011.
17. Baban, S.M.J. *et al.*, Developing and applying a GIS-assisted approach to locating wind farms in the UK. 63, 59–71, 2001, https://doi.org/10.1016/S0960-1481(00)00169-5.
18. Hofer, T. *et al.*, Wind farm sitting using a spatial analytic Hierarchy Process approach a case study of Stadteregion Aachen. 229, 222–243, 2016, https://doi.org/10.1016/j.apenergy.2015.10.138.
19. Villacreses, G. *et al.*, Wind farms suitability location using geographical information system GIS based on multi-criteria decision making MCDM

methods the case of continental Ecuador. 280, 275–286, 2017, https://doi.org/10.1016/j.renene.2017.03.041.

20. Amrani, S.-E., Merrouni, A.A., Touili, S., Dekhissi, H., A multi-scenario site suitability analysis to assess the installation of large scale photovoltaic-hydrogen production units. Case study: Eastern Morocco. *Energy Convers. Manage.*, 295, 117615, nov. 2023, doi: 10.1016/j.enconman.2023.117615.
21. Rahmouni, S., Negrou, B., Settou, N., Dominguez, J., Gouareh, A., Prospects of hydrogen production potential from renewable resources in Algeria. *Int. J. Hydrogen Energy*, 42, 2, 13831395, janv. 2017, doi: 10.1016/j.ijhydene.2016.07.214.
22. Nikolaidis, P. and Poullikkas, A., A comparative overview of hydrogen production processes. *Renew. Sustain. Energy Rev.*, 67, 597611, janv. 2017, doi: 10.1016/j.rser.2016.09.044.
23. Touili, S., Merrouni, A.A., Hassouani, Y.E., Amrani, A., Azouzoute, A., Techno-economic investigation of electricity and hydrogen production from wind energy in Casablanca, Morocco. *IOP Conf. Ser.: Mater. Sci. Eng.*, 948, 1, 012012, nov. 2020, doi: 10.1088/1757-899X/948/1/012012.
24. Touili, S., Merrouni, A.A., Azouzoute, A., El Hassouani, Y., Amrani, A., A technical and economical assessment of hydrogen production potential from solar energy in Morocco. *Int. J. Hydrogen Energy*, 43, 51, 2277722796, déc. 2018, doi: 10.1016/j.ijhydene.2018.10.136.
25. Boudries, R., Techno-economic study of hydrogen production using CSP technology. *Int. J. Hydrogen Energy*, 43, 6, 34063417, févr. 2018, doi: 10.1016/j.ijhydene.2017.05.157.
26. Diaf, S. and Notton, G., Technical and economic analysis of large-scale wind energy conversion systems in Algeria. *Renew. Sustain. Energy Rev.*, 19, 3751, mars 2013, doi: 10.1016/j.rser.2012.11.026.
27. Ayodele, T.R., Ogunjuyigbe, A.S.O., Amusan T, O., Wind power utilization assessment and economic analysis of wind turbines across fifteen locations in the six geographical zones of Nigeria. *J. Cleaner Prod.*, 129, 341349, août 2016, doi: 10.1016/j.jclepro.2016.04.060.
28. Gouareh, A. *et al.*, GIS-based analysis of hydrogen production from geothermal electricity using CO2 as working fluid in Algeria. *Int. J. Hydrogen Energy*, 40, 44, 1524415253, nov. 2015, doi: 10.1016/j.ijhydene.2015.05.105.
29. Touili, S., Merrouni, A.A., El Hassouani, Y., Amrani, A., Rachidi, S., Analysis of the yield and production cost of large-scale electrolytic hydrogen from different solar technologies and under several Moroccan climate zones. *Int. J. Hydrogen Energy*, 45, 51, 2678526799, oct. 2020, doi: 10.1016/j.ijhydene.2020.07.118.

10

Bifunctional Electrocatalyst–Driven Hybrid Water Splitting for Energy-Saving Coproduction of Green H_2 and Valuable Chemicals

Hui Jiang, Guoliang Mei and Bo You*

Key Laboratory of Material Chemistry for Energy Conversion and Storage, Ministry of Education, Hubei Key Laboratory of Material Chemistry and Service Failure, School of Chemistry and Chemical Engineering, Huazhong University of Science and Technology, Wuhan, Hubei, China

Abstract

Electrocatalytic water splitting to generate green H_2 with renewable energy inputs is attractive to relieve the environmental contamination and energy dilemma. Restricted by the sluggish kinetics of oxygen evolution reaction (OER) and single functionality of cathode/anode catalysts, conventional overall water splitting (OWS) suffers from high overpotentials and respective synthesis of monofunctional electrocatalysts, leading to high cost. Moreover, the generated O_2 bears low value and its possible mixing with H_2 poses safety issues. In response, hybrid water splitting (HWS) that integrates hydrogen evolution reaction (HER) with favorable oxidative upgrading reactions over bifunctional electrocatalysts can generate H_2 and value-added chemicals simultaneously with lower energy input and avoid the explosive H_2/O_2 mixtures. Hence, this chapter focuses on the recent progress in bifunctional electrocatalyst–driven HWS for energy-saving coproduction of highly valuable chemicals and green H_2. Firstly, the fundamentals of HER, OER, OWS, and HWS are introduced, followed by a discussion on the electrochemical reconstruction of bifunctional electrocatalysts during OWS and HWS. Next, the HWS based on bifunctional electrocatalysts is summarized, wherein the alternative oxidation reactions include upgrading of alcohol, aldehydes, amines, biomass,

**Corresponding author*: youbo@hust.edu.cn

Mehmet Sankir and Nurdan Demirci Sankir (eds.) Towards Green Hydrogen Generation, (293–358)

and plastic waste. Finally, a brief overview of current challenges and perspective for bifunctional electrocatalyst–driven HWS is discussed.

Keywords: Water splitting, electrocatalysis, bifunctional, green hydrogen, oxidative upgrading

10.1 Introduction

The environmental contamination and energy dilemma caused by the combustion of nonrenewable fossil fuels accelerate the exploration of renewable and clean energy sources [1–9]. Due to the zero pollution and high energy density after combustion, hydrogen is a promising alternative, and it has been widely employed in the petrochemical industry, ammonia production, metal smelting, and fuel desulfurization [10–15]. To achieve global net zero emissions by 2050, the Hydrogen Council estimates that the yearly production capacity of hydrogen may reach 16 to 24 million tons (Mt) with an annual demand of 100 Mt by 2030 [16–20]. Note that the global hydrogen market is anticipated to reach $209 billion in 2027. To satisfy the increasing demand for hydrogen energy, green, highly efficient, and low-cost technologies to generate hydrogen will play an important role in the future hydrogen economy. Compared with traditional H_2 production based on fossil fuels, such as the gasification of coal and reforming of steam-methane [21–25], electrocatalytic water splitting driven by a sustainable energy source is deemed as the most promising sustainable and environmentally friendly technique [26–30].

Two half-reactions related to water electrolysis are hydrogen evolution reaction (HER) and oxygen evolution reaction (OER). There are three typical electrochemical water splitting technologies: solid oxide electrolysis (SOE), alkaline water electrolysis (AWE), and proton exchange membrane water electrolysis (PEMWE). Among them, AWE, reported by Troostwijk and Diemann in 1789 [31], uses the concentrated alkali electrolyte (generally 20 to 30 wt% KOH or NaOH) and operates at 50°C to 80°C to endure maximum electrical conductivity. Due to the usage of non-noble Ni-based electrodes, low-cost cell components, and high durability, it has been extensively implemented worldwide [32–34]. While the Ni-based materials only show modest activities, the involved asbestos diaphragms may result in high resistance for ion migration and possible gas crossover [33]. These issues can be partially addressed with the development of anion exchange membranes. In contrast, PEMWE, which mostly employs Ru/Ir-based materials as anode catalysts and Pt-based materials as cathode catalysts, benefits from high proton conductivity, low gas permeability, and thin thickness of the PEM [31], offering advantages of compact design, high operating current density (above 2000 mA cm^{-2}), high energy conversion efficiency, high dynamic range, and high purity of produced gases. Nonetheless, these merits also come with a cost of the requirement of noble metal–based electrocatalysts to drive the two half-reactions, leading to excessive costs and limited large-scale applications [35]. SOE is a high-temperature electrolysis technology with oxide ion–conducting ceramics as the solid electrolyte and separator. At the elevated working temperature (800°C to 1000°C), the water vapor is initially reduced into H_2 and O^{2-} at the cathode [31, 36–38]. The generated O^{2-} ions migrate across the solid electrolyte to the anode, leading to the formation of oxygen (O_2). SOE uses industrial waste heat at elevated temperatures to effectively reduce the amount of input electrical energy. Moreover, the ceramic membrane, with excellent airtightness, makes it possible to generate hydrogen with high purity. However, SOE is still at its early stage with significant stability issues under harsh conditions [39–41]. By contrast, both AEM and PEMWE can work in relatively mild circumstances, exhibiting promising potential for hydrogen generation *via* water electrolysis. Considering the sluggish kinetics of two half-reactions, as well as the relatively high cost associated with the separate and multistep synthesis of two different electrocatalysts, the design and preparation of bifunctional electrocatalysts with excellent electrocatalytic performance and inexpensive price for water splitting have attracted a large amount of attention.

From the perspective of energy conversion efficiency, the generation of H_2 from water electrolysis is predominantly determined by the OER at the anode due to its multiple proton-coupled electron transfer nature. Additionally, the generated O_2 byproduct is low value and the reactive oxygen species (ROS) can degrade PEM, leading to possible gas crossover and thus posing safety issues [42, 43]. To alleviate these issues, researchers are motivated by exploring alternative anodic reactions that are more thermodynamically favorable and economically attractive to replace OER [44], such that a novel paradigm termed hybrid water splitting (HWS) is developed. Based on the functionalities of the substrates, we can categorize these reactions into three main groups: sacrificing agent oxidation which consumes valuable substrates, pollutants oxidative degradation which uses environmental pollutants as oxidative substrates, and oxidative upgrading reaction which obtains value-added products. All these three types of oxidation reactions allow a much lower cell voltage for water splitting when hybridizing with HER, benefiting from their low oxidative potential compared to OER. The drawback of sacrificing agent oxidation is the transformation of valuable agents (e.g., hydrazine, ammonia, and urea) to worthless products, which increases the cost of the system. The pollutant's oxidative degradation, when coupled into the HWS system, helps with pollutant (e.g., formic acid/formate, phenol, thiourea/sodium thiocyanate, and sulfion) removal from wastewater streams. The oxidative upgrading reaction is of the most interest because of its ability to produce value-added chemicals in the anode, and we will focus on this type of HWS in the following discussion. Possible reactions in this category include alcohol oxidation, aldehyde oxidation, amine oxidation, biomass oxidation, and plastic waste oxidation.

In this chapter, we offer a systematic summary of recent achievements in HWS coupled with oxidative upgrading driven by bifunctional electrocatalysts. We first clarify the mechanisms of HER, OER, overall water splitting (OWS), and HWS. Next, the reconstruction of electrocatalysts during both reduction (HER) and oxidation (OER and alternative oxidation reactions) was discussed in detail. After that, we elaborate on several bifunctional electrocatalyst–driven HWS systems coupled with different oxidative upgrading reactions, which could generate H_2 and highly valuable chemicals at both electrodes. The final section provides a concise conclusion with future challenges pertaining to the development of bifunctional electrocatalysts for more large-scale HWS applications.

10.2 Fundamentals of Hybrid Water Splitting

10.2.1 Mechanism of HER

As shown in Figure 10.1a, the HER is a two-electron process. Based on the pH of the electrolyte, the proton needed to produce H_2 comes from hydronium ion (H_3O^+) in acidic conditions or H_2O in alkaline/neutral conditions. Specifically, H_3O^+ in acidic media could combine with an electron to form the adsorbed hydrogen atom (H*) at active sites (*) of the electrocatalyst which is known as the Volmer step (Equation 10.1). Subsequently, the absorbed H* could convert to H_2 *via* the Volmer-Heyrovský or Volmer-Tafel step. Specifically, the H* combines with an electron and another H_3O^+ (Heyrovský step, Equation 10.2) or with another adjacent H* (Tafel mechanism, Equation 10.3) to form H_2.

$$\text{Volmer step: } H_3O^+ + e^- + * \rightarrow H^* + H_2O \tag{10.1}$$

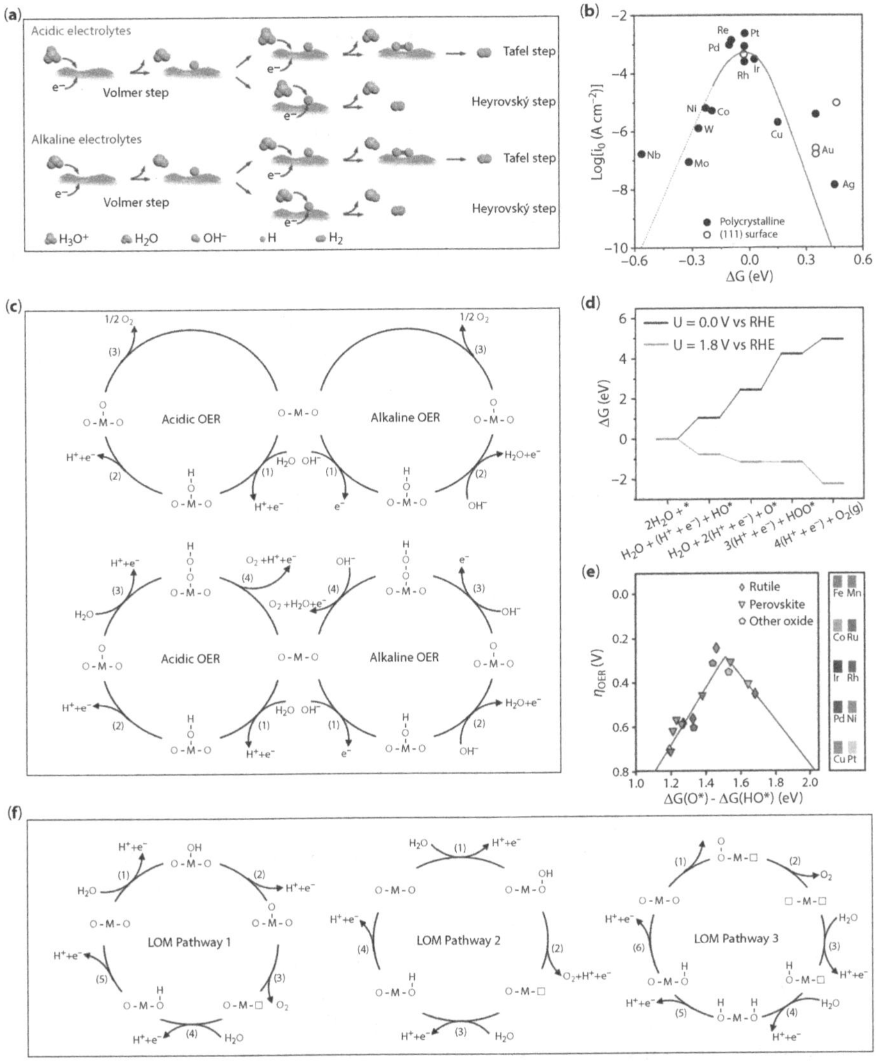

Figure 10.1 (a) Mechanism of HER electrocatalysis in acidic and alkaline solutions. Reproduced with permission from [69]. Copyright 2020 John Wiley & Sons. (b) The relationship between activity and ΔG of hydrogen adsorption. The curve and circles are from microkinetic theory and experimental data, respectively. Reproduced with permission from [70]. Copyright 2019 John Wiley & Sons. (c) Schematic of the AEM mechanism for OER. Reproduced with permission from [71]. Copyright 2022 John Wiley & Sons. (d) The free energy diagrams for AEM-based OER at different potentials. (e) OER volcano plot for various transition metal oxides against $\Delta G_{O*}-\Delta G_{HO*}$. Reproduced with permission from [65]. Copyright 2017 Springer Nature. (f) Schematic of the LOM mechanism for OER. Reproduced with permission from [71]. Copyright 2022 John Wiley & Sons.

$$\text{Heyrovsky step: } H_3O^+ + e^- + H^* \rightarrow H_2 + H_2O \tag{10.2}$$

$$\text{Tafel step: } H^* + H^* \rightarrow H_2 + 2* \tag{10.3}$$

Similarly, the HER in alkaline/neutral media could also occur by either the Volmer-Heyrovský or Volmer-Tafel steps. Owing to the lack of H^+, the *H originated from the water molecule, which could accept an electron and dissociate to H* and OH^- (Volmer step, Equation 10.4). Afterward, the generated H* captures a water molecule and an electron to produce H_2 (Heyrovský step, Equation 10.5), or two adjacent H* could combine to generate H_2 (Tafel step, Equation 10.6).

$$\text{Volmer step: } H_2O + e^- + * \rightarrow H^* + OH^- \tag{10.4}$$

$$\text{Heyrovsky step: } H_2O + e^- + H^* \rightarrow H_2 + OH^- \tag{10.5}$$

$$\text{Tafel step: } H^* + H^* \rightarrow H_2 + 2* \tag{10.6}$$

These elementary steps reveal that electron transfer and hydrogen adsorption play an important role in hydrogen evolution. Meanwhile, it is apparent that acidic media are generally more advantageous for the HER due to the greater challenge of breaking the H-OH bond to produce H* in alkaline and neutral electrolytes. In principle, any of these three elementary steps could serve as the rate-determining step (RDS) of HER over the real electrocatalysts, and the calculated value of Tafel slope is generally considered as the indicator of RDS. Specifically, the Tafel slopes of Tafel, Heyrovský, and Volmer steps are approximately 30, 39, and 118 mV dec^{-1}, respectively [45–51].

As for the HER, amounts of reports proposed that the interaction between electrode materials (mainly metals) and H* plays a critical role in HER kinetics [52–54] and the volcano plot describing the relationship between ΔG (hydrogen binding free energy at the metal surface) and i_0 (exchange current density) has been constructed (Figure 10.1b) [55]. It is revealed that optimal H* adsorption is beneficial to improve HER kinetics [56]. In various computational technologies, density functional theory (DFT) has gathered a lot of attention and has been recognized as a cogent tool to explore the electrochemical mechanism at the atomic level. Based on the DFT calculations, the computational energetics of elementary steps from the perspectives of both thermodynamics and kinetics can be obtained, making it possible to distinguish potential active centers and RDS in a specific reaction process. Regarding the acidic HER, the ΔG of H* (ΔG_{H*}) is generally employed as the descriptor to evaluate the activity of electrocatalysts. Meanwhile, the theoretical predictions based on DFT calculation coincide well with the experimental results as depicted in Figure 10.1b.

10.2.2 Mechanism of OER

The mechanism of OER refers to a complicated four-electron transfer process, which contains a series of successive elementary steps. In general, two possible mechanisms are involved in the OER: the traditional adsorbate evolution mechanism (AEM) (Figure 10.1, c to e) and the lattice oxygen-mediated mechanism (LOM) (Figure 10.1f) [57, 58].

For AEM in acidic electrolytes,

$$H_2O + * \rightarrow HO^* + H^+ + e^- \tag{10.7}$$

$$HO^* \rightarrow O^* + H^+ + e^- \tag{10.8}$$

$$2O^* \rightarrow 2* + O_2 \tag{10.9}$$

$$\text{or } O^* + H_2O \rightarrow HOO^* + H^+ + e^- \tag{10.10}$$

$$HOO^* \rightarrow * + O_2 + H^+ + e^- \tag{10.11}$$

For AEM in alkaline electrolytes,

$$OH^- + * \rightarrow HO^* + e^- \tag{10.12}$$

$$HO^* + OH^- \rightarrow O^* + H_2O + e^- \tag{10.13}$$

$$2O^* \rightarrow 2* + O_2 \tag{10.14}$$

$$\text{or } O^* + OH^- \rightarrow HOO^* + e^- \tag{10.15}$$

$$HOO^* + OH^- \rightarrow * + O_2 + H_2O + e^- \tag{10.16}$$

The initial step for AEM involves the generation of HO* on the active site (*). In acidic electrolytes, the breaking of strong covalent HO–H bonds in H_2O is necessary to commence the first step due to the absence of free OH^- (Equation 10.7), which could result in slower kinetics, whereas the free OH^- in alkaline electrolytes could be directly absorbed on the active sites *via* a one-electron oxidation process (Equation 10.12). After another one-electron oxidation process (Equations 10.8 and 10.13), the formed O* intermediates could form the oxygen (O_2) through two different pathways. The direct reaction involves the coupling of O* intermediates to produce O_2 (Equations 10.9 and 10.14), or the O* intermediate could convert into HOO* intermediate by reacting with H_2O (in acidic media) (Equation 10.10) or OH^- (in alkaline media) (Equation 10.15), and then the HOO* intermediate decomposes to form O_2 (Equations 10.11 and 10.16). It is worth noting that the later pathway possesses a lower thermodynamic barrier than that of the direct coupling of two O* [59]. From the aspects of both thermodynamics and kinetics, the elementary step that possesses the highest energy barrier is typically considered as the RDS. It has been verified by Bockris and co-workers that the obtained Tafel slope based on the electrochemical tests intensely relies on the RDS within five elementary steps [60], and thus, the Tafel slope can also be used as a key descriptor to identify the RDS in OER and clarifying the in-depth mechanism [61–64].

The Gibbs free energies (ΔG) of all four elementary steps could be calculated based on DFT-based simulations. Ideally, the value of ΔG for each elementary step is expected to be 1.23 eV, where the overpotentials are zero [62]. However, the linear relations of the binding energies of intermediates imply that the real electrocatalyst could not satisfy the ideal requirements. For example, as shown in Figure 10.1d, the linear relationship between the ΔG_{HO*} and ΔG_{HOO*} could be observed under varied potentials [65]. Fortunately, DFT calculations can be simplified by taking advantage of this scaling relation, such as the difference between ΔG_{O*} and ΔG_{HO*} ($\Delta G_{O*}-\Delta G_{HO*}$), which typically serves as the key indicator to evaluate the OER activities of electrocatalysts. Therefore, a volcano plot based on DFT calculations with η_{OER} as a function of $\Delta G_{O*}-\Delta G_{HO*}$ can be established, which is beneficial for comprehending the OER mechanism in detail (Figure 10.1e). Collectively, the ΔG analysis based on DFT calculations and the Tafel slope investigations could provide complementary information about the thermodynamic and kinetic aspects of OER.

Though the AEM has achieved many remarkable successes, a few issues remain. For instance, there are plenty of OER electrocatalysts that possess a smaller onset potential than the minimum (0.37 V) calculated in AEM pathways [66–68]. Moreover, a plurality of experimental reports found that OER electrocatalysts could suffer from the electrochemical reconstruction, and thus, the surface of electrocatalysts would not be a static platform under high oxidative potential. In order to figure out these mysteries, LOM pathways have been put forward where the lattice oxygen could participate in the OER process [66–68]. Figure 10.1f shows three possible LOM mechanisms for acidic OER, and the main difference is the binding between the lattice oxygen and the intermediate species. In pathway 1, the adsorbed O* intermediate could couple with a lattice of oxygen to liberate O_2 and an oxygen vacancy could be left on the surface of electrocatalysts. Pathway 2 involves the direct binding of HO* with the lattice oxygen of the electrocatalyst to form the HOO* intermediate, which subsequently decomposes to generate O_2 and an oxygen vacancy, as well as H^+. For pathway 3, O_2 could be produced by the combination of two lattice oxygen atoms generated in pathways 1 and 2, and two oxygen vacancies could be left. As we can see, all three LOM pathways for the acidic OER process require lattice oxygen participation to produce O_2, and thus, lattice oxygen activation would be essential for the underlying OER electrocatalysts following the LOM mechanisms.

10.2.3 Overall Water Splitting

It has been mentioned that OWS includes cathodic HER and anodic OER. The theoretical thermodynamic voltage of OWS under standard conditions is 1.23 V and is independent of the mechanism of the two half-reactions [72].

$$\text{Overall reaction: } H_2O \rightarrow H_2 + \frac{1}{2}O_2 \quad E^0 = 1.23\text{ V} \tag{10.17}$$

In acidic electrolytes,

$$\text{HER}: 2\text{H}^+ + 2\text{e}^- \rightarrow \text{H}_2 \tag{10.18}$$

$$\text{OER}: \text{H}_2\text{O} \rightarrow 2\text{H}^+ + \frac{1}{2}\text{O}_2 + 2\text{e}^- \tag{10.19}$$

In neutral and alkaline electrolytes,

$$\text{HER}: 2\text{H}_2\text{O} + 2\text{e}^- \rightarrow \text{H}_2 + 2\text{OH}^- \tag{10.20}$$

$$\text{OER}: 2\text{OH}^- \rightarrow \text{H}_2\text{O} + \frac{1}{2}\text{O}_2 + 2\text{e}^- \tag{10.21}$$

To overcome intrinsic energy barriers at both cathodic HER (η_{HER}) and anodic OER (η_{OER}), as well as other resistances (η_{other}) (e.g., solution and contact resistances), the additional voltage, referred to as overpotential (η), is required. Thus, the practical applied voltage to power OWS is always higher than the thermodynamic value. Consequently, the actual voltage required for OWS is given by 10.22 [73]:

$$E = 1.23\ \text{V} + \eta_{\text{HER}} + \eta_{\text{OER}} + \eta_{\text{other}} \tag{10.22}$$

Clearly, to minimize the applied voltage of OWS, in addition to optimizing the cell configuration to reduce η_{other}, the HER and OER electrocatalysts with prominent electrocatalytic performance are vital to reduce η_{HER} and η_{OER}, respectively. At present, Pt-based electrocatalysts exhibit excellent HER activity, while Ru/Ir-based oxides possess high activity for OER.

10.2.4 Hybrid Water Splitting

Compared to OWS, HWS couples cathodic HER with alternative oxidation reactions with relatively low oxidation potentials to replace the sluggish OER (Equations 10.23 and 10.24), allowing for lower cell voltages and thus higher energy efficiency for hydrogen production. In particular, the oxidative upgrading reaction can also produce value-added products at the anode and thus maximize the return on energy investment, making it a highly promising candidate. Typical reactions within this category include the upgrading of alcohol, aldehydes, amines, biomass, and plastic waste.

$$\text{HMF} + 6\text{OH}^- = \text{FDCA} + 4\text{H}_2\text{O} + 6\text{e}^- \ \ \text{E}^0 = -0.53\ \text{V} \quad (10.23)$$

$$\text{CH}_3\text{OH} + 5\text{OH}^- = \text{HCOO}^- + 4\text{H}_2\text{O} + 4\text{e}^- \ \ \text{E}^0 = 0.103\ \text{V} \quad (10.24)$$

Benefitting from the advantages mentioned above, HWS for energy-saving coproduction of green H_2 and valuable chemicals is currently garnering significant attention, and a multitude of bifunctional electrocatalysts have been explored to drive both HER and anodic alternative oxidation reactions.

10.3 Electrochemical Reconstruction

The electrochemical reconstruction means variations of physicochemical properties of electrocatalysts, including the component, structure, and crystallinity, and it could connect the real electrochemical surface under the actual catalytic reaction process. It is strongly correlated to electrocatalytic activity and stability, and serves as an effective strategy to engineer and synthesize high-performance electrocatalysts with bifunctionality nowadays [74–80]. Benefiting from the advanced *in situ* physiochemical characterization technologies, the reconstruction phenomenon has been ascertained in the vast majority of electrocatalysts during the electrochemical process, especially for the bifunctional electrocatalysts for OWS and HWS [81–83], which leads to significantly altered interfacial properties and thus electrochemical activity [84–90]. It has been verified that both HER and OER electrocatalysts undergo electrochemical reconstruction such as self-reduction/oxidation, components leaching/deposition, phase separation/transition, or even structure collapse under operating conditions [91–93]. Various factors such as the intrinsic properties of pristine electrocatalysts like components, microstructures, valence states, and structural defects and the practical reaction conditions (e.g., operating potentials, electrolyte pH, ion, and gaseous atmosphere) can determine the reconstruction rate and degree [91]. Commonly, appropriate reconstruction can create amorphous/low-crystalline and defects-rich active species, leading to improved electrocatalytic activity [91–93]. Meanwhile, there are also several negative consequences that could result from electrochemical reconstruction. For instance, low crystallinity and abundant structural defects caused by electrochemical reconstruction could result in reduced conductivity and poor structural stability, respectively [94, 95]. Furthermore, the dissociated species may obstruct the diaphragm of the electrolyzer during reconstruction, raising the internal resistance [95]. Therefore, the *in situ* observation of electrochemical reconstruction is crucial to clarify the activity origin and failure mechanisms for electrocatalyst design and synthesis toward energy-saving HWS [91–121].

It has been discussed above that the electrochemical reconstruction of electrocatalysts may exist in both the reduction and oxidation processes, and it could result in variations of physicochemical properties [96–107]. For instance, Zhang and collaborators reported that Mo in Ni_4Mo electrocatalysts under HER operating potential could suffer from oxidative dissolution, re-adsorption, and polymerization on the Ni_4Mo surface because of the more negative redox potential of Mo/MoO_4^{2-} than that of HER, leading to enhanced HER activity and durability [96]. Xie and collaborators found that the introduction of P in $CoSe_2$ results in a phase transition from cubic to orthorhombic phase, and the P-doped orthorhombic $CoSe_2$ was unstable for alkaline HER [122]. Recently, Zhang and colleagues monitored the phase and valence state evolution of pyrite-type NiS_2 during the alkaline HER process *via* a variety of *ex situ/operando* physicochemical characterization technologies [101]. They found the surface sulfur vacancy–mediated dynamic reconstruction of NiS_2 propagated from surface to bulk with terminated Ni_3S_2 under HER operating potential. Besides, a similar reduction reconstruction was also observed for α-NiS, β-NiS, and Ni_3S_4, and appropriate Fe substitution can facilitate the dynamic evolution process. Corroborated by DFT calculations, they proposed that the introduction of Fe promotes the S 3p band of NiS_2 to the Fermi level, facilitating the occurrence of sulfur vacancies, and increases the Ni 3d–S 3p band overlap to accelerate the sulfur release, jointly responsible for the reconstruction of NiS_2 into Ni_3S_2.

By comparison, electrochemical oxidation reconstruction is more typical and has attracted much more interest. Generally, the applied anodic potential drives the original electrochemical reconstruction of OER electrocatalysts, and local microenvironments (e.g., pH, ion, and temperature) also play non-negligible roles in reconstruction behavior [112, 123–133]. The electrochemical reconstruction process of OER electrocatalysts can be triggered from surface to bulk, particularly under higher anodic potential and harsher circumstances. The intrinsic properties of the pristine electrocatalysts could also determine the reconstruction behavior [108, 134–141]. For instance, Lim and collaborators have realized the surface reconstruction of layered $LiCoO_{2-x}Cl_x$ during OER by a cationic redox–tuning method. They discovered that chlorine doping results in a lower triggered potential of cobalt oxidation than that of chlorine-free $LiCoO_2$ counterpart, and the surface reconstruction for $LiCo_{1.8}Cl_{0.2}$ could be induced by lithium leaching to form amorphous (oxy)hydroxide [142]. Physiochemical characterizations along with DFT calculations revealed that the introduction of chlorine into $LiCoO_2$ facilitates the delithiation process and the formation of thermodynamically stable (oxy)hydroxide structure. A similar metal ion dissolution and lattice oxygen oxidation reconstruction of perovskites can be observed under OER oxidation potentials [91, 143–152]. Recently, Xu and co-workers found that biased metal-oxygen covalency between MO_4 (tetrahedron, M_T–O) and MO_6 (octahedron, M_O–O) controls the reconstruction direction and degree of spinel $Li_xCo_{1-x}Co_2O_4$. Specifically, a stronger M_O–O covalency compared to M_T–O covalency favors a more complete restructuring toward oxyhydroxides [93].

In contrast to transition metal oxides, metal nonoxides like carbides, phosphides, and nitrides, generally have more positive enthalpy values [133, 153]. Accordingly, surface reconstruction of these species to (oxy)(hydr)oxide under anodic oxidative conditions is thermodynamically favorable [154–161]. Intriguingly, the reconstruction behavior is sometimes reversible. For instance, Strasser and collaborators found that the subnanometer shell of crystalline Co_3O_4 can be reconstructed into amorphous $CoO_x(OH)_y$ during OER and the formed electrocatalytic active layer can be reverted by recrystallization when the applied potential went back to the non-OER region [162]. Lately, Wei and co-workers reported the dynamic and reversible coupling of O on the Ir center over AD-HM-Ir electrocatalysts (atomically dispersed hetero-nitrogen-configured Ir), and they found that one oxygen atom can combine on the Ir site to obtain the active O-hetero-Ir-N_4 for OER, and this dynamically adsorbed oxygen would vanish after OER [163].

In short, electrochemical reconstruction is a common phenomenon in both reduction and oxidation processes, which would have a crucial effect on the electrocatalytic activity of bifunctional electrocatalysts for HWS [91, 101, 164–178]. Various advanced characterization techniques have been developed to monitor this dynamic electrochemical reconstruction, including electrochemistry (e.g., cycle voltammetry, chronopotentiometry, electrochemical impedance spectroscopy, underpotential deposition, and CO stripping), spectroscopy (e.g., FTIR, Raman, UV-vis, XRD, XPS, XAS, and XRF), scanning probe microscopy (e.g., AFM, STM, SECCM, and SECM), electron microscopy (e.g., TEM and SEM), ion scattering, atom probe tomography, Mössbauer spectroscopy, and mass spectrometry.

10.4 Hybrid Water Splitting with Oxidative Upgrading

In a traditional electrolyzer for OWS, the electrocatalysts with either OER or HER activities are usually employed to reduce overpotentials for efficient hydrogen generation. While there are still several challenges that need to be overcome, (i) most of the reported electrocatalysts are single functional, such that different synthetic methods and/or instruments are needed, which increase the overall synthetic cost; (ii) traditionally, OER and HER are coupled, and H_2 and O_2 can be obtained at the same time, which results in the possible mixing due to gas crossover and thus poses safety issues. Moreover, the cooccurrence of electrocatalysts, O_2, and H_2 can generate ROS that would degrade the membrane and therefore reduce the life spans of the water electrolysis device and (iii) the cathodic hydrogen production rate is generally restricted by anodic OER due to its slower reaction kinetics which require much higher overpotentials than those of HER and lower overall energy conversion efficiency for H_2 generation. Meanwhile, the obtained O_2 at the anode during the OER process possesses a low value. In response, developing bifunctional electrocatalyst–driven HWS that couples HER with favorable alternative oxidative upgrading reactions relative to OER can fully address the above issues. These alternative oxidative reactions include oxidative upgrading of alcohol, aldehydes, amines, biomass, and plastic waste.

10.4.1 Alcohols Oxidation

Electrochemical oxidation of various alcohols has been explored due to their low toxicity, great promise for carbon neutrality processes, and low oxidative potentials compared to that of OER. Additionally, the resulting products are value-added aldehyde and carboxylic acid. For example, formate, acetate, and glycerate can be yielded from methanol, ethanol, and glycerol, respectively.

10.4.1.1 Methanol Oxidation

Methanol (CH_3OH) is the simplest alcohol with the virtue of easy availability, high-density hydrogen, and abundance [179]. Methanol oxidation reaction (MOR) has garnered significant interest owing to the low thermodynamic potential (0.11 V) and the high value of formate or formic acid

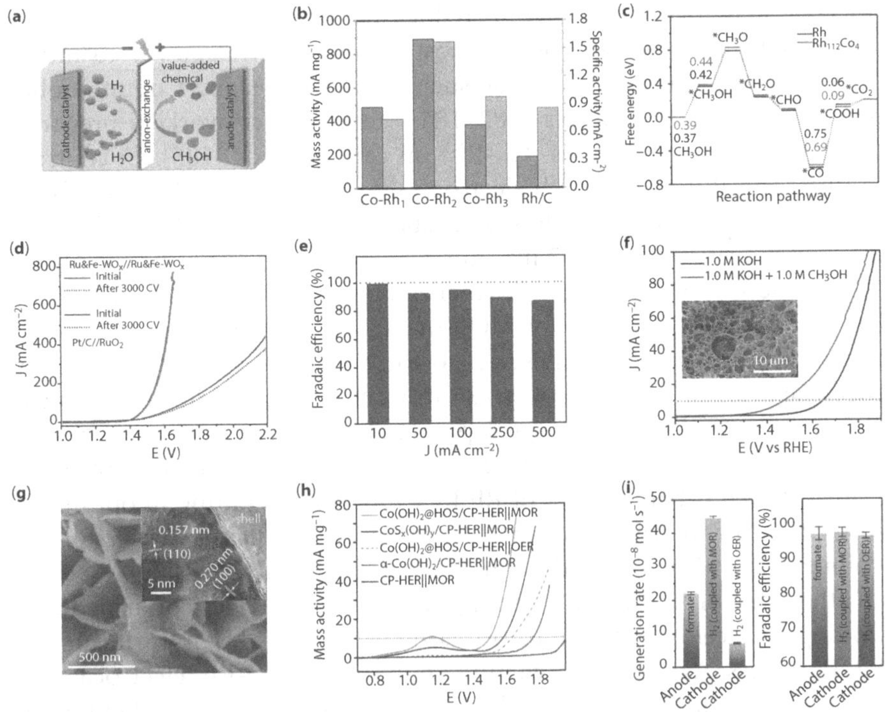

Figure 10.2 (a) Scheme of methanol-assisted hybrid water splitting (HWS) that combines HER with value-added methanol oxidation reaction (MOR). Reproduced with permission from [192]. Copyright 2019 John Wiley & Sons. (b) Specific activity and mass activity for Co-Rh electrocatalysts and Rh/C for MOR at the peak potential. (c) Gibbs free energy diagrams for MOR on the Rh and $Rh_{112}Co_4$ models. Reproduced with permission from [185]. Copyright 2020 John Wiley & Sons. (d) Pt/C||IrO_2 and Ru&Fe-WO_x||Ru&Fe-WO_x couples for the MOR-assisted HWS. (e) The corresponding FE of MOR at different current densities for Ru&Fe-WO_x couple. Reproduced with permission from [186]. Copyright 2021 Elsevier. (f) Polarization curve of OER//HER and MOR//HER for $Ni_{50}Co_{15}Fe_{30}Cu_5$ MEAAs. Inset: SEM image of $Ni_{50}Co_{15}Fe_{30}Cu_5$ MEAAs. Reproduced with permission from [189]. Copyright 2022 John Wiley & Sons. (g) SEM and HRTEM images of $Co(OH)_2$@HOS. (h) LSV curves for MOR-assisted HWS and OWS in a two-electrode configuration. Reproduced with permission from [190]. Copyright 2020 John Wiley & Sons. (i) FEs of formate and H_2 during in HER||MOR), and H_2 in HER||OER systems. Reproduced with permission from [191]. Copyright 2020 John Wiley & Sons.

products (Figure 10.2a), which are important chemicals in pharmaceutical, textile, rubber, and pesticide industries [180].

Noble metals like Rh-, Ru-, Ir-, Pd-, and Pt-based electrocatalysts have long been known to deliver exceptional electrocatalytic activities for HER and/or MOR [181]. However, in view of their high cost and limited reserves, it is highly desirable to reduce their usage without sacrificing the catalytic activity. One of the most promising strategies is to introduce a second nonprecious metal, such as Ni, Co, Fe, and Cu, into noble metal–based electrocatalysts [182–184]. Yang and co-workers prepared Co-doped Rh nanoparticles (Co-Rh) on carbon black [185]. In contrast to Rh/C and Pt/C, the Co-Rh electrocatalyst with the atomic ratio of Co/Rh of 3.71/96.29 and particle size of about 1.94 nm (Co-Rh_2) possessed the lowest overpotential of 2 mV to reach 10 mA cm^{-2} and the calculated Tafel slope was 33 mV dec^{-1}. Furthermore, Co-Rh_2 also exhibited excellent electrocatalytic activity for MOR in 1.0 M KOH with 1.0 M methanol. Specifically, the specific activity of Co-Rh_2 was calculated to be 1.57 mA cm^{-2}, superior to that of Rh/C (0.86 mA cm^{-2}). Meanwhile, a high mass activity of 889 mA mg^{-1} at the peak potential for Co-Rh_2 can be achieved as well, far better than that of Rh/C (Figure 10.2b). DFT calculations revealed a downshifted d-band center of Rh by Co doping, resulting in optimized adsorption for intermediates in HER to lower the energy barriers of the Volmer step. Meanwhile, the smaller energy barrier of $Rh_{112}Co_4$ (0.69 eV) than that of Rh (0.75 eV) during MOR (Figure 10.2c) confirmed that Co-doping accelerates the transformation of *CO into *COOH during the MOR process. As expected, when the bifunctional Co-Rh_2 electrocatalyst was employed for MOR-assisted HWS in a two-electrode configuration, a cell voltage to obtain 10 mA cm^{-2} was merely 1.545 V, superior to that of Pt/C||IrO_2 couple. Feng and colleagues fabricated a ternary electrocatalyst of Ru&Fe-WO_x with urchin-like morphology as a bifunctional electrocatalyst for MOR-assisted HWS in 1.0 M KOH with 3.0 M methanol. The two-electrode configuration needed a voltage of merely 1.50/1.62 V to deliver 100/500 mA cm^{-2} for cogeneration of H_2 and formate with high faradaic efficiency (FE > 90%) as well as robust stability (Figure 10.2, d and e) [186]. Combined experiments and DFT computations confirmed that WO_x host and Ru&Fe-based dopants synergistically optimize the adsorption of hydrogen on lattice oxygen during the hydrogen evolution process and *COOH desorption on Ru site for MOR. Recently, NiIr-MOF/NF electrocatalysts (NiIr-based MOF) were reported as bifunctional electrocatalysts for MOR and HER with excellent performance, which merely requires a voltage of 1.39 V to obtain 10 mA cm^{-2} [187].

Besides non-noble metal–doped noble metal catalysts, medium-entropy alloys (MEAs) were also reported for electrocatalytic applications because of their high tolerance, regulable composition, and complicated surface. Compared to high-entropy alloys (HEAs), the MEAs with fewer components benefit from correlating the relationship between catalytically active sites and activity [188]. However, it is still challenging to prepare phase-pure MEAs aerogels (MEAAs) due to the immiscibility of different metals. Guo and collaborators successfully addressed this challenge *via* a combined auto-combustion and following a low-temperature reduction strategy to obtain 3D $Ni_{50}Co_{15}Fe_{30}Cu_5$ MEAAs with the highest porosity and lowest density for MOR-coupled HWS [189]. SEM image of the obtained $Ni_{50}Co_{15}Fe_{30}Cu_5$ MEAAs demonstrated a porous structure (inset of Figure 10.2f), with the 8.1 m^2 g^{-1} specific surface area as determined by N_2 sorption experiments. Compared with other low-entropy controls, the quaternary $Ni_{50}Co_{15}Fe_{30}Cu_5$ MEAAs exhibited the highest mass activity (1.62 A mg^{-2}) for MOR in 1.0 M KOH with 1.0 M methanol and lowest overpotential (113 mV) to afford 10 mA^{-2} for HER. Importantly, as shown in Figure 10.2f, the two-electrode MOR-assisted HWS employing $Ni_{50}Co_{15}Fe_{30}Cu_5$ MEAAs as the bifunctional electrocatalysts delivered 10 mA cm^{-2} at a low cell voltage of 1.476 V, much lower than that of traditional OWS system at the same current density (1.649 V). Also, the formate was the main anodic product and H_2 was the dominant cathodic product.

Although metal-based electrocatalysts with excellent electrocatalytic performance have gathered more attention for OER and alternative oxidation reactions, the function of metalloid elements (e.g., O, S, N, and P) could not be ignored. Fu and co-workers prepared a core-shell $Co(OH)_2$@HOS (cobalt hydroxide@hydroxysulfide) nanosheets on carbon paper (CP) and employed it as both cathodic and anodic electrocatalysts for MOR-assisted HWS in 1.0 M KOH containing 1.0 M methanol to obtain high-value chemicals (formate and H_2) under low voltages [190]. SEM image (Figure 10.2g) revealed that $Co(OH)_2$@HOS/CP retains the nanosheet-assembled structure of pristine precursor after surface sulfurization. The porous microstructure and open space were beneficial to mass transfer and gas diffusion. As depicted in the HRTEM image (inset of Figure 10.2g), the peripheral thin layer of $Co(OH)_2$@HOS with a thickness of 2 to 2.5 nm was amorphous hydroxysulfide, and the lattice fringes (0.270 and 0.157 nm) in crystalline core originated from (100) and (110) planes of $Co(OH)_2$, respectively. The $Co(OH)_2$@HOS/CP-based MOR-assisted HWS attained a voltage of 1.497 V to achieve 10 mA cm^{-2} with more than 98% FE for co-production of value-added chemicals (formate and H_2), superior to that of OWS (1.631 V). Afterward, Fu and co-workers also prepared the NiSe hollow nanocrystals (h-NiSe) on a carbon nanotube (CNT) network for methanol oxidation in HWS [191]. The small size (15 nm) of hollow NiSe nanocrystals and CNT substrate could increase the exposure of active sites and augment the quantity/density of edge sites. When the h-NiSe/CNT bifunctional electrocatalyst was employed for MOR-assisted HWS in 1.0 M KOH with 1.0 M methanol, the high activity for HWS can be retained for more than 20 h at 1.62 V with high current density (>300 mA cm^{-2}). Besides, the FEs of formate and H_2 were 97.97% and 98.33%, respectively (Figure 10.2i). Physicochemical characterizations and DFT studies indicated that the d band center of h-NiSe/CNTs was adjusted by the synergistic effect between non-negligible SeO_x and *in situ* formed Ni-OOH, resulting in optimized adsorption for methanol oxidation and thus boosting the transformation of methanol to high-value formate, plus H_2 with lower energy consumption relative to OWS.

10.4.1.2 Ethanol Oxidation

Owing to the easy availability and storage, ethanol (EtOH) is a promising alternative for oxidation upgrading to replace anodic OER for energy-saving coproduction of high value-added chemicals and H_2 (Figure 10.3a) [193]. The thermodynamic potential for ethanol oxidation reaction (EOR) to acetate is only 0.057 V, remarkably lower than that of OER [180].

Noble metal–based materials with high activities for both EOR and HER are ideal bifunctional electrocatalysts for EOR-coupled HWS albeit with high cost. For example, Pt-based catalysts have displayed superior electrocatalytic activity for hydrogen evolution and/or small molecule oxidations [194–196]. Usually, single Pt has mediocre or even poor selectivity for EOR due to the weak substrates' adsorption (ethanol and H_2O) and C–H bond dissociation capacities [197]. Introducing other elements to form Pt-based alloys is a promising method to improve their EOR and HER activities. Liu and co-workers fabricated the bifunctional PtCu NF (3D PtCu nanoframe) with high-index facets as the bifunctional electrocatalysts for EOR-assisted HWS in 1.0 M KOH containing 0.5 M ethanol [198]. The TEM and HRTEM images (Figure 10.3b) demonstrated that PtCu NF possesses multi-channels, and plenty of high-index facets of (331), which provide superior accessibility to active sites. After loading on VulcanXC-72 carbon (PtCu NF/C), it exhibited a low overpotential of 76 mV to obtain 10 mA cm^{-2} for HER (Figure 10.3c), much lower than that of Pt/C. Also, it possessed a smaller EOR potential than Pt/C (0.754 vs. 0.794 V vs. RHE) with acetate as the end anodic product. Meanwhile, the two-electrode system fabricated with the bifunctional PtCu NF/C couple can deliver 10 mA cm^{-2} at 0.58 V, 170 mV lower than that of the control couple (Pt/C). The prominent electrochemical performance of PtCu NF/C for HER and EOR was ascribed to the synergistic effect of high-index facets, Cu doping, and open 3D structure, collectively resulting in optimized adsorption for EOR, and improved C–H bond dissociation of ethanol (Figure 10.3d). Similarly, introducing Cu to Pd can result in enhanced EOR performance as well.

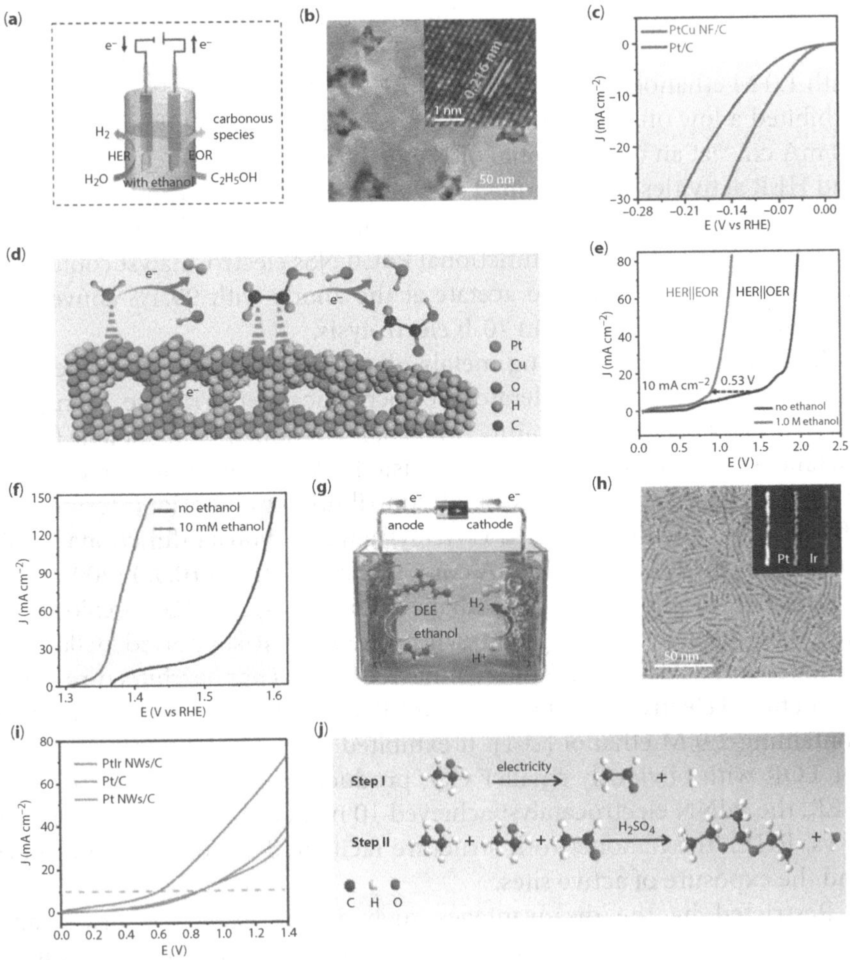

Figure 10.3 (a) Schematic of two-electrode EOR-assisted HWS system. Reproduced with permission from [200]. Copyright 2023 Royal Society of Chemistry. (b) TEM images of PtCu NF. (c) LSV curves of Pt/C and PtCu NF/C for HER in 1.0 M KOH. (d) Schematic diagram for EOR and HER pathway on PtCu NF/C. Reproduced with permission from [198]. Copyright 2022 American Chemical Society. (e) LSV curves of a two-electrode electrolyzer using PdSP metallene as the bifunctional electrocatalysts for HER||EOR (1.0 M KOH + 1.0 M ethanol) and HER||OER (1.0 M KOH). Reproduced with permission from [200]. Copyright 2023 Royal Society of Chemistry. (f) LSV curves of Ni_3S_2/NF for EOR and OER in 1.0 M KOH with and without 10 mM ethanol. Reproduced with permission from [44]. Copyright 2016 American Chemical Society. (g) Schematic of EOR toward value-added DEE coupled with HER. (h) TEM image and its corresponding elemental mappings of PtIr NWs. (i) LSV curves of the PtIr NWs/C couple, Pt NWs/C couple, and Pt/C couple in 0.5 M H_2SO_4 with ethanol solution. (j) Possible mechanism for the transformation of ethanol to DEE. Reproduced with permission from [203]. Copyright 2021 American Chemical Society.

Zheng and co-workers prepared hierarchical 3D PdCu alloy nanosheets (NSs) *via* a simple CO-assisted method to catalyze EOR in 1.0 M KOH with 1.0 M ethanol and HER in 1.0 M KOH [199]. The 3D PdCu alloy NSs exhibited a low onset potential of 330 mV vs. RHE for EOR and obtained 10 mA cm^{-2} at an overpotential of 106 mV for HER. These enhanced EOR and HER activities were ascribed to synergistic effects (Cu and Pd) and the ultrathin 3D structure. Driven by the solar cell (1.0 V), the two-electrode electrolyzer based on the bifunctional PdCu NSs electrocatalyst could produce H_2 at the cathode and acetate at the anode with 99.4% conversion efficiency and 98% FE within 70-h electrolysis.

In addition to introducing metals, doping with nonmetal elements is also a valid method to accelerate electrochemical performance. Wang and collaborators prepared the bifunctional S, P-co–doped Pd metallene (PdSP metallene) electrocatalyst for EOR-assisted HWS. The co-doping of S and P regulated the electronic state of Pd, contributing to superior intermediate affinity. Thus, the EOR-assisted HWS system based on a bifunctional PdSP metallene couple reached 10 mA cm^{-2} at 0.88 V (Figure 10.3e) [200].

Apart from alloying/doping, shape engineering has been explored as well. For instance, Jeon's group synthesized a self-supported palladium nanonetwork (PdNN) with interconnected porous nanoarchitecture as the bifunctional electrocatalyst for HER in 1.0 M KOH and EOR in 1.0 M KOH containing 1.0 M ethanol [201]. It exhibited 2.6-fold higher mass activity for EOR with obviously smaller CO_2 production compared to Pt/C. For HER, the PdNN electrocatalyst achieved 10 mA cm^{-2} at a potential of 0.11 V vs. RHE. The nanonetwork structure facilitated the electron transport and the exposure of active sites.

Restricted by the disadvantages such as the expensive prices and non-abundant resources of noble metal electrocatalysts, the substitution with non-noble and earth-abundant metal electrocatalysts is of great significance. Sun's group prepared the hierarchically porous Ni_3S_2/NF *via* the sulfurization treatment of commercial NF as a bifunctional electrocatalyst for both EOR in 1.0 M KOH with 10 mM ethanol and HER in 1.0 M KOH [44]. As shown in the LSV curve of Ni_3S_2/NF, a low onset potential of ~1.35 V vs. RHE was required for Ni_3S_2/NF during the EOR process, 150 mV smaller than that of OER, and the current density for EOR raised rapidly within 1.40 V vs. RHE, implying the higher conversion efficiency (Figure 10.3f). ^{1}H NMR analysis demonstrated the almost complete conversion of ethanol and near unity FE for acetate production. Physiochemical characterizations including SEM and XPS revealed that Ni_3S_2/NF is partially oxidized to active Ni species, which served as the real catalytic sites for

EOR. In addition to the remarkable EOR activity, only a low overpotential of 116 mV was required for the Ni_3S_2/NF to deliver 10 mA cm^{-2} for HER. Afterward, Cao and co-workers proposed that further phosphorization of transition-metal sulfides like CoS_2 results in enhanced EOR performance [202]. The EOR-assisted HWS electrolyzer fabricated with P-doped CoS_2 nanosheet arrays (Co-S-P) on carbon cloth (CC) couple reached 10 mA^{-2} at a voltage of 1.63 V, 140 mV smaller than that for OWS.

Besides acetate, 1,1-diethoxyethane (DEE) can also be obtained by EOR, which is employed as a versatile precursor and liquid fuel. In terms of energy efficiency, direct conversion of ethanol to DEE coupled with HER is an attractive strategy for future green hydrogen generation (Figure 10.3g) [203, 204]. Guo's group realized the cogeneration of DEE at the anode and H_2 at the cathode from ethanol and water in 0.5 M H_2SO_4 ethanol electrolyte by using PtIr alloy nanowire (NWs) as the bifunctional electrocatalyst [203]. As shown in Figure 10.3h, the TEM image demonstrated that the PtIr NWs possess ultrafine 1D morphology with 30-nm length and 10-nm diameter. Before electrochemical tests, the PtIr NWs were loaded on commercial carbon supports followed by acetic acid and potential cycling pretreatments. For the selective EOR toward DEE, the lowest onset potential can be observed for the as-prepared PtIr NWs/C, and an applied potential of only 0.45 V vs. SCE was needed to afford 30 mA cm^{-2}, in sharp contrast to those of Pt NWs/C (~0.93 V vs. SCE) and commercial Pt/C (~0.92 V vs. SCE). Impressively, a high mass activity of 55.6 mA ${mg_{Pt+Ir}}^{-1}$ can be achieved for the PtIr NWs/C, almost 6.3- and 4.7-fold higher than those of Pt/C (8.7 mA ${mg_{Pt}}^{-1}$) and Pt NWs/C (11.7 mA ${mg_{Pt}}^{-1}$), respectively. In addition to the remarkable EOR activity, PtIr NWs/C also showed the highest electrocatalytic performance for HER, reaching 10 mA cm^{-2} at a potential of −150 mV vs. SCE, superior to those of Pt/C (−300 mV vs. SCE) and Pt NWs/C (−280 mV vs. SCE). As expected, the EOR-assisted HWS electrolyzer catalyzed by bifunctional PtIr NWs/C attained 10 mA cm^{-2} at a voltage of 0.61 V and reached an FE of 77.8% for DEE product at 1.8 V (Figure 10.3i). Moreover, affected by acetalation rate and extent, the highest activity for PtIr NWs/C couple can be obtained in $HClO_4$ ethanol solution, and the FE for selective ethanol oxidation could reach 85% at 10 mA cm^{-2}, much higher than those in 0.5 M H_2SO_4- (70%) or HNO_3-contained (52%) ethanol. The mechanism analysis based on *in situ* FTIR spectroscopy studies revealed that the conversion of ethanol to DEE can be divided into two steps: Firstly, the ethanol was dehydrogenated and converted to acetaldehyde. Subsequently, acid-catalyzed acetalation resulted in the conversion of acetaldehyde to DEE (Figure 10.3j).

10.4.1.3 Ethylene Glycol Oxidation

Ethylene glycol (EG) is the simplest dibasic alcohol, with low toxicity, affordable cost, and convenient storage [205]. During the process of EG oxidation reaction (EGOR), the electrons could be extracted from the hydroxyl groups to form various products (e.g., oxalic acid, carbonate, glyoxylic acid, glycolic acid, glyoxal, glycolaldehyde, or their mixtures) [206]. Among them, glycolic acid possesses high value and has a wide application for health-care products, textiles, leather processing, metal cleaning, and adhesives. More importantly, the corresponding thermodynamic potential of EGOR (0.57 V vs. RHE) toward glycolic acid is smaller than that of OER as well [207].

Two-dimensional (2D) metallene with superior electron transport properties, atomic-level thickness, and large electrochemical active area exhibited excellent electrocatalytic activities [208–211]. Noble metals like Rh and Pt have been widely employed for EGOR because of their outstanding catalytic electrochemical performance and high inherent corrosion resistance. For example, Wang's group prepared Rh/RhOOH metallene electrocatalyst (the partially hydroxylated Rh metallene) *via* a straightforward and simple hydrothermal method for HER and EGOR (Figure 10.4a) [212]. HAADF-STEM and HRTEM images in Figure 10.4b revealed a 2D ultrathin nanosheet, amorphous/crystalline, and defect-rich structure for Rh/RhOOH metallene with numerous pores and vacancies on the surface. These features provided abundant active sites and accelerated mass diffusion/transport for reactants to the active centers. FTIR spectroscopy displayed a peak at ~3700 cm^{-1}, implying the presence of surface RhOOH species for Rh/RhOOH metallene, beneficial for the stabilization of the unique 2D structure (Figure 10.4c). Notably, the Rh/RhOOH metallene displayed prominent and comparable electrochemical performance for HER both in alkaline water (1.0 M KOH) and seawater conditions (1.0 M KOH with seawater) with an overpotential of 29 mV to deliver 10 mA cm^{-2}. For EGOR, Rh/RhOOH metallene also exhibited superior specific activity and mass activity (1.051 mA cm^{-2} and 0.633 mA μg^{-1}_{Rh}), higher than those of control samples. Inspired by its prominent HER and EGOR activities, the authors further assembled the two-electrode electrolyzer for EGOR-assisted HWS based on bifunctional Rh/RhOOH electrocatalysts to increase the energy efficiency of H_2 generation. As depicted in Figure 10.4d, EGOR-assisted HWS can attain 10 mA cm^{-2} at 0.678V, in sharp contrast to that for OWS (1.548 V). Encouragingly, the Rh/RhOOH metallene couple also demonstrated superior activity and durability for EGOR-assisted HWS in alkaline seawater conditions (1.0 M KOH with 6 M EG

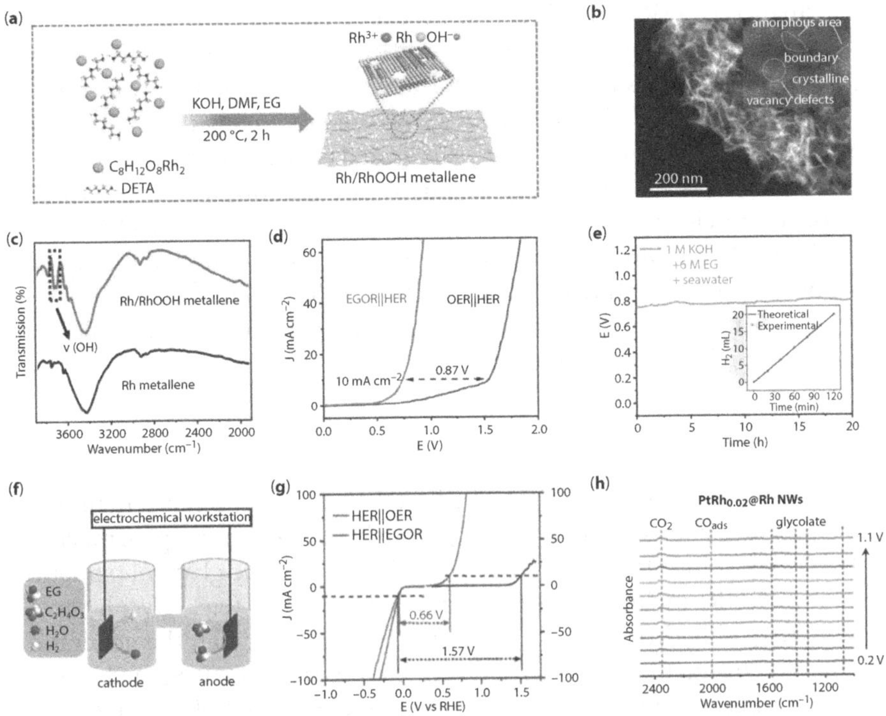

Figure 10.4 (a) Schematic for the preparation of Rh/RhOOH metallene. (b) HAADF-STEM and HRTEM images of Rh/RhOOH metallene. (c) FTIR spectra of Rh/RhOOH metallene and Rh metallene. (d) LSV curves of Rh/RhOOH-based two-electrode electrolyzer for EGOR||HER (1.0 M KOH with 6.0 M EG) and OER||HER couple (1.0 M KOH). (e) Voltage-time curve at 10 mA cm^{-2} of Rh/RhOOH metallene couple for EGOR-assisted HWS in alkaline seawater condition (1.0 M KOH with seawater and 6.0 M EG). Inset: The theoretical and measured H_2 quantity at cathodic HER. Reproduced with permission from [212]. Copyright 2022 John Wiley & Sons. (f) Schematic of the EGOR-coupled HWS system. (g) LSV curves of $PtRh_{0.02}$@Rh NWs-based two-electrode electrolyzer for HWS (EGOR|| HER) and OWS (OER||HER). (h) *In situ* FTIR spectra of $PtRh_{0.02}$@Rh in 0.1 M KOH with 0.5 M EG solution. Reproduced with permission from [213]. Copyright 2022 Royal Society of Chemistry.

and seawater), maintaining a low cell voltage at 10 mA cm^{-2} for 20 h with ~100% FE (Figure 10.4e).

Subsequently, Wang and co-workers synthesized $PtRh_{0.02}$@Rh NWs electrocatalyst (PtRh nanowires with Rh modification) *via* a facile one-step solvothermal reduction for EGOR-coupled HWS (Figure 10.4f) [213]. Doping Rh with a low concentration gave rise to lattice strain and adjusted electronic structures, contributable to enhanced electrocatalytic performance of $PtRh_{0.02}$@Rh NWs. As expected, only low overpotentials of 30.6 and 45.8 mV were required for $PtRh_{0.02}$@Rh NWs to attain 10 mA cm^{-2} for HER in alkaline water and seawater electrolytes, respectively. Additionally, the EGOR-coupled HWS system employing the $PtRh_{0.02}$@Rh NWs as a bifunctional electrocatalyst revealed remarkable activity in 0.1 M KOH with 0.5 M EG for energy-saving H_2 production, with a cell voltage of 0.66 V to deliver 10 mA cm^{-2}, in sharp contrast to that for OWS (1.57 V) (Figure 10.4g). Moreover, a high FE of 90% can be obtained for glycolic acid production. As shown in Figure 10.4f, $PtRh_{0.02}$@Rh NWs possessed stronger infrared signals corresponding to C_2 intermediates, beneficial to producing high-value glycolic acid.

10.4.1.4 Glycerol Oxidation

Glycerol ($C_3H_8O_3$) is a typical polyhydroxyl alcohol and a cheap byproduct during biodiesel production, which is widely utilized in the pharmaceutical industry [214, 215]. The products of electrocatalytic glycerol oxidation reaction (GOR) include C_1 (e.g., formic acid, FA), C_2 (e.g., glycolic acid), and C_3 (e.g., dihydroxyacetone, DHA) [216]. In particular, the generation of the C_1 product of FA ($C_3H_8O_3 + 8OH^- \rightarrow 3HCOOH + 8e^- + 5H_2O$) is appealing because of the low thermodynamic potential of only 0.69 V in alkaline media, making it possible to replace OER for GOR-assisted HWS [217].

Considering the critical role of hydroxyl adsorption for GOR, it is desirable to explore the (oxy)(hydr)oxides-based electrocatalysts with high oxophilicity [218]. Xie and co-workers synthesized ternary NiVRu LDH nanosheet arrays on NF (NiVRu-LDHs NAs/NF) for H_2 generation in 1.0 M KOH and selective oxidation of glycerol in 1.0 M KOH with 0.1 M glycerol [217]. The SEM images (Figure 10.5a) showed that NiVRu-LDHs nanosheet arrays with thin thickness (7 to 8 nm) are vertically grown on the NF. For the GOR toward FA (HCOOH), the potentials for NiVRu-LDHs NAs/NF to achieve the current density of 10 (1.24 V vs. RHE) and 300 mA cm^{-2} (1.41 V vs. RHE) were more superior to that for OER (>1.5 V vs. RHE). Besides the prominent GOR activity, the superior electrocatalytic HER activity also be observed for the NiVRu-LDHs NAs/NF, and a low overpotential of 35 mV can be observed to achieve 10 mA cm^{-2}, comparable to that of Pt/C. To investigate the underlying GOR mechanism, operando Raman spectroscopy analysis was conducted. It can be observed that there are two peaks at 555 and 476 cm^{-1} during the OER, corresponding to the $\nu(Ni^{III}\text{-}O)$ and $\delta(Ni^{III}\text{-}O)$ vibrations of Ni-OOH, respectively (Figure 10.5b). However, these signals completely disappeared upon the

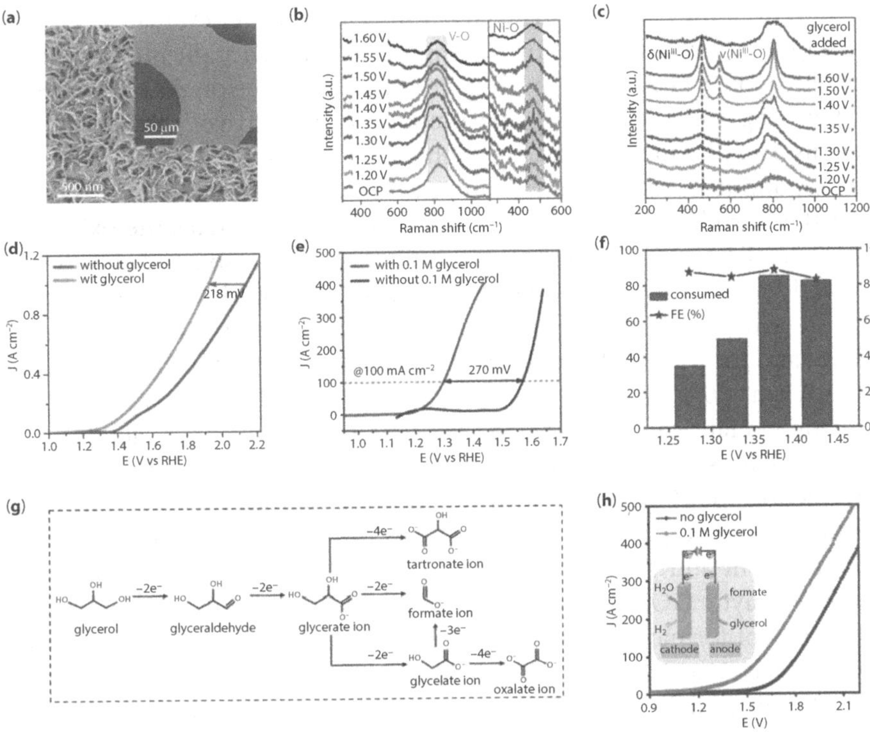

Figure 10.5 (a) SEM images of NiVRu-LDHs NAs/NF. *In situ* Raman spectra for (b) GOR (1.0 M KOH with 0.1 M glycerol) and (c) OER (1.0 M KOH) at different potentials over the NiVRu-LDHs NAs/NF. (d) LSV curves of the flow electrolyzer for OWS and GOR-assisted HWS catalyzed by NiVRu-LDHs NAs/NF couple. Reproduced with permission from [217]. Copyright 2023 John Wiley & Sons. (e) LSV curves for OER (1.0 M KOH) and GOR (1.0 M KOH with 0.1 M glycerol) over Pt_{SA}-NiCo LDH/NF. (f) Glycerol FE and conversion efficiency of formate at specific potentials. Reproduced with permission from [220]. Copyright 2023 Elsevier. (g) Possible mechanisms of electrocatalytic GOR to formate in alkaline medium. (h) LSV curves of two-electrode electrolyzer based on Ni_3N/Co_3N-NWs electrocatalyst for GOR-coupled HWS and OWS. The inset shows the schematic of a two-electrode cell. Reproduced with permission from [222]. Copyright 2023 John Wiley & Sons.

addition of glycerol (Figure 10.5c). The authors proposed that glycerol is oxidized by Ni^{III} with the participation of the hydroxyl group, and thus, the Ni^{III} is reduced into Ni^{II}, accountable for the disappearance of the typical peaks of Ni^{III}-OOH. Benefitting from the prominent GOR and HER performance of the NiVRu-LDHs NAs/NF, the GOR-assisted HWS electrolyzer required cell voltages of 1.3 and 1.62 V to deliver 10 and 100 mA cm^{-2}, respectively, with FEs of >80% for formate generation at 1.35 to 1.75

V. Moreover, they also constructed a flow cell to further evaluate the practical application value. As shown in Figure 10.5d, even at the industry-level current density (1 A cm^{-2}), the voltage of the flow electrolyzer is only 1.933 V, lowering 10.1% electricity consumption compared with that for OWS (Figure 10.5d). Furthermore, the high FEs of almost 80% and 96% for the corresponding formate and H_2 can be achieved. DFT calculations verified that the introduced Ru optimizes ΔG_{H*} and thus facilitates the oxidation of Ni^{II} to Ni^{III} and C–C bond cleavage. Similarly, Johnsson and collaborate located Pt nanoparticles at oxygen vacancies (O_v) of MoO_x nanosheets to obtain advanced MoO_x/Pt bifunctional electrocatalysts for GOR-assisted HWS, and the MoO_x/Pt couple exhibited a cell voltage of 0.70 V to achieve 10 mA cm^{-2}, 0.90 V lower than that for OWS [219]. Recently, Wang and co-workers fabricated Pt_{SA}-NiCo LDHs/NF by employing oxygen O_v-rich NiCo LDHs as a substrate to capture single Pt for energy-saving hydrogen production in 1.0 M KOH with 0.1 M glycerol [220]. XAS results confirmed that Pt–O_5 coordination can be formed by the combination of the oxygen atom and Pt_{SA}. The Pt_{SA}-NiCo LDHs/NF demonstrated superior electrocatalytic GOR activity (Figure 10.5e) with high FE and conversion efficiency of formate (Figure 10.5f) in the glycerol-containing electrolyte. When serving as the bifunctional electrode in a GOR-assisted HWS system, the Pt_{SA}-NiCo LDHs/NF couple exhibited only a cell voltage of 1.37 V to achieve 100 mA cm^{-2}. The glycerol conversion was calculated to be 83.4% at 1.47 V and the FE of H_2 was close to 100% (Figure 10.5f).

Several nonprecious metal-based HER electrocatalysts were also proved to be highly active for GOR, showing promise for GOR-coupled HWS. For instance, Shi and collaboration synthesized the bifunctional Ni-Mo-N electrocatalyst (Ni-Mo nitride) on carbon fiber cloth (CFC) for cogeneration of formate and H_2 [221]. The Ni-Mo-N/CFC–based GOR-coupled HWS electrolyzer revealed a cell voltage of 1.36 V to attain 10 mA cm^{-2}, with high FEs of 99.7% for H_2 and 95.0% for formate. Recently, Zhang and co-workers prepared biphasic Ni_3N/Co_3N heterostructure nanowires (Ni_3N/Co_3N-NWs) bifunctional electrocatalysts for GOR-coupled HWS [222]. As shown in Figure 10.5g, the primary reaction mechanism for GOR on Ni_3N/Co_3N-NWs was verified by operando synchrotron radiation FTIR and NMR analysis. The Ni_3N/Co_3N-NWs demonstrated a potential of 1.18 V vs. RHE to obtain 20 mA cm^{-2} for GOR in 1.0 M KOH with 0.1 M glycerol, and an overpotential of 69.0 mV to achieve 10 mA cm^{-2} for HER in 1.0 M KOH. The GOR-coupled HWS merely needed a cell voltage of 1.47 V at 50 mA cm^{-2}, superior to that of the OWS system (1.72 V) (Figure 10.5h). Impressively, a 1 A cm^{-2} industry-level current density can be realized in the flow electrolyzer at a voltage of 2.01 V with robust stability over 200 h, enabling electricity savings of 9.7% and electricity consumption of 4.82 kWh $m^{-3}{}_{H2}$. Additionally, the co-production rates for H_2 and formate were 21.4 and 11 mmol cm^{-2} h^{-1}, respectively. DFT calculations revealed that the ΔG_{H*} value for HER and glycerol dehydrogenation kinetics are simultaneously optimized by the regulated electron distribution on Ni_3N/Co_3N heterointerfaces, beneficial for the enhanced bifunctional electrochemical activity.

10.4.1.5 Benzyl Alcohol Oxidation

Electrocatalytic benzyl alcohol oxidation reaction (BAOR) is another ideal alternative to OER, and its oxidation products are benzaldehyde or benzoic acid. Both are extensively utilized in fine chemicals and pharmaceuticals [223, 224].

Sun's group reported that a bifunctional hp-Ni electrocatalyst (3D hierarchically porous nickel framework) prepared by facile dynamic electrodeposition can serve as a bifunctional electrocatalyst for BAOR-coupled HWS (Figure 10.6a) [223]. As shown in Figure 10.6b, SEM images of hp-Ni indicated the existence of interconnected macroporous walls stacked with numerous Ni nanoparticles forming plentiful smaller macropores (about 10 μm). The hierarchical porosity was believed to favor mass transport and gas diffusion for accelerated BAOR in 1.0 M KOH with 10 mM benzyl alcohol (BA) and HER in 1.0 M KOH. LSV curves of hp-Ni revealed that the onset potential decreases from 1.51 V to 1.35 V vs. RHE after adding BA. Particularly, the hp-Ni exhibited excellent stability for BAOR with high FE (96% to 98%) under five successive cycles. Additionally, in comparison to BA, the similar LSV curves for (MBA (4-methylbenzyl alcohol) and NBA (4-nitrobenzyl alcohol) oxidation (Figure 10.6c) indicated that the oxidation state of Ni species in electrocatalysts, rather than the intrinsic thermodynamics of substrates, predominantly determines their catalytic oxidation. After stability tests, the corresponding hp-Ni (named post-BA hp-Ni) maintained the integral 3D hierarchical porous structure (Figure 10.6d). Compared with the fresh counterpart, XPS spectra of Ni $2p_{3/2}$ for post-BA hp-Ni indicated a strengthened peak at 856.0 eV and a weakened peak at 852.6 eV originating from oxidized Ni species and metallic Ni, respectively, as well as increased O 1s peak, suggestive of the surface reconstruction of hp-Ni during BAOR (Figure 10.6e). Impressively, the hp-Ni also revealed a low overpotential of 219 mV for HER at 50 mA cm^{-2}. As expected, BAOR-assisted HWS based on hp-Ni couple attained low cell voltages of 1.50, 1.54, 1.60, and 1.66 V to achieve 10, 20, 50, and 100 mA cm^{-2}, respectively, superior to those of OWS (Figure 10.6f). Also, the FEs

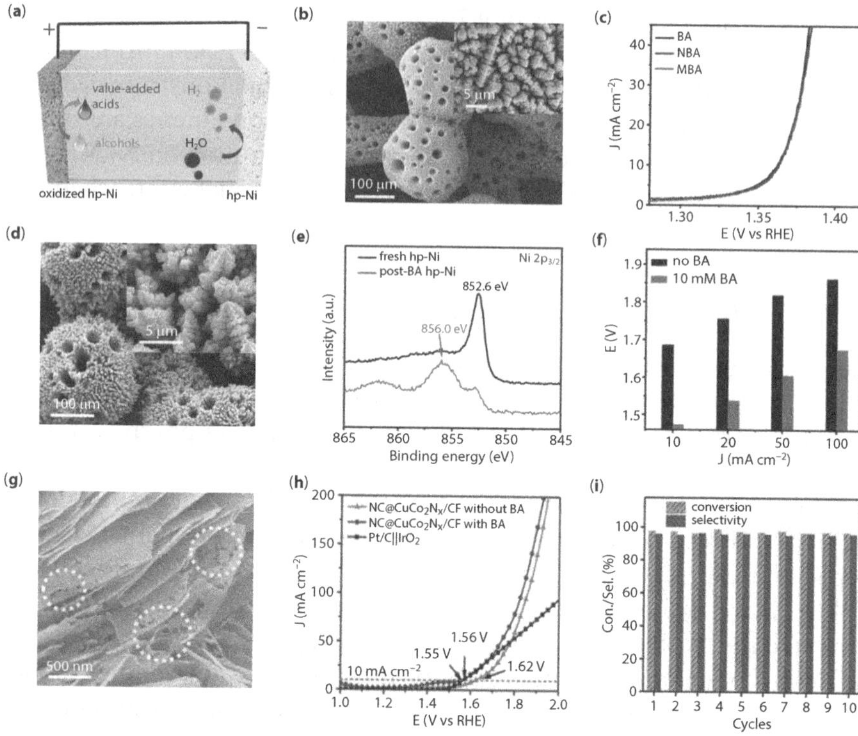

Figure 10.6 (a) Schematic of BAOR-coupled HWS over bifunctional hp-Ni. (b) SEM images of hp-Ni. (c) LSV curves in 1.0 M KOH with 10 mM NBA, MBA, or BA for hp-Ni. (d) SEM images of post-BA hp-Ni. (e) XPS spectra of post-BA hp-Ni and fresh counterpart electrocatalysts for Ni $2p_{3/2}$. (f) The recorded voltages for hp-Ni couple at 10, 20, 50, and 10 mA^{-2} for HWS (1.0 M KOH with 10 mM BA) and OWS (1.0 M KOH). Reproduced with permission from [223]. Copyright 2017 American Chemical Society. (g) SEM image of NC@$CuCo_2N_x$/CF. (h) LSV curves for Pt/C||IrO_2 and NC@$CuCo_2N_x$/CF||NC@$CuCo_2N_x$/CF. (i) Conversion and selectivity of BA for BAOR-assisted HWS under 10 cycles. Reproduced with permission from [228]. Copyright 2017 John Wiley & Sons.

of benzoic acid at the anode instead of low-value O_2 and H_2 at the cathode reached 97% and nearly 100%, respectively. Apart from hp-Ni, the bifunctional Ni_3S_2/NF electrocatalyst was also reported by Sun's group to catalyze HER and BAOR, with ~190 mV voltage saving at 100 mA cm^{-2} relative to OWS [225].

In addition to benzoic acid, benzaldehyde is also a highly valuable oxidation product from BAOR with wide applications in the pharmaceuticals and resin industry [226, 227]. The partial oxidation of BA into benzaldehyde can also be coupled with HER. By employing combined hydrothermal, carbon coating, and NH_3 calcination, Wang and co-workers obtained a hierarchical porous, self-supported N-doped carbon-coated $CuCo_2N_x$ (NC@$CuCo_2N_x$) on carbon fiber (CF) as bifunctional electrocatalysts for BAOR-coupled HWS for energy-efficient coproduction of benzaldehyde and H_2 [228]. The obtained nanosheets consisted of numerous interconnected $CuCoN_x$ nanoparticles (size: 20 to 30 nm) and holes with various sizes (Figure 10.6g). The unique porosity benefited mass transport for rapid electrochemical kinetics. For BAOR in 1.0 M KOH with 15 mM BA, a low potential of 1.25 V vs. RHE for NC@$CuCo_2N_x$/CF was required to reach 10 mA cm^{-2}, 210 mV lower than for OER. Meanwhile, the NC@$CuCo_2N_x$/CF possessed promising HER performance in 1.0 M KOH as well, with an overpotential of approximately 105 mV at 10 mA cm^{-2}. When the bifunctional NC@$CuCo_2N_x$/CF was employed in a two-electrode system for BAOR-assisted HWS, the NC@$CuCo_2N_x$/CF couple required a voltage of 1.55 V to afford 10 mA cm^{-2} in 1.0 M KOH with 15 mM BA, lower than that in 1.0 M KOH for OWS (1.67 V) (Figure 10.6h). The NC@$CuCo_2N_x$/CF couple exhibited prominent electrocatalytic activity in 1.0 M KOH with 15 mM BA for selective BAOR toward benzaldehyde with a high conversion of 97.25% and FE of 81.3% for benzaldehyde at 15 mA. Impressively, the NC@$CuCo_2N_x$/CF couple retained the high selectivity (≥96%) and activity (≥97%) for 10 cycles (Figure 10.6i). DFT calculations demonstrated that the optimized ΔG of key intermediates and modified electronic structure can be achieved by the robust conjugation between CuN and CoN.

10.4.2 Aldehyde Oxidation

Typical aldehydes include formaldehyde (HCHO) and acetaldehyde. Efficient production of H_2 and simultaneous upgrading of aldehydes into high-value chemicals like carboxylic acids are attractive. Owing to the lower price of formaldehyde relative to acetaldehyde, electrocatalytic formaldehyde oxidation reaction (FAOR) was usually employed to replace OER for H_2 production from water, and its oxidative product, carbon monoxide, is a vital C_1 feedstock to produce hydrocarbons in the chemical industry [229]. Although noble metal–based electrocatalysts feature excellent performance in HER, CO produced from anodic FAOR will poison them due to the strong adsorption. Therefore, nonprecious metal–based electrocatalysts, like Ni- and Co-based nanomaterials, were explored for FAOR-assisted HWS.

With their unique electronic properties, well-defined active centers, and ultimate atomic utilization, single-atom catalysts (SACs) have become a rising star in the field of catalysis. Fan and co-worker fabricated Co-N_x-C@Co containing abundant atomically dispersed Co-N_4 sites and Co nanoparticles as both cathodic and anodic electrocatalysts for HER and FAOR in 1.0 M KOH with 3.2 M HCHO *via* pyrolysis of the Zn/Co bimetal ZIF (Figure 10.7a) [230]. As shown in Figure 10.7b, the SEM image of Co-N_x-C@Co revealed that the high-temperature annealing process results in the collapse of the dodecahedron wall inward (Figure 10.7b). HAADF-STEM images of Co-N_x-C@Co showed the presence of some bright spots of Co nanoparticles loaded at the center of carbon and a few cobalt single-atom

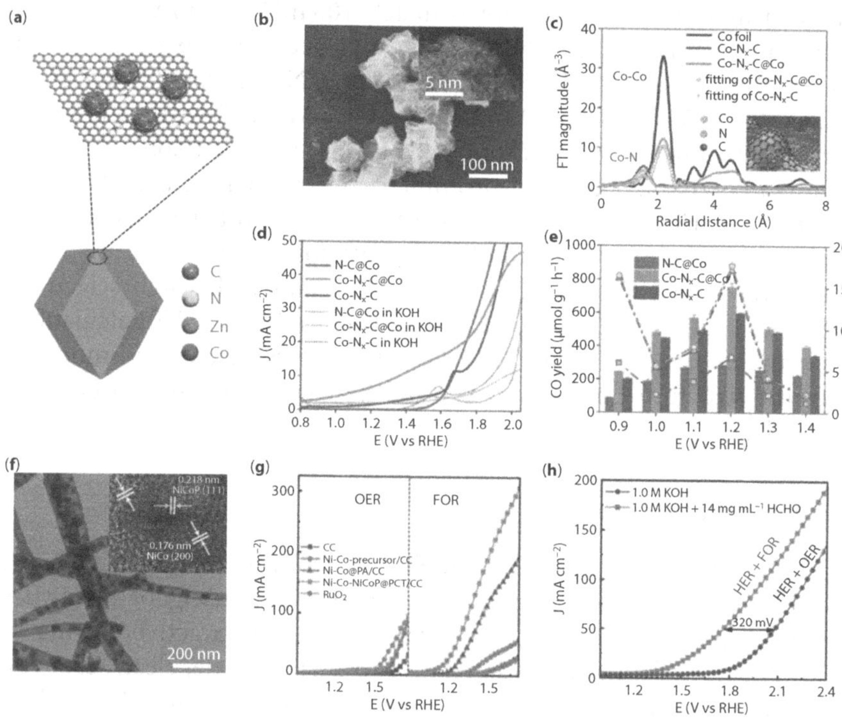

Figure 10.7 (a) Schematic of Co-N_x-C@Co. (b) SEM and HAADF-STEM images of Co-N_x-C@Co. (c) FT-EXAFS spectra of Co-N_x-C@Co, Co-N_x-C, and Co foil. (d) LSV curves of N-C@Co, Co-N_x-C@Co, and Co-N_x-C couples in 1.0 M KOH electrolyte with and without HCHO for FAOR-assisted HWS and OWS. (e) CO yield and FE for N-C@Co, Co-N_x-C@Co, and Co-N_x-C at different potentials in HCHO-containing 1.0 M KOH. Reproduced with permission from [230]. Copyright 2021 American Chemical Society. (f) TEM images of NiCo-NiCoP@PCT. (g) LSV curves of RuO_2, NiCo-NiCoP@PCT/CC, Ni-Co@PA/CC, Ni-Co-precursor/CC, and CC for OER (1.0 M KOH) and FAOR (1.0 M KOH with HCHO). (h) LSV curves for FAOR-assisted HWS. Reproduced with permission from [231]. Copyright 2021 Elsevier.

sites marked with red circles (inset Figure 10.7b). As depicted in the Fourier transform EXAFS (FT-EXAFS) spectra of Co-N_x-C@Co, the main peak at about 2.21 Å was ascribed to the Co–Co bond, confirming that the dominant component for Co element was metallic cobalt and another broad peak at around 1.47 Å corresponded to the Co–N coordination (Figure 10.7c). The resulting Co-N_x-C@Co exhibited excellent electrocatalytic activity for FAOR with a low potential of 1.00 V to deliver 10 mA^{-2}, much lower than that of OER (1.73 V vs. RHE), and the high CO yield of 1993 μmol g^{-1} h^{-1} can be realized at 1.2 V vs. RHE. In addition to the prominent performance for FAOR, Co-N_x-C@Co also exhibited the lowest overpotential for HER to reach 50 mA cm^{-2} compared to Co-N_x-C and N-C@Co. To realize H_2 and CO production simultaneously, a Co-N_x-C@Co–based FAOR-assisted HWS electrolyzer was constructed, and it delivered 10 mA cm^{-2} at 1.43 V vs. RHE with the highest CO yield of 754 μmol g^{-1} h^{-1} at 1.2 V vs. RHE (Figure 10.7e).

Similarly, Zhou's group fabricated the linearly distributed NiCo-NiCoP@PCT (NiCo-NiCoP inlaid into P-doped CNTs) on CC for FAOR-assisted HWS (Figure 10.7f). The NiCo-NiCoP@PCT displayed excellent electrochemical activities for both HER with a low overpotential of 135 mV and FAOR with an ultralow potential of 1.18 V vs. RHE to afford 10 mA cm^{-2} (Figure 10.7g) [231]. Besides, the bifunctional NiCo-NiCoP@PCT/CC–based two-electrode electrolyzer acquired 50 mA cm^{-2} at 1.76 V, 320 mV smaller than that for OWS (Figure 10.7h).

In addition, formaldehyde could also be partially oxidized at the anode, resulting in the co-production of formate and the H_2 evolution with a minimal thermodynamic potential of −0.22 V vs. RHE ($HCHO+2OH^- \rightarrow HCOO^- + 1/2H_2$) [232]. The ultralow theoretical potential makes it possible to achieve industry-level current densities (>500 mA cm^{-2}) at a cell voltage of less than 1 V. Although H_2 generation *via* partial FAOR has been reported on several metal materials, the bifunctional electrocatalyst for FAOR-assisted HWS for cogeneration of H_2 at both cathode and anode has rarely been explored.

10.4.3 Amine Oxidation

Amine represents a large family of nitrogen-containing organic compounds derived from ammonia (NH_3). The oxidation of selected amines can obtain several kinds of typical precursors of fine chemicals [233–236]. Moreover, the amine oxidation reactions (AORs) are thermodynamically more favorable than OER, making them excellent alternatives for OER to achieve HWS. In this section, we mainly introduce the AORs based on two types of substrates: 1,2,3,4-tetrahydroisoquinoline (THIQ) and benzylamine (BAm). The corresponding reactions are denoted as THIQ semi-dehydrogenation (THIQ-SDH) and BAm oxidation reaction (BAmOR), respectively.

The mild oxidation of THIQ can generate a highly valuable chemical of dihydroisoquinolines (DHIQ), which demonstrates potential applications in the pharmaceutical industry for the development of drugs in antitumor, antifungal, vasodilation, and monoamine oxidase inhibition [237, 238]. The problem is how to regulate the degree of oxidation to selectively acquire THIQ *via* a two-electron SDH pathway instead of a four-electron pathway toward dehydrogenated isoquinoline (IQ). Zhang and co-workers successfully demonstrated the THIQ-SDH using a bifunctional Ni_2P nanosheet electrode and coupled it with HER to boost H_2 evolution in water (Figure 10.8a) [239]. The SEM image (Figure 10.8b) showed that Ni_2P possesses porous nanosheet-assembled morphology with an open network. With 0.5 mM THIQ added in 1.0 M KOH, the current density rose rapidly at the potential beyond 1.32 V vs. RHE without oxygen emerging until the potential reached 1.48 V vs. RHE (Figure 10.8c). The FE of about 96% and selectivity of about 99% can be obtained at the potential range of 1.30 to 1.45 V vs. RHE. The FE of THIQ-SDH reduced to about 50% at 1.70 V vs. RHE, while the selectivity of about 96% for the DHIQ product could still be achieved. The mechanism for electrocatalytic THIQ-SDH over a Ni_2P anode was explored by CV curves and *in situ* Raman spectroscopy. The controllable SDH of THIQ to DHIQ can originate from the redox active species (*in situ* formed Ni^{II}/Ni^{III}). The Ni^{III}-OOH species were first generated on the Ni_2P surface due to the surface reconstruction and then quickly reduced by THIQ to Ni^{II}. At the same time, the THIQ was oxidized to DHIQ. Under high potentials (>1.69 V), the secondary dehydrogenation of THIQ was more difficult than OER, inhibiting further oxidation of DHIQ to dehydrogenated isoquinoline (IQ) which guaranteed the high selectivity of THIQ-SDH. Considering the remarkable HER performance of Ni_2P, with a low overpotential of 95 mV at 10 mV cm^{-2}, the two-electrode THIQ-SDH–assisted HWS electrolyzer based on the Ni_2P was assembled. LSV curve (Figure 10.8e) demonstrated that THIQ-SDH–assisted HWS required a 300-mV smaller voltage than that of OWS to deliver 20 mA cm^{-2}, revealing higher energy efficiency. The Ni_2P couple exhibited excellent durability during long-term electrolysis at 1.50 V with high FEs of about 100% and 96% for cathodic H_2 production and anodic THIQ-SDH (Figure 10.8f), respectively.

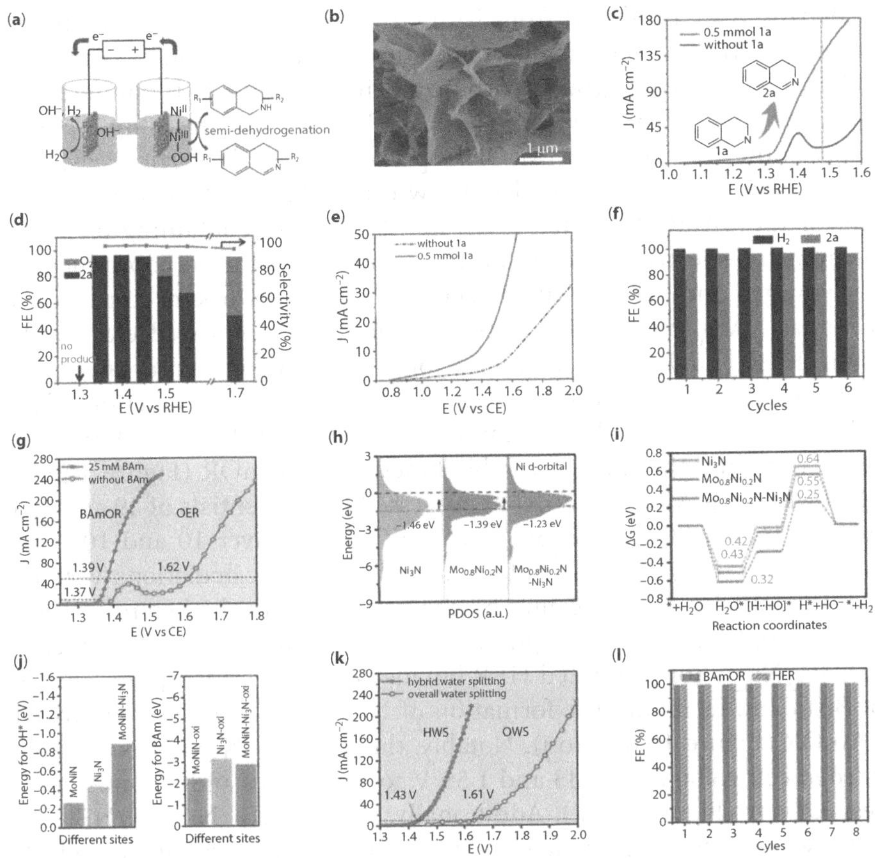

Figure 10.8 (a) Scheme of electrocatalytic THIQ-SDH–assisted HWS. (b) SEM image of the bifunctional Ni_2P nanosheet electrode. (c) LSV curves of the Ni_2P electrocatalyst in 1.0 M KOH with and without THIQ for THIQ-SDH and OER. (d) FEs and selectivities of DHIQ production at selected potentials. (e) LSV curves of a $Ni_2P||Ni_2P$ couple–based electrolyzer for THIQ-SDH–assisted HWS (1.0 M KOH with 0.5 mmol THIQ) and OWS (1.0 M KOH). (f) FEs of a Ni_2P couple electrolyzer with six cycles for both H_2 and dihydroisoquinolines production. Reproduced with permission from [239]. Copyright 2019 John Wiley & Sons. (g) LSV curves of $Mo_{0.8}Ni_{0.2}N$-Ni_3N/NF in 0.5 M Na_2SO_4/0.1 M KOH with and without 25 mM BAm. (h) DOS with H* adsorption, and (i) free energy diagram of HER for $Mo_{0.8}Ni_{0.2}N$-Ni_3N, $Mo_{0.8}Ni_{0.2}N$, and Ni_3N. (j) HO* adsorption energy and BAm adsorption energy of oxidized $Mo_{0.8}Ni_{0.2}N$-Ni_3N/NF, $Mo_{0.8}Ni_{0.2}N$/NF, and Ni_3N. (k) LSV curves for OWS and BAm-assisted HWS. (l) Cycle-dependent FEs of H_2 and benzonitrile products over $Mo_{0.8}Ni_{0.2}N$-Ni_3N/NF couple. Reproduced with permission from [242]. Copyright 2023 John Wiley & Sons.

Benzonitrile, produced from BAmOR, is also a useful precursor for the preparation of various pharmaceuticals, natural products, and pesticides [240]. Chen's group synthesized the ultrathin NiP nanomeshes with 1.9 nm thickness on NF (Ni_2P-UNMs/NF) *via* the simple phosphorization treatment of $Ni(OH)_2$ precursor [241]. Particularly, the Ni_2P-UNMs/NF couple revealed a low cell voltage of 1.41 V at 10 mA cm for the BAmOR-assisted HWS in 1.0 M KOH containing 0.125 M BAm with respective FEs of about 96.2% and 95.3% for the coproduction of hydrogen and benzonitrile. Recently, Fu and co-workers constructed a biphasic $Mo_{0.8}Ni_{0.2}N$-Ni_3N heterojunction with improved electrocatalytic HER and BAmOR activities for energy-saving hydrogen generation in 0.5M Na_2SO_4/0.1 M KOH [242]. After the introduction of 25 mM BAm, the potential significantly decreased from 1.62 V to 1.39 V vs. RHE, revealing the more favorable kinetics of BAmOR (Figure 10.8g). Meanwhile, the $Mo_{0.8}Ni_{0.2}N$-Ni_3N possessed overpotentials of 49 and 133 mV for HER in 0.5 M Na_2SO_4/0.1 M KOH to deliver 10 and 100 mA cm^{-2}, respectively. Experimental results and DFT calculations verified that the tailored d-band center by charge transfer in the heterojunction not only optimizes H* adsorption and decreases water activation energy on $Mo_{0.8}Ni_{0.2}N$ for promoted HER but also optimizes adsorption energy of BAm and facilitates the formation of NiOOH on Ni_3N for improved BAmOR (Figure 10.8, h to j). Notably, the $Mo_{0.8}Ni_{0.2}N$-Ni_3N/NF couple needed cell voltages of 1.43 and 1.59 V to acquire 10 and 220 mA cm^{-2}, respectively (Figure 10.8k). Additionally, the $Mo_{0.8}Ni_{0.2}N$-Ni_3N/NF couple also showed prominent stability, with FEs of benzonitrile and H_2 approaching 100% for 10 cycles (Figure 10.8l).

10.4.4 Biomass Oxidation

Biomass, a renewable carbon source coming from plants and animals, is emerging as a promising alternative to crude oil for the synthesis of fuels and other value-added products [243–246]. The challenge lies in the development of affordable and effective conversion technology to obtain biomass-based chemical products that are comparable or even superior to petroleum-based chemical products. Particularly, electrocatalytic biomass oxidation represents a green route for biomass upgrading and it has the potential to be coupled with HER to realize the coproduction of chemicals and H_2 hydrogen in an energy-efficient manner.

10.4.4.1 *Furans Oxidation*

Furanic compounds, including 5-hydroxymethylfurfural (HMF) and furfural (FF), are exemplary primary biomass-derived building block chemicals, which can be produced by dehydration of the relevant carbohydrates. They are widely utilized as universal precursors for fuels, polymers, and plastics [247–249]. For example, FDCA (2,5-furandicarboxylic acid) acquired from HMF oxidation reaction (HMFOR) is a value-added alternative to terephthalic acid for the synthesis of polyesters, polyurethanes, and polyamides (Figure 10.9a) [250]. The overall electrolysis reaction of the HMFOR-assisted HWS electrolyzer is shown below: $HMF+2H_2O \rightarrow FDCA+3H_2$, and the thermodynamic potential of HMFOR toward FDCA is approximately 0.3 V, much smaller than that of conventional OWS [180]. Since HMFOR toward FDCA involves a six-electron transfer, the performance of HMF oxidation is highly reliant on the electrocatalyst.

As expected, the noble metal–based electrocatalysts, like Ir/Co_3O_4 [251], $Pt/Ni(OH)_2$ [252], and PbO-CuO [253], showed excellent electrocatalytic activities for HMFOR. Lately, Guo and colleagues fabricated bifunctional Rh-SA/NiFe NMLDH by employing nanoporous mesh-type NiFe LDH to anchor $Rh\text{-}O_5/Ni(Fe)$ atomic site for HMFOR-assisted HWS (Figure 10.9b) [254]. Electrochemical tests in 1.0 M KOH with 0.05 M HMF revealed that Rh-SA/NiFe NMLDH delivers 50 mA cm^{-2} at small potentials of 1.30 V vs. RHE (Figure 10.9c). When coupled with HER for HWS, the cell voltages required to afford 10, 50, and 100 mA cm^{-2} were 1.32, 1.42, and 1.48 V, respectively, lower than those of OWS (Figure 10.9, d and e). During the chronoamperometric measurement, the Rh-SA/NiFe NMLDH showed outstanding catalytic performance for more than 100 h with an average FE of 96.8% to FDCA. Theoretical calculations elucidated that the respective electronic structures of Ni 3d and Rh 4d were optimized by the interaction between Rh and neighboring Ni in $Rh\text{-}O_5/Ni(Fe)$, resulting in reduced energy barriers of key intermediates/substrates like HMF and HO* conversion (Figure 10.9, f and g).

Compared to the noble metal–based electrocatalysts, low-cost transition metal-based materials, especially the Ni- and Co-based materials, receive more attention due to the cost advantage. Sun and co-workers synthesized a variety of Ni- and Co-based bifunctional electrocatalysts including Co-P/CF [255], Ni_2P NPA/NF (Ni_2P nanoparticle arrays on NF) [256], Ni_3S_2/NF [225], and hp-Ni [223] for HWS assisted by oxidation of various biomass like HMF, FF, furfuryl alcohol (FFA). For example, the hierarchically

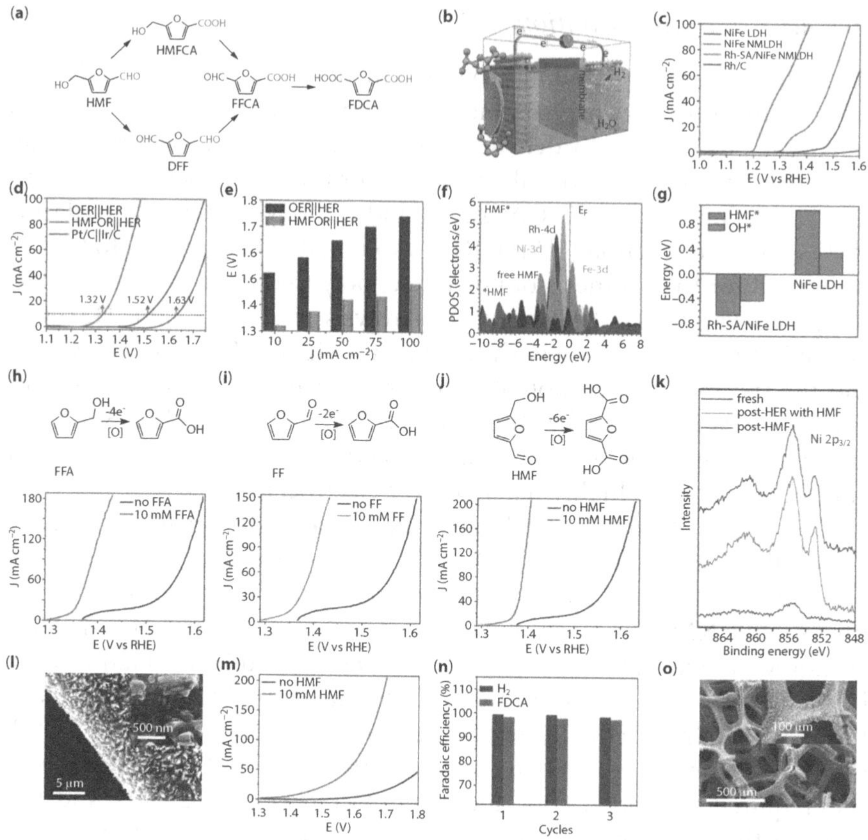

Figure 10.9 (a) The reaction mechanism of HMF oxidation. (b) Schematic of HMFOR-coupled HWS. (c) Polarization curves of four electrocatalysts in 1.0 M KOH with 0.05 M HMF. (d) LSV curves of HMFOR-assisted HWS and OWS for Rh-SA/NiFe NMLDH couple and Pt/C||Ir/C couple. (e) Summary of the cell voltages at selected current densities for HMFOR-assisted HWS and OWS over Rh-SA/NiFe NMLDH couple. (f) PDOS for the adsorption of HMF on Rh-SA/NiFe LDH. (g) Comparations of the adsorption energy of HO* and HMF* on NiFe LDH and Rh-SA/NiFe LDH. Reproduced with permission from [254]. Copyright 2023 American Chemical Society. (h-j) Reaction equations for selective oxidation of organics and the polarization curves in 1.0 M KOH with and without 10 mM substrates (HMF, FF, and FFA) for Ni_3S_2/NF. (k) XPS spectra of Ni $2p_{3/2}$ for the post-HMF, post-HER with HMF, and fresh Ni_3S_2/NF samples. Reproduced with permission from [225]. Copyright 2016 American Chemical Society. (l) SEM images of Ni_2P NPA/NF. (m) LSV curves for OWS (1.0 M KOH) and HMFOR-coupled HWS (1.0 M KOH with 10 mM HMF) in a Ni_2P NPA/NF-based two-electrode configuration. (n) FEs of Ni_2P NPA/NF-based two-electrode system for co-production of FDCA and H_2 in 1.0 M KOH with 10 mM HMF for three cycles. Reproduced with permission from [256]. Copyright 2016 John Wiley & Sons. (o) SEM images of Co-P/CF. Reproduced with permission from [255]. Copyright 2016 American Chemical Society.

porous Ni_3S_2/NF with the morphology of stacked nanoparticles was prepared by direct sulfurization of NF and showed superior electrocatalytic activities for HER and oxidation of a series of biomass substrates. After introducing 10 mM FFA (Figure 10.9h), FF (Figure 10.9i), and HMF (Figure 10.9j) in 1.0 M KOH, all LSV curves for furans oxidation exhibited lower potentials to deliver the same current density, implying more rapid kinetics for biomass oxidation than OER. Using HMF oxidation reaction (HMFOR) as a model, the systematic physicochemical characterizations, including XRD, SEM, and XPS (Figure 10.9k), revealed that although the Ni_3S_2/NF maintains the whole 3D hierarchically porous morphology and main phase of Ni_3S_2 after stability testing, the partial Ni_3S_2 is restructured to oxidized Ni species (nickel oxyhydroxides/hydroxides/oxides), different from that for HER. Moreover, Ni_3S_2/NF displayed almost overlapped LSV curves for HER even when 10 HMF was added and required an overpotential of 116 mV to afford 10 mA cm^{-2}. The Ni_3S_2/NF couple acquired 100 mA cm^{-2} at 1.84 V, 200 mV smaller than that for OWS, and achieved the cogeneration of H_2 and FDCA with FEs of 100% and 98%, respectively.

Similarly, Sun and co-workers also fabricated the Ni_2P NPA/NF bifunctional electrocatalyst *via* facile phosphorization of NF for HMFOR-assisted HWS [256]. As shown in Figure 10.9l, the SEM image revealed that Ni_2P NPA/NF possesses a porous and rough morphology consisting of Ni_2P nanoparticles. Introducing 10 mM HMF into 1.0 M KOH decreased the onset potential of Ni_2P NPA/NF from 1.50 to 1.35 V vs. RHE, with a dramatic current rising within 1.40 V vs. RHE. Besides, the HER-HMFOR two-electrode system exhibited a low cell voltage of 1.58 V to attain 50 mA cm^{-2}, 220 mV smaller than that for OWS (Figure 10.9m). Additionally, the high FEs of 100% and 98% FEs were obtained for the coproduction of H_2 and FDCA within three cycles, respectively (Figure 10.9n). Afterward, they also synthesized the bifunctional Co-P/CF electrocatalyst (Co-P on copper foam) with a porous morphology containing numerous Co-P particles on CF for HMFOR and HER (Figure 10.9o) and found that the Co-P/CF–based two-electrode configuration exhibits a cell voltage of 1.44 V to afford 20 mA cm^{-2} for HMFOR-coupled HWS in 1.0 M KOH with 50 mM HMF [255].

More recently, a carbon-encapsulated porous MoO_2-FeP@C electrocatalyst (MoO_2-FeP heterojunction) with spindle-like morphology was constructed by Fu and co-workers for HER and HMFOR [257]. The electron transfer from MoO_2 to FeP regulated the adsorption of H_2O, H^*, and HMF for improved HER and HMFOR activities. Consequently, the MoO_2-FeP@C–based two-electrode system achieved 10 mA cm^{-2} at 1.486 V for H_2 and FDCA cogeneration. Particularly, the high selectivity of 98.6% and

FE of 97.8% can be achieved for FDCA, and the conversion efficiency of HMF was high at 99.4%.

10.4.4.2 Carbohydrates Oxidation

Carbohydrates like glucose and chitin are the most widespread organic substances in nature. For example, glucose is a widely distributed monosaccharide and it can be easily obtained from biomass [258]. Glucose oxidation reaction (GlOR) can generate value-added glucaric acid (GRA), gluconate, gluconic acid (GNA), and so on (Figure 10.10a), making it promising to couple with HER for HWS [259].

Zhang and co-workers prepared a bifunctional sandwich-structured Fe_3O_4/Au/CoFe-LDH for the GlOR-assisted HWS (Figure 10.10b) [260]. The resulting Fe_3O_4/Au/CoFe-LDH demonstrated decent alkaline HER performance with an overpotential of 141 mV to deliver 10 mA cm^{-2} and superior GlOR performance with a potential of 0.404 V vs. RHE to afford 100 mA cm^{-2} in 1.0 M KOH containing 0.5 M glucose. Structural and electrochemical characterizations verified that the sandwiched Au facilitates electron transfer between Fe_3O_4 and CoFe-LDH, responsible for improved activities. Further electrocatalytic evaluations in a two-electrode electrolyzer confirmed that the Fe_3O_4/Au/CoFe-LDH couple needs a low voltage of 0.48 V to acquire 10 mA cm^{-2} (Figure 10.10c) and achieves a high selectivity of 89.7% for GlOR even at 1 A cm^{-2} (Figure 10.10d). Likewise, Yang and co-workers integrated Ru nanoparticles with nickel boride grown on NF (Ru@Ni-B/NF) as the bifunctional electrocatalyst for GlOR-assisted HWS, with a cell voltage of 1.24 and 1.56 V to afford 10 mA cm^{-2} and 50 mA cm^{-2}, respectively [261].

To further decrease the cost of electrocatalysts, nonprecious metal-based electrocatalysts have attracted wide attention and were used for HER and GlOR. For example, Wen and collaborators synthesized bifunctional Fe-doped $CoSe_2$ nanowires ($Fe_{0.1}$-$CoSe_2$) on CC for GlOR-assisted HWS (Figure 10.10e) [262]. For HER in 0.5 M H_2SO_4, the as-prepared $Fe_{0.1}$-$CoSe_2$/CC required a low overpotential of 270 mV to drive 100 mA cm^{-2}. For GlOR in 1.0 M KOH containing 0.5 M glucose, the $Fe_{0.1}$-$CoSe_2$/CC revealed an onset potential of 1.12 V vs. RHE, 530 mV smaller than that for OER, implying the favorable kinetics of GlOR relative to OER. In view of the excellent electrocatalytic performance of HER and GlOR, the $Fe_{0.1}$-$CoSe_2$/CC–based GlOR-HER electrolyzer was constructed for the cogeneration of gluconate and H_2, and a cell voltage of only 0.72 V was needed to deliver 10 mA cm^{-2}, much lower than that for OWS (1.34 V vs. RHE) (Figure 10.10f). To gain in-depth mechanical insight into $Fe_{0.1}$-$CoSe_2$

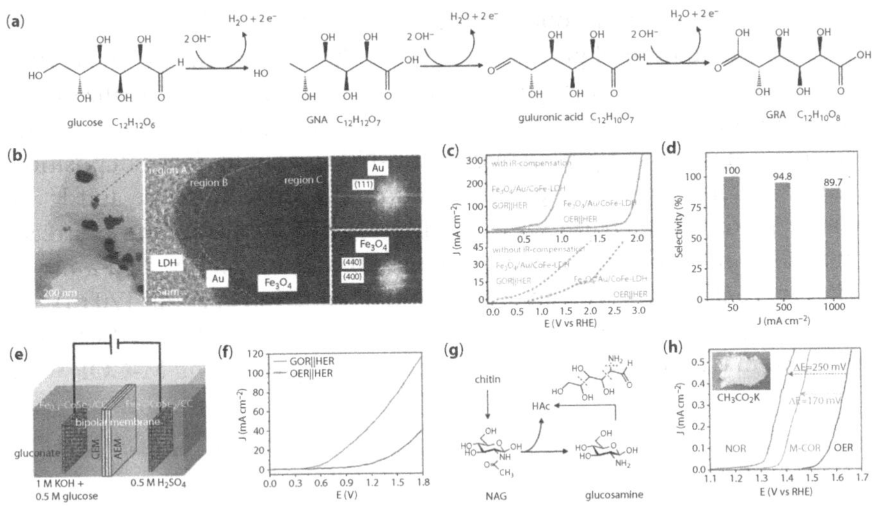

Figure 10.10 (a) Possible mechanisms for GlOR toward GRA and GNA. Reproduced with permission from [259] under the terms of a CC-BY 4.0 license. Copyright 2020 The Authors, published by Springer Nature. (b) TEM and HRTEM images of Fe_3O_4/Au/CoFe-LDH-1000, along with selected area electron diffraction. (c) LSV curves of Fe_3O_4/Au/CoFe-LDH in an asymmetrical electrolytic cell with/without iR-compensation for OER (1.0 M KOH) and GlOR (1.0 M KOH with glucose). (d) Selectivity of GlOR at different current densities. Reproduced with permission from [260]. Copyright 2021 John Wiley & Sons. (e) Schematic of asymmetric-electrolyte electrolyzer for GOR-assisted HWS based on the bifunctional $Fe_{0.1}$-$CoSe_2$/CC catalyst with 0.5 M H_2SO_4 as the cathodic electrolyte and 1.0 M KOH containing 0.5 M glucose as the anodic electrolyte, separated by a bipolar membrane. (f) LSV curves for the asymmetric-electrolyte cell with ΔpH = 14 based on $Fe_{0.1}$-$CoSe_2$/CC couple. Reproduced with permission from [262]. Copyright 2020 Elsevier. (g) Main reaction pathway for electrochemical reforming of chitin selectively to acetic acid (HAc). (h) LSV curves for OER (1.0 M KOH; black curve), M-COR (1.0 M KOH with 3.5 g L^{-1} milled chitin; blue curves), and NOR (1.0 M KOH with 2.2 g L^{-1} N-acetyl glucosamine; red curve). Inset is the photograph of the oxidation product potassium acetate (CH_3CO_2K). Reproduced with permission from [246] under the terms of a CC-BY 4.0 license. Copyright 2021 The Authors, published by Springer Nature.

for improved electrocatalytic GlOR and HER, DFT calculations were performed. They found that the Co in $Fe_{0.1}$-$CoSe_2$ are the primary active centers for both GlOR and HER, with more neutral ΔG_{H^*} for HER and a lower thermodynamic barrier for GlOR.

Apart from glucose, chitin, consisting of β-(1,4)-linked *N*-acetyl-β-D-glucosamine, is another abundant natural carbohydrate in crustacean shells and insect skeletons. Figure 10.10g depicts the electrochemical reforming of chitin selectivity to acetic acid (HAc). Firstly, the chitin was depolymerized to form N-acetylglucosamine (NAG) monomer. The resulting NGA was then decomposed to obtain glucosamine and HAc by deacetylation. The further oxidation of glucosamine through breaking the pyranose ring can generate more HAc (Figure 10.10g). Li and co-workers realized the integration of chitin oxidation (COR) reaction and HER with three-dimensional hierarchical porous nickel (hp-Ni) *via* facile electrodeposition on bubbles for energy-saving coproduction of acetate at anode and H_2 at cathode [246]. To increase the solubility of chitin, ball-milling pretreatment was conducted. As shown in Figure 10.11h, hp-Ni revealed a much higher current density with an overpotential reduction of about 170 mV by introducing 3.5 g L^{-1} milled chitin (G-COR) in 1.0 M KOH. Furthermore, about 250 mV potential saving can be achieved by replacing M-COR with

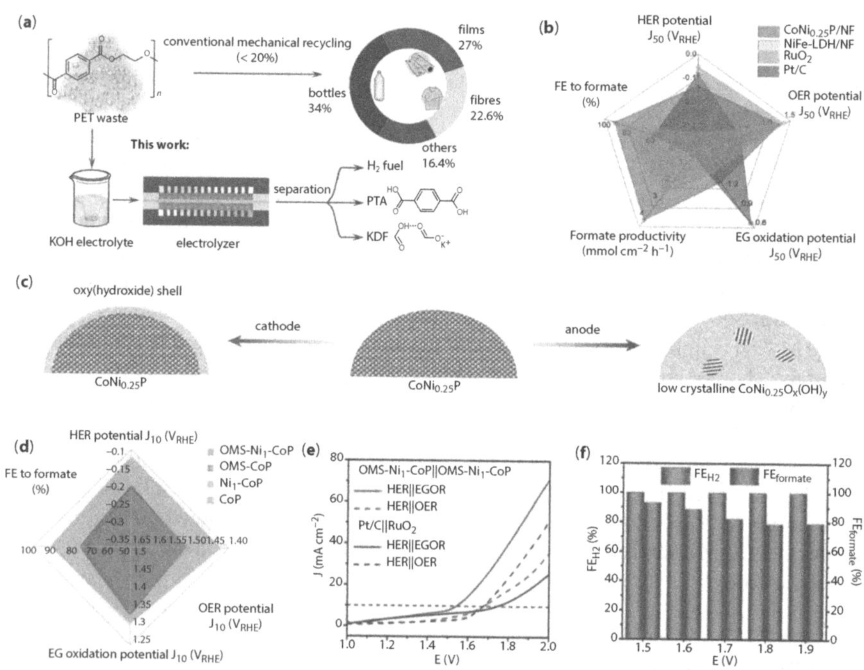

Figure 10.11 (a) Conventional and proposed electrocatalysis routes for PET recycling. (b) Summary of the electrocatalytic performances of $CoNi_{0.25}P/NF$ and known catalysts. (c) Schematic of the structure changes of $CoNi_{0.25}P$ catalyst under reaction conditions. Reproduced with permission from [206] under the terms of a CC-BY 4.0 license. Copyright 2021 The Authors, published by Springer Nature. (d) Comparison of the catalytic performances of OMS-Ni_1-CoP and controls. (e) LSV curves of OMS-Ni_1-CoP couple and Pt/C||RuO_2 couple for EGOR-assisted HWS and OWS. (f) FE of formic acid and H_2 production for EGOR-assisted HWS over OMS-Ni_1-CoP couple. Reproduced with permission from [267]. Copyright 2022 Elsevier.

NAG oxidation reaction (NOR) in 1.0 M KOH with 2.2 g L^{-1} NAG, and the potassium acetate powder can be obtained by precipitation extraction of the resulting anolyte (inset of Figure 10.10h).

10.4.5 Plastic Waste Oxidation

Currently, plenty of plastic wastes are thrown away and stored in garbage and aquatic environments [263–266]. Therefore, converting low-value plastic waste into value-added products is extremely meaningful, especially incorporating it with HER.

Polyethylene terephthalate (PET) is one of the most used polymers worldwide with an annual production of around 70 million tons, yet only a small fraction (<20%) of them are recycled (Figure 10.11a) [206, 265]. Considering the read degradation of PET into its monomers including EG and terephthalic acid (TPA) under alkaline conditions, Duan and co-workers integrated the oxidative upcycling of PET waste with HER using the bifunctional $CoNi_{0.25}P$/NF electrocatalyst for the cogeneration of potassium diformate (KDF)/terephthalic acid (PTA) and H_2 [206]. Compared with known Pt/C, RuO_2, and NiFe-LDH/NF electrocatalysts, $CoNi_{0.25}P$/NF displayed the highest FE for formate production (91.3%) with excellent productivity of 4.01 mmol cm^{-2} h^{-1} at 1.7 V vs. RHE for EGOR, indicating the crucial role of phosphorization for oxidative C–C cleavage (Figure 10.11b). Besides, the $CoNi_{0.25}P$/NF also exhibited excellent HER activity (Figure 10.11b). After introducing 0.3 M EG, the potential for $CoNi_{0.25}P$/NF shifted from 1.50 to 0.70 V vs. RHE to reach 50 mA cm^{-2}. Furthermore, they constructed a membrane-electrode assembly (MEA) flow cell with zero gaps based on the bifunctional $CoNi_{0.25}P$/NF. The $CoNi_{0.25}P$/NF couple required a low cell voltage of only 1.8 V to attain an industrial-grade current density of 500 mA cm^{-2} with formate productivity of 5.1 mmol cm^{-2} h^{-1}, much higher than that for traditional H-type electrolyzer under same voltage due to the smaller ohmic resistance in MEA. To deeply understand the prominent performance of $CoNi_{0.25}P$/NF for both HER and EGOR, the authors conducted systematic structural and electrochemical characterizations including HRTEM, Raman spectroscopy, and XAS. For the anodic EGOR and OER, the $CoNi_{0.25}P$ sample underwent complete reconstruction to the metal oxy(hydroxide) with low crystallinity after EGOR. For the cathodic HER, $CoNi_{0.25}P$ maintained the crystalline structure, with slight surface oxidation and dissolution of transition metal phosphide toward an amorphous $CoNi_{0.25}P/CoNi_{0.25}O_x(OH)_y$ core-shell structure (Figure 10.11c).

Similarly, Zhu and co-workers developed an atomic Ni-modified CoP with ordered macroporous superstructure and abundant mesopores (OMS-Ni_1-CoP) for energy-saving H_2 production and PET plastic recycling [267]. As shown in Figure 10.11d, a low overpotential of only 119 mV was needed for the HER in 1.0 M KOH over OMS-Ni_1-CoP to achieve 10 mA cm^{-2}. Moreover, the required potentials for EGOR over OMS-Ni_1-CoP were only 1.30 and 1.38 V vs. RHE to deliver 10 and 50 mA cm^{-2}, respectively (Figure 10.11d). The assembled two-electrode electrolyzer using bifunctional OMS-Ni_1-CoP electrocatalyst as both cathode and anode exhibited a cell voltage of 1.52 V at 10 mA cm^{-2} (Figure 10.11e) and showed the FEs of 100% and 93.2% for the coproduction of H_2 and value-add formate, respectively (Figure 10.11f). Even in the electrolyte containing PET plastic hydrolysate, the OMS-Ni_1-CoP couple also showed a low voltage of 1.57 V at 10 mA cm^{-2}. Other transition metal–based electrocatalysts, like Ni_3N/W_5N_4 [268], exhibited excellent electrochemical performance for HER/EGOR as well, implying the significant potential for plastic waste oxidation-assisted HWS.

The thermoplastic polylactic acid (PLA) plastic is widely used to produce disposable goods. The theoretical thermodynamic potential of PLA electro-oxidation reaction (PAOR) toward acetic acid (HAc) is much smaller than that of OER, such that it is possible to use PAOR to replace OER for energy-efficient water splitting. Gao and co-workers prepared the bifunctional cobalt selenide nanosheets on nickel foam ($CoSe_2$/NF) electrocatalyst for PAOR-assisted HWS in 1.0 M KOH and about 170 mV voltage saving can be achieved relative OWS [269]. Also, the electrooxidation conversion rate of PLA toward HAc was about 87%. Moreover, only 11-mV degradation can be observed after 10-h electrolysis at 50 mA cm^{-2}, revealing the outstanding stability of the $CoSe_2$/NF couple–based PAOR-assisted HWS system.

10.5 Conclusions and Outlook

This chapter focuses on the current progress in bifunctional electrocatalyst–driven HWS for energy-saving coproduction of green H_2 and valuable chemicals. With the achievements in synthetic methodologies, various noble metals (e.g., Ru, Ir, Au, Pd, Pt, and Rh) and nonprecious metals (e.g., Mo, Ni, Co, Fe, Cu, Mn, and W) based bifunctional electrocatalysts with diverse morphologies have been prepared. To address the issues regarding OER in traditional OWS like sluggish kinetics, low-value O_2, and the possible presence of explosive H_2/O_2 mixture and ROS, alternative oxidation upgrading reactions have been introduced to replace OER for HWS catalyzed by the bifunctional electrocatalysts. These alternative reactions with favorable kinetics compared to OER include the upgrading of alcohols, aldehydes, amines, biomass, and plastic waste toward valuable chemicals, along with the cogeneration of green H_2 under lower energy input. Despite significant progress has been made, several challenges remain.

First, current synthetic strategies for bifunctional electrocatalysts in the laboratory based on solvothermal technology, electrodeposition, templating, annealing, photochemistry, and so on, may fail for commercialization due to the variations in mass/heat transfer, operability, and cost at large scale. In the same case, toxic gases like NH_3, PH_3, and S are involved, making them neither economical nor environmentally benign. It is necessary to develop green, scalable, affordable, and rapid routes. In this regard, novel methods based on microwave, joule heat, plasma, and laser as well as electroshock and 3D printing can be introduced to synthesize bifunctional electrocatalysts. To improve the mass/heat transport, multichannel flow reactors may be desirable. To heighten the electrochemical activity and durability of the resulting bifunctional electrocatalysts, creating 3D hierarchical porosity, plentiful defects, and adjustable local microenvironment is welcome.

Second, electrochemical reconstruction endows the resulting electrocatalysts' bifunctionality for HWS, while a full understanding of the reconstruction process is still not yet achieved due to the convolution of chemical reactions and mass transport. Currently, the time/space resolution of *in situ/operando* characterization techniques is still insufficient to accurately track the dynamic evolution of electrochemical reconstruction, and/or only the average or bulk information can be obtained. For example, FTIR, Raman, and XPS cannot monitor the fast electrooxidation reconstruction on the order of picoseconds to nanoseconds in a real-time fashion. Additionally, due to the limitations in detection range and resolution, it is impossible that one or a few characterization techniques can give us the full picture regarding electrochemical reconstruction. Consequently, developing combined characterization techniques with high spatiotemporal resolution and broad detection range is strongly recommended. Note that composition detaching usually happens during electrochemical reconstruction, such that the detached species may dissolve in electrolytes and change the local microenvironments at the catalytic interface, interfacing with the investigation of the activity-structure relationship. Intentionally adding analogs to electrolytes to mimic these electrocatalytic interfaces can offer some insightful information.

Third, theoretical computations like DFT, combined with recent achievements in computing power, algorithms, and big data, have gained great success in understanding the mechanisms of OWS and HWS, as well as accelerating the discovery of advanced bifunctional electrocatalysts. Meanwhile, current models for DFT calculations are too simple to simulate the real electrocatalytic processes because electric field, electrochemical double-layer, solvation, and dissolved ions in realistic electrochemical processes are usually not considered. Moreover, considering the intricacy and significance of electrochemical reconstruction, more attention should be paid to the DFT simulation of the dynamic reconstruction process. In addition, the alternative oxidations reactions usually involve multiple electron transfers/intermediates/adsorption sites, such that experimentally identifying the precise stereo-configuration of catalytic interface and measuring the binding energies of intermediates are often challenging, extending DFT modeling for these oxidation reactions beyond HER and OER is urgent and necessary.

Fourth, considering the substantial demand for green H_2 in the future, it is more appealing to explore the oxidative upgrading of commodity chemicals with high solubility. Moreover, to obtain the H_2 with high purity or avoid gaseous mixture separation, the gas-free alternative oxidation reactions have a higher priority. In other words, the techno-economic analysis should be first performed when deciding on replacing OER with an alternative oxidation reaction.

References

1. Chu, S. and Majumdar, A., Opportunities and Challenges for a Sustainable Energy Future. *Nature*, 488, 294–303, 2012.
2. Lewis, N.S., Research Opportunities to Advance Solar Energy Utilization. *Science*, 351, aad1920, 2016.
3. Dotan, H., Landman, A., Sheehan, S.W., Malviya, K.D., Shter, G.E., Grave, D.A., Arzi, Z., Yehudai, N., Halabi, M., Gal, N., Decoupled Hydrogen and Oxygen Evolution by a Two-Step Electrochemical-Chemical Cycle for Efficient Overall Water Splitting. *Nat. Energy*, 4, 786–795, 2019.
4. Muhich, C.L., Evanko, B.W., Weston, K.C., Lichty, P., Liang, X., Martinek, J., Musgrave, C.B., Weimer, A.W., Efficient Generation of H_2 by Splitting Water with an Isothermal Redox Cycle. *Science*, 341, 540–542, 2013.
5. Wang, J., Yang, H., Li, F., Li, L., Wu, J., Liu, S., Cheng, T., Xu, Y., Shao, Q., Huang, X., Single-Site Pt-Doped RuO_2 Hollow Nanospheres With Interstitial C for High-Performance Acidic Overall Water Splitting. *Sci. Adv.*, 8, eabl9271, 2022.

6. Turner, J.A., Sustainable Hydrogen Production. *Science*, 305, 972–974, 2004.
7. Jaramillo, T.F., Jorgensen, K.P., Bonde, J., Nielsen, J.H., Horch, S., Chorkendorff, I., Identification of Active Edge Sites for Electrochemical H_2 Evolution from MoS_2 Nanocatalysts. *Science*, 317, 100–102, 2007.
8. Jiang, H., Sun, Y., You, B., Dynamic Electrodeposition on Bubbles: An Effective Strategy toward Porous Electrocatalysts for Green Hydrogen Cycling. *Acc. Chem. Res.*, 56, 1421–1432, 2023.
9. Zhao, X., He, D., Xia, B.Y., Sun, Y., You, B., Ambient Electrosynthesis toward Single-Atom Sites for Electrocatalytic Green Hydrogen Cycling. *Adv. Mater.*, 35, 2210703, 2023.
10. Zhao, H. and Yuan, Z.-Y., Progress and Perspectives for Solar-Driven Water Electrolysis to Produce Green Hydrogen. *Adv. Energy Mater.*, 13, 2300254, 2023.
11. Kovac, A., Paranos, M., Marcius, D., Hydrogen in Energy Transition: A Review. *Int. J. Hydrogen Energy*, 46, 10016–10035, 2021.
12. Barbir, F., PEM Electrolysis for Production of Hydrogen from Renewable Energy Sources. *Sol. Energy*, 78, 661–669, 2005.
13. Dodds, P.E., Staffell, L., Hawkes, A.D., Li, F., Grunewald, P., Mcdowall, W., Ekins, P., Hydrogen and Fuel Cell Technologies for Heating: A Review. *Int. J. Hydrogen Energy*, 40, 2065–2083, 2015.
14. Moradi, R. and Groth, K.M., Hydrogen Storage and Delivery: Review of the State of the Art Technologies and Risk and Reliability Analysis. *Int. J. Hydrogen Energy*, 44, 12254–12269, 2019.
15. Wang, M., Wang, Z., Gong, X., Guo, Z., The Intensification Technologies to Water Electrolysis for Hydrogen Production - A Review. *Renew. Sustain. Energy Rev.*, 29, 573–588, 2014.
16. Luo, M., Yi, Y., Wang, S., Wang, Z., Du, M., Pan, J., Wang, Q., Review of Hydrogen Production Using Chemical-Looping Technology. *Renew. Sustain. Energy Rev.*, 81, 3186–3214, 2018.
17. Parra, D., Valverde, L., Pino, F.J., Patel, M.K.A., Review On The Role, Cost and Value of Hydrogen Energy Systems for Deep Decarbonisation. *Renew. Sustain. Energy Rev.*, 101, 279–294, 2019.
18. Acar, C. and Dincer, I., The Potential Role of Hydrogen as a Sustainable Transportation Fuel to Combat Global Warming. *Int. J. Hydrogen Energy*, 45, 3396–3406, 2020.
19. Elam, C.C., Padró, C.E.G., Sandrock, G., Luzzi, A., Lindblad, P., Hagen, E.F., Realizing the Hydrogen Future: The International Energy Agency's Efforts to Advance Hydrogen Energy Technologies. *Int. J. Hydrogen Energy*, 28, 601–607, 2003.
20. Najjar, Y.S.H., Hydrogen Safety: The Road Toward Green Technology. *Int. J. Hydrogen Energy*, 38, 10716–10728, 2013.
21. Posso, F., Pulido, A., Acevedo-Páez, J.C., Towards the Hydrogen Economy: Estimation of Green Hydrogen Production Potential and the Impact of Its

Uses in Ecuador as a Case Study. *Int. J. Hydrogen Energy*, 48, 11922–11942, 2023.
22. Wu, T. and Wang, J., Reliability Evaluation for Integrated Electricity-Gas Systems Considering Hydrogen. *IEEE Trans. Sustain. Energy*, 14, 920–934, 2023.
23. Mueller-Langer, F., Tzimas, E., Kaltschmitt, M., Peteves, S., Techno-Economic Assessment of Hydrogen Production Processes for The Hydrogen Economy for the Short and Medium Term. *Int. J. Hydrogen Energy*, 32, 3797–3810, 2007.
24. Barreto, L., Makihira, A., Riahi, K., The Hydrogen Economy in the 21st Century: A Sustainable Development Scenario. *Int. J. Hydrogen Energy*, 28, 267–284, 2003.
25. Marbán, G. and Valdás-Solís, T., Towards the Hydrogen Economy? *Int. J. Hydrogen Energy*, 32, 1625–1637, 2007.
26. Schmidt, O., Gambhir, A., Staffell, I., Hawkes, A., Nelson, J., Few, S., Future Cost and Performance of Water Electrolysis: An Expert Elicitation Study. *Int. J. Hydrogen Energy*, 42, 30470–30492, 2017.
27. De Luna, P., Hahn, C., Higgins, D., Jaffer, S.A., Jaramillo, T.F., Sargent, E.H., What Would It Take for Renewably Powered Electrosynthesis to Displace Petrochemical Processes? *Science*, 364, eaav3506, 2019.
28. Shi, Y. and Zhang, B., Recent Advances in Transition Metal Phosphide Nanomaterials: Synthesis and Applications in Hydrogen Evolution Reaction. *Chem. Soc. Rev.*, 45, 1529–1541, 2016.
29. Abbasi, T. and Abbasi, S.A., 'Renewable' Hydrogen: Prospects and Challenges. *Renew. Sustain. Energy Rev.*, 15, 3034–3040, 2011.
30. Sharma, S. and Ghoshal, S.K., Hydrogen the Future Transportation Fuel: From Production to Applications. *Renew. Sustain. Energy Rev.*, 43, 1151–1158, 2015.
31. Shiva Kumar, S. and Himabindu, V., Hydrogen Production by PEM Water Electrolysis - A Review. *Mater. Sci. Energy Technol.*, 2, 442–454, 2019.
32. Wan, L., Xu, Z., Xu, Q., Pang, M., Lin, D., Liu, J., Wang, B., Key Components and Design Strategy of the Membrane Electrode Assembly for Alkaline Water Electrolysis. *Energy Environ. Sci.*, 16, 1384–1430, 2023.
33. Yu, Z.-Y., Duan, Y., Feng, X.-Y., Yu, X., Gao, M.-R., Yu, S.-H., Clean and Affordable Hydrogen Fuel from Alkaline Water Splitting: Past, Recent Progress, and Future Prospects. *Adv. Mater.*, 33, 2007100, 2021.
34. Zeng, K. and Zhang, D., Recent Progress in Alkaline Water Electrolysis for Hydrogen Production and Applications. *Prog. Energy Combust. Sci.*, 36, 307–326, 2010.
35. Mccrory, C.C.L., Jung, S., Peters, J.C., Jaramillo, T.F., Benchmarking Heterogeneous Electrocatalysts for the Oxygen Evolution Reaction. *J. Am. Chem. Soc.*, 135, 16977–16987, 2013.

36. Biswas, S., Kulkarni, A.P., Giddey, S., Bhattacharya, S.A., Review on Synthesis of Methane as a Pathway for Renewable Energy Storage with a Focus on Solid Oxide Electrolytic Cell-Based Processes. *Front. Energy Res.*, 8, 570112, 2020.
37. Hua, B., Li, M., Sun, Y.-F., Li, J.-H., Luo, J.-L., Enhancing Perovskite Electrocatalysis of Solid Oxide Cells through Controlled Exsolution of Nanoparticles. *ChemSusChem*, 10, 3333–3341, 2017.
38. Tucker, M.C., Progress in Metal-Supported Solid Oxide Electrolysis Cells: A Review. *Int. J. Hydrogen Energy*, 45, 24203–24218, 2020.
39. Ogumerem, G.S. and Pistikopoulos, E.N., Parametric Optimization and Control for a Smart Proton Exchange Membrane Water Electrolysis (PEMWE) System. *J. Process Control*, 91, 37–49, 2020.
40. Chen, Z., Guo, L., Pan, L., Yan, T., He, Z., Li, Y., Shi, C., Huang, Z.-F., Zhang, X., Zou, J.-J., Advances in Oxygen Evolution Electrocatalysts for Proton Exchange Membrane Water Electrolyzers. *Adv. Energy Mater.*, 12, 2103670, 2022.
41. Laguna-Bercero, M.A., Recent Advances in High Temperature Electrolysis Using Solid Oxide Fuel Cells: A Review. *J. Power Sources*, 203, 4–16, 2012.
42. Zhang, Z., Li, X., Zhong, C., Zhao, N., Deng, Y., Han, X., Hu, W., Spontaneous Synthesis of Silver-Nanoparticle-Decorated Transition-Metal Hydroxides for Enhanced Oxygen Evolution Reaction. *Angew. Chem. Int. Ed.*, 59, 7245–7250, 2020.
43. Du, J., Li, F., Sun, L., Metal-Organic Frameworks and Their Derivatives as Electrocatalysts for the Oxygen Evolution Reaction. *Chem. Soc. Rev.*, 50, 2663–2695, 2021.
44. You, B., Liu, X., Jiang, N., Sun, Y.A., General Strategy for Decoupled Hydrogen Production from Water Splitting by Integrating Oxidative Biomass Valorazation. *J. Am. Chem. Soc.*, 138, 13639, 2016.
45. Tian, X., Zhao, P., Sheng, W., Hydrogen Evolution and Oxidation: Mechanistic Studies and Material Advances. *Adv.Mater.*, 31, 1808066, 2019.
46. Zhu, J., Hu, L., Zhao, P., Lee, L.Y.S., Wong, K.-Y., Recent Advances in Electrocatalytic Hydrogen Evolution Using Nanoparticles. *Chem. Rev.*, 120, 851–918, 2020.
47. Mahmood, N., Yao, Y., Zhang, J.-W., Pan, L., Zhang, X., Zou, J.-J., Electrocatalysts for Hydrogen Evolution in Alkaline Electrolytes: Mechanisms, Challenges, and Prospective Solutions. *Adv. Sci.*, 5, 1700464, 2018.
48. Cheng, T., Wang, L., Merinov, B.V., Goddard, W.A., Explanation of Dramatic pH-Dependence of Hydrogen Binding on Noble Metal Electrode: Greatly Weakened Water Adsorption at High pH. *J. Am. Chem. Soc.*, 140, 7787–7790, 2018.
49. Giles, S.A., Wilson, J.C., Nash, J., Xu, B., Vlachos, D.G., Yan, Y., Recent Advances in Understanding the pH Dependence of the Hydrogen Oxidation and Evolution Reactions. *J. Catal.*, 367, 328–331, 2018.

50. Ledezma-Yanez, I., Wallace, W.D.Z., Sebastian-Pascual, P., Climent, V., Feliu, J.M., Koper, M.T.M., Interfacial Water Reorganization as a pH-Dependent Descriptor of the Hydrogen Evolution Rate on Platinum Electrodes. *Nat. Energy*, 2, 17031, 2017.
51. Lu, S.Q. and Zhuang, Z.B., Investigating the Influences of the Adsorbed Species on Catalytic Activity for Hydrogen Oxidation Reaction in Alkaline Electrolyte. *J. Am. Chem. Soc.*, 139, 5156, 2017.
52. Quaino, P., Juarez, F., Santos, E., Schmickler, W., Volcano Plots in Hydrogen Electrocatalysis-Uses and Abuses. *Beilstein J. Nanotech.*, 5, 846–854, 2014.
53. Parsons, R., The Rate of Electrolytic Hydrogen Evolution and the Heat of Adsorption of Hydrogen. *Trans. Faraday Soc.*, 54, 1053–1063, 1958.
54. Xiao, P., Sk, M.A., Thia, L., Ge, X., Lim, R.J., Wang, J.-Y., Lim, K.H., Wang, X., Molybdenum phosphide as an efficient electrocatalyst for the hydrogen evolution reaction. *Energy Environ. Sci.*, 7, 2624, 2014.
55. Trasatti, S., Work Function, Electronegativity, and Electrochemical Behaviour of Metals: III. Electrolytic Hydrogen Evolution in Acid Solutions. *J. Electroanal. Chem. Interfacial Electrochem.*, 39, 163–184, 1972.
56. Liu, H., Wang, R., Wei, J., Cheng, C., Zheng, Y., Pan, Y., He, X., Ding, M., Tan, H., Fu, Q., Conformation-Directed Micelle-to-Vesicle Transition of Cholesterol Decorated Polypeptide Triggered by Oxidation. *J. Am. Chem. Soc.*, 140, 6604–6610, 2018.
57. Tahir, M., Pan, L., Idrees, F., Zhang, X., Wang, L., Zou, J.-J., Wang, Z.L., Electrocatalytic Oxygen Evolution Reaction for Energy Conversion and Storage: A Comprehensive Review. *Nano Energy*, 37, 136–157, 2017.
58. Kim, J.S., Kim, B., Kim, H., Kang, K., Recent Progress On Multimetal Oxide Catalysts for the Oxygen Evolution Reaction. *Adv. Energy Mater.*, 8, 1702774, 2018.
59. Li, S., Hao, X., Abudula, A., Guan, G., Nanostructured Co-Based Bifunctional Electrocatalysts for Energy Conversion and Storage: Current Status and Perspectives. *J. Mater. Chem. A*, 7, 18674–18707, 2019.
60. Bockris, J.O.M., Kinetics of Activation Controlled Consecutive Electrochemical Reactions: Anodic Evolution of Oxygen. *J. Chem. Phys.*, 24, 817–827, 1956.
61. Iqbal, B., Saleem, M., Arshad, S.N., Rashid, J., Hussain, N., Zaheer, M., One-Pot Synthesis of Heterobimetallic Metal-Organic Frameworks (MOFs) for Multifunctional Catalysis. *Chem. Eur. J.*, 25, 10490–10498, 2019.
62. Man, I.C., Su, H.-Y., Calle-Vallejo, F., Hansen, H.A., Martinez, J., II, Inoglu, N.G., Kitchin, J., Jaramillo, T.F., Nørskov, J.K., Rossmeisl, J., Universality in Oxygen Evolution Electrocatalysis on Oxide Surfaces. *ChemCatChem*, 3, 1159–1165, 2011.
63. Rossmeisl, J., Dimitrievski, K., Siegbahn, P., Nørskov, J.K., Comparing Electrochemical and Biological Water Splitting. *J. Phys. Chem. C*, 111, 18821–18823, 2007.

64. Zhang, J., Yang, H.B., Zhou, D., Liu, B., Adsorption Energy in Oxygen Electrocatalysis. *Chem. Rev.*, 122, 17028–17072, 2022.
65. Montoya, J.H., Seitz, L.C., Chakthranont, P., Vojvodic, A., Jaramillo, T.F., Nørskov, J.K., Materials for Solar Fuels and Chemicals. *Nat. Mater.*, 16, 70–81, 2017.
66. Lei, Z., Wang, T., Zhao, B., Cai, W., Liu, Y., Jiao, S., Li, Q., Cao, R., Liu, M., Recent Progress in Electrocatalysts for Acidic Water Oxidation. *Adv. Energy Mater.*, 10, 2000478, 2020.
67. Zhang, N. and Chai, Y., Lattice Oxygen Redox Chemistry in Solid-State Electrocatalysts for Water Oxidation. *Energy Environ. Sci.*, 14, 4647–4671, 2021.
68. Lin, Y., Dong, Y., Wang, X., Chen, L., Electrocatalysts for the Oxygen Evolution Reaction in Acidic Media. *Adv. Mater.*, 35, 2210565, 2023.
69. Weng, C.-C., Ren, J.-T., Yuan, Z.-Y., Transition Metal Phosphide-Based Materials for Efficient Electrochemical Hydrogen Evolution: A Critical Review. *ChemSusChem*, 13, 3357–3375, 2020.
70. You, B., Tang, M.T., Tsai, C., Abild-Pedersen, F., Zheng, X., Li, H., Enhancing Electrocatalytic Water Splitting by Strain Engineering. *Adv. Mater.*, 31, 1807001, 2019.
71. Zhou, B., Gao, R., Zou, J.-J., Yang, H., Surface Design Strategy of Catalysts for Water Electrolysis. *Small*, 18, 2202336, 2022.
72. Liao, J.-Y., Lei, B.-X., Chen, H.-Y., Kuang, D.-B., Su, C.-Y., Oriented Hierarchical Single Crystalline Anatase TiO_2 Nanowire Arrays on Ti-Foil Substrate for Efficient Flexible Dye-Sensitized Solar Cells. *Energy Environ. Sci.*, 5, 5750–5757, 2012.
73. Zou, X. and Zhang, Y., Noble Metal-Free Hydrogen Evolution Catalysts for Water Splitting. *Chem. Soc. Rev.*, 44, 5148–5180, 2015.
74. Zhou, H., Li, Z., Xu, S.-M., Lu, L., Xu, M., Ji, K., Ge, R., Yan, Y., Ma, L., Kong, X., Zheng, L., Duan, H., Selectively Upgrading Lignin Derivatives to Carboxylates through Electrochemical Oxidative C(OH)-C Bond Cleavage by a Mn-Doped Cobalt Oxyhydroxide Catalyst. *Angew. Chem. Int. Ed.*, 60, 8976–8982, 2021.
75. Zhao, S., Yang, Y., Tang, Z., Insight into Structural Evolution, Active Sites, and Stability of Heterogeneous Electrocatalysts. *Angew. Chem. Int. Ed.*, 61, 202110186, 2022.
76. Singh, B., Mannu, P., Huang, Y.C., Prakash, R., Shen, S., Dong, C.L., Indra, A., Deciphering Ligand Controlled Structural Evolution of Prussian Blue Analogues and Their Electrochemical Activation during Alkaline Water Oxidation. *Angew. Chem. Int. Ed.*, 61, 202211585, 2022.
77. Nguyen, T.X., Su, Y.H., Lin, C.C., Ting, J.M., Self-Reconstruction of Sulfate-Containing High Entropy Sulfide for Exceptionally High-Performance Oxygen Evolution Reaction Electrocatalyst. *Adv. Funct. Mater.*, 31, 2106229, 2021.

78. Guo, T., Li, L., Wang, Z., Recent Development and Future Perspectives of Amorphous Transition Metal-Based Electrocatalysts for Oxygen Evolution Reaction. *Adv. Energy Mater.*, 12, 2200827, 2022.
79. Liu, X., Guo, R., Huang, W., Zhu, J., Wen, B., Mai, L., Advances in Understanding the Electrocatalytic Reconstruction Chemistry of Coordination Compounds. *Small,* 17, 2100629, 2021.
80. Ding, H., Liu, H., Chu, W., Wu, C., Xie, Y., Structural Transformation of Heterogeneous Materials for Electrocatalytic Oxygen Evolution Reaction. *Chem. Rev.*, 121, 13174–13212, 2021.
81. Zhong, H., Wang, X., Sun, G., Tang, Y., Tan, S., He, Q., Zhang, J., Xiong, T., Diao, C., Yu, Z., Optimization of Oxygen Evolution Activity by Tuning e^*_g Band Broadening in Nickel Oxyhydroxide. *Energy Environ. Sci.*, 16, 641–652, 2023.
82. Hong, Q., Wang, Y., Wang, R., Chen, Z., Yang, H., Yu, K., Liu, Y., Huang, H., Kang, Z., Menezes, P.W., *In Situ* Coupling of Carbon Dots with Co-ZIF Nanoarrays Enabling Highly Efficient Oxygen Evolution Electrocatalysis. *Small,* 19, 2206723, 2023.
83. Li, G.L., Qiao, X.Y., Miao, Y.Y., Wang, T.Y., Deng, F., Synergistic Effect of N-$NiMoO_4$ /Ni Heterogeneous Interface with Oxygen Vacancies in N-$NiMoO_4$/Ni/CNTS for Superior Overall Water Splitting. *Small,* 19, 2207196, 2023.
84. Wang, S., Jiang, Q., Ju, S., Hsu, C.S., Chen, H.M., Zhang, D., Song, F., Identifying the Geometric Catalytic Active Sites of Crystalline Cobalt Oxyhydroxides for Oxygen Evolution Reaction. *Nat.Commun.*, 13, 6650, 2022.
85. Zhang, L., Wang, J., Jiang, K., Xiao, Z., Gao, Y., Lin, S., Chen, B., Self-Reconstructed Metal-Organic Framework Heterojunction for Switchable Oxygen Evolution Reaction. *Angew. Chem. Int. Ed.*, 61, e202214794, 2022.
86. Chen, R.R., Chen, G., Ren, X., Ge, J., Ong, S.J.H., Xi, S., Wang, X., Xu, Z.J., $SmCo_5$ with a Reconstructed Oxyhydroxide Surface for Spin-Selective Water Oxidation at Elevated Temperature. *Angew. Chem. Int. Ed.*, 60, 25884–25890, 2021.
87. An, L., Zhang, H., Zhu, J., Xi, S., Huang, B., Sun, M., Peng, Y., Xi, P., Yan, C.H., Balancing Activity and Stability in Spinel Cobalt Oxides through Geometrical Sites Occupation towards Efficient Electrocatalytic Oxygen Evolution. *Angew. Chem. Int. Ed.*, 62, e202214600, 2023.
88. Cai, Z., Wang, P., Zhang, J., Chen, A., Zhang, J., Yan, Y., Wang, X., Reinforced Layered Double Hydroxide Oxygen-Evolution Electrocatalysts: A Polyoxometallic Acid Wet-Etching Approach and Synergistic Mechanism. *Adv. Mater.*, 34, 2110696, 2022.
89. Liu, Y., Li, X., Zhang, S., Wang, Z., Wang, Q., He, Y., Huang, W.H., Sun, Q., Zhong, X., Hu, J., Molecular Engineering of Metal-Organic Frameworks as Efficient Electrochemical Catalysts for Water Oxidation. *Adv. Mater.*, 35, 2300945, 2023.

90. Chang, G., Zhou, Y., Wang, J., Zhang, H., Yan, P., Wu, H.B., Yu, X.Y., Dynamic Reconstructed RuO_2/NiFeOOH With Coherent Interface for Efficient Seawater Oxidation. *Small*, 19, 2206768, 2023.
91. Zeng, Y., Zhao, M., Huang, Z., Zhu, W., Zheng, J., Jiang, Q., Wang, Z., Liang, H., Surface Reconstruction of Water Splitting Electrocatalysts. *Adv.Energy Mater.*, 12, 2201713, 2022.
92. Yang, Y., Luo, M., Zhang, W., Sun, Y., Chen, X., Guo, S., Metal Surface and Interface Energy Electrocatalysis: Fundamentals, Performance Engineering, and Opportunities. *Chem*, 4, 2054–2083, 2018.
93. Sun, Y., Wang, J., Xi, S., Shen, J., Luo, S., Ge, J., Sun, S., Chen, Y., Hanna, J.V., Li, S., Wang, X., Xu, Z.J., Navigating Surface Reconstruction of Spinel Oxides for Electrochemical Water Oxidation. *Nat. Commun.*, 14, 2467, 2023.
94. Zhai, W., Chen, Y., Liu, Y., Sakthivel, T., Ma, Y., Guo, S., Qu, Y., Dai, Z., Bimetal-Incorporated Black Phosphorene with Surface Electron Deficiency for Efficient Anti-Reconstruction Water Electrolysis. *Adv. Funct. Mater.*, 33, 2301565, 2023.
95. Sun, H., Tung, C.W., Qiu, Y., Zhang, W., Wang, Q., Li, Z., Tang, J., Chen, H.C., Wang, C., Chen, H.M., Atomic Metal-Support Interaction Enables Reconstruction-Free Dual-Site Electrocatalyst. *J. Am. Chem. Soc.*, 144, 1174–1186, 2022.
96. Du, W., Shi, Y., Zhou, W., Yu, Y., Zhang, B., Unveiling the *In Situ* Dissolution and Polymerization of Mo in Ni_4Mo Alloy For Promoting The Hydrogen Evolution Reaction. *Angew. Chem. Int. Ed.*, 60, 7051–7055, 2021.
97. Enkhtuvshin, E., Yeo, S., Choi, H., Kim, K.M., An, B.S., Biswas, S., Lee, Y., Nayak, A.K., Jang, J.U., Na, K.-H., Choi, W.-Y., Ali, G., Chae, K.H., Akbar, M., Chung, K.Y., Yoo, K., Chung, Y.-C., Shin, T.H., Kim, H., Chung, C.-Y., Han, H., Surface Reconstruction of Ni–Fe Layered Double Hydroxide Inducing Chloride Ion Blocking Materials for Outstanding Overall Seawater Splitting. *Adv. Funct. Mater.*, 33, 2214069, 2023.
98. Jin, J., Yin, J., Liu, H., Huang, B., Hu, Y., Zhang, H., Sun, M., Peng, Y., Xi, P., Yan, C.-H., Atomic Sulfur Filling Oxygen Vacancies Optimizes H Absorption and Boosts the Hydrogen Evolution Reaction in Alkaline Media. *Angew. Chem. Int. Ed.*, 60, 14117–14123, 2021.
99. Lin, G., Zhang, Z., Ju, Q., Wu, T., Segre, C.U., Chen, W., Peng, H., Zhang, H., Liu, Q., Liu, Z., Zhang, Y., Kong, S., Mao, Y., Zhao, W., Suenaga, K., Huang, F., Wang, J., Bottom-Up Evolution of Perovskite Clusters into High-Activity Rhodium Nanoparticles toward Alkaline Hydrogen Evolution. *Nat. Commun.*, 14, 280, 2023.
100. Liu, J., Qiao, W., Zhu, Z., Hu, J., Xu, X., Chameleon-Like Reconstruction on Redox Catalysts Adaptive to Alkali Water Electrolysis. *Small*, 18, 2202434, 2022.
101. Sun, Y., Wu, J., Zhang, Z., Liao, Q., Zhang, S., Wang, X., Xie, Y., Ma, K., Kang, Z., Zhang, Y., Phase Reconfiguration of Multivalent Nickel Sulfides in Hydrogen Evolution. *Energy Environ. Sci.*, 15, 633–644, 2022.

102. Wang, L., Xie, L., Zhao, W., Liu, S., Zhao, Q., Oxygen-facilitated dynamic active-site generation on strained MoS_2 during photo-catalytic hydrogen evolution. *Chem. Eng. J.*, 405, 127028, 2021.
103. Yang, C., Zhong, W., Shen, K., Zhang, Q., Zhao, R., Xiang, H., Wu, J., Li, X., Yang, N., Electrochemically Reconstructed Cu-FeOOH/Fe_3O_4 Catalyst for Efficient Hydrogen Evolution in Alkaline Media. *Adv. Energy Mater.*, 12, 2200077, 2022.
104. Luo, X., Ji, P., Wang, P., Tan, X., Chen, L., Mu, S., Spherical Ni_3S_2/Fe-NiP_x Magic Cube with Ultrahigh Water/Seawater Oxidation Efficiency. *Adv. Sci.*, 9, 2104846, 2022.
105. Zhao, Y., Dongfang, N., Triana, C.A., Huang, C., Erni, R., Wan, W., Li, J., Stoian, D., Pan, L., Zhang, P., Dynamics and Control of Active Sites in Hierarchically Nanostructured Cobalt Phosphide/Chalcogenide-Based Electrocatalysts for Water Splitting. *Energy Environ. Sci.*, 15, 727–739, 2022.
106. Ji, P., Yu, R., Wang, P., Pan, X., Jin, H., Zheng, D., Chen, D., Zhu, J., Pu, Z., Wu, J., Ultra-Fast and In-Depth Reconstruction of Transition Metal Fluorides in Electrocatalytic Hydrogen Evolution Processes. *Adv. Sci.*, 9, e2103567, 2022.
107. Banerjee, S., Kakekhani, A., Wexler, R.B., Rappe, A.M., Relationship Between the Surface Reconstruction of Nickel Phosphides and Their Activity toward the Hydrogen Evolution Reaction. *ACS Catal.*, 13, 4611–4621, 2023.
108. Zhou, D., Li, F., Zhao, Y., Wang, L., Zou, H., Shan, Y., Fu, J., Ding, Y., Duan, L., Liu, M., Mechanistic Regulation by Oxygen Vacancies in Structural Evolution Promoting Electrocatalytic Water Oxidation. *ACS Catal.*, 13, 4398–4408, 2023.
109. Hu, Y., Zheng, Y., Jin, J., Wang, Y., Peng, Y., Yin, J., Shen, W., Hou, Y., Zhu, L., An, L., Lu, M., Xi, P., Yan, C.-H., Understanding the Sulphur-Oxygen Exchange Process of Metal Sulphides Prior to Oxygen Evolution Reaction. *Nat. Commun.*, 14, 1949, 2023.
110. Liu, C., Shi, X.R., Yue, K., Wang, P., Zhan, K., Wang, X., Xia, B.Y., Yan, Y., S-Species-Evoked High-Valence $Ni^{2+\delta}$ of the Evolved Beta-$Ni(OH)_2$ Electrode for Selective Oxidation of 5-Hydroxymethylfurfural. *Adv. Mater.*, 35, 2211177, 2023.
111. Liu, H., Xie, R., Wang, Q., Han, J., Han, Y., Wang, J., Fang, H., Qi, J., Ding, M., Ji, W., He, B., Lü, W., Enhanced OER Performance and Dynamic Transition of Surface Reconstruction in $LaNiO_3$ Thin Films With Nanoparticles Decoration. *Adv. Sci.*, 10, 2207128, 2023.
112. Reith, L., Hausmann, J.N., Mebs, S., Mondal, I., Dau, H., Driess, M., Menezes, P.W., *In Situ* Detection of Iron in Oxidation States ≥ IV in Cobalt-Iron Oxyhydroxide Reconstructed during Oxygen Evolution Reaction. *Adv. Energy Mater.*, 13, 2203886, 2023.
113. Tajuddin, A.A.H., Wakisaka, M., Ohto, T., Yu, Y., Fukushima, H., Tanimoto, H., Li, X., Misu, Y., Jeong, S., Fujita, J., II, Corrosion-Resistant and High-Entropic Non-Noble-Metal Electrodes for Oxygen Evolution in Acidic Media. *Adv. Mater.*, 35, 2207466, 2023.

114. Wang, C., Zhai, P., Xia, M., Liu, W., Gao, J., Sun, L., Hou, J., Identification of the Origin for Reconstructed Active Sites on Oxyhydroxide for Oxygen Evolution Reaction. *Adv. Mater.*, 35, 2209307, 2023.
115. Wu, T., Sun, Y., Ren, X., Wang, J., Song, J., Pan, Y., Mu, Y., Zhang, J., Cheng, Q., Xian, G., Xi, S., Shen, C., Gao, H.-J., Fisher, A.C., Sherburne, M.P., Du, Y., Ager, J.W., Gracia, J., Yang, H., Zeng, L., Xu, Z.J., Reconstruction of Thiospinel to Active Sites and Spin Channels for Water Oxidation. *Adv. Mater.*, 35, 2207041, 2023.
116. Zhang, R., Pan, L., Guo, B., Huang, Z.F., Chen, Z., Wang, L., Zhang, X., Guo, Z., Xu, W., Loh, K.P., Tracking the Role of Defect Types in Co_3O_4 Structural Evolution and Active Motifs during Oxygen Evolution Reaction. *J. Am. Chem. Soc.*, 145, 2271–2281, 2023.
117. Zhao, C.-X., Liu, J.-N., Wang, C., Wang, J., Song, L., Li, B.-Q., Zhang, Q., An Anionic Regulation Mechanism for the Structural Reconstruction of Sulfide Electrocatalysts under Oxygen Evolution Conditions. *Energy Environ. Sci.*, 15, 3257–3264, 2022.
118. Kuai, C., Xi, C., Hu, A., Zhang, Y., Xu, Z., Nordlund, D., Sun, C.J., Cadigan, C.A., Richards, R.M., Li, L., Revealing the Dynamics and Roles of Iron Incorporation in Nickel Hydroxide Water Oxidation Catalysts. *J. Am. Chem. Soc.*, 143, 18519–18526, 2021.
119. Tesch, M.F., Bonke, S.A., Jones, T.E., Shaker, M.N., Xiao, J., Skorupska, K., Mom, R., Melder, J., Kurz, P., Knop-Gericke, A., Evolution of Oxygen-Metal Electron Transfer and Metal Electronic States during Manganese Oxide Catalyzed Water Oxidation Revealed with *In Situ* Soft X-Ray Spectroscopy. *Angew. Chem. Int. Ed.*, 58, 3426–3432, 2019.
120. Jin, Z. and Bard, A.J., Surface Interrogation of Electrodeposited MnO_x and $CaMnO_3$ Perovskites by Scanning Electrochemical Microscopy: Probing Active Sites and Kinetics for the Oxygen Evolution Reaction. *Angew. Chem. Int. Ed.*, 60, 794–799, 2021.
121. Mefford, J.T., Akbashev, A.R., Kang, M., Bentley, C.L., Gent, W.E., Deng, H.D., Alsem, D.H., Yu, Y.S., Salmon, N.J., Shapiro, D.A., Correlative Operando Microscopy of Oxygen Evolution Electrocatalysts. *Nature*, 593, 67–73, 2021.
122. Sun, Y., Li, X., Zhang, T., Xu, K., Yang, Y., Chen, G., Li, C., Xie, Y., Nitrogen-Doped Cobalt Diselenide with Cubic Phase Maintained for Enhanced Alkaline Hydrogen Evolution. *Angew. Chem. Int. Ed.*, 60, 21575–2152, 2021.
123. Jiang, H., He, Q., Zhang, Y., Song, L., Structural Self-Reconstruction of Catalysts in Electrocatalysis. *Acc. Chem. Res.*, 51, 2968–2977, 2018.
124. Hao, J., Luo, W., Wang, S., Zhao, K., Hou, J., Li, L., Ge, B., Yang, W., Shi, W., Discharge-Induced Enhancement of the Oxygen Evolution Reaction. *Angew. Chem. Int. Ed.*, 60, 20042–20048, 2021.
125. Villalobos, J., González-Flores, D., Urcuyo, R., Montero, M.L., Schuck, G., Beyer, P., Risch, M., Requirements for Beneficial Electrochemical Restructuring: A Model Study on a Cobalt Oxide in Selected Electrolytes. *Adv. Energy Mater.*, 11, 2101737, 2021.

126. Wang, Q., Cheng, Y., Tao, H.B., Liu, Y., Ma, X., Li, D.S., Yang, H.B., Liu, B., Long-Term Stability Challenges and Opportunities in Acidic Oxygen Evolution Electrocatalysis. *Angew. Chem. Int. Ed.*, 62, e202216645, 2023.
127. Clament Sagaya Selvam, N., Kwak, S.J., Choi, G.H., Oh, M.J., Kim, H., Yoon, W.-S., Lee, W.B., Yoo, P.J., Unveiling the Impact of Fe Incorporation on Intrinsic Performance of Reconstructed Water Oxidation Electrocatalyst. *ACS Energy Lett.*, 6, 4345–4354, 2021.
128. Peng, C.K., Lin, Y.C., Chiang, C.L., Qian, Z., Huang, Y.C., Dong, C.L., Li, J.F., Chen, C.T., Hu, Z., Chen, S.Y., Zhang-Rice Singlets State Formed by Two-Step Oxidation for Triggering Water Oxidation under Operando Conditions. *Nat. Commun.*, 14, 529, 2023.
129. Ye, C., Liu, J., Zhang, Q., Jin, X., Zhao, Y., Pan, Z., Chen, G., Qiu, Y., Ye, D., Gu, L., Activating Metal Oxides Nanocatalysts for Electrocatalytic Water Oxidation by Quenching-Induced Near-Surface Metal Atom Functionality. *J. Am. Chem. Soc.*, 143, 14169–14177, 2021.
130. Lin, X., Huang, Y.C., Hu, Z., Li, L., Zhou, J., Zhao, Q., Huang, H., Sun, J., Pao, C.W., Chang, Y.C., 5f Covalency Synergistically Boosting Oxygen Evolution of $UCoO_4$ Catalyst. *J. Am. Chem. Soc.*, 144, 416–423, 2022.
131. Hunter, B.M., Gray, H.B., Muller, A.M., Earth-Abundant Heterogeneous Water Oxidation Catalysts. *Chem. Rev.*, 116, 14120–14136, 2016.
132. Kuznetsov, D.A., Han, B., Yu, Y., Rao, R.R., Hwang, J., Roman-Leshkov, Y., Shao-Horn, Y., Tuning Redox Transitions via Inductive Effect in Metal Oxides and Complexes, and Implications in Oxygen Electrocatalysis. *Joule*, 2, 225–244, 2018.
133. Gao, L., Cui, X., Sewell, C.D., Li, J., Lin, Z., Recent Advances in Activating Surface Reconstruction for the High-Efficiency Oxygen Evolution Reaction. *Chem. Soc. Rev.*, 50, 8428–8469, 2021.
134. Sun, Y., Li, R., Chen, X., Wu, J., Xie, Y., Wang, X., Ma, K., Wang, L., Zhang, Z., Liao, Q., A-Site Management Prompts the Dynamic Reconstructed Active Phase of Perovskite Oxide OER Catalysts. *Adv. Energy Mater.*, 11, 2003755, 2021.
135. Lei, H., Ma, L., Wan, Q., Tan, S., Yang, B., Wang, Z., Mai, W., Fan, H.J., Promoting Surface Reconstruction of NiFe Layered Double Hydroxide for Enhanced Oxygen Evolution. *Adv. Energy Mater.*, 12, 2202522, 2022.
136. Bai, J., Mei, J., Liao, T., Sun, Q., Chen, Z.G., Sun, Z., Molybdenum-Promoted Surface Reconstruction in Polymorphic Cobalt for Initiating Rapid Oxygen Evolution. *Adv. Energy Mater.*, 12, 2103247, 2022.
137. Abdelhafiz, A., Wang, B., Harutyunyan, A.R., Li, J., Carbothermal Shock Synthesis of High Entropy Oxide Catalysts: Dynamic Structural and Chemical Reconstruction Boosting the Catalytic Activity and Stability toward Oxygen Evolution Reaction. *Adv. Energy Mater.*, 12, 2200742, 2022.
138. Xu, X., Pan, Y., Zhong, Y., Shi, C., Guan, D., Ge, L., Hu, Z., Chin, Y.Y., Lin, H.J., Chen, C.T., New Undisputed Evidence and Strategy for Enhanced

Lattice-Oxygen Participation of Perovskite Electrocatalyst through Cation Deficiency Manipulation. *Adv. Sci.*, 9, 2200530, 2022.
139. Zegeye, T.A., Chen, W.-T., Hsu, C.-C., Valinton, J.A.A., Chen, C.-H., Activation Energy Assessing Potential-Dependent Activities and Site Reconstruction for Oxygen Evolution. *ACS Energy Lett.*, 7, 2236–2243, 2022.
140. Weng, Z., Huang, H., Li, X., Zhang, Y., Shao, R., Yi, Y., Lu, Y., Zeng, X., Zou, J., Chen, L., Coordination Tailoring of Epitaxial Perovskite-Derived Iron Oxide Films for Efficient Water Oxidation Electrocatalysis. *ACS Catal.*, 13, 2751–2760, 2023.
141. He, Q., Wan, Y., Jiang, H., Pan, Z., Wu, C., Wang, M., Wu, X., Ye, B., Ajayan, P.M., Song, L., Nickel Vacancies Boost Reconstruction in Nickel Hydroxide Electrocatalyst. *ACS Energy Lett.*, 3, 1373–1380, 2018.
142. Wang, J., Kim, S.-J., Liu, J., Gao, Y., Choi, S., Han, J., Shin, H., Jo, S., Kim, J., Ciucci, F., Redirecting Dynamic Surface Restructuring of a Layered Transition Metal Oxide Catalyst for Superior Water Oxidation. *Nat. Catal.*, 4, 212–222, 2021.
143. Selvam, N.C.S., Du, L., Xia, B.Y., Yoo, P.J., You, B., Reconstructed Water Oxidation Electrocatalysts: The Impact of Surface Dynamics on Intrinsic Activities. *Adv. Funct. Mater.*, 31, 2008190, 2021.
144. Chen, Y., Sun, Y., Wang, M., Wang, J., Li, H., Xi, H., Wei, C., Xi, P., Sterbinsky, G.E., Freelandet, J.W., Fisher, A.C., Ager III, J.W., Feng, Z., Xu, Z.J., Lattice Site–Dependent Metal Leaching in Perovskites toward a Honeycomb-Like Water Oxidation Catalyst. *Sci. Adv.*, 7, eabk1788, 2021.
145. Lu, M., Zheng, Y., Hu, Y., Huang, B., Ji, D., Sun, M., Li, J., Peng, Y., Si, R., Xi, P., Yan, C., Artificially Steering Electrocatalytic Oxygen Evolution Reaction Mechanism by Regulating Oxygen Defect Contents in Perovskites. *Sci. Adv.*, 8, eabq3563, 2022.
146. Wu, T., Sun, S., Song, J., Xi, S., Du, Y., Chen, B., Sasangka, W.A., Liao, H., Gan, C.L., Scherer, G.G., Iron-Facilitated Dynamic Active-Site Generation on Spinel $CoAl_2O_4$ with Self-Termination of Surface Reconstruction for Water Oxidation. *Nat. Catal.*, 2, 763–772, 2019.
147. Shen, T.H., Spillane, L., Peng, J., Shao-Horn, Y., Tileli, V., Switchable Wetting of Oxygen-Evolving Oxide Catalysts. *Nat. Catal.*, 5, 30–36, 2022.
148. Lopes, P.P., Chung, D.Y., Rui, X., Zheng, H., He, H., Martins, P.F.B.D., Strmcnik, D., Stamenkovic, V.R., Zapol, P., Mitchell, J.F., Dynamically Stable Active Sites from Surface Evolution of Perovskite Materials during the Oxygen Evolution Reaction. *J. Am. Chem. Soc.*, 143, 2741–2750, 2021.
149. Zhao, J.-W., Zhang, H., Li, C.-F., Zhou, X., Wu, J.-Q., Zeng, F., Zhang, J., Li, G.-R., Key Roles of Surface Fe Sites and Sr Vacancies in the Perovskite for an Efficient Oxygen Evolution Reaction via Lattice Oxygen Oxidation. *Energy Environ. Sci.*, 15, 3912–3922, 2022.
150. Xiao, K., Wang, Y., Wu, P., Hou, L., Liu, Z.Q., Activating Lattice Oxygen in Spinel $ZnCo_2O_4$ through Filling Oxygen Vacancies with Fluorine for

Electrocatalytic Oxygen Evolution. *Angew. Chem. Int. Ed.*, 62, e202301408, 2023.

151. Duan, Y., Lee, J.Y., Xi, S., Sun, Y., Ge, J., Ong, S.J.H., Chen, Y., Dou, S., Meng, F., Diao, C., Anodic Oxidation Enabled Cation Leaching for Promoting Surface Reconstruction in Water Oxidation. *Angew. Chem. Int. Ed.*, 60, 7418–7425, 2021.
152. Sun, Y., Wu, J., Xie, Y., Wang, X., Ma, K., Tian, Z., Zhang, Z., Liao, Q., Zheng, W., Kang, Z., Dynamics of Both Active Phase and Catalysis Pathway for Spinel Water-Oxidation Catalysts. *Adv. Funct. Mater.*, 32, 2207116, 2022.
153. Rasmussen, F.A. and Thygesen, K.S., Computational 2D Materials Database: Electronic Structure of Transition-Metal Dichalcogenides and Oxides. *J. Phys. Chem. C*, 119, 13169–13183, 2015.
154. Mei, J., Shang, J., He, T., Qi, D., Kou, L., Liao, T., Du, A., Sun, Z., 2D/2D Black Phosphorus/Nickel Hydroxide Heterostructures for Promoting Oxygen Evolution via Electronic Structure Modulation and Surface Reconstruction. *Adv. Energy Mater.*, 12, 2201141, 2022.
155. Li, D., Wan, W., Wang, Z., Wu, H., Wu, S., Jiang, T., Cai, G., Jiang, C., Ren, F., Self-Derivation and Surface Reconstruction of Fe-Doped Ni_3S_2 Electrode Realizing High-Efficient and Stable Overall Water and Urea Electrolysis. *Adv. Energy Mater.*, 12, 2201913, 2022.
156. Zhang, H., Geng, S., Ouyang, M., Yadegari, H., Xie, F., Riley, D.J., A Self-Reconstructed Bifunctional Electrocatalyst of Pseudo-Amorphous Nickel Carbide@Iron Oxide Network for Seawater Splitting. *Adv. Sci.*, 9, 2200146, 2022.
157. Sun, Y., Wang, J., Qi, Y., Li, W., Wang, C., Efficient Electrooxidation of 5-Hydroxymethylfurfural Using Co-Doped Ni_3S_2 Catalyst: Promising for H_2 Production Under Industrial-Level Current Density. *Adv. Sci.*, 9, 2200957, 2022.
158. Zhang, L., Cai, W., Bao, N., Yang, H., Implanting an Electron Donor to Enlarge The D-P Hybridization of High-Entropy (Oxy)Hydroxide: A Novel Design to Boost Oxygen Evolution. *Adv.Mater.*, 34, 2110511, 2022.
159. Zhang, L., Cai, W., Bao, N., Top-Level Design Strategy to Construct an Advanced High-Entropy Co-Cu-Fe-Mo (Oxy)Hydroxide Electrocatalyst for the Oxygen Evolution Reaction. *Adv. Mater.*, 33, 2100745, 2021.
160. Chen, R., Zhang, Z., Wang, Z., Wu, W., Du, S., Zhu, W., Lv, H., Cheng, N., Constructing Air-Stable and Reconstruction-Inhibited Transition Metal Sulfide Catalysts via Tailoring Electron-Deficient Distribution for Water Oxidation. *ACS Catal.*, 12, 13234–13246, 2022.
161. Liu, J., Du, W., Guo, S., Pan, J., Hu, J., Xu, X., Iron-Locked Hydr(Oxy)Oxide Catalysts via Ion-Compensatory Reconstruction Boost Large-Current-Density Water Oxidation. *Adv. Sci.*, 10, 2300717, 2023.
162. Bergmann, A., Martinez-Moreno, E., Teschner, D., Chernev, P., Gliech, M., De Araujo, J.F., Reier, T., Dau, H., Strasser, P., Reversible Amorphization and

the Catalytically Active State of Crystalline Co_3O_4 During Oxygen Evolution. *Nat. Commun.*, 6, 8625, 2015.

163. Su, H., Zhou, W., Zhou, W., Li, Y., Zheng, L., Zhang, H., Liu, M., Zhang, X., Sun, X., Xu, Y., *In-Situ* Spectroscopic Observation of Dynamic-Coupling Oxygen on Atomically Dispersed Iridium Electrocatalyst for Acidic Water Oxidation. *Nat. Commun.*, 12, 6118, 2021.
164. Ning, M., Zhang, F., Wu, L., Xing, X., Wang, D., Song, S., Zhou, Q., Yu, L., Bao, J., Chen, S., Boosting Efficient Alkaline Fresh Water and Seawater Electrolysis via Electrochemical Reconstruction. *Energy Environ. Sci.*, 15, 3945–3957, 2022.
165. Lin, R., Kang, L., Zhao, T., Feng, J., Celorrio, V., Zhang, G., Cibin, G., Kucernak, A., Brett, D.J.L., Corà, F., Identification and Manipulation of Dynamic Active Site Deficiency-Induced Competing Reactions in Electrocatalytic Oxidation Processes. *Energy Environ. Sci.*, 15, 2386–2396, 2022.
166. Wang, Y., Jiao, Y., Yan, H., Yang, G., Tian, C., Wu, A., Liu, Y., Fu, H., Vanadium-Incorporated CoP_2 with Lattice Expansion for Highly Efficient Acidic Overall Water Splitting. *Angew. Chem. Int. Ed.*, 61, e202116233, 2022.
167. Sun, J., Xue, H., Guo, N., Song, T., Hao, Y.R., Sun, J., Zhang, J., Wang, Q., Synergetic Metal Defect and Surface Chemical Reconstruction Into $NiCo_2S_4$/ZnS Heterojunction to Achieve Outstanding Oxygen Evolution Performance. *Angew. Chem. Int. Ed.*, 60, 19435–19441, 2021.
168. Marelli, E., Gazquez, J., Poghosyan, E., Muller, E., Gawryluk, D.J., Pomjakushina, E., Sheptyakov, D., Piamonteze, C., Aegerter, D., Schmidt, T.J., Correlation between Oxygen Vacancies and Oxygen Evolution Reaction Activity for a Model Electrode: $PrBaCo_2O_{5+\delta}$. *Angew. Chem. Int. Ed.*, 60, 14609–14619, 2021.
169. Chong, X., Liu, C., Wang, C., Yang, R., Zhang, B., Integrating Hydrogen Production and Transfer Hydrogenation with Selenite Promoted Electrooxidation of α-Nitrotoluenes to *E*-Nitroethenes. *Angew. Chem. Int. Ed.*, 60, 22010–22016, 2021.
170. Zhuang, L., Li, J., Wang, K., Li, Z., Zhu, M., Xu, Z., Structural Buffer Engineering on Metal Oxide for Long-Term Stable Seawater Splitting. *Adv. Funct. Mater.*, 32, 2201127, 2022.
171. Liu, X., Xia, F., Guo, R., Huang, M., Meng, J., Wu, J., Mai, L., Ligand and Anion Co-Leaching Induced Complete Reconstruction of Polyoxomolybdate-Organic Complex Oxygen-Evolving Pre-Catalysts. *Adv. Funct. Mater.*, 31, 2101792, 2021.
172. Liao, H., Luo, T., Tan, P., Chen, K., Lu, L., Liu, Y., Liu, M., Pan, J., Unveiling Role of Sulfate Ion in Nickel-Iron (Oxy)Hydroxide with Enhanced Oxygen-Evolving Performance. *Adv. Funct. Mater.*, 31, 2102772, 2021.
173. Hu, C., Dong, H., Li, Y., Sinha, S., Wang, C., Xu, W., Song, L., Suenaga, K., Geng, H., Wang, J., Huang, Q., Tan, Y., Huang, X., Self-Reconstruction of Single-Atom-Thick A Layers in Nanolaminated MAX Phases for Enhanced Oxygen Evolution. *Adv. Funct. Mater.*, 33, 2211530, 2023.

174. Zhao, Y., Wen, Q., Huang, D., Jiao, C., Liu, Y., Liu, Y., Fang, J., Sun, M., Yu, L., Operando Reconstruction toward Dual-Cation-Defects Co-Containing NiFe Oxyhydroxide for Ultralow Energy Consumption Industrial Water Splitting Electrolyzer. *Adv. Energy Mater.*, 13, 2203595, 2023.
175. Solomon, G., Landström, A., Mazzaro, R., Jugovac, M., Moras, P., Cattaruzza, E., Morandi, V., Concina, I., Vomiero, A., $NiMoO_4$@Co_3O_4 Core–Shell Nanorods: *In Situ* Catalyst Reconstruction toward High Efficiency Oxygen Evolution Reaction. *Adv. Energy Mater.*, 11, 2101324, 2021.
176. Zhang, Y., Fu, Q., Song, B., Xu, P., Regulation Strategy of Transition Metal Oxide-Based Electrocatalysts for Enhanced Oxygen Evolution Reaction. *Acc. Mater. Res.*, 3, 1088–1100, 2022.
177. Zhang, T., Liu, Y., Tong, L., Yu, J., Lin, S., Li, Y., Fan, H.J., Oxidation State Engineering in Octahedral Ni by Anchored Sulfate to Boost Intrinsic Oxygen Evolution Activity. *ACS Nano*, 17, 6770–6780, 2023.
178. Wu, T., Ren, X., Sun, Y., Sun, S., Xian, G., Scherer, G.G., Fisher, A.C., Mandler, D., Ager, J.W., Grimaud, A., Spin Pinning Effect to Reconstructed Oxyhydroxide Layer on Ferromagnetic Oxides for Enhanced Water Oxidation. *Nat. Commun.*, 12, 3634, 2021.
179. Cloutier, C.R. and Wilkinson, D.P., Electrolytic Production of Hydrogen from Aqueous Acidic Methanol Solutions. *Int. J. Hydrogen Energy*, 35, 3967–3984, 2010.
180. Deng, C., Toe, C.Y., Li, X., Tan, J., Yang, H., Hu, Q., He, C., Earth-Abundant Metal-Based Electrocatalysts Promoted Anodic Reaction in Hybrid Water Electrolysis for Efficient Hydrogen Production: Recent Progress and Perspectives. *Adv. Energy Mater.*, 12, 2201047, 2022.
181. Wang, S., Mao, Q., Ren, H., Wang, W., Wang, Z., Xu, Y., Li, X., Wang, L., Wang, H., Liquid Metal Interfacial Growth and Exfoliation to Form Mesoporous Metallic Nanosheets for Alkaline Methanol Electroreforming. *ACS Nano*, 16, 2978–2987, 2022.
182. Sheng, J., Kang, J., Ye, H., Xie, J., Zhao, B., Fu, X.-Z., Yu, Y., Sun, R., Wong, C.-P., Porous Octahedral PdCu Nanocages as Highly Efficient Electrocatalysts for the Methanol Oxidation Reaction. *J. Mater. Chem. A*, 6, 3906–3912, 2018.
183. Wang, Z., Ren, X., Luo, Y., Wang, L., Cui, G., Xie, F., Wang, H., Xie, Y., Sun, X., An Ultrafine Platinum-Cobalt Alloy Decorated Cobalt Nanowire Array with Superb Activity toward Alkaline Hydrogen Evolution. *Nanoscale*, 10, 12302–12307, 2018.
184. Xu, W., Chang, J., Cheng, Y., Liu, H., Li, J., Ai, Y., Hu, Z., Zhang, X., Wang, Y., Liang, Q.A., Multi-Step Induced Strategy to Fabricate Core-Shell Pt-Ni Alloy as Symmetric Electrocatalysts for Overall Water Splitting. *Nano Res.*, 15, 965–971, 2022.
185. Guo, Y., Yang, X., Liu, X., Tong, X., Yang, N., Coupling Methanol Oxidation with Hydrogen Evolution on Bifunctional Co-Doped Rh Electrocatalyst for Efficient Hydrogen Generation. *Adv. Funct. Mater.*, 33, 2209134, 2023.

186. Yang, Q., Zhang, C., Dong, B., Cui, Y., Wang, F., Cai, J., Jin, P., Feng, L., Synergistic Modulation of Nanostructure and Active Sites: Ternary Ru&Fe-WO_x Electrocatalyst for Boosting Concurrent Generations of Hydrogen and Formate over 500 mA cm^{-2}. *Appl. Catal., B*, 296, 120359, 2021.
187. Xu, Y., Liu, M., Wang, M., Ren, T., Ren, K., Wang, Z., Li, X., Wang, L., Wang, H., Methanol Electroreforming Coupled to Green Hydrogen Production over Bifunctional NiIr-Based Metal-Organic Framework Nanosheet Arrays. *Appl. Catal., B*, 300, 120753, 2022.
188. Katiyar, N.K., Biswas, K., Yeh, J.-W., Sharma, S., Tiwary, C.S.A., Perspective on the Catalysis Using the High Entropy Alloys. *Nano Energy*, 88, 106261, 2021.
189. Han, G., Li, M., Liu, H., Zhang, W., He, L., Tian, F., Liu, Y., Yu, Y., Yang, W., Guo, S., Short-Range Diffusion Enables General Synthesis of Medium-Entropy Alloy Aerogels. *Adv. Mater.*, 34, 2202943, 2022.
190. Xiang, K., Wu, D., Deng, X., Li, M., Chen, S., Hao, P., Guo, X., Luo, J.-L., Fu, X.-Z., Boosting H_2 Generation Coupled with Selective Oxidation of Methanol into Value-Added Chemical over Cobalt Hydroxide@Hydroxysulfide Nanosheets Electrocatalysts. *Adv.Funct. Mater.*, 30, 1909610, 2020.
191. Zhao, B., Liu, J., Xu, C., Feng, R., Sui, P., Wang, L., Zhang, J., Luo, J.L., Fu, X.Z., Hollow NiSe Nanocrystals Heterogenized with Carbon Nanotubes for Efficient Electrocatalytic Methanol Upgrading to Boost Hydrogen Co-Production. *Adv. Funct. Mater.*, 31, 2008812, 2020.
192. Li, M., Deng, X., Xiang, K., Liang, Y., Zhao, B., Hao, J., Luo, J.-L., Fu, X.-Z., Value-Added Formate Production from Selective Methanol Oxidation as Anodic Reaction to Enhance Electrochemical Hydrogen Cogeneration. *ChemSusChem*, 13, 914–921, 2020.
193. Dai, L., Qin, Q., Zhao, X., Xu, C., Hu, C., Mo, S., Wang, Y.O., Lin, S., Tang, Z., Zheng, N., Electrochemical Partial Reforming of Ethanol into Ethyl Acetate Using Ultrathin Co_3O_4 Nanosheets as a Highly Selective Anode Catalyst. *ACS Cent.Sci.*, 2, 538–544, 2016.
194. Jeong, S., Mai, H.D., Nam, K.-H., Park, C.-M., Jeon, K.-J., Self-Healing Graphene-Templated Platinum-Nickel Oxide Heterostructures for Overall Water Splitting. *ACS Nano*, 16, 930–938, 2022.
195. Kim, T., Roy, S.B., Moon, S., Yoo, S., Choi, H., Parale, V.G., Kim, Y., Lee, J., Jun, S.C., Kang, K., Chun, S.-H., Kanamori, K., Park, H.-H., Highly Dispersed Pt Clusters on F-Doped Tin(IV) Oxide Aerogel Matrix: An Ultra-Robust Hybrid Catalyst for Enhanced Hydrogen Evolution. *ACS Nano*, 16, 1625–1638, 2022.
196. Li, M., Duanmu, K., Wan, C., Cheng, T., Zhang, L., Dai, S., Chen, W., Zhao, Z., Li, P., Fei, H., Zhu, Y., Yu, R., Luo, J., Zhang, K., Lin, Z., Ding, M., Huang, J., Sun, H., Guo, J., Pan, X., Goddard, W., Sautet, P., Huang, Y., Duan, X., Single-Atom Tailoring of Platinum Nanocatalysts for High-Performance Multifunctional Electrocatalysis. *Nat. Catal.*, 2, 495–503, 2019.

197. Dai, C., Yang, Y., Zhao, Z., Fisher, A., Liu, Z., Cheng, D., From Mixed to Three-Layer Core/Shell PtCu Nanoparticles: Ligand-Induced Surface Segregation to Enhance Electrocatalytic Activity. *Nanoscale*, 9, 8945–8951, 2017.
198. Fu, H., Zhang, N., Lai, F., Zhang, L., Chen, S., Li, H., Jiang, K., Zhu, T., Xu, F., Liu, T., Surface-Regulated Platinum-Copper Nanoframes in Electrochemical Reforming of Ethanol for Efficient Hydrogen Production. *ACS Catal.*, 12, 11402–11411, 2022.
199. Zhao, X., Dai, L., Qin, Q., Pei, F., Hu, C., Zheng, N., Self-Supported 3D PdCu Alloy Nanosheets as a Bifunctional Catalyst for Electrochemical Reforming of Ethanol. *Small*, 13, 1602970, 2017.
200. Wang, H., Guo, Y., Mao, Q., Yu, H., Deng, K., Wang, Z., Li, X., Xu, Y., Wang, L., Sulfur and Phosphorus Co-Doping Optimized Electronic Structure and Modulated Intermediate Affinity on PdSP Metallene for Ethanol-Assisted Energy-Saving H_2 Production. *Nanoscale*, 15, 7765–7771, 2023.
201. Begum, H., Ahmed, M.S., Jeon, S., Highly Efficient Dual Active Palladium Nanonetwork Electrocatalyst for Ethanol Oxidation and Hydrogen Evolution. *ACS Appl. Mater. Interfaces*, 9, 39303–39311, 2017.
202. Sheng, S., Ye, K., Sha, L., Zhu, K., Gao, Y., Yan, J., Wang, G., Cao, D., Rational Design of Co-S-P Nanosheet Arrays as Bifunctional Electrocatalysts for both Ethanol Oxidation Reaction and Hydrogen Evolution Reaction. *Inorg. Chem. Front.*, 7, 4498–4506, 2020.
203. Yin, K., Chao, Y., Lv, F., Tao, L., Zhang, W., Lu, S., Li, M., Zhang, Q., Gu, L., Li, H., Guo, S., One Nanometer PtIr Nanowires as High-Efficiency Bifunctional Catalysts for Electrosynthesis of Ethanol into High Value-Added Multicarbon Compound Coupled with Hydrogen Production. *J. Am. Chem. Soc.*, 143, 10822–10827, 2021.
204. Li, J.-Y., Li, Y.-H., Zhang, F., Tang, Z.-R., Xu, Y.-J., Visible-Light-Driven Integrated Organic Synthesis and Hydrogen Evolution over 1D/2D CdS-$Ti_3C_2T_x$ Mxene Composites. *Appl. Catal., B*, 269, 118783, 2020.
205. Zhou, H., Ren, Y., Li, Z., Xu, M., Wang, Y., Ge, R., Kong, X., Zheng, L., Duan, H., Electrocatalytic Upcycling of Polyethylene Terephthalate to Commodity Chemicals and H_2 Fuel. *Nat. Commun.*, 12, 4679, 2021.
206. Yan, Y., Zhou, H., Xu, S.-M., Yang, J., Hao, P., Cai, X., Ren, Y., Xu, M., Kong, X., Shao, M., Duan, H., Electrocatalytic Upcycling of Biomass and Plastic Wastes to Biodegradable Polymer Monomers and Hydrogen Fuel at High Current Densities. *J. Am. Chem. Soc.*, 145, 6144, 2023.
207. Si, D., Xiang, B., Chen, L., Shi, J., Highly Selective and Efficient Electrocatalytic Synthesis of Glycolic Acid in Coupling with Hydrogen Evolution. *Chem. Catal.*, 1, 941–955, 2021.
208. Fan, J., Wu, J., Cui, X., Gu, L., Zhang, Q., Meng, F., Lei, B.-H., Singh, D.J., Zheng, W., Hydrogen Stabilized RhPdH 2D Bimetallene Nanosheets for Efficient Alkaline Hydrogen Evolution. *J. Am. Chem. Soc.*, 142, 3645–3651, 2020.

209. Luo, M., Zhao, Z., Zhang, Y., Sun, Y., Xing, Y., Lv, F., Yang, Y., Zhang, X., Hwang, S., Qin, Y., Ma, J.-Y., Lin, F., Su, D., Lu, G., Guo, S., PdMo Bimetallene for Oxygen Reduction Catalysis. *Nature*, 574, 81–85, 2019.
210. Prabhu, P. and Lee, J.-M., Metallenes as Functional Materials in Electrocatalysis. *Chem. Soc. Rev.*, 50, 6700–6719, 2021.
211. Yu, H., Zhou, T., Wang, Z., Xu, Y., Li, X., Wang, L., Wang, H., Defect-Rich Porous Palladium Metallene for Enhanced Alkaline Oxygen Reduction Electrocatalysis. *Angew. Chem. Int. Ed.*, 60, 12027–12031, 2021.
212. Mao, Q., Deng, K., Yu, H., Xu, Y., Wang, Z., Li, X., Wang, L., Wang, H., *In Situ* Reconstruction of Partially Hydroxylated Porous Rh Metallene for Ethylene Glycol-Assisted Seawater Splitting. *Adv. Funct. Mater.*, 32, 2201081, 2022.
213. Jiang, X., Dong, Z., Zhang, Q., Xu, G.-R., Lai, J., Li, Z., Wang, L., Decoupled Hydrogen Evolution from Water/Seawater Splitting by Integrating Ethylene Glycol Oxidation on $PtRh_{0.02}$@Rh Nanowires with Rh Atom Modification. *J. Mater. Chem. A*, 10, 20571–20579, 2022.
214. Li, Y., Bao, X., Chen, D., Wang, Z., Dewangan, N., Li, M., Xu, Z., Wang, J., Kawi, S., Zhong, Q., A Minireview on Nickel-Based Heterogeneous Electrocatalysts for Water Splitting. *ChemCatChem*, 11, 5913–5928, 2019.
215. Selembo, P.A., Perez, J.M., Lloyd, W.A., Logan, B.E., High Hydrogen Production from Glycerol or Glucose by Electrohydrogenesis Using Microbial Electrolysis Cells. *Int. J. Hydrogen Energy*, 34, 5373–5381, 2009.
216. Li, T. and Harrington, D.A., An Overviewof Glycerol Electrooxidation Mechanisms on Pt, Pd and Au. *ChemSusChem*, 14, 1472–1495, 2021.
217. Qian, Q., He, X., Li, Z., Chen, Y., Feng, Y., Cheng, M., Zhang, H., Wang, W., Xiao, C., Zhang, G., Xie, Y., Electrochemical Biomass Upgrading Coupled with Hydrogen Production under Industrial-Level Current Density. *Adv. Mater.*, 35, 2300935, 2023.
218. Yang, Y. and Mu, T., Electrochemical Oxidation of Biomass Derived 5-Hydroxymethylfurfural (HMF): Pathway, Mechanism, Catalysts and Coupling Reactions. *Green Chem.*, 23, 4228, 2021.
219. Yu, X., dos Santos, E.C., White, J., Salazar-Alvarez, G., Pettersson, L.G.M., Cornell, A., Johnsson, M., Electrocatalytic Glycerol Oxidation with Concurrent Hydrogen Evolution Utilizing an Efficient MoO_x/Pt Catalyst. *Small*, 17, 2104288, 2021.
220. Yu, H., Wang, W., Mao, Q., Deng, K., Wang, Z., Xu, Y., Li, X., Wang, H., Wang, L., Pt single atom captured by oxygen vacancy-rich NiCo layered double hydroxides for coupling hydrogen evolution with selective oxidation of glycerol to formate. *Appl. Catal., B*, 330, 122617, 2023.
221. Li, Y., Wei, X., Chen, L., Shi, J., He, M., Nickel-Molybdenum Nitride Nanoplate Electrocatalysts for Concurrent Electrolytic Hydrogen and Formate Productions. *Nat. Commun.*, 10, 5335, 2019.
222. Zhu, Y., Qian, Q., Chen, Y., He, X., Shi, X., Wang, W., Li, Z., Feng, Y., Zhang, G., Cheng, F., Biphasic Transition Metal Nitride Electrode Promotes Nucleophile

Oxidation Reaction for Practicable Hybrid Water Electrocatalysis. *Adv. Functi. Mater.*, 33, 2300547, 2023.
223. You, B., Liu, X., Liu, X., Sun, Y., Efficient H_2 Evolution Coupled with Oxidative Refining of Alcohols via a Hierarchically Porous Nickel Bifunctional Electrocatalyst. *ACS Catal.*, 7, 4564–4570, 2017.
224. Cheung, K., Wong, W., Ma, D., Lai, T., Wong, K., Transition Metal Complexes as Electrocatalysts-Development and Applications in Electro-oxidation Reactions. *Coord. Chem. Rev.*, 251, 2367, 2007.
225. You, B., Liu, X., Jiang, N., Sun, Y.A., General Strategy for Decoupled Hydrogen Production from Water Splitting by Integrating Oxidative Biomass Valorization. *J. Am. Chem. Soc.*, 138, 13639–13646, 2016.
226. Zhu, P., Shen, Y., Dai, L., Yu, Q., Zhang, Z.-M., An, C., Accelerating Anode Reaction with Electro-oxidation of Alcohols over Ru Nanoparticles to Reduce the Potential for Water Splitting. *ACS Appl. Mater. Interface*, 14, 1452, 2021.
227. Chen, X., Zhong, X., Yuan, B., Li, S., Gu, Y., Zhang, Q., Zhuang, G., Li, X., Deng, S., Wang, J.-g., Defect Engineering of Nickel Hydroxide Nanosheets by Ostwald Ripening for Enhanced Selective Electrocatalytic Alcohol Oxidation. *Green Chem.*, 21, 578, 2019.
228. Zheng, J., Chen, X., Zhong, X., Li, S., Liu, T., Zhuang, G., Li, X., Deng, S., Mei, D., Wang, J.-G., Hierarchical Porous NC@CuCo Nitride Nanosheet Networks: Highly Efficient Bifunctional Electrocatalyst for Overall Water Splitting and Selective Electrooxidation of Benzyl Alcohol. *Adv. Funct. Mater.*, 27, 1704169, 2017.
229. Sharpe, H.R., Geer, A.M., Taylor, L.J., Gridley, B.M., Blundell, T.J., Blake, A.J., Davies, E.S., Lewis, W., Mcmaster, J., Robinson, D., Kays, D.L., Selective Reduction and Homologation of Carbon Monoxide by Organometallic Iron Complexes. *Nat. Commun.*, 9, 3757, 2018.
230. Qin, M., Fan, S., Li, X., Yin, Z., Wang, L., Chen, A., Double Active Sites in Co-N_x-C@Co Electrocatalysts for Simultaneous Production of Hydrogen and Carbon Monoxide. *ACS Appl. Mater. Interfaces*, 13, 38256–38265, 2021.
231. Zhou, Z., Zeng, L., Xiong, G., Yang, L., Yuan, H., Yu, J., Xu, S., Wang, D., Zhang, X., Liu, H., Zhou, W., Multifunctional Electrocatalyst of NiCo-NiCoP Nanoparticles Embedded into P-Doped Carbon Nanotubes for Energy-Saving Hydrogen Production and Upgraded Conversion of Formaldehyde. *Chem. Eng. J.*, 426, 129214, 2021.
232. Li, G., Han, G., Wang, L., Cui, X., Moehring, N.K., Kidambi, P.R., Jiang, D.-e., Sun, Y., Dual Hydrogen Production from Electrocatalytic Water Reduction Coupled with Formaldehyde Oxidation via a Copper-Silver Electrocatalyst. *Nat. Commun.*, 14, 525, 2023.
233. Jagadeesh, R.V., Junge, H., Beller, M., "Nanorust"-Catalyzed Benign Oxidation of Amines for Selective Synthesis of Nitriles. *ChemSusChem*, 8, 92–96, 2015.

234. Kim, J., Golime, G., Kim, H.Y., Oh, K., Copper(II)-Catalyzed Aerobic Oxidation of Amines: Divergent Reaction Pathways by Solvent Control to Imines and Nitriles. *Asian J. Org. Chem.*, 8, 1674–1679, 2019.
235. Xu, B., Hartigan, E.M., Feula, G., Huang, Z., Lumb, J.-P., Arndtsen, B.A., Simple Copper Catalysts for the Aerobic Oxidation of Amines: Selectivity Control by the Counterion. *Angew. Chem. Int. Ed.*, 55, 15802–15806, 2016.
236. Yamaguchi, K. and Mizuno, N., Efficient Heterogeneous Aerobic Oxidation of Amines by a Supported Ruthenium Catalyst. *Angew. Chem. Int. Ed.*, 42, 1480–1483, 2003.
237. Chaumontet, M., Piccardi, R., Baudoin, O., Synthesis of 3,4-Dihydroisoquinolines by a C(sp^3)-H Activation/Electrocyclization Strategy: Total Synthesis of Coralydine. *Angew. Chem. Int. Ed.*, 48, 179–182, 2009.
238. Zheng, B., Trieu, T.H., Li, F.-L., Zhu, X.-L., He, Y.-G., Fan, Q.-Q., Shi, X.-X., Copper-Catalyzed Benign and Efficient Oxidation of Tetrahydroisoquinolines and Dihydroisoquinolines Using Air as a Clean Oxidant. *ACS Omega*, 3, 8243–8252, 2018.
239. Huang, C., Huang, Y., Liu, C., Yu, Y., Zhang, B., Integrating Hydrogen Production with Aqueous Selective Semi-Dehydrogenation of Tetrahydroisoquinolines over a Ni_2P Bifunctional Electrode. *Angew. Chem. Int. Ed.*, 58, 12014–12017, 2019.
240. Galli, C., Radical Reactions of Arenediazonium Ions: An Easy Entry into the Chemistry of the Aryl Radical. *Chem. Rev.*, 88, 765–792, 1988.
241. Ding, Y., Miao, B.-Q., Li, S.-N., Jiang, Y.-C., Liu, Y.-Y., Yao, H.-C., Chen, Y., Benzylamine Oxidation Boosted Electrochemical Water-Splitting: Hydrogen and Benzonitrile Co-Production at Ultra-Thin Ni_2P Nanomeshes Grown on Nickel Foam. *Appl. Catal., B*, 268, 118393, 2020.
242. Li, Y., Jiao, Y., Yan, H., Yang, G., Liu, Y., Tian, C., Wu, A., Fu, H., Mo-Ni-based Heterojunction with Fine-customized d-Band Centers for Hydrogen Production Coupled with Benzylamine Electrooxidation in Low Alkaline Medium. *Angew. Chem. Int. Ed.*, 62, 202306640, 2023.
243. Luterbacher, J.S., Rand, J.M., Alonso, D.M., Han, J., Youngquist, J.T., Maravelias, C.T., Pfleger, B.F., Dumesic, J.A., Nonenzymatic Sugar Production from Biomass Using Biomass-Derived Gamma-Valerolactone. *Science*, 343, 277–280, 2014.
244. Xia, Q., Chen, Z., Shao, Y., Gong, X., Wang, H., Liu, X., Parker, S.F., Han, X., Yang, S., Wang, Y., Direct Hydrodeoxygenation of Raw Woody Biomass into Liquid Alkanes. *Nat. Commun.*, 7, 11162, 2016.
245. Zhang, Z. and Deng, K., Recent Advances in the Catalytic Synthesis of 2, 5-Furandicarboxylic Acid and its Derivatives. *ACS Catal.*, 5, 6529–6544, 2015.
246. Zhao, H., Lu, D., Wang, J., Tu, W., Wu, D., Koh, S.W., Gao, P., Xu, Z.J., Deng, S., Zhou, Y., You, B., Li, H., Raw Biomass Electroreforming Coupled to Green Hydrogen Generation. *Nat. Commun.*, 12, 2008, 2021.

247. Hutchings, G.J., Selective Oxidation Using Supported Gold Bimetallic and Trimetallic Nanoparticles. *Catal. Today*, 238, 69–73, 2014.
248. Lv, G., Wang, H., Yang, Y., Deng, T., Chen, C., Zhu, Y., Hou, X., Graphene Oxide: A Convenient Metal-Free Carbocatalyst for Facilitating Aerobic Oxidation of 5-Hydroxymethylfurfural into 2, 5-Diformylfuran. *ACS Catal.*, 5, 5636–5646, 2015.
249. Wang, T., Nolte, M.W., Shanks, B.H., Catalytic Dehydration of C_6 Carbohydrates for the Production of Hydroxymethylfurfural (HMF) as a Versatile Platform Chemical. *Green Chem.*, 16, 548–572, 2014.
250. van Putten, R.-J., van der Waal, J.C., de Jong, E., Rasrendra, C.B., Heeres, H.J., de Vries, J.G., Hydroxymethylfurfural, A Versatile Platform Chemical Made from Renewable Resources. *Chem. Rev.*, 113, 1499–1597, 2013.
251. Lu, Y., Liu, T., Dong, C.L., Huang, Y.C., Li, Y., Chen, J., Zou, Y., Wang, S., Tuning the Selective Adsorption Site of Biomass on Co_3O_4 by Ir Single Atoms for Electrosynthesis. *Adv. Mater.*, 33, 2007056, 2021.
252. Zhou, B., Li, Y., Zou, Y., Chen, W., Zhou, W., Song, M., Wu, Y., Lu, Y., Liu, J., Wang, Y., Wang, Y., Platinum Modulates Redox Properties and 5-Hydroxymethylfurfural Adsorption Kinetics of $Ni(OH)_2$ for Biomass Upgrading. *Angew. Chem. Int. Ed.*, 60, 22908–22914, 2021.
253. Zhou, P., Lv, X., Tao, S., Wu, J., Wang, H., Wei, X., Wang, T., Zhou, B., Lu, Y., Frauenheim, T., Fu, X., Wang, S., Zou, Y., Heterogeneous-Interface-Enhanced Adsorption of Organic and Hydroxyl for Biomass Electrooxidation. *Adv. Mater.*, 34, 2204089, 2022.
254. Zeng, L., Chen, Y., Sun, M., Huang, Q., Sun, K., Ma, J., Li, J., Tan, H., Li, M., Pan, Y., Liu, Y., Luo, M., Huang, B., Guo, S., Cooperative Rh-O_5/Ni(Fe) Site for Efficient Biomass Upgrading Coupled with H_2 Production. *J. Am. Chem. Soc.*, 145, 17577–17587, 2023.
255. Jiang, N., You, B., Boonstra, R., Rodriguez, I.M.T., Sun, Y., Integrating Electrocatalytic 5-Hydroxymethylfurfural Oxidation and Hydrogen Production via Co-P-Derived Electrocatalysts. *ACS Energy Lett.*, 1, 386–390, 2016.
256. You, B., Jiang, N., Liu, X., Sun, Y., Simultaneous H_2 Generation and Biomass Upgrading in Water by an Efficient Noble-Metal-Free Bifunctional Electrocatalyst. *Angew. Chem. Int. Ed.*, 55, 9913–9917, 2016.
257. Yang, G., Jiao, Y., Yan, H., Xie, Y., Wu, A., Dong, X., Guo, D., Tian, C., Fu, H., Interfacial Engineering of MoO_2-FeP Heterojunction for Highly Efficient Hydrogen Evolution Coupled with Biomass Electrooxidation. *Adv. Mater.*, 32, 2000455, 2020.
258. Du, P., Zhang, J., Liu, Y., Huang, M., Hydrogen Generation from Catalytic Glucose Oxidation by Fe-Based Electrocatalysts. *Electrochem. Commun.*, 83, 11–15, 2017.
259. Liu, W.-J., Xu, Z., Zhao, D., Pan, X.-Q., Li, H.-C., Hu, X., Fan, Z.-Y., Wang, W.-K., Zhao, G.-H., Jin, S., Huber, G.W., Yu, H.-Q., Efficient Electrochemical

Production of Glucaric Acid and H_2 via Glucose Electrolysis. *Nat. Commun.*, 11, 265, 2020.
260. Sun, F., Zhou, Y., You, Z., Xia, H., Tuo, Y., Wang, S., Jia, C., Zhang, J., Bi-Functional Fe_3O_4/Au/CoFe-LDH Sandwich-Structured Electrocatalyst for Asymmetrical Electrolyzer with Low Operation Voltage. *Small*, 17, 2103307, 2021.
261. Liu, H., Zhang, R., Chen, L., Wang, L., Guo, Y., Yang, Y., Synergistic Coupling of Nickel Boride with Ru Cluster as a Highly Active Multifunctional Electrocatalyst for Overall Water Splitting and Glucose Electrolysis. *Adv. Sustain. Syst.*, 5, 2000184, 2021.
262. Zheng, D., Li, J., Ci, S., Cai, P., Ding, Y., Zhang, M., Wen, Z., Three-Birds-with-One-Stone Electrolysis for Energy-Efficiency Production of Gluconate and Hydrogen. *Appl. Catal., B*, 277, 119178, 2020.
263. Geyer, R., Jambeck, J.R., Law, K.L., Production, Use, and Fate of All Plastics Ever Made. *Sci. Adv.*, 3, e1700782, 2017.
264. Kane, I.A., Clare, M.A., Miramontes, E., Wogelius, R., Rothwell, J.J., Garreau, P., Pohl, F., Seafloor Microplastic Hotspots Controlled by Deep-Sea Circulation. *Science*, 368, 1140–1145, 2020.
265. Li, L., Luo, Y., Li, R., Zhou, Q., Peijnenburg, W.J.G.M., Yin, N., Yang, J., Tu, C., Zhang, Y., Effective Uptake of Submicrometre Plastics by Crop Plants via a Crack-Entry Mode. *Nat. Sustain.*, 3, 929–937, 2020.
266. Uekert, T., Kasap, H., Reisner, E., Photoreforming of Nonrecyclable Plastic Waste over a Carbon Nitride/Nickel Phosphide Catalyst. *J. Am. Chem. Soc.*, 141, 15201–15210, 2019.
267. Wang, N., Li, X., Hu, M.-K., Wei, W., Zhou, S.-H., Wu, X.-T., Zhu, Q.-L., Ordered Macroporous Superstructure of Bifunctional Cobalt Phosphide with Heteroatomic Modification for Paired Hydrogen Production and Polyethylene Terephthalate Plastic Recycling. *Appl. Catal., B*, 316, 121667, 2022.
268. Ma, F., Wang, S., Gong, X., Liu, X., Wang, Z., Wang, P., Liu, Y., Cheng, H., Dai, Y., Zheng, Z., Huang, B., Highly Efficient Electrocatalytic Hydrogen Evolution Coupled with Upcycling of Microplastics in Seawater Enabled via Ni_3N/W_5N_4 Janus Nanostructures. *Appl. Catal., B*, 307, 121198, 2022.
269. Li, Y., Zhao, Y., Zhao, H., Wang, Z., Li, H., Gao, P., A Bifunctional Catalyst of Ultrathin Cobalt Selenide Nanosheets for Plastic-Electroreforming-Assisted Green Hydrogen Generation. *J. Mater. Chem. A*, 10, 20646–20646, 2022.

11

Electrolyzers for H_2 Production: A Review

Edisson Villa-Ávila[1,2], Paul Arévalo[1,2], Marcos Tostado-Véliz[1] and Francisco Jurado[1*]

[1]*Department of Electrical Engineering, University of Jaen, EPS Linares, Jaen, Spain*
[2]*Department of Electrical, Electronics and Telecommunications Engineering (DEET), Universidad de Cuenca, Balzay Campus, Cuenca, Ecuador*

Abstract
The growing interest in sustainable energy has driven significant innovation in energy production, with water electrolysis standing out as a promising technology for carbon-free hydrogen production from renewable sources. This study presents a bibliometric analysis of recent and highly cited research on hydrogen electrolyzers using the Web of Science database. The main objective is to identify current trends and potential future applications in this dynamically evolving field. This analysis will leverage case studies from global scientific literature, enabling a comprehensive assessment of the current state-of-the-art in terms of electrolyzer technologies, operational models, key challenges facing the industry, highlighted themes, and future research directions. Given the continually expanding market for renewable energy and the increasing investment in research and development, it is crucial to have an up-to-date review of this rapidly advancing field. Understanding the latest trends and advancements in hydrogen electrolyzers is essential for guiding future research and facilitating the effective implementation of sustainable technologies. Furthermore, this bibliometric review will provide valuable insights into emerging applications and potential challenges to overcome in the production and utilization of hydrogen as a clean energy source.

***Keywords*:** Sustainable energy, water electrolysis, hydrogen production, bibliometric analysis

**Corresponding author*: fjurado@ujaen.es

Mehmet Sankir and Nurdan Demirci Sankir (eds.) Towards Green Hydrogen Generation, (359–406)

11.1 Introduction

Since the emergence of the Industrial Revolution in the 18th century, humanity has undergone an unprecedented transformation in terms of economic and technological development. This era of innovation brought about a growing demand for energy, initially met by the relentless burning of coal and later by the discovery and exploitation of vast oil and natural gas reserves [1]. While this reliance on fossil fuels has propelled global economic growth, it has also triggered a series of crucial challenges [2].

The depletion of finite reserves, price instability, and, most notably, the exacerbation of climate change due to greenhouse gas emissions have prompted a fundamental reassessment of our energy approach [3]. In this context, renewable energies have emerged as the vanguard of a new era in energy production. The capture and conversion of natural and sustainable sources such as solar, wind, hydroelectric, and geothermal energy offer a promising and necessary alternative [4].

As countries worldwide intensify their efforts to establish renewable energy infrastructures and address the challenge of climate change, a critical need arises to efficiently store and utilize the generated energy. In this context, hydrogen production through electricity stands out as a central pillar, especially as the world moves toward the 2050 horizon with net-zero emissions [5]. Its significance in this transition toward more sustainable energy is such that the International Energy Agency (IEA) has highlighted it as a key player in its roadmap to achieve zero emissions [6]. This gas is expected to have a crucial role in the upcoming scenario: It is predicted that by 2030, there will be 15 million hydrogen-powered vehicles globally, and by 2050, the maritime sector will rely significantly on hydrogen as its main source of fuel [6].

As energy and transportation systems globally undergo a significant evolution marked by a substantial increase in the use of renewable energy sources, energy-saving measures, and the growing adoption of electric vehicles, the landscape decisively tilts toward a cleaner energy matrix. In this context, hydrogen emerges as a promising energy vector. Its status as a clean and sustainable gas, emitting no carbon dioxide (CO_2), positions it as an alternative with significant economic potential for future economies [7, 8]. Being the most abundant element and present in water and various organic substances, hydrogen stands out as the simplest and lightest element, composed of one electron and one proton. It is a colorless, tasteless, and highly flammable gas, with a remarkable energy density per unit mass, surpassing most conventional fuels by more than double [5, 8–10].

Its nontoxic nature and environmental compatibility, combined with the versatility of being produced from both renewable and nonrenewable sources, reinforce its position as a highly advantageous energy carrier [8]. The production of hydrogen from renewable sources such as wind, solar, hydroelectric, wave, geothermal, and biomass energy underscores its potential in the future energy scenario [8].

This is where electrolyzers play an essential role. These devices enable the conversion of electrical energy into hydrogen, a highly versatile and clean energy vector. Should the electricity utilized in the electrolysis process be derived from renewable energy, the resulting hydrogen would be characterized as "green hydrogen" with no linked CO_2 emissions. At present, electrolysis constitutes merely 2% of the overall global hydrogen production [6], and a mere 0.1% originates from water electrolysis. This technology encounters elevated costs, and the transition from gray to green hydrogen necessitates a substantial amount of electricity, often reaching the TWh range [6].

Global trends are analyzed, with highlighted examples from leading countries in the transition to a hydrogen economy. Additionally, the economic aspect will be addressed, examining prices and market projections for electrolyzers and produced hydrogen. The sophisticated electrolyzer technology of today is transforming the global energy landscape, opening new possibilities for large-scale hydrogen production, storage, and utilization. The growing interest in hydrogen production as a clean and sustainable energy source has driven significant advances in electrolyzer technology [11].

These devices play a crucial role in converting electricity into hydrogen through highly efficient electrochemical processes. In this chapter, a comprehensive review of the latest developments and perspectives in the field of electrolyzers is presented, with a focus on high-purity and high-performance hydrogen production. As the world faces the urgency of transitioning to cleaner and more sustainable energy sources, hydrogen production from renewable sources emerges as a promising solution.

Technological advances in electrolyzers open up new possibilities for industrial-scale hydrogen generation, with applications ranging from the chemical industry to sustainable mobility. In this context, three types of electrolyzers will be explored in detail: solid oxide electrolysis cells (SOEC) [12], proton exchange membrane (PEM) cells [12], and alkaline electrolyzers [13].

Each of these systems has unique characteristics that influence its efficiency, costs, and potential applications. This review will provide a comprehensive assessment of these technologies, highlighting both their advantages and areas that require further research and development. Throughout the chapter, current challenges, emerging trends, and insights into how electrolyzers are shaping the hydrogen production landscape will be addressed. Case studies and specific examples illustrating the successful implementation of these technologies in various industrial contexts will also be examined.

Ultimately, this work aims to provide a comprehensive overview of advances and prospects in electrolyzer technology, emphasizing its crucial role in transitioning to a more sustainable and clean energy economy. Through this exploration, readers will be better equipped to understand and contribute to the exciting future of hydrogen production.

11.2 Materials and Methods

The extensive literature review is conducted following the methodology proposed by [14] to gather and analyze datasets published on the Web of Science (WoS). The primary objective is to access articles and select those providing insights into electrolyzers for hydrogen production. Articles are collected from open-access databases for the comprehensive study. The selection process considers titles, keywords, abstracts, highlights, and document types. Various keywords are employed for searches, including "electrolyzers," "hydrogen production," "electrolysis technologies," "renewable hydrogen," "renewable energies and hydrogen," "electrochemistry and hydrogen production," "energy sustainability," "hydrogen economy," "energy storage," and "energy conversion technologies."

Only the most relevant articles for the topics of this review are selected and grouped based on research quality, responding to a set of research-related questions and describing factual information based on real-world scenarios. The analysis of documents is conducted based on electrolyzers for hydrogen production. Research questions are formulated based on various studies and article analysis. These include the following:

- What are the most significant advancements in electrolyzer technology for hydrogen production?
- How do technological advancements influence the efficiency and profitability of electrolyzer systems?
- What are the key economic challenges facing the implementation of large-scale electrolyzer technologies?
- To what extent has progress been made in integrating electrolyzers into broader energy systems?

- How are economic and financial aspects evaluated in the implementation of industrial-scale electrolyzer systems?
- What impact does technological maturity have on the economic viability of hydrogen production from electrolyzers?
- What role do electrolyzers play in transitioning toward a more sustainable hydrogen economy?
- How do electrolyzers contribute to reducing greenhouse gas emissions compared to other hydrogen production technologies?
- What prospects are envisioned for the future integration of electrolyzers into the global energy landscape?
- How do the short and long-term economic costs and benefits of electrolyzer systems compare to other hydrogen production technologies?

11.2.1 Choice of Search Database

WoS and Scopus are globally recognized research indexing databases. They are critical components of the current global research ecosystem, providing data for global university rankings and bibliometric analyses for global research assessments [15]. WoS is the oldest multidisciplinary database with a wide variety of journals, journal titles, conference proceedings, and monographs covering a vast range of disciplines [16]. WoS was chosen for the search for several reasons [17]:

- Comprehensive coverage: WoS covers a broad range of scientific disciplines, including natural sciences, social sciences, health sciences, and humanities. Its extensive coverage ensures that researchers have access to a wide variety of sources and research areas.
- Recognized citation metrics: WoS is known for its citation system and metrics, such as the impact factor (IF), which provides a quantitative measure of the influence and relevance of scientific publications. These metrics are widely used to assess the quality and impact of research.
- Rigorous selection process: WoS has a rigorous journal selection process for inclusion in the database. Journals must meet certain quality criteria and adhere to high editorial standards. This ensures that publications indexed in WoS are reliable and of high quality.

- Interface and advanced search tools: WoS offers an intuitive interface and advanced search tools that make navigation and retrieval of relevant information easy. Researchers can conduct precise searches and obtain reliable results.
- Connections with other databases: WoS is connected with other databases and resources, allowing researchers to access a wide range of additional information, such as citations, patents, and research data.

11.2.2 Literature Review Process

After selecting the database, an exhaustive examination of recent research outcomes on hydrogen electrolyzers for sustainable energy production from the past decade is conducted. Utilizing the WoS database, this study employs bibliometric methods to scrutinize articles published between 2013 and 2023. The search date is set as the second week of June 2023, considering that the number of articles may have evolved with time due to additional publications. As of the search date, WoS exhibits a higher number of publications compared to Scopus, prompting the decision to work with this database. WoS stands out as the largest peer-reviewed and interdisciplinary database globally, consistently updated and expanded. It offers a comprehensive view of research production across various disciplines, including engineering, sciences, social sciences, technology, arts, medicine, humanities, and others [18], making it the ideal choice for a bibliometric review.

The literature search begins by using keywords, with the word "electrolyzer" yielding 10,930 results (see Figure 11.1), and the phrase "hydrogen electrolyzer" yielding 7854 results (see Figure 11.2) [19]. This illustrates the broad spectrum of research paths, with over 100 keywords directly related to the study area. The most frequently used keywords in titles and abstracts of all articles, with frequencies ranging from 4 to 93, are highlighted. Different font sizes and colors indicate keyword frequency and rank. In Figure 11.1, the favorable keyword is "electrolyzer," followed by "electrolyzer," "model," "voltage," "operation," and "performance." According to Figure 11.2, "hydrogen" is the most frequently used favorable keyword, with "water electrolyzer," "study," "electricity," and "economic" also prominently featured. It is noteworthy to observe that the keyword statistics presented in Figure 11.2 have been employed in abstracts or titles.

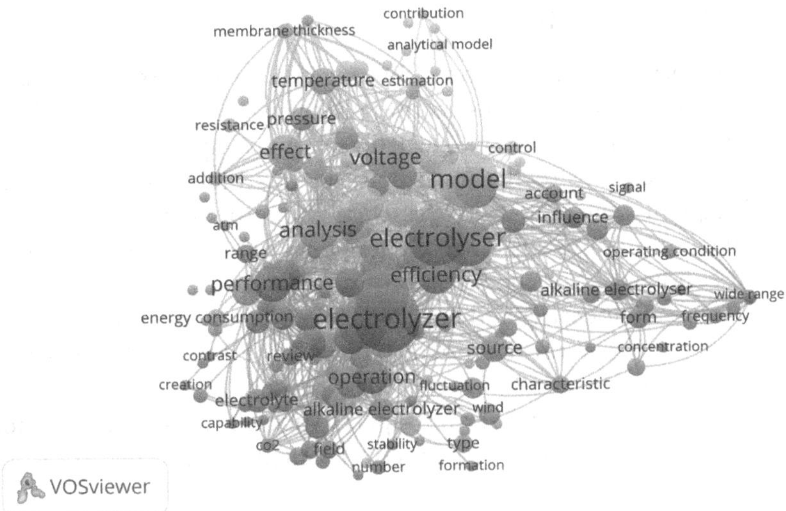

Figure 11.1 Results from WoS core collection for "electrolyzer".

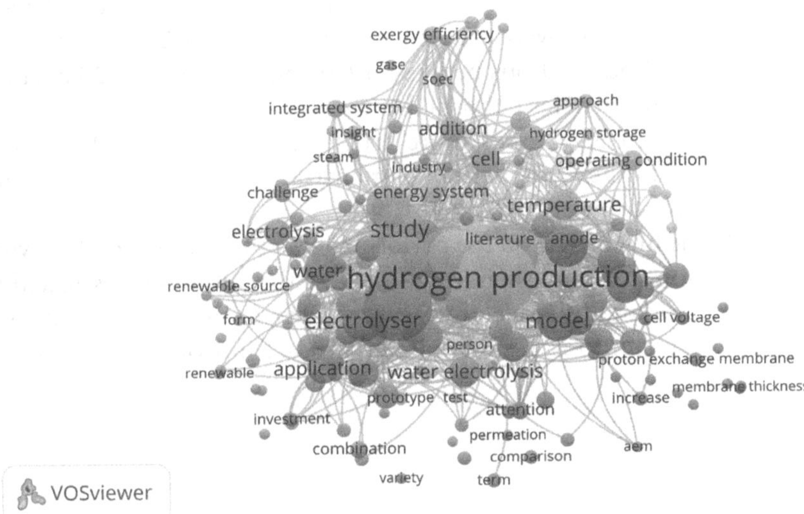

Figure 11.2 Results from WoS core collection for "hydrogen electrolyzer".

11.2.3 Study Selection

During the primary exploration of hydrogen electrolyzers, a total of 7854 results emerged in the initial selection phase covering 1970 to 2023 in the WoS database. The keywords "electrolyzer" and "hydrogen electrolyzer" proved most relevant to the focal point of this bibliometric analysis. In the subsequent phase, 360 articles were pinpointed using refined keywords like control and sustainable energy, within the time frame of 2013 to 2023. Following this, the selection was narrowed down to 360 articles related to the engineering and energy field, and subsequently, 328 articles were extracted based on document type and the English language criterion. Ultimately, about 184 articles were chosen, prioritizing those with the highest citations, considering aspects such as titles, abstracts, focus, citations, content relevance, and practical contributions.

11.2.4 Trend and Review Analysis

Figure 11.3 depicts annual trends publications trends related to electrolyzers from 1970 to 2023. Between 1970 and 2023, 7894 articles related to hydrogen electrolyzers were published, with a total of 297,003 citations. There were 1129 publications between 1970 and 1913 with 1129 citations, increasing to 6765 articles published in the last 10 years with 285,533 citations, resulting in 5.99 times more articles published with 252.9 times more citations in the same period, demonstrating the growth of research in hydrogen electrolyzers.

In 1970, the first article on the performance of an electrolyzer for hydrogen production was published in [20], where the authors investigated the performance of a modified electrolyzer for hydrogen production under specific working conditions, resulting in an increase in working temperature,

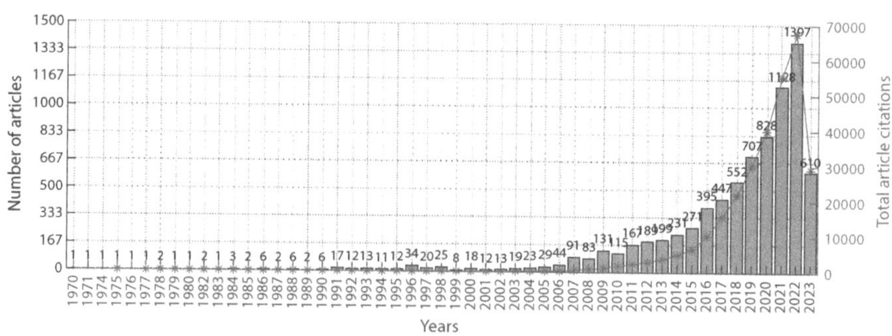

Figure 11.3 Number of articles and citations per year.

a reduction in electrolyte temperature fluctuations, a significant decrease in the carbonate content growth rate, and a significant increase in hydrogen production.

The highest number of publications and citations so far was in 2022 with articles related to the study area, totaling 1397 articles and 67,948 citations. As illustrated in Figure 11.4, where 25 countries provide highly cited documents with citation counts ranking high. Each country contributed a different number of documents between 2013 and 2023. In terms of countries, China, the United States, Germany, South Korea, and Canada are the most productive countries in terms of publications in bibliometric studies. China leads in scientific paper creation based on hydrogen electrolyzers.

Figure 11.5 delves into the assessment of journals by editors, highlighting 15 distinct editors associated with the leading journals. Notably, Elsevier emerged as the publisher for 80% of the top articles, showcasing a commitment to high-quality publications. The International Journal of Hydrogen Energy is frequently mentioned (45%). Then, Energy Conversion and Management (9.17%), Energy, and Journal of Energy Storage, contributed 5% of the total each. Other journals such as Applied Energy, Renewable Energy, International Journal of Electrical Power and Energy Systems, Journal of Cleaner Production, Renewable and Sustainable Energy Reviews, Energy and Buildings, Applied Thermal Engineering, Journal of Power Sources, Energy Reports, and Results in Engineering contributed to less than 5% of the overall published papers. This underscores Elsevier's dominance as a publisher in hydrogen electrolyzer research [21].

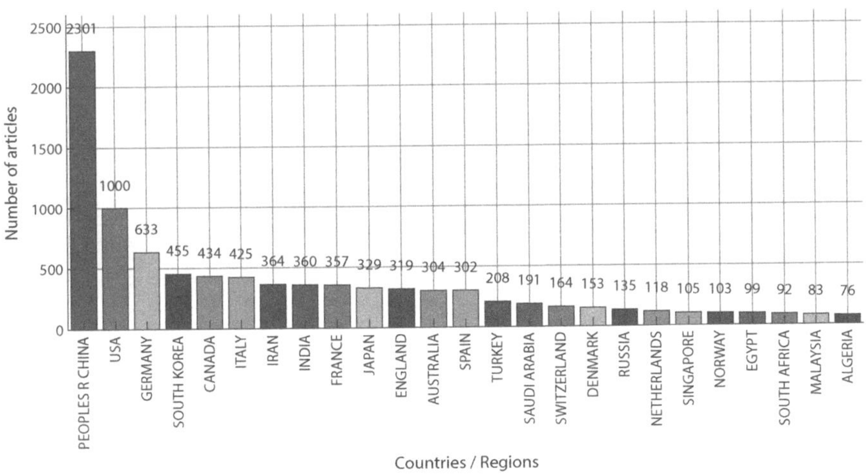

Figure 11.4 Number of articles for countries.

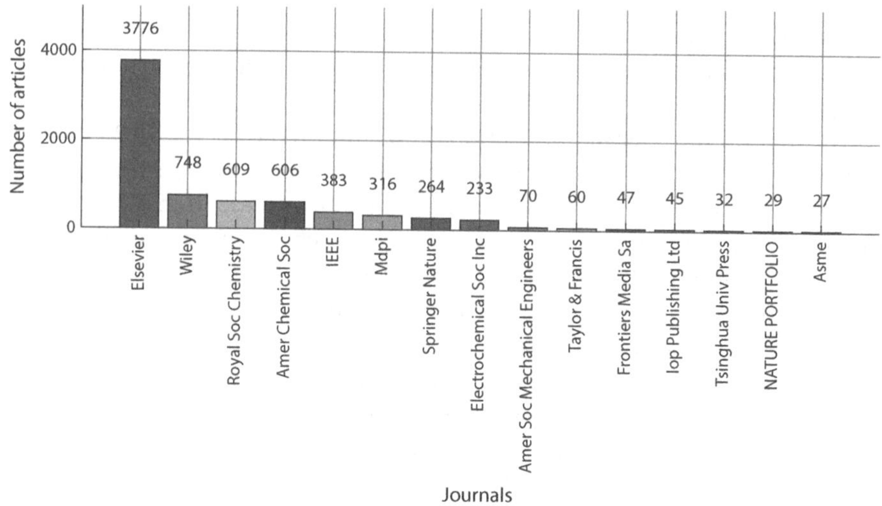

Figure 11.5 Editor of most-cited articles on hydrogen electrolyzers.

Figure 11.6 spotlights the most impactful articles curated by 10 different editors, featured across 25 journals with IF values spanning from 0.7 to 32.5. The IF is commonly employed to gauge a journal's standing in relation to others within the same field. However, its use as a metric for individual

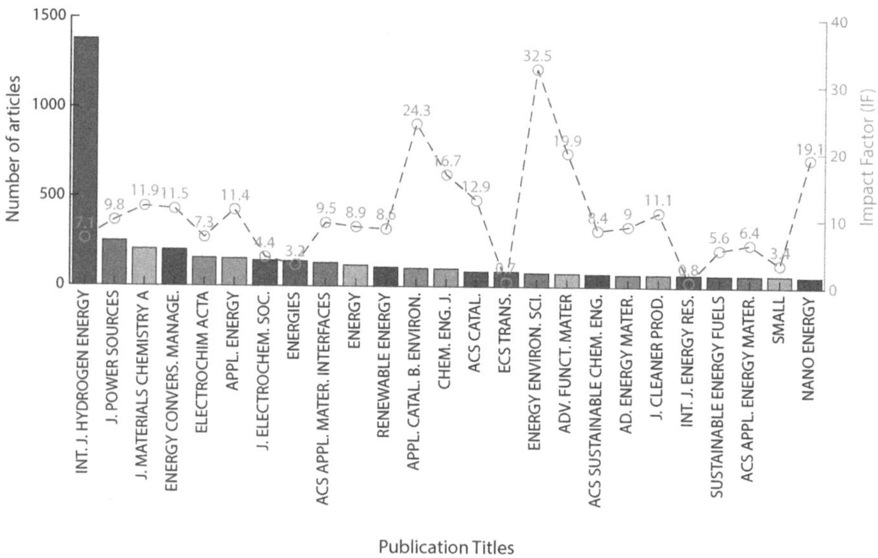

Figure 11.6 Articles for journal evaluation and impact factors.

article quality may inadvertently elevate lower-quality papers while potentially downplaying high-quality ones [21]. According to Figure 11.6, the International Journal of Hydrogen Energy from Elsevier stands out with the highest frequency of article publications. The second-largest journal is Energy Conversion and Management. The third most frequent journal is Energy, and Journal of Energy Storage. Renewable and Sustainable Energy Reviews, Energy Conversion and Management, and Journal of Cleaner Production. The frequency of articles on hydrogen electrolyzers published in Renewable and Sustainable Energy Reviews, Energy Conversion and Management, and Journal of Cleaner Production. However, the most popular focus is on the most cited articles published in the International Journal of Hydrogen Energy.

Upon scrutinizing the keywords, it becomes evident that contingent on the type of electrolyte employed, there exist mainly three categories of hydrogen-producing electrolyzers: alkaline, solid polymer, and solid oxide [22]. Among these, alkaline electrolysis (AEL) stands out as the technology that is the most established and commercially developed, owing to its outstanding technical performance, and established market presence. The use of demineralized water is pivotal in ensuring module durability, mitigating corrosion and electrochemical mechanisms, and enhancing overall process efficiency [22]. Various parameters, such as power, operating temperatures, hydrogen, and pressure flow factors, play crucial roles in these electrolyzers [23].

Key articles have employed specific keywords related to electrolyzers, such as "alkaline water electrolyzer" [22, 24], "hydrogen generative aqua electrolyzer (HAE)" [25], "methanol electrolyzer" [26], "PEM electrolyzer" [12, 27], "electrolyzer" [24, 28, 29], "PEV electrolyzer system" [30], and "SOE electrolyzer" [12, 31].

11.2.5 State of the Art in Hydrogen Electrolyzers

The production of hydrogen from renewable sources is essential for addressing energy and environmental concerns efficiently and economically [32]. Among the prominent technologies for hydrogen production from water are solid oxide electrolysis (SOE), AEL, and PEM electrolysis [33]. Large-scale AEL is an established industrial process that can produce up to 200 Nm^3 of hydrogen per hour and is known for its reliability and safety [34, 35]. On the other hand, PEM uses a solid and expensive membrane instead of a corrosive liquid electrolyte, and although efficient, its current production involves high costs [34, 35]. Lastly, SOE operates at very high temperatures, resulting in even higher efficiency than AEL or PEM technologies [35].

The SOE technology is currently undergoing research and development, primarily at the laboratory scale, with a production capacity of up to 5.7 Nm3/h and 18 kW. In this process, oxygen is generated at the anode as oxide anions traverse the solid electrolyte [35]. Despite hydrogen's high gravimetric energy density, its volumetric energy density is constrained, posing challenges for its application in fuel cell vehicles [5, 34]. Efficient hydrogen storage is pivotal for electricity production through fuel cells, necessitating systems that do not significantly increase the vehicle's weight while providing an adequate driving range. This aspect holds critical importance in the hydrogen economy, and the development of storage techniques that are safe, reliable, affordable, and effective presents a considerable challenge [34]. Three primary storage methods exist compressed gas, cryogenic liquid hydrogen, and solid-state storage, with the first two being the most prevalent. Solid-state storage is currently under development and relies on advanced materials [34]. While compressed hydrogen gas tanks are the prevailing choice in fuel cell vehicles, liquid hydrogen storage introduces additional challenges, including the potential for evaporation even in well-insulated tanks [36].

Researchers are focused on developing economical and practical methods for hydrogen storage and production, especially for their application in fuel cells, the most promising technique for obtaining energy from this element. Models of hydrogen production and storage units driven by renewable energy sources, such as the wind-photovoltaic model [37], which uses an imperial competitive algorithm to optimize its operation, have been proposed. Olivier *et al.* [38] combined renewable energies with hydrogen production and storage through simulations. Han *et al.* [39] examined electrolysis by considering various factors such as pressure, current density, temperature, and ohmic loss of the PEM electrolyte. Zhang *et al.* [40] evaluated a PV-H_2/battery storage system connected to the grid, comparing the efficiency between hydrogen and battery storage. Gonzatti and Farret [41] conducted simulations of a system that includes AEL, metal hydride storage, and a PEM fuel cell, confirming the accuracy of the results with direct measurements. Additionally, Kavadias *et al.* [37] optimized and sized a hydrogen production and storage system based on AEL, calculating the nominal power to achieve optimal storage and production.

In the literature review, numerous works addressing hydrogen production, electrolyzers, and storage have been published, including references [42–45]. Nikolaidis and Poullikkas [46] conducted a thorough analysis of major technologies for hydrogen production and their technical attributes.

Maggio *et al.* [43] examined recent literature on the production and application of hydrogen, with a specific focus on the power-to-hydrogen concept, emphasizing key economic, social, and environmental aspects. Thema *et al.* [47] investigated European power-to-gas (P2G) plants that produce hydrogen or a sustainable substitute for natural gas. Hirscher *et al.* [45] provided a review of various hydrogen production methods, assessing their current status and exploring future possibilities. Egeland-Eriksen *et al.* [44] analyzed continuously operating hydrogen energy storage systems, ranging from laboratory scale to full scale, and assessed the concepts and performances of 15 initiatives. Yue *et al.* [42] examined current and future developments in hydrogen generation technologies, electrification, and storage. Additionally, Asghari *et al.* [48] conducted a review of metal-free seawater electrolytic catalysts, highlighting advances and challenges in this field.

11.3 Advances in Electrolyzer Technology

This section provides a detailed overview of the most recent advances in electrolyzer technology, highlighting the most significant achievements in this ever-evolving field. From efficiency enhancements to innovations in materials and design, we will delve into the most promising developments that are currently driving the effectiveness and viability of electrolyzers.

Electrolysis involves an electrochemical process that breaks down water into hydrogen (H_2) and oxygen (O_2) by introducing an electric current, as illustrated in Figure 11.7. This technology stands as the most advanced and commercialized to date. Networked electrolysis stands out as the swiftest and most cost-effective approach for H_2 production through conventional electricity, thereby augmenting the hydrogen economy. The electrolyzer is connected to the electrical grid, and this limitation can be overcome through wind, solar, or biomass electrolysis, where the electrolyzer is linked to wind turbines, photovoltaic solar panels, or biogas engines to generate electricity. Thanks to this more environmentally friendly method of H_2 production, greenhouse gas emissions were reduced by incorporating clean energy into the electrical grid [49].

Solar electrolysis, a method wherein solar water splitting triggers H_2 production, is regarded as a sophisticated and economically efficient procedure, providing a viable opportunity for electricity generation [28]. This approach serves as an effective solution to significant challenges like H_2 storage and transportation, while also mitigating the production of harmful and toxic substances [50]. In [22], a 2D water electrolyzer

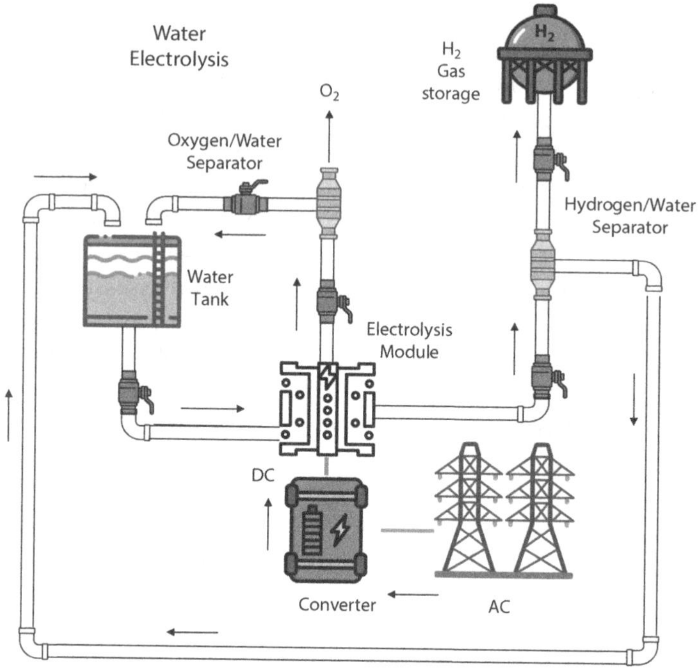

Figure 11.7 Water electrolysis.

model was developed and applied to the PEM fuel system integrated with multi-junction photovoltaic solar cells. The conclusion drawn was that system integration enhances H_2 production. Wulf *et al.* [51] conducted a prototype study to generate H_2 gas using a PEM electrolyzer, and the results indicate that this system can achieve approximately 4.5 times greater efficiency compared to other systems. The magnet-enhanced PEM electrolyzer demonstrated higher efficiency than the simple electrolyzer. For hydrogen production through electrolysis, AWE (alkaline water electrolysis) was recommended as a cost-effective and well-established electrolysis technology [52].

11.3.1 Polymer Electrolyte Membrane Water Electrolysis (PEMWE)

PEMWE stands as an effective and environmentally friendly technique for producing hydrogen H_2 from water through electrolysis. PEMWE exhibits substantial potential for H_2 production from RES, showcasing advantages that play a pivotal contribution to mitigating greenhouse gas emissions within the hydrogen sector. According to studies, it is projected that by 2050, PEMWE could replace conventional methane steam reforming in the Hy-Pro field [11]. In contrast to alkaline electrolyzers, PEMWE swiftly responds to RES fluctuations, accommodates a wide range of temperatures, demonstrates strong performance under varying conditions and energy input scenarios, operates at elevated current densities, and boasts notable energy conversion efficiencies [11]. Additionally, PEMWE is simpler compared to the alkaline type, providing high-quality gas, and demanding lower maintenance. Therefore, operating, and maintenance costs are reduced with use, offering the advantage of ultra-pure hydrogen production with a purity exceeding 99.999%, capable of fueling a fuel cell. The benefits of PEMWE include lower corrosiveness [53], flexibility, high protonic conductivity, and thin PEMs [54]. Several articles elaborate on factors such as a relatively low operating temperature [54], and low computational complexity [12]. Additionally, aspects like decarbonization, rapid response [55], and quick cold start [56] are also notable within this context.

11.3.2 Proton Exchange Membrane Electrolyzer

The PEM represents a significant innovation in hydrogen production. It uses a solid membrane to conduct ions, allowing efficient operation and rapid response to intermittent RES such as sun and wind. Although it requires expensive materials like platinum and ruthenium to prevent corrosion, it offers considerable advantages in terms of efficiency and maintenance. Additionally, it has promising potential in renewable energy storage and provides opportunities to enhance the overall system efficiency [6]. The PEM is an efficient technology for industrial hydrogen production, with efficiencies comparable to alkaline technology, ranging between 50% and 68% in terms of lower heating value (LHV) [57]. However, PEM presents significant advantages in its operating range. While alkaline technology has a minimum operating limit of 20% of its nominal capacity, PEM can effectively operate even at only 5% of its nominal capacity. Furthermore, PEM can operate beyond its nominal capacity, a feature not achievable with alkaline electrolyzers. Another notable advantage of PEM is its ability to operate at higher current densities, reaching up to 2 A/m^2, compared to the 0.8 A/m^2 of alkaline technology. This feature translates into a significant improvement in hydrogen production [6].

11.3.3 Anion Exchange Membrane Electrolyzer (AEM)

AEM represents an emerging technology that amalgamates the merits of PEM while functioning in alkaline conditions. This implies that the underlying reactions driving the process align with those observed in alkaline technology [58]. A distinctive feature of AEM is its operation in an alkaline environment, eliminating the need for scarce materials like iridium or platinum. The structural composition of the AEM electrolyzer mirrors that of PEM, featuring a membrane situated between two electrodes. However, the differentiator lies in the separator: The AEM membrane exclusively facilitates the conduction of hydroxide ions (OH-) [28], whereas, in alkaline technology, the separator consists of a porous material through which ions traverse its pores [58].

AEM has the flexibility to be supplied with pure or demineralized water, akin to PEM. Nevertheless, the performance of the system may suffer due to the limited availability of hydroxide ions in the membrane, potentially leading to electrode degradation if exclusively supplied to the anode. One viable solution involves the addition of an electrolyte, such as KOH, mirroring practices in alkaline technology [28, 59]. Despite being in a developmental phase, commercial products based on AEM have already entered the market, exemplified by electrolyzers manufactured by HydroLite or Enapter [60, 61]. Nevertheless, AEM currently exhibits lower efficiency than both alkaline and PEM technology (67%, based on the LHV for a unit of a few kW), blending characteristics from both. It can operate at the same maximum current density as PEM (2 A/m^2) but is constrained by the nominal capacity, akin to alkaline technology [62].

11.3.4 Alkaline Electrolysis

AEL has reached a well-developed stage as a technology for hydrogen (H_2) production, particularly at a scale of up to several megawatts, and currently stands as the most widely employed electrolytic technology on a global commercial scale [63–65]. The AEL cell is constructed with two electrodes, an anode and a cathode, submerged in a significantly concentrated aqueous alkaline solution containing approximately 20% to 30% by mass of KOH. In traditional AEL, the anode and cathode are commonly fabricated from low-cost steel or nickel-coated alloy steel [63]. These electrodes are configured in a zero-gap formation (or quasi-zero-gap) using a thin diaphragm, facilitating the separation of the gases produced. The diaphragm is permeable to hydroxide ions and water [63].

Table 11.1 Características típicas de las principales tecnologías de electrolisis.

	Low-temperature electrolysis			**High-temperature electrolysis**		
	Alkaline (OH^-) electrolysis		Proton exchange (H^+) electrolysis		Oxygen ion (O^{2-}) electrolysis	
	Liquid	Polymer electrolyte membrane		Solid oxide electrolysis (SOE)		
	Conventional	Solid alkaline	H^+-PEM	H^+-SOE	O^{2-}-SOE	Co-electrolysis
Operation principles	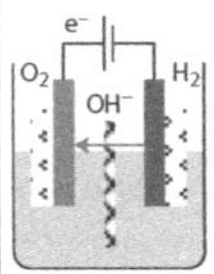	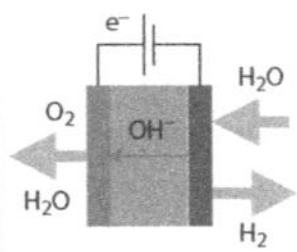	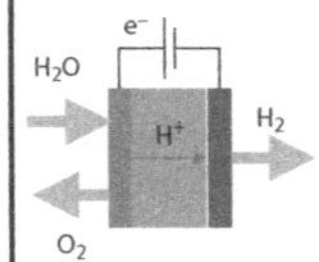	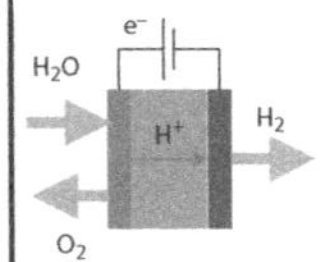	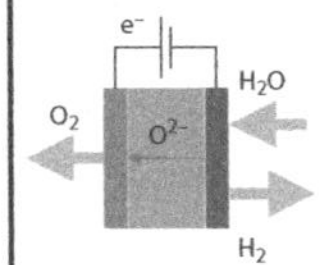	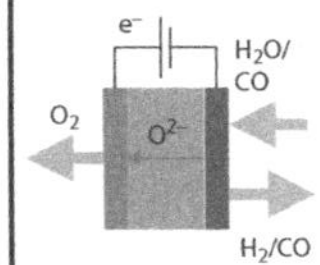
Charge carrier	OH^-	OH^-	H^+	H^+	O^{2-}	O^{2-}
Temperature (°C)	20–80	20–200	20–200	500–1000	500–1000	750–900
Electrolyte	Liquid	Solid (polymeric)	Solid (polymeric)	Solid (ceramic)	Solid (ceramic)	Solid (ceramic)
Anodic reaction (OER)	$4OH \rightarrow 2H_2O+O_2+4e^-$	$4OH \rightarrow 2H_2O+O_2+4e^-$	$2H_2O \rightarrow 4H^++O_2+4e^-$	$2H_2O \rightarrow 4H^++O_2+4e^-$	$O^{2-} \rightarrow 1/2O_2+2e^-$	$O^{2-} \rightarrow 1/2O_2+2e^-$
Anodes	Ni>Co>Fe (oxides) perovskites: $Ba_{0.5}Sr_{0.5}Co_{0.8}Fe_{0.2}O_3$, $LaCoO_3$	Ni-based	IrO_2, RuO_2, $Ir_xRu_{1-x}O_2$ supports: TiO_2, ITO, TiC	Perovskites with protonic-electronic conductivity	$La_xSr_{1-x}MnO_3$ +Y-stabilized ZrO_2 (LSM-YSZ)	$La_xSr_{1-x}MnO_3$ +Y-stabilized ZrO_2 (LSM-YSZ)
Cathodic reaction (HER)	$2H_2O+4e^- \rightarrow 4OH^-+2H_2$	$2H_2O+4e^- \rightarrow 4OH^-+2H_2$	$4H^++4e^- \rightarrow 2H_2$	$4H^++4e^- \rightarrow 2H_2$	$H_2O+2e^- \rightarrow H_2+O^{2-}$	$H_2O+2e^- \rightarrow H_2+O^{2-}$ $CO_2+2e^- \rightarrow CO+O^{2-}$
Cathodes	Ni alloys	Ni, Ni-Fe, $NiFe_2O_4$	Pt/C MoS_2	Ni-cermets	Ni-YSZ Subst. $LaCrO_3$	Ni-YSZ perovskites

(Continued)

Table 11.1 Características típicas de las principales tecnologías de electrolisis. (*Continued*)

	Low-temperature electrolysis			**High-temperature electrolysis**		
	Alkaline (OH^-) electrolysis		Proton exchange (H^+) electrolysis		Oxygen ion (O^{2-}) electrolysis	
	Liquid	Polymer electrolyte membrane		Solid oxide electrolysis (SOE)		
	Conventional	Solid alkaline	H^+-PEM	H^+-SOE	O^{2-}-SOE	Co-electrolysis
Efficiency (%)	59–70		65–82	Up to 100	Up to 100	-
Applicability	Commercial	Laboratory scale	Near-term commercialization	Laboratory Scale	Demonstration	Laboratory scale
Advantages	Low cost, stable, mature	Combination of alkaline and H^+-PEM features	Compact, fast response /high-purity H_2	Improved kinetics, lower demands, and capital		Direct syngas production
Disadvantage	Corrosive electrolyte, slow dynamics	Low OH^- conductivity in membranes	High-cost membranes, safety issues	Unstable electrodes, safety concerns		
Challenges	Improve durability/ oxygen evolution	Enhance electrolyte	Reduce noble metal use	Microstructural changes in the electrodes: delamination, blocking of TPBs, passivation		C deposition, microstructural change electrodes

Key difficulties linked to AELs include managing the corrosive nature of the electrolyte and addressing constrained current densities because of the moderate mobility of OH- [63]. Additionally, the diaphragm does not completely block the passage of gas transfer from one half-cell to the other. Oxygen diffusion entering the cathode section diminishes electrolysis efficiency by reacting with the hydrogen on the cathode side to form water. Moreover, substantial mixing occurs, particularly the diffusion of hydrogen into the O_2 evolution half-cell, and it needs to be minimized for safety reasons. This issue becomes especially troublesome under low loads (<40%), where the O_2 production rate decreases, leading to an increase in the hydrogen crossover concentration to potentially hazardous levels (lower explosion limit >4 mol% H_2) [63].

Water electrolysis offers various technologies, each with its own advantages and disadvantages (summarized in Table 11.1). AEL is the most commercially established, being used by companies, e.g., De Nora SAP, Norsk Hydro, and General Electric, due to its low cost and long-term stability thanks to the utilization of non-precious electrodes [66]. On the other hand, PEM electrolysis faces challenges due to its acidic environment, requiring expensive noble metal catalysts and specific materials. Thermodynamic analysis suggests that high-pressure electrolysis has lower energy requirements [65]. Although alkaline electrolyzers can operate at high pressures, they do so with higher degradation rates [28]. Operating at different pressures on the two sides of the electrolyzer could facilitate hydrogen storage and oxygen production [67]. SOE and PEM technologies, which use solid electrolytes, enable a more condensed design and operation under varying pressures. Additionally, their solid nature provides a quicker response to changes in power load in contrast to liquid alkaline electrolyzers [68–71].

11.4 Future Perspectives and Trends

This section provides a glimpse into the future of electrolyzer technology, exploring emerging trends and long-term development prospects. We will examine potential disruptive advances that could redefine the field of electrolyzers, as well as promising research directions on the horizon. Hydrogen has the potential to serve as a chemical carrier for the storage of surplus energy produced by renewable sources [72]. This stored hydrogen can then be utilized in diverse processes, playing a role in realizing a carbon-neutral energy landscape [73]. It can serve as an on-demand fuel in applications such as engines powered by internal combustion and fuel cells, offering a clean energy option without post-combustion emissions [74].

Additionally, hydrogen can function as a reducing agent for specific CO_2 recycling and valorization processes, transforming carbon dioxide into valuable fine chemicals or fuels. This approach contributes to an overall decrease in the consumption of fossil fuels.

It is crucial to acknowledge that water electrolysis yields oxygen as a valuable chemical in its pure form. Anticipated advancements, for example, the electrification of the chemical manufacturing sector, may create scenarios in situations where a segment of the produced oxygen finds application in production processes. Another sector where both hydrogen and oxygen play crucial roles belongs to the coal-to-liquids sector, experiencing rapid growth in China [64]. Through the integration of by combining renewable electricity with coal technology, oxygen can be utilized in the gasification process, and hydrogen can be introduced into the syngas. This integration minimizes the water-gas shift process, consequently reducing CO_2 production. Thus, the synergy of electrolysis with coal technology presents significant potential to mitigate the adverse environmental effects linked to coal utilization [64].

Water electrolysis stands out as a promising technology for H_2 production, given its high efficiency and minimal environmental impact. Despite the demonstration of the water electrolysis concept two centuries ago, widespread implementation for renewable hydrogen production faces challenges due to the limited availability of electrocatalysts that are both active and stable and can adapt to fluctuating conditions [75]. Current research predominantly focuses on the development of electrocatalysts that are either free of or contain lower quantities of precious metals. The European Union's energy roadmap envisions the establishment of an integrated network of electrolyzers working in tandem with renewable energy sources. This vision requires consistent performance during intermittent operation on a daily cycle and an extended life span [76]. Electrolysis technologies are categorized based on the electrolyte employed, including alkaline liquids, polymer membranes, and solid oxide ceramics. Low-temperature electrolysis offers advantages such as a straightforward cell design, rapid startup, and safe operation. In contrast, high-temperature SOE can achieve higher efficiencies using cost-effective materials. Additionally, thermodynamics enables the simultaneous co-electrolysis of CO_2 and H_2O, resulting in the direct production of syngas (H_2 + CO), a process with significant potential for reducing CO_2 emissions.

This review concentrates on recent progress in the creation of cost-effective electrocatalysts that are both active and stable across the three water electrolysis technologies (alkaline, PEM, and SOE), specifically within the realm of heterogeneous electrocatalysis. It is important to note that both structural and functional features are shared between homogeneous and heterogeneous catalytic systems for water splitting, although this article limits its discussion to concepts in heterogeneous electrocatalysis.

11.4.1 Polymer Electrolyte Membrane Water Electrolysis (PEMWE)

The importance of the polymer electrolyte membrane water electrolysis in hydrogen production from water electrolysis is emphasized. The increased efficiency of PEMWE depends on advances in various scientific disciplines. The relevance of the chemistry of new catalytic compounds and the importance of materials science in designing more efficient electrolyzers using modern materials, such as nanomaterials, is mentioned. Independent case studies are recommended to better understand the optimal operating conditions of PEMWE in various situations. Additionally, the potential of PEMWE in renewable energy storage is highlighted, suggesting the investigation of hybrid systems for continuous hydrogen production. The International Journal of Hydrogen Energy is recommended for valuable insights into current advances and future perspectives on electrolyzer technology and hydrogen production, as seen in articles like "Innovation trends in electrolyzers for hydrogen production" [13, 77, 78].

11.4.2 Proton Exchange Membrane Electrolyzer

The utilization of solid polymer membranes featuring proton conductivity in electrolysis below 100°C offers advantages, allowing for designs of cells that are compact, robust, and stackable with minimal safety concerns. However, the acidic environment in PEM electrolyzers poses challenges to redox reaction kinetics, necessitating the use of costly noble metal catalysts, which in turn hinders the commercialization of this technology [12]. In PEM electrolysis, the primary challenges are associated with the oxygen evolution reaction (OER), as the hydrogen evolution reaction (HER) occurs much more rapidly. While RuO_2 and IrO_2 exhibit high activity for the OER, RuO_2 corrodes/dissolves during oxygen evolution in acidic environments, and IrO_2 is more stable but less abundant and, therefore, expensive [48]. Efforts to develop efficient and cost-effective electrocatalysts for the OER are progressing in three research directions. The first direction employs relatively less expensive Ru-based materials, incorporating other elements like Ir or Sn to enhance chemical stability. The emphasis in the second and third directions is on reducing the loading of expensive catalysts (Ir).

The second direction explores high surface area supports (such as TiO_2, TiC, and TaC) to introduce the catalytic phase, while the third approach seeks to decrease the use of noble metal catalysts through alternative preparation methods (e.g., spray deposition) and new catalyst structures (e.g., core-shell structures) [8].

Concerning the cathode, where the HER occurs [7], the latest-generation electrocatalyst comprises Pt nanoparticles highly dispersed on carbon supports. Despite Pt's known high cost, low Pt loads are typically required for HER in PEM electrolysis, making this catalyst suitable even for large-scale applications. Ongoing efforts aim to develop alternative inorganic materials as HER electrocatalysts, such as sulfides, phosphides, carbides, and nitrides. Among them, MoS_2 shows promise, although its performance currently lags behind that of state-of-the-art Pt-based electrocatalysts due to low stability and electron conductivity. Introducing MoS_2 into new nanostructures is under investigation for potential advancements in this field [54].

11.4.3 Alkaline Electrolysis

Water electrolysis in liquid alkaline media occurs at temperatures below 100°C and represents the most straightforward and developed electrolysis technology among the three primary methods. Despite being the least efficient, AEL has already found practical applications in various industries [37]. In alkaline electrolyzers, cost-effective electrocatalysts, primarily based on nickel alloys, are utilized in both the anodic and cathodic compartments. Current research endeavors concentrate on two key aspects: (i) enhancing durability and reliability, as both cathodic and anodic electrocatalysts undergo deactivation over prolonged operation; and (ii) improving electrocatalytic activity, particularly for the OER, whose kinetics are less favorable in alkaline environments [13].

To address these challenges, research is pursuing two primary directions. The first seeks to develop innovative materials that are both efficient and stable, achieved through approaches such as doping Ni-based electrocatalysts with other elements or exploring alternative materials, with perovskites demonstrating promise in the latter case [79]. Theoretical studies aim to identify relevant descriptors for the activity of hydrogen and OERs, considering specific characteristics like reaction intermediate binding energies and the occupation of 3D orbitals with, e.g., symmetry, to maximize electrocatalytic activity. Exploring the magnetic properties of the catalyst in facilitating O_2 formation during the OER is a novel aspect that warrants attention. Fundamental insights derived from descriptors must complement kinetic studies to comprehensively examine the mechanistic pathway under realistic conditions, facilitating the systematic creation of electrocatalysts with exceptional efficiency, specially tailored for the OER [80]. The second direction involves recent strides in nanotechnology. The evolution of various 2D or 3D nanostructures composed of nickel- and cobalt-based materials has led to the creation of electrocatalysts with characteristics that have been optimized, including surface area, porosity, and conductivity, resulting in optimal efficiency and prolonged stability. Progress in material nanostructuring is anticipated to significantly enhance electrocatalyst performance, addressing aspects such as stability under chemical conditions, surface area, utilization of the catalyst, and bubble management [81].

11.4.4 Solid Oxide Electrolysis

High-temperature SOE can provide high efficiency and lower overall costs compared to conventional low-temperature electrolysis due to favorable thermodynamics and kinetics. Moreover, SOE operating temperatures open the possibility of simultaneous CO_2 and H_2O electrolysis for syngas production. However, in viable systems, heat must be generated renewably (PV or nuclear power plant) or provided by waste heat generated from exothermic processes [82]. SOE technology hinges on the utilization of solid oxide ceramics as electrolytes. Traditionally, these oxides demonstrate the capacity to conduct oxygen ions (O_2−). However, recent advancements have led to the development of oxides with proton conductivity (H+). Although the kinetics are advantageous within the operating temperature range of SOE, challenges arise in the selection of appropriate materials for electrode catalysts. These challenges stem from various degradation and stability issues, constituting the primary reason why this technology remains in the research and development (R&D) stage [83].

Typically, composite materials are employed, wherein the active electrocatalytic phase is blended in conjunction with the ceramic phase of the electrolyte. This approach not only maximizes the length of triple-phase boundaries, thereby enhancing electrocatalytic activity but also enhances enhanced mechanical stability by promoting improved adhesion between the electrode and the electrolyte. Perovskite oxides are commonly utilized as oxygen electrode electrocatalysts, leveraging their advantageous characteristics, including high catalytic activity, excellent thermal compatibility with ceramic oxide electrocatalysts, and high electronic conductivity. Despite the identification of high-performance materials, ongoing research seeks alternative perovskites with mixed ionic-electronic conductivity to further extend the reaction zone and improve chemical compatibility with electrolyte oxides and interconnection materials [84].

In the realm of the hydrogen electrode, nickel-based compounds stand out as the predominant materials used in SOE. The utilization of proton-conducting SOEs presents a promising avenue, as hydrogen is produced in a separate chamber, mitigating nickel reoxidation issues associated with water presence. Proton-conducting SOEs have already demonstrated sufficient performance, requiring optimization primarily related to microstructure for further improvements in catalytic properties. Conversely, in O_2^--conducting SOE, nickel-based compounds experience deactivation not only following intermittent operation in steam electrolysis but also throughout H_2O/CO_2 co-electrolysis. This necessitates the exploration and development of alternative materials, an area where perovskite materials are anticipated to play a significant role in the future [51].

11.5 Current Challenges and Obstacles

This section delves into the challenges and obstacles currently restraining the full exploitation of the potential of electrolyzer technology in hydrogen production. Technical, economic, and environmental issues are explored, demanding attention and innovative solutions. The H_2 economy faces various barriers hindering its successful development and commercialization, especially in large-scale production through different processes, underscoring the need to address these challenges from a techno-economic perspective [85].

Despite significant advances in electrolyzer technology, several challenges persist, requiring attention for widespread adoption. One key challenge is optimizing energy efficiency. As the goal is to enhance the conversion of electricity into hydrogen, minimizing losses and maximizing the efficiency of electrochemical processes are essential. Furthermore, the durability and life span of electrolyzers are critical focal points. Factors such as corrosion, material degradation, and impurity accumulation can significantly impact performance over time. Advanced design strategies and materials are necessary to address these challenges. The effective integration of electrolyzers into larger energy systems also presents challenges, including managing fluctuations in the availability of RES and adapting to the variable demands of the electrical system [86].

11.5.1 Consideration of Technical, Economic, and Environmental Aspects

The viability and success of electrolyzer technology depend on a comprehensive evaluation across multiple dimensions. From a technical perspective, addressing the interoperability and compatibility of electrolyzers with other components of the energy system, such as storage systems and conversion devices, is crucial [87]. Economically, cost reduction is essential to enhance the competitiveness of hydrogen produced through electrolysis.

This involves optimizing manufacturing processes, using affordable materials, and maximizing production efficiency. Finally, from an environmental perspective, it is fundamental to assess and mitigate the environmental impact of electrolyzers throughout their life cycle [88]. This includes sustainable material management, emission minimization, and considering the carbon footprint associated with hydrogen production. Effective understanding and addressing of these challenges and considerations collectively will significantly contribute to the evolution and successful adoption of electrolyzer technology in global hydrogen production [89, 90].

11.6 Applications and Industrial Potential

In this section, we address the broad spectrum of practical applications of electrolyzers in various industrial sectors, as well as the potential impact this technology can have on hydrogen production and utilization. Electrolyzers have the potential to revolutionize multiple industrial sectors by providing a clean and sustainable source of hydrogen.

One of the most promising sectors is the chemical industry, where hydrogen is a key raw material in the production of ammonia, methanol, and various chemicals [91]. Moreover, electrolyzers can play a crucial role in the mobility sector. Hydrogen production from renewable sources has the potential to drive the adoption of hydrogen vehicles, offering an alternative to fossil fuel–powered vehicles [92]. The energy sector also benefits greatly from electrolyzers, facilitating large-scale energy storage and the integration of intermittent sources such as solar and wind into the electrical grid [93].

Table 11.2 Hydrogen demand by end-use sector.

H_2 use	Description	Ref.
Refining	Application of hydrogen in the refining process of oil products.	[55]
Ammonia	Utilization of hydrogen in the production of ammonia, encompassing its role as a chemical feedstock or fuel.	[96]
Methanol	Application of hydrogen in the production of methanol, serving as a chemical feedstock or fuel.	[97]
Iron and steel	Application of hydrogen in steelmaking processes, including iron that has undergone direct reduction, introduction into blast furnaces, hot finishing, and other applications involving high temperatures. procedures.	[55]
Other Ind	Application of hydrogen in high-temperature heat processes, except for the industrial processes of oil refining, ammonia production, methanol production, and steelmaking.	[98]
Mobility	Application of hydrogen in various vehicles, including those for road, off-road, rail, maritime, or aviation purposes.	[99]
Power	The integration of hydrogen as a clean energy source into the electricity grid through the utilization of various power generation technologies, including gas turbines, reciprocating engines, and fuel cells.	[100]
Grid inj.	Introducing hydrogen into the natural gas distribution network	[101]
CHP	Utilizing hydrogen for heat and power generation through combined heat and power systems, such as in fuel cells or turbines.	[102]
Domestic heat	Utilizing hydrogen directly in buildings for heating water and spaces.	[103]
Biofuels	Incorporating hydrogen in the production of biofuels.	[104]
Synfuels	Manufacturing synthetic liquid fuels, excluding methanol.	[51]
CH_4 grid inj.	Introducing synthetic methane into the natural gas distribution network.	[51]
CH_4 mobility	Utilizing synthetic methane as a fuel for vehicles.	[105]

11.6.1 Assessment of Potential Impact on Hydrogen Production and Use

The potential impact of electrolyzers on hydrogen production and use is considerable and spans multiple dimensions. In terms of sustainability, the widespread adoption of electrolyzers powered by RES can significantly reduce greenhouse gas emissions associated with hydrogen production.

Economically, hydrogen production through electrolysis offers opportunities for job creation and the development of a new industry based on clean technologies [94]. Globally, the adoption of electrolyzers has the potential to transform energy and economic landscapes, driving the transition toward a hydrogen economy and accelerating the decarbonization of numerous industrial sectors [95]. Table 11.2 presents hydrogen demand by sector for end use. Electrolyzers represent a fundamental component in the transition toward a cleaner and more sustainable energy future, with a potentially transformative impact on a wide range of industrial sectors and on the global production and use of hydrogen.

11.7 Case Studies

This section showcases case studies and concrete examples illustrating the successful implementation of electrolyzer technologies, and analyzing the results obtained. One of the most notable examples of a successful electrolyzer implementation is found in the renewable energy sector in Denmark. The "Hy-Synergy" project has demonstrated the technical and economic viability of large-scale hydrogen production from wind and solar sources, with the produced hydrogen being used in the chemical industry and as fuel for transportation [45]. Another relevant case is the "Refhyne" project in Germany, which has established one of the world's largest alkaline water electrolysis plants. This plant, powered by renewable energy, aims to produce hydrogen for use in the chemical industry and refineries [51]. In Spain, a $150 million investment initiative was launched in 2021 to operate a 20-MW electrolyzer using PV. The generated hydrogen will function as a raw material for the production of ammonia [106]. This highlights the importance of near-term renewable hydrogen generation, supported by substantial investments from major oil-producing nations. Saudi Arabia, a prominent global oil producer, unveiled a $5 billion project to exclusively produce hydrogen from renewable sources. By 2030, it is anticipated that the cost of hydrogen from this venture will be lower than conventional fossil fuel–based production methods [107], which currently contribute to an annual emission of 830 million tons of CO_2 [108].

In Europe, regional trains powered by hydrogen have initiated operations, with expectations of capturing a more significant market share in the future, potentially displacing 30% of the current diesel fleets [109]. Moreover, rather than introducing hydrogen into the gas distribution network, incorporating hydrogen into onboard systems presents an opportunity for hydrogen producers to resell it at a premium. The National Renewable Energy Laboratory (NREL) has disclosed that the potential cost of hydrogen ranges from approximately 3 to 10 \$/kg, whereas the prevalent hydrogen fuel price is 13.99 \$/kg [110]. Additionally, the development of onboard hydrogen applications can contribute to diminishing the necessity for expanding infrastructure capacity for vehicle charging. Given that hydrogen refueling infrastructure can be deemed comparable to existing gasoline and diesel stations, there might be less demand for refueling stations for alternative battery-equipped electric vehicles. The IEA has compiled a database of large-scale electrolysis projects [108]. Figure 11.8 shows the worldwide distribution of infrastructure for hydrogen projects, while Figure 11.9 displays the global distribution of projects in construction that will exclusively generate hydrogen through renewable sources.

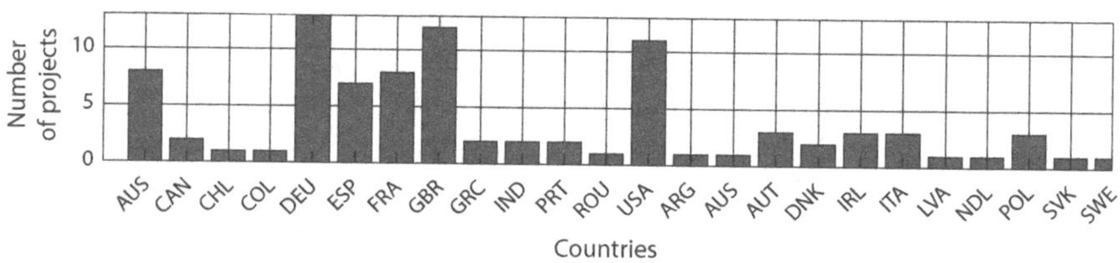

Figure 11.8 Worldwide distribution of infrastructure for hydrogen projects.

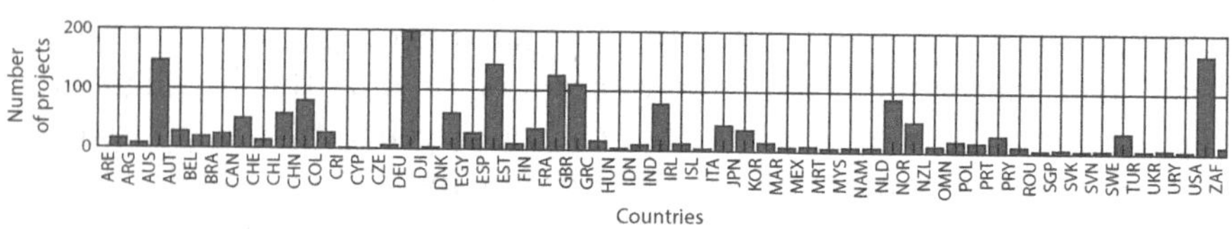

Figure 11.9 Global dispersion of electrolysis projects aimed at hydrogen production.

11.8 Economic and Sustainability Considerations

Europe has taken a leading role in advancing the development of hydrogen-powered energy systems, showcasing substantial market potential for further lucrative initiatives. In the European Union, over 300 companies actively participate in the fuel cell and hydrogen sectors, and an even greater number operate within related supply chains [51, 111]. Various techno-economic assessments of hydrogen facility operations have affirmed the economic feasibility of energy systems based on hydrogen, accounting for both direct and indirect costs. The commercialization of hydrogen-based fuels carries a twofold impact—integrating renewable energy into the existing fuel infrastructure while mitigating greenhouse gas emissions. To foster a more conducive environment, several measures could be implemented:

- Electricity Price: The price of electricity is a major concern. Taxes represent a significant portion of the electricity bill in European hydrogen energy applications, making it challenging to reduce costs. Tax exemptions and rates ought to be deliberated, incorporating specific regulations, such as discouraging power usage during peak periods.
- Gas Pipeline Systems: Gas pipeline systems should be implemented for hydrogen transport soon. Studies have demonstrated the feasibility of distributing and transmitting hydrogen through the current infrastructure of the natural gas distribution network. However, hydrogen injection is currently largely limited by national permit regulations. Authorities should consider legal pathways to support the development of gas pipeline networks, thereby providing more opportunities for energy-to-gas applications.
- Limited Renewable Energy Production: European countries have experienced constrained electricity production from renewable sources. The advancement of Renewable Energy Sources (RES) surpasses the capacity to integrate them into the local grid, resulting in congestion and restricting the utilization of renewable energy. A viable solution involves implementing regulations that prioritize the connection and dispatch of renewable energy sources. In instances of limitation, policymakers should contemplate providing appropriate compensation.

11.8.1 Economic Aspects Related to the Implementation of Electrolyzer Technologies

The current capital costs related to alkaline water electrolyzers, compressed tanks, and fuel cells have been compiled in [112], and recent projects and research have undertaken various profitability analyses [113–115]. The capital expenditure for the existing hydrogen production system employing alkaline electrolyzers spans from 1000 to 1500 €/kW, encompassing installation expenses. In contrast, the figures for PEMs are twice as high, ranging from 2000 to 3000 €/kW [112]. Despite the well-established nature of alkaline water electrolysis, production volumes remain relatively modest. This is attributed to electrolyzer suppliers predominantly manufacturing low-volume electrolyzers for a niche market, concurrently elevating the costs of auxiliary equipment. Consequently, the potential reduction of the cost for alkaline electrolyzers hinges on more streamlined production processes, while achieving further cost reduction for PEM electrolyzers requires advancements in technological development [116].

Various configurations of stationary fuel cell systems are implemented to cater to diverse needs, ranging from residential buildings to industrial applications. Micro-combined heat and power (CHP) fuel cell systems designed for households and small buildings, boasting an installed capacity of 0.3 to 5 kW, currently incur costs of around 10,000 $/kW. Medium-sized installations for larger buildings ranging from 5 to 400 kW carry costs ranging from 4500 to 7500 $/kW, while large-scale installations with capacities of 0.4 to 30 MW command costs from 2000 to 3000 $/kW, specifically for certain industrial applications [117]. Capital expenses possess the possibility for a substantial reduction in the near future due to becoming more economical through the advancement of installation technologies and the realization of economies of scale. Anticipated capital expenditure (CAPEX) values for 2030 are projected to reach 3500 $/kW, 1500 to 4000 $/kW, and 1200 to 1750 $/kW for micro-CHP, medium-sized, and applications on a large scale, respectively [117, 118].

The cost of hydrogen production through steam methane reforming (SMR) is primarily influenced by gas prices, currently ranging between 1.4 and 1.8 $/kg with CO_2 capture [42]. However, driven by fossil fuel depletion and decreasing costs of renewable electricity, electrolytic hydrogen is becoming competitive and is positioned for continued deployment in the near future. The current cost of producing hydrogen with alkaline electrolyzers ranges from 3.2 to 5.2 $/kg, while with PEMs, the cost is 4.1 to 6 $/kg [75]. As for hydrogen compression, storage, and dispensing costs, they are 1.8 to 2.6 $/kg in the pipeline scenario and 2.1 to 3 $/kg in the distributed scenario. This cost is expected to decrease to the ultimate goal of 1.6 \$/kg [112].

Although hydrogen production *via* the process of water electrolysis presents a hopeful and viable resolution, electricity consumption is a critical consideration. Assuming a 60% efficiency in the production of hydrogen through water electrolysis, meeting the current specific demand for hydrogen necessitates an electricity consumption of 3600 TWh, surpassing the total annual electricity generation in Europe [119]. A promising avenue to mitigate electricity costs is the generation of electricity from renewable or nuclear sources. As renewable energy sources, such as solar and wind, with declining costs, gain traction, renewable electricity becomes more cost-effective. While hydrogen produced with renewable energy may incur high transmission and distribution costs due to remote locations, the ultimate benefits are substantial. An evaluation of costs and benefits associated with an integrated wind-hydrogen system on Corvo Island demonstrated that local renewable energy could meet 80% of the island's electricity demand [120]. Initiatives to install electrolyzers in locations having exceptional solar and wind resources worldwide have been initiated, with large-scale industrial deployments on the horizon. With the gradual maturation of technologies, capital costs for both electrolyzer and fuel cell systems are expected to witness significant reductions by 2030, particularly in fuel cell costs. Technological breakthroughs are imperative to expand the active area of the fuel cell, thereby reducing the required number of cells to produce a specific amount of hydrogen and subsequently lowering costs.

The implementation of electrolyzer technologies involves several crucial economic considerations. Firstly, the initial acquisition and installation costs of electrolyzers, as well as the associated infrastructure and electricity supply costs, must be assessed. Additionally, it is essential to consider operating costs, including energy consumption, maintenance, and plant management. Research, such as that by Parissis *et al.* [120], has highlighted the need for comprehensive life cycle analyses to evaluate the total costs and long-term profitability of electrolyzer systems. Another economic aspect is the possibility of revenue from the sale of produced hydrogen. Depending on specific markets and applications, hydrogen can have significant value as a clean energy source and as a raw material in various industries. Therefore, a detailed analysis of hydrogen prices and demand in the regional and sectoral context is crucial [120]. In the evaluation of hydrogen production costs through electrolysis, the consideration extends beyond production expenses to encompass storage and transportation costs as well. A study conducted by Farràs *et al.* [121] presented an economic analysis of advanced water electrolysis (AWE) technology at various European locations, including Germany, Austria, and Spain. The reported levelized cost of hydrogen stands at approximately 3.64 €/kg in Germany, while the Austrian and Spanish sites exhibit slightly higher costs (15% to 18%), attributed to elevated electricity expenses. Another research endeavor by Farràs *et al.* [121] estimated the production cost of hydrogen for AWE at approximately 6 €/kg and for PEM-based water electrolyzers at 7 €/kg. This study explored the economic feasibility of producing high-purity H_2 gas using a concentrated photovoltaic (CPV) electrolyzer system. The findings indicate that hydrogen production increases with elevated electrolysis temperature and solar concentration, resulting in a decrease in the hydrogen production rate's cost with increased concentration.

Despite advancements in non–platinum group metal (PGM)–based electrocatalysts and real-world applications necessitate pretreatment (such as alkaline treatment or reverse osmosis), incurring costs comparable to those associated with currently employed PGM-based freshwater electrolyzers. Presently, commercial PEM systems utilize PGM-based electrodes; however, emerging anion exchange membrane water electrolyzers (AEMWEs) present themselves as promising contenders for future applications due to their compatibility with low-cost electrocatalysts. In the realm of hydrogen production, two economic approaches are suggested [121]: (a) A single-step approach utilizing treating seawater directly without any pre-treatment, focusing research on designing selective, stable, and cost-effective materials. (b) A two-step approach involving desalination plants, predominantly reverse osmosis. Through this approach, research endeavors and expenditures are directed toward reducing desalination expenses. Recent insights from [121] indicate that, from an energetic and economic standpoint, the two-step strategy is currently more practical. However, given the introduction of emerging technologies aimed at developing stable and low-cost electrocatalysts and innovative cell configurations, routine energy and economic analyses are imperative. Strategies should be subject to regular reviews to align with evolving technological landscapes.

11.8.2 Financial Viability Analysis

Financial viability and long-term sustainability are critical aspects of the success and scalability of electrolyzer systems. It is important to consider not only short-term economic costs and benefits but also the long-term impact in terms of profitability and environmental sustainability. The significance of incorporating sensitivity analyses to assess the robustness of electrolyzer projects in the face of changing energy prices and investment costs is highlighted [122]. Long-term sustainability involves evaluating the environmental impact of H_2 production, including greenhouse gas emissions and natural resource consumption. Additionally, it is crucial to consider the potential for job creation and local economic development associated with electrolyzer project implementation [49]. Rigorous evaluation of economic and sustainability aspects plays a fundamental role in making informed decisions about the implementation of electrolyzer technologies, contributing to the construction of a more sustainable and economically viable energy future.

Both thermochemical conversion and water electrolysis are key technologies in hydrogen production. However, significant technical and economic challenges exist in integrating them into the energy system and operating them commercially on a large scale. In economic terms, the cost of electricity represents a substantial portion, ranging from 40% to 57% of the levelized cost of hydrogen. To overcome these challenges, strategies such as selecting optimal geographical locations and implementing intelligent operations to reduce costs can be considered. Additionally, minimizing CO_2 emissions can be achieved by considering the carbon footprint of the electricity source used. The commercialization of H_2 production faces challenges like reduced production efficiency compared to alternative methods and increased production expenses [123]. Fermentative H_2 production is still in its early stages and undergoing gradual technological changes [52].

Enhancing yield and hydrogen production rates necessitate diverse strategies at both the fermentative and genetic levels. The utilization of genetic engineering plays a pivotal role in crafting strains geared toward efficient hydrogen production [124]. Table 11.3 delineates the production costs of hydrogen through electrolysis, gleaned from selected literature, juxtaposed with costs related to SMR and other methodologies [125]. In certain instances, hydrogen supply prices, encompassing costs associated with storage and delivery, are also delineated. For improved comparability, all values are standardized to hydrogen mass units (cost/kg). Accurately estimating hydrogen costs poses a formidable challenge, given the multitude of variables with uncertain values and the requisite assumptions underlying the calculations.

Hydrogen, positioned as a clean energy carrier, emerges as a pivotal consideration for prospective advancements in global energy systems.

Table 11.3 Hydrogen cost comparison (H_2/kg).

Production cost	Supply cost	Notes	Reference
	4.5 $	Electrolyzer efficiency 80% Electricity cost 0.07 $/kWh	[126]
	10.9 $ (from wind)	Grid-mix refers to the Korean electricity mix	[127]
	9.1 $ (from grid - mix)		
	9.4 $ (SMR)		
	4.7 $ (from wind)	No taxes on hydrogen	
	3.3 $ (from grid - mix)		
	2.9 $ (SMR)		
	6.7 €	On-site production	[125]
	5.0 €		
	4.8–8.3 $ (untaxed)		[128]
	4.4–7.2 $ (bioethanol reforming)		
	3.3–4.3 $ (SMR)		
4–11 € (from wind grid-connected)		Alkaline electrolysis	[129]
9–20 € (from wind isolated)		The lowest cost (4.0 e/kg) is for a 2000 barrels per day BtL station grid connected	
10 €		The higher costs refer to electrolyzer utilization	[130]
2–6 €			
1.1 $ (2025?)	The cost of compression, storage, and dispensing is $2.47 /kg.	Decentralized electrolysis production and the industrial electricity tariff in Iceland.	[131]

(*Continued*)

Table 11.3 Hydrogen cost comparison (H_2/kg). (*Continued*)

Production cost	Supply cost	Notes	Reference
8.0 €		Costs refer to EU-28	[132]
2.9 € (SMR)			
3.6 € (SMR + CCS)			
4.0 € (SMR on-site)			
6.0 € (SMR with biogas)			
4.4 € (2055)			
3.3 € (SMR, 2055)			
4.1 € (SMR + CCS, 2055)			
3.0 € (SMR on-site, 2055)			
4.0 € (SMR with biogas, 2055)			
	10 €		[103]
	6-8 € (future)		
	5-10 € (untaxed)		[133]
2–3 €	9.5 € (untaxed)	The lower production cost is obtained by combining FC and electrolyzer and participating in the SCR market	[134]
1.1 €			
	4.9 - 11.3 € (2020)	On-site electrolysis from produced RES	[135]
	4.6 - 11.6 € (2030)		
	5.1 - 13.4 € (2050)		

Presently, the primary imperatives revolve around cost reduction and the exploration of extensive applications. The following recommendations are proffered within the social and political domains:

- Advocating for large-scale implementation is paramount. Given the readiness of technologies, expansive applications are deemed a cost-effective avenue, entailing minimal infrastructure investments compared to high operational capacity. Particularly in regions abundant in renewable energy, applications ranging from a hundred megawatts to gigawatts are anticipated, contingent on the support of pertinent policies.

- The forthcoming decades should witness the development of new hydrogen markets. Acknowledged as a promising solution for the energy transition, hydrogen ought to be factored into all sectors grappling with resource depletion and pollution, encompassing industry, freight transportation, aviation, shipping, and chemical domains.
- Governments assume a pivotal role in expediting the hydrogen economy. The absence of specific regulations poses impediments to industrial operations, introducing both temporal and cost-related impacts. Therefore, governments must champion coordination, regulation, and standardization to facilitate the seamless integration of hydrogen-powered energy systems into the market.

11.9 Conclusions and Recommendations

This work has thoroughly addressed the current state and future prospects of electrolyzer technology for hydrogen production. From the initial attempts to use electricity to decompose water to today's sophisticated electrolyzer technology, a historical timeline has been traced, highlighting the evolution and significant impact of this technology on the global energy landscape. The importance of electrolyzers in the context of transitioning toward a more sustainable and clean energy economy has been analyzed in detail.

One key finding from this chapter is the critical relevance of electrolyzers in the context of renewable energies and hydrogen production from sustainable sources. It has been emphasized how these devices play a crucial role in converting electricity into hydrogen through highly efficient electrochemical processes. Additionally, the need to address current challenges, such as water impurity sensitivity and system optimization for increased efficiency and profitability, has been underscored.

To advance in electrolyzer technology, several areas of future research are proposed. Firstly, the exploration of new materials and design techniques that can enhance the efficiency and durability of electrolyzers is suggested. Research into advanced catalysts and selective membranes could lead to significant breakthroughs in this regard [62].

Furthermore, conducting detailed studies on the integration of electrolyzers into larger energy systems is recommended, including considerations for electric load management and interaction with other energy storage technologies. The industrial-scale implementation of electrolyzer systems also requires a thorough analysis of economic and financial aspects, as well as an assessment of social and environmental impacts throughout the entire system's life cycle [136].

Ultimately, researchers and professionals in the field are encouraged to collaborate on interdisciplinary initiatives addressing the complex and promising challenges posed by electrolyzer technology. The synergy between academic research, industry, and policy formulation will play a crucial role in advancing toward a more sustainable and hydrogen-centric energy future.

Acknowledgments

The author Edisson Villa Ávila expresses his sincere gratitude for the opportunity to partially present the findings of his research, conducted as part of his doctoral studies in the Ph.D. program in Advances in Engineering of Sustainable Materials and Energies at the University of Jaen, Spain.

The author Paul Arévalo thanks the Call for Grants for the Requalification of the Spanish University System for 2021–2023 and Margarita Salas Grants for the training of young doctors awarded by the Ministry of Universities and financed by the European Union – NextGenerationEU.

Finally, the authors thank Universidad de Cuenca, Ecuador, for easing access to the facilities of the Micro-Grid Laboratory of the Centro Científico Tecnológico y de Investigación Balzay (CCTI-B), for allowing the use of its equipment, and for authorizing members of its staff the provision of technical support necessary to carry out the experiments described in this article.

References

1. Candelaresi, D. and Spazzafumo, G., Production of Substitute Natural Gas Integrated with Allam Cycle for Power Generation. *Energies*, 16, 5, 2162, 2023, doi: 10.3390/en16052162.
2. Ibagon, N., Munoz, P., Correa, G., Techno economic analysis tool for the sizing and optimization of an off-grid hydrogen hub. *J. Energy Storage*, 73, 659–670, 2023, doi: 10.1016/j.est.2023.108787.
3. Morales Ramos, A., Pérez Figueroa, M., Pérez Gallardo, J., De León Almaraz, S., Energías renovables y el hidrógeno: un par prometedor en la transición

energética de México. *Investig. y Cienc. la Univ. Autónoma Aguascalientes*, 25, 70, 92–101, 2017.

4. Yujing, H., Jie, C., Xueqin, T., Yafei, W., Integrated photovoltaic storage joint smoothing strategy based on simultaneous perturbation stochastic approximation algorithm. *Front. Energy Res.*, 10, 1–12, Jan. 2023, doi: 10.3389/FENRG.2022.1014027.
5. Hwang, H.T. and Varma, A., Hydrogen storage for fuel cell vehicles. *Curr. Opin. Chem. Eng.*, 5, 42–48, 2014, doi: 10.1016/j.coche.2014.04.004.
6. Lopez, V.A.M., Ziar, H., Haverkort, J.W., Zeman, M., Isabella, O., Dynamic operation of water Electrolysers: A review for applications in photovoltaic systems integration. *Renew. Sustain. Energy Rev.*, 182, 113407, 2023, doi: 10.1016/j.rser.2023.113407.
7. Kumar, S.S. and Himabindu, V., Hydrogen production by PEM water electrolysis – A review. *Mater. Sci. Energy Technol.*, 2, 3, 442–454, 2019, doi: 10.1016/j.mset.2019.03.002.
8. Abe, J.O., Popoola, A.P., II, Ajenifuja, E., Popoola, O.M., Hydrogen energy, economy and storage: Review and recommendation. *Int. J. Hydrogen Energy*, 44, 29, 15072–15086, 2019, doi: 10.1016/j.ijhydene.2019.04.068.
9. Arsad, A.Z. *et al.*, Hydrogen energy storage integrated hybrid renewable energy systems: A review analysis for future research directions. *Int. J. Hydrogen Energy*, 47, 39, 17285–17312, 2022, doi: 10.1016/j.ijhydene.2022.03.208.
10. Gielen, D., Boshell, F., Saygin, D., Bazilian, M.D., Wagner, N., Gorini, R., The role of renewable energy in the global energy transformation. *Energy Strategy Rev.*, 24, January, 38–50, 2019, doi: 10.1016/j.esr.2019.01.006.
11. Makhsoos, A., Kandidayeni, M., Pollet, B.G., Boulon, L., A perspective on increasing the efficiency of proton exchange membrane water Electrolysers-a review. *Int. J. Hydrogen Energy*, 48, 41, 15341–15370, 2023, doi: 10.1016/j.ijhydene.2023.01.048.
12. Chandrasekar, A., Flynn, D., Syron, E., Operational challenges for low and high temperature Electrolysers exploiting curtailed wind energy for hydrogen production. *Int. J. Hydrogen Energy*, 46, 57, 28900–28911, 2021, doi: 10.1016/j.ijhydene.2020.12.217.
13. Kumar, S.S. and Lim, H., An overview of water electrolysis technologies for green hydrogen production. *Energy Rep.*, 8, 13793–13813, 2022, doi: 10.1016/j.egyr.2022.10.127.
14. Thürer, M., Tomašević, I., Stevenson, M., Qu, T., Huisingh, D., A systematic review of the literature on integrating sustainability into engineering curricula. *J. Clean. Prod.*, 181, 608–617, 2018, doi: 10.1016/j.jclepro.2017.12.130.
15. Asubiaro, T.V. and Onaolapo, S., A comparative study of the coverage of African journals in Web of Science, Scopus, and CrossRef. *J. Assoc. Inf. Sci. Technol.*, March, 74, 745–758, 2023, doi: 10.1002/asi.24758.
16. Zychová, K., Šímová, T., Fejfarová, M., A bibliometric analysis of team autonomy research. *Cogent Bus. Manage.*, 10, 1, 2195024, 2023, doi: 10.1080/23311975.2023.2195024.

17. Worrall, J.L. and Cohn, E.G., Citation Data and Analysis: Limitations and Shortcomings. *J. Contemp. Crim. Justice*, 39, 3, 327–340, 2023, doi: 10.1177/10439862231170972.
18. Mokhnacheva, Y.V., Document Types Indexed in WoS and Scopus: Similarities, Differences, and Their Significance in the Analysis of Publication Activity. *Sci. Tech. Inf. Process.*, 50, 1, 40–46, 2023, doi: 10.3103/S0147688223010033.
19. VOSviewer - Visualizing scientific landscapes, https://www.vosviewer.com/ (accessed May 21, 2023).
20. S., N., Performance Studies On An Electrolyser For The Production Of Hydroge. *Indian J. Technol.*, 8, 2, 65–70, 1970.
21. Li, W. and Zhao, Y., Bibliometric analysis of global environmental assessment research in a 20-year period. *Environ. Impact Assess. Rev.*, 50, 158–166, 2015, doi: 10.1016/j.eiar.2014.09.012.
22. Ursúa, A., San Martín, I., Barrios, E.L., Sanchis, P., Stand-alone operation of an alkaline water electrolyser fed by wind and photovoltaic systems. *Int. J. Hydrogen Energy*, 38, 35, 14952–14967, 2013, doi: 10.1016/j.ijhydene.2013.09.085.
23. Clarke, D.P., Al-Abdeli, Y.M., Kothapalli, G., The impact of renewable energy intermittency on the operational characteristics of a stand-alone hydrogen generation system with on-site water production. *Int. J. Hydrogen Energy*, 38, 28, 12253–12265, 2013, doi: 10.1016/j.ijhydene.2013.07.031.
24. Arsad, A.Z. *et al.*, Hydrogen electrolyser technologies and their modelling for sustainable energy production: A comprehensive review and suggestions. *Int. J. Hydrogen Energy*, 48, 72, 27841–27871, 2023, doi: 10.1016/j.ijhydene.2023.04.014.
25. Francis, R. and Chidambaram, I.A., Optimized PI+ load-frequency controller using BWNN approach for an interconnected reheat power system with RFB and hydrogen electrolyser units. *Int. J. Electr. Power Energy Syst.*, 67, 381–392, 2015, doi: 10.1016/j.ijepes.2014.12.012.
26. Christensen, E., Berg, R.W., Kruger, R., Bjerrum, N.J., Methanol Synthesis from CO_2 and H_2O Using a Solid Phosphate Electrolyser at 240 °C and 28 Bar. *J. Electrochem. Soc.*, 170, 1, 014502, 2023, doi: 10.1149/1945-7111/acb296.
27. Ramirez, M.A.P., Fogel, S., Reinecke, S.F., Hampel, U., Techno-Economic Assessment of PEM Electrolysis for O_2 Supply in Activated Sludge Systems-A Simulation Study Based on the BSM2 Wastewater Treatment Plant. *Processes*, 11, 6, 1639, 2023, doi: 10.3390/pr11061639.
28. Li, C. and Baek, J.B., The promise of hydrogen production from alkaline anion exchange membrane Electrolysers. *Nano Energy*, 87, May, 106162, 2021, doi: 10.1016/j.nanoen.2021.106162.
29. Krishnan, S. *et al.*, Present and future cost of alkaline and PEM electrolyser stacks. *Int. J. Hydrogen Energy*, 48, 83, 32313–32330, 2023, doi: 10.1016/j.ijhydene.2023.05.0310360-3199.

30. Sathre, R., Scown, C.D., Kavvada, O., Hendrickson, T.P., Energy and climate effects of second-life use of electric vehicle batteries in California through 2050. *J. Power Sources*, 288, 82–91, Mar. 2015, doi: 10.1016/j.jpowsour.2015.04.097.
31. Nejadian, M.M., Ahmadi, P., Houshfar, E., Comparative optimization study of three novel integrated hydrogen production systems with SOEC, PEM, and alkaline Electrolyser. *Fuel*, 336, 126835, 2023, doi: 10.1016/j.fuel.2022.126835.
32. Yoshida, A., Nakazawa, H., Kenmotsu, N., Amano, Y., Economic analysis of a proton exchange membrane electrolyser cell for hydrogen supply scenarios in Japan. *Energy*, 251, 123943, 2022, doi: 10.1016/j.energy.2022.123943.
33. Luo, Z., Wang, X., Wen, H., Pei, A., Hydrogen production from offshore wind power in South China. *Int. J. Hydrogen Energy*, 47, 58, 24558–24568, 2022, doi: 10.1016/j.ijhydene.2022.03.162.
34. Zhang, F., Zhao, P., Niu, M., Maddy, J., The survey of key technologies in hydrogen energy storage. *Int. J. Hydrogen Energy*, 41, 33, 14535–14552, 2016, doi: 10.1016/j.ijhydene.2016.05.293.
35. Bhandari, R., Trudewind, C.A., Zapp, P., Life cycle assessment of hydrogen production via electrolysis - A review. *J. Clean. Prod.*, 85, 151–163, 2014, doi: 10.1016/j.jclepro.2013.07.048.
36. Zheng, J., Liu, X., Xu, P., Liu, P., Zhao, Y., Yang, J., Development of high pressure gaseous hydrogen storage technologies. *Int. J. Hydrogen Energy*, 37, 1, 1048–1057, 2012, doi: 10.1016/j.ijhydene.2011.02.125.
37. Kavadias, K.A., Apostolou, D., Kaldellis, J.K., Modelling and optimisation of a hydrogen-based energy storage system in an autonomous electrical network. *Appl. Energy*, 227, August 2017, 574–586, 2018, doi: 10.1016/j.apenergy.2017.08.050.
38. Olivier, P., Bourasseau, C., Bouamama, B., Modelling, simulation and analysis of a PEM electrolysis system. *IFAC-PapersOnLine*, 49, 12, 1014–1019, 2016, doi: 10.1016/j.ifacol.2016.07.575.
39. Han, B., Steen, S.M., Mo, J., Zhang, F.Y., Electrochemical performance modeling of a proton exchange membrane Electrolyser cell for hydrogen energy. *Int. J. Hydrogen Energy*, 40, 22, 7006–7016, 2015, doi: 10.1016/j.ijhydene.2015.03.164.
40. Zhang, Y., Campana, P.E., Lundblad, A., Yan, J., Comparative study of hydrogen storage and battery storage in grid connected photovoltaic system: Storage sizing and rule-based operation. *Appl. Energy*, 201, 397–411, 2017, doi: 10.1016/j.apenergy.2017.03.123.
41. Gonzatti, F. and Farret, F.A., Mathematical and experimental basis to model energy storage systems composed of Electrolyser, metal hydrides and fuel cells. *Energy Convers. Manage.*, 132, 241–250, 2017, doi: 10.1016/j.enconman.2016.11.035.
42. Yue, M., Lambert, H., Pahon, E., Roche, R., Jemei, S., Hissel, D., Hydrogen energy systems: A critical review of technologies, applications, trends

and challenges. *Renew. Sustain. Energy Rev.*, 146, May, 111180, 2021, doi: 10.1016/j.rser.2021.111180.

43. Maggio, G., Nicita, A., Squadrito, G., How the hydrogen production from RES could change energy and fuel markets: A review of recent literature. *Int. J. Hydrogen Energy*, 44, 23, 11371–11384, 2019, doi: 10.1016/j.ijhydene.2019.03.121.
44. Egeland-Eriksen, T., Hajizadeh, A., Sartori, S., Hydrogen-based systems for integration of renewable energy in power systems: Achievements and perspectives. *Int. J. Hydrogen Energy*, 46, 63, 31963–31983, 2021, doi: 10.1016/j.ijhydene.2021.06.218.
45. Hirscher, M. *et al.*, Materials for hydrogen-based energy storage – past, recent progress and future outlook. *J. Alloys Compd.*, 827, 153548, 2020, doi: 10.1016/j.jallcom.2019.153548.
46. Nikolaidis, P. and Poullikkas, A., A comparative overview of hydrogen production processes. *Renew. Sustain. Energy Rev.*, 67, 597–611, 2017, doi: 10.1016/j.rser.2016.09.044.
47. Thema, M., Bauer, F., Sterner, M., Power-to-Gas: Electrolysis and methanation status review. *Renew. Sustain. Energy Rev.*, 112, January, 775–787, 2019, doi: 10.1016/j.rser.2019.06.030.
48. Asghari, E., Abdullah, M., II, Foroughi, F., Lamb, J.J., Pollet, B.G., Advances, opportunities, and challenges of hydrogen and oxygen production from seawater electrolysis: An electrocatalysis perspective. *Curr. Opin. Electrochem.*, 31, 100879, 2022, doi: 10.1016/j.coelec.2021.100879.
49. Kannah, R.Y. *et al.*, Techno-economic assessment of various hydrogen production methods – A review. *Bioresour. Technol.*, 319, September 2020, 124175, 2021, doi: 10.1016/j.biortech.2020.124175.
50. Martins, M.E., Peretti, H.A., Spinadel, E., Zinola, C.F., Visintiny, A., Energy Conversion And Storage Based On Hydrogen. State of Art And Strategized Proposals In Mercosur, Conversión Y Almacenamiento De Energía Basados En Hidrógeno. Estado Del Arte Y Propuesta De Estrategias En El Mercosur. *Rev. Tec. la Fac. Ing. Univ. del Zulia*, 31, 2, 1–18, 2008.
51. Wulf, C., Zapp, P., Schreiber, A., Review of Power-to-X Demonstration Projects in Europe. *Front. Energy Res.*, 8, September, 1–12, 2020, doi: 10.3389/fenrg.2020.00191.
52. Becker, H., Murawski, J., Shinde, D.V., Stephens, I.E.L., Hinds, G., Smith, G., Impact of impurities on water electrolysis: a review. *Sustain. Energy Fuels*, 7, 7, 1565–1603, 2023, doi: 10.1039/d2se01517j.
53. Lee, C.Y., Chen, C.H., Bin Jung, G., Zheng, Y.X., Liu, Y.C., Persistent effect test and internal microscopic monitoring for pem water Electrolyser. *Micromachines*, 12, 5, 1–14, 2021, doi: 10.3390/mi12050494.
54. Kang, Z. *et al.*, Novel thin/tunable gas diffusion electrodes with ultra-low catalyst loading for hydrogen evolution reactions in proton exchange membrane Electrolyser cells. *Nano Energy*, 47, March, 434–441, 2018, doi: 10.1016/j.nanoen.2018.03.015.

55. Griffiths, S., Sovacool, B.K., Kim, J., Bazilian, M., Uratani, J.M., Industrial decarbonization via hydrogen: A critical and systematic review of developments, socio-technical systems and policy options. *Energy Res. Soc. Sci.*, 80, July, 102208, 2021, doi: 10.1016/j.erss.2021.102208.
56. Kim, J., Qi, M., Kim, M., Lee, J., Lee, I., Moon, I., Biogas reforming integrated with PEM electrolysis via oxygen storage process for green hydrogen production: From design to robust optimization. *Energy Convers. Manage.*, 251, November 2021, 115021, 2022, doi: 10.1016/j.enconman.2021.115021.
57. Feng, X., Ma, T., Yamaguchi, Y., Peng, J., Dai, Y., Ji, D., Potential of residential building integrated photovoltaic systems in different regions of China. *Energy Sustain. Dev.*, 72, August 2022, 19–32, 2023, doi: 10.1016/j.esd.2022.11.006.
58. Santoro, C. *et al.*, What is Next in Anion-Exchange Membrane Water Electrolysers? Bottlenecks, Benefits, and Future. *ChemSusChem*, 15, 8, 1–17, 2022, doi: 10.1002/cssc.202200027.
59. Miller, H.A. *et al.*, Green hydrogen from anion exchange membrane water electrolysis: A review of recent developments in critical materials and operating conditions. *Sustain. Energy Fuels*, 4, 5, 2114–2133, 2020, doi: 10.1039/c9se01240k.
60. Electrolyser Technology AEM and Use of Hydrogen - Hydrolite, https://www.hydrolite-h2.com/aem-Electrolyser-technology/ (accessed Nov. 01, 2023).
61. Enapter AEM Electrolysers: Your Green Hydrogen Production Starts Now - Green Hydrogen - Enapter, https://www.enapter.com/ (accessed Nov. 01, 2023).
62. Sharshir, S.W., Joseph, A., Elsayad, M.M., Tareemi, A.A., Kandeal, A.W., Elkadeem, M.R., A review of recent advances in alkaline Electrolyser for green hydrogen production: Performance improvement and applications. *Int. J. Hydrogen Energy*, 49, 458–488, 2023, doi: 10.1016/j.ijhydene.2023.08.107.
63. Ursúa, A., Barrios, E.L., Pascual, J., San Martín, I., Sanchis, P., Integration of commercial alkaline water electrolysers with renewable energies: Limitations and improvements. *Int. J. Hydrogen Energy*, 41, 30, 12852–12861, 2016, doi: 10.1016/j.ijhydene.2016.06.071.
64. Sapountzi, F.M., Gracia, J.M., Weststrate, C. J. (K. J.), Fredriksson, H.O.A., Niemantsverdriet, J. W. (H.), Electrocatalysts for the generation of hydrogen, oxygen and synthesis gas. *Prog. Energy Combust. Sci.*, 58, 1–35, 2017, doi: 10.1016/j.pecs.2016.09.001.
65. Schalenbach, M., Kasian, O., Mayrhofer, K.J.J., An alkaline water Electrolyser with nickel electrodes enables efficient high current density operation. *Int. J. Hydrogen Energy*, 43, 27, 11932–11938, 2018, doi: 10.1016/j.ijhydene.2018.04.219.
66. Gatto, I., Capri, A., Lo Vecchio, C., Zignani, S., Patti, A., Baglio, V., Optimal operating conditions evaluation of an anion-exchange-membrane Electrolyser based on FUMASEP? FAA3-50 membrane. *Int. J. Hydrogen Energy*, 48, 32, 11914–11921, 2023, doi: 10.1016/j.ijhydene.2022.04.176.

67. Dang, J., Li, Y., Liu, B., Hu, S., Yang, F., Ouyang, M., Design and economic analysis of high-pressure proton exchange membrane electrolysis for renewable energy storage. *Int. J. Hydrogen Energy*, 48, 28, 10377–10393, 2023, doi: 10.1016/j.ijhydene.2022.11.250.
68. Li, H. *et al.*, Highly hydrophilic polybenzimidazole/Zirconia composite separator with reduced gas crossover for alkaline water electrolysis. *J. Membr. Sci.*, 683, 121844, 2023, doi: 10.1016/j.memsci.2023.121844.
69. Ferriday, T.B., Middleton, P.H., Kolhe, M.L., Van Herle, J., Raising the temperature on electrodes for anion exchange membrane electrolysis-activity and stability aspects. *Chem. Eng. J. Adv.*, 16, 100525, 2023, doi: 10.1016/j.ceja.2023.100525.
70. Lee, J.K., Lau, G.Y., Sabharwal, M., Weber, A.Z., Peng, X., Tucker, M.C., Titanium porous-transport layers for PEM water electrolysis prepared by tape casting. *J. Power Sources*, 559, 232606, 2023, doi: 10.1016/j.jpowsour.2022.232606.
71. Cheng, G. *et al.*, Analysis of Hydrogen Production Potential from Water Electrolysis in China. *Energy Fuels*, 37, 13, 9220–9232, 2023, doi: 10.1021/acs.energyfuels.3c00859.
72. Ibagon, N., Munoz, P., Diaz, V., Teliz, E., Correa, G., Techno-economic analysis for off-grid green hydrogen production in Uruguay. *J. Energy Storage*, 67, 107604, 2023, doi: 10.1016/j.est.2023.107604.
73. Cisterna, L., Améstica, L., Piderit, M., Proyectos Fotovoltaicos en Generación Distribuida ¿Rentabilidad Privada o Sustentabilidad Ambiental? *Rev. Politécnica*, 45, 2, 31–40, 2020, doi: 10.33333/rp.vol45n2.03.
74. Reynoso, W.E.C., Aguilar, L.J., Féliz, C.A., Propuesta de una Matriz energética sostenible a partir de la innovación en las tecnologías de las Celdas de Combustible para hidrógeno y biogás. *Lat. Am. Caribb. Conf. Eng. Technol.*, 11, 1–10, 2013.
75. Li, Y., Chen, D.W., Liu, M., Wang, R.Z., Life cycle cost and sensitivity analysis of a hydrogen system using low-price electricity in China. *Int. J. Hydrogen Energy*, 42, 4, 1899–1911, 2017, doi: 10.1016/j.ijhydene.2016.12.149.
76. Nuevas medidas europeas para que las baterías sean más sostenibles y éticas, Noticias, Parlamento Europeo, Apr. 03, 2022, [Online], Available: https://www.europarl.europa.eu/news/es/headlines/economy/20220228STO24218/nuevas-medidas-europeas-para-que-las-baterias-sean-mas-sostenibles-y-eticas.
77. IRENA, Innovation landscape brief: Pay-as-you-go models, *International Renewable Energy Agency*, pp. 1–22, Abu Dhabi, 2020.
78. IRENA, Innovation trends in electrolysers for hydrogen production, p. 37, 2021, [Online], Available: https://www.irena.org/-/media/Files/IRENA/Agency/Publication/2022/May/IRENA_EPO_Electrolysers_H2_production_2022.pdf%0Ahttps://www.irena.org/publications/2022/May/Innovation-Trends-in-Electrolysers-for-Hydrogen-Production.

79. Wang, L. *et al.*, Sustainability evaluations on material consumption for terawatt-scale manufacturing of silicon-based tandem solar cells. *Prog. Photovoltaics Res. Appl.*, January, 31, 1–13, 2023, doi: 10.1002/pip.3687.
80. Wang, Y., Robson, M.J., Manzotti, A., Ciucci, F., High-entropy perovskites materials for next-generation energy applications. *Joule*, 7, 5, 848–854, 2023, doi: 10.1016/j.joule.2023.03.020.
81. Zheng, H., Riegraf, M., Sata, N., Costa, R., A double perovskite oxygen electrode in Zr-rich proton conducting ceramic cells for efficient electricity generation and hydrogen production. *J. Mater. Chem. A*, 11, 20, 10955–10970, 2023, doi: 10.1039/d3ta01298k.
82. Xia, Q., Zhao, J., Chen, C., Jin, W., Modeling of CO2/H2O Co-electrolysis using solar-driven SOEC coupled with ammonia-based chemical heat pump. *Renew. Energy*, 212, 128–137, 2023, doi: 10.1016/j.renene.2023.05.041.
83. Wu, I. *et al.*, Effects of the Hydrophobic Block Length Ratio of Poly(vinylbenzyl *N*-methylpiperidinium carbonate)-*b*-polyethylene-*b*-poly(vinylbenzyl *N*-methylpiperidinium carbonate) Block Copolymers for Anion Exchange Membrane Electrolysis. *ACS Appl. Polym. Mater.*, 5, 8, 5834–5845, 2023, doi: 10.1021/acsapm.3c00414.
84. Santiago, P.V.B. *et al.*, Perovskite Oxides as an Opportunity to Systematically Study the Electrooxidation of Alcohols and Polyols on Materials Based on Abundant Elements: Learning from the Experience Using Pure Metals and Metallic Oxides in Electrocatalysis. *ACS Appl. Energy Mater.*, 6, 13, 7025–7051, 2023, doi: 10.1021/acsaem.3c00492.
85. Kojima, H., Nagasawa, K., Todoroki, N., Ito, Y., Matsui, T., Nakajima, R., Influence of renewable energy power fluctuations on water electrolysis for green hydrogen production. *Int. J. Hydrogen Energy*, 48, 12, 4572–4593, 2023, doi: 10.1016/j.ijhydene.2022.11.018.
86. Singla, M.K., Gupta, J., Alsharif, M.H., Jahid, A., Optimizing Integration of Fuel Cell Technology in Renewable Energy-Based Microgrids for Sustainable and Cost-Effective Energy. *Energies*, 16, 4482, 2023. https://doi.org/10.3390/en16114482
87. IRENA, Green Hydrogen Cost Reduction: Scaling up Electrolysers to Meet the 1.5°C Climate Goal, *International Renewable Energy Agency*, Abu Dhabi, 2020.
88. Patonia, A. and Poudineh, R., *Cost-competitive green hydrogen: how to lower the cost of electrolysers?*, Oxford Institute for Energy Studies, locatio: Wellington Square, Oxford OX1 2JD, United Kingdom, January, 2022.
89. Zhao, G. *et al.*, Life cycle assessment of H2O electrolysis technologies. *Int. J. Hydrogen Energy*, 45, 43, 23765–23781, 2020, doi: 10.1016/j.ijhydene.2020.05.282.
90. Weidner, T., Tulus, V., Guillén-Gosálbez, G., Environmental sustainability assessment of large-scale hydrogen production using prospective life cycle analysis. *Int. J. Hydrogen Energy*, 48, 22, 8310–8327, 2023, doi: 10.1016/j.ijhydene.2022.11.044.

91. Wang, K. *et al.*, Operando analysis of through-plane interlayer temperatures in the PEM Electrolyser cell under various operating conditions. *Appl. Energy*, 348, 121588, 2023, doi: 10.1016/j.apenergy.2023.121588.
92. Farahani, S.S., Bleeker, C., Van Wijk, A., Lukszo, Z., Hydrogen-based integrated energy and mobility system for a real-life office. *Appl. Energy*, 264, February, 114695, 2020, doi: 10.1016/j.apenergy.2020.114695.
93. Liu, Z., Gao, Y., Liu, B., An artificial intelligence-based electric multiple units using a smart power grid system. *Energy Rep.*, 8, 13376–13388, 2022, doi: 10.1016/j.egyr.2022.09.138.
94. IEA, World Energy Outlook 2018, IEA, Paris https://www.iea.org/reports/world-energy-outlook-2018, Licence: CC BY 4.0, 2018.
95. Huertas, O.P., Opciones de descarbonizacion del sector residencial, comercial e institucional en el municipio de Madrid, 2020. [Online]. Available: https://oa.upm.es/63460/1/TFM_OSCAR_PEREZ_HUERTAS.pdf
96. Palone, O., Gagliardi, G.G., Mechelli, M., Cedola, L., Borello, D., Techno-economic analysis of sustainable methanol and ammonia production by chemical looping hydrogen generation from waste plastic. *Energy Convers. Manage.*, 292, 117389, 2023, doi: 10.1016/j.enconman.2023.117389.
97. Lin, H. *et al.*, Comprehensive analysis of a novel power and methanol coproduction process using landfill gas reforming and renewable hydrogen generation by an alkaline Electrolyser. *J. Clean. Prod.*, 418, 138131, 2023, doi: 10.1016/j.jclepro.2023.138131.
98. Dunn, S., Hydrogen futures: Toward a sustainable energy system. *Int. J. Hydrogen Energy*, 27, 3, 235–264, 2002, doi: 10.1016/S0360-3199(01)00131-8.
99. Low, J., Haszeldine, R.S., Mouli-Castillo, J., Comparative evaluation of battery electric and hydrogen fuel cell electric vehicles for zero carbon emissions road vehicle fuel in Scotland. *Eng. Archive*, 1, 1–29, July, 2020, doi: 10.31224/osf.io/dcjrt.
100. Babu, M.K. and Ray, P., Sensitivity analysis, optimal design, cost and energy efficiency study of a hybrid forecast model using HOMER pro. *J. Eng. Res.*, 11, 2, 100033, 2023, doi: 10.1016/j.jer.2023.100033.
101. Mostert, C., Ostrander, B., Bringezu, S., Kneiske, T.M., Comparing electrical energy storage technologies regarding their material and carbon footprint. *Energies*, 11, 12, 3386, 2018, doi: 10.3390/en11123386.
102. Pazouki, S. and Haghifam, M.R., "Optimal planning and scheduling of energy hub in presence of wind, storage and demand response under uncertainty. *Int. J. Electr. Power Energy Syst.*, 80, 219–239, 2016, doi: 10.1016/j.ijepes.2016.01.044.
103. Alanne, K. and Cao, S., Zero-energy hydrogen economy (ZEH2E) for buildings and communities including personal mobility. *Renew. Sustain. Energy Rev.*, 71, October 2016, 697–711, 2017, doi: 10.1016/j.rser.2016.12.098.
104. Chu, S. and Majumdar, A., Opportunities and challenges for a sustainable energy future. *Nature*, 488, 7411, 294–303, 2012, doi: 10.1038/nature11475.

105. Choe, C., Kim, H., Lim, H., Feasibility study of power-to-gas as simultaneous renewable energy storage and CO_2 utilization: Direction toward economic viability of synthetic methane production. *Sustain. Energy Technol. Assess.*, 57, 103261, 2023, doi: 10.1016/j.seta.2023.103261.
106. Green hydrogen, solar and storage: Iberdrola plans 'Europe's largest' industrial facility - Energy-Storage.News, https://www.energy-storage.news/green-hydrogen-solar-and-storage-iberdrola-plans-europes-largest-industrial-facility/ (accessed Nov. 01, 2023).
107. Saudi Arabia's Plan to Rule the $700 Billion Hydrogen Market from a Marine Perspective — Mr. Sustainability, https://www.mr-sustainability.com/stories/2021/saudi-arabias-plan-to-rule-700-billion-hydrogen-market-maritime (accessed Nov. 01, 2023).
108. Hydrogen Production and Infrastructure Projects Database - Data product - IEA, https://www.iea.org/data-and-statistics/data-product/hydrogen-production-and-infrastructure-projects-database (accessed Nov. 01, 2023).
109. F. C. and H. 2 J. Undertaking *et al.*, *Study on the use of fuel cells and hydrogen in the railway environment*, Publications Office, 2019, doi: doi/10.2881/495604.
110. Eichman, J., Townsend, A., Melaina, M., Economic Assessment of Hydrogen Technologies Participating in California Electricity Markets Economic Assessment of Hydrogen Technologies Participating in California Electricity Markets. *Natl. Renewable Energy Lab.*, 2016, [Online], Available: www.nrel.gov/publications.%0Ahttps://www.nrel.gov/docs/fy16osti/65856.pdf%0Ahttps://www.nrel.gov/docs/fy16osti/65856.pdf.
111. Kraus, N. and Damman, S., *HyLAW*. 2271647, 2019, March, 2019.
112. Parra, D., Valverde, L., Pino, F.J., Patel, M.K., A review on the role, cost and value of hydrogen energy systems for deep decarbonisation,". *Renew. Sustain. Energy Rev.*, 101, November 2018, 279–294, 2019, doi: 10.1016/j.rser.2018.11.010.
113. Zuberi, M.J.S., Bless, F., Chambers, J., Arpagaus, C., Bertsch, S.S., Patel, M.K., Excess heat recovery: An invisible energy resource for the Swiss industry sector. *Appl. Energy*, 228, June, 390–408, 2018, doi: 10.1016/j.apenergy.2018.06.070.
114. Lund, H. *et al.*, 4th Generation District Heating (4GDH). Integrating smart thermal grids into future sustainable energy systems. *Energy*, 68, 1–11, 2014, doi: 10.1016/j.energy.2014.02.089.
115. Luo, X., Wang, J., Dooner, M., Clarke, J., Overview of current development in electrical energy storage technologies and the application potential in power system operation. *Appl. Energy*, 137, 511–536, 2015, doi: 10.1016/j.apenergy.2014.09.081.
116. Zheng, N., Zhang, H., Duan, L., Wang, Q., Bischi, A., Desideri, U., Techno-economic analysis of a novel solar-driven PEMEC-SOFC-based multi-generation system coupled parabolic trough photovoltaic thermal collector and thermal energy storage. *Appl. Energy*, 331, 120400, 2023, doi: 10.1016/j.apenergy.2022.120400.

117. Parra, D., Norman, S.A., Walker, G.S., Gillott, M., Optimum community energy storage for renewable energy and demand load management. *Appl. Energy*, 200, 358–369, 2017, doi: 10.1016/j.apenergy.2017.05.048.
118. Nojavan, S., Zare, K., Mohammadi-Ivatloo, B., Application of fuel cell and Electrolyser as hydrogen energy storage system in energy management of electricity energy retailer in the presence of the renewable energy sources and plug-in electric vehicles. *Energy Convers. Manage.*, 136, 404–417, 2017, doi: 10.1016/j.enconman.2017.01.017.
119. Zhong, Q., Nelson, J.R., Tong, D., Grubesic, T.H., A spatial optimization approach to increase the accuracy of rooftop solar energy assessments. *Appl. Energy*, 316, April, 119128, 2022, doi: 10.1016/j.apenergy.2022.119128.
120. Parissis, O.S. *et al.*, Integration of wind and hydrogen technologies in the power system of Corvo island, Azores: A cost-benefit analysis. *Int. J. Hydrogen Energy*, 36, 13, 8143–8151, 2011, doi: 10.1016/j.ijhydene.2010.12.074.
121. Farràs, P., Strasser, P., Cowan, A.J., Water electrolysis: Direct from the sea or not to be? *Joule*, 5, 8, 1921–1923, 2021, doi: 10.1016/j.joule.2021.07.014.
122. Montenegro Camacho, Y.S., Bensaid, S., Piras, G., Antonini, M., Fino, D., Techno-economic analysis of green hydrogen production from biogas autothermal reforming. *Clean Technol. Environ. Policy*, 19, 5, 1437–1447, 2017, doi: 10.1007/s10098-017-1341-1.
123. Liu, W. *et al.*, Multi-branched AgAuPt nanoparticles for efficient electrocatalytic hydrogen evolution: Synergism of tip-enhanced electric field effect and local electric field effect. *J. Energy Chem.*, 81, 339–348, 2023, doi: 10.1016/j.jechem.2023.02.032.
124. Bartolucci, L., Cordiner, S., Mulone, V., Santarelli, M., Ortenzi, F., Pasquali, M., PV assisted electric vehicle charging station considering the integration of stationary first- or second-life battery storage. *J. Clean. Prod.*, 383, 135426, Mar. 2023, doi: 10.1016/j.jclepro.2022.135426.
125. Shayegan, S., Pearson, P.J.G., Hart, D., Hydrogen for buses in London: A scenario analysis of changes over time in refuelling infrastructure costs. *Int. J. Hydrogen Energy*, 34, 19, 8415–8427, 2009, doi: 10.1016/j.ijhydene.2009.06.084.
126. Stevens, M.B., Fowler, M.W., Elkamel, A., Elhedhli, S., Macro-level optimized deployment of an electrolyser-based hydrogen refuelling infrastructure with demand growth. *Eng. Optim.*, 40, 10, 955–967, Oct. 2008, doi: 10.1080/03052150802236061.
127. Lee, J.Y., Yoo, M., Cha, K., Lim, T.W., Hur, T., Life cycle cost analysis to examine the economical feasibility of hydrogen as an alternative fuel. *Int. J. Hydrogen Energy*, 34, 10, 4243–4255, 2009, doi: 10.1016/j.ijhydene.2009.03.012.
128. Hwang, J.J., Sustainability study of hydrogen pathways for fuel cell vehicle applications. *Renew. Sustain. Energy Rev.*, 19, 220–229, 2013, doi: 10.1016/j.rser.2012.11.033.
129. Le Duigou, A. *et al.*, Hydrogen pathways in France: Results of the HyFrance3 Project. *Energy Policy*, 62, 1562–1569, 2013, doi: 10.1016/j.enpol.2013.06.094.

130. Quarton, C.J. and Samsatli, S., Power-to-gas for injection into the gas grid: What can we learn from real-life projects, economic assessments and systems modelling? *Renew. Sustain. Energy Rev.*, 98, September, 302–316, 2018, doi: 10.1016/j.rser.2018.09.007.
131. Shafiei, E., Davidsdottir, B., Leaver, J., Stefansson, H., Asgeirsson, E., II, Comparative analysis of hydrogen, biofuels and electricity transitional pathways to sustainable transport in a renewable-based energy system. *Energy*, 83, 614–627, 2015, doi: 10.1016/j.energy.2015.02.071.
132. Cantuarias-Villessuzanne, C., Weinberger, B., Roses, L., Vignes, A., Brignon, J.M., Social cost-benefit analysis of hydrogen mobility in Europe. *Int. J. Hydrogen Energy*, 41, 42, 19304–19311, 2016, doi: 10.1016/j.ijhydene.2016.07.213.
133. Gis, W., Kruczynski, S., Menes, M., Waskiewicz, J., Motor transport in Poland – hydrogenization. 00019, 2017, 1–6, 2020.
134. Grueger, F., Möhrke, F., Robinius, M., Stolten, D., Early power to gas applications: Reducing wind farm forecast errors and providing secondary control reserve. *Appl. Energy*, 192, 551–562, 2017, doi: 10.1016/j.apenergy.2016.06.131.
135. Viesi, D., Crema, L., Testi, M., The Italian hydrogen mobility scenario implementing the European directive on alternative fuels infrastructure (DAFI 2014/94/EU). *Int. J. Hydrogen Energy*, 42, 44, 27354–27373, 2017, doi: 10.1016/j.ijhydene.2017.08.203.
136. Hosseinzadeh, A., Zhou, J.L., Li, X., Afsari, M., Altaee, A., Techno-economic and environmental impact assessment of hydrogen production processes using bio-waste as renewable energy resource. *Renew. Sustain. Energy Rev.*, 156, December 2021, 111991, 2022, doi: 10.1016/j.rser.2021.111991.

12

Green Hydrogen Generation by Water Electrolysis

Weizhe Zhang[1], Yixiang Shi[1,2*], Shuang Li[1,2] and Ningsheng Cai[1]

[1]Key Laboratory for Thermal Science and Power Engineering of Ministry of Education, Department of Energy and Power Engineering, Tsinghua University, Beijing, China

[2]Shanxi Research Institute for Clean Energy, Tsinghua University, Taiyuan, China

Abstract

Water electrolysis, which exploits electrical power to split H_2O, is an effective and rapidly progressing strategy for renewable energy consumption and large-scale green H_2 generation. Nevertheless, various electrolysis techniques confront certain advantages and challenges, including prolonged stability but inadequate performance, and enhanced reaction thermodynamics and kinetics whereas attendant accelerated degradation or increased cost for dependence on precious metals, and are at their specific stage of development. In this chapter, a brief introduction is first proposed, and the fundamental insights of water electrolysis are presented. Subsequently, evaluations of alkaline electrolysis cell (AEC), polymer electrolyte membrane electrolysis cell (PEMEC), together with the frontier research focusing on elevated temperature AEC (ET-AEC) and ET-PEMEC, as well as protonic ceramic electrolysis cell (PCEC) and solid oxide electrolysis cell (SOEC) are sequentially performed, which cover studies concentrating on electrodes and electrolytes, analysis of performance and stability, as well as opportunities and challenges for development and operation of stacks and systems.

Keywords: Green hydrogen, water electrolysis

**Corresponding author*: shyx@tsinghua.edu.cn

Mehmet Sankir and Nurdan Demirci Sankir (eds.) Towards Green Hydrogen Generation, (407–462)

12.1 Introduction

Electrolysis, which consumes electrical power to split H_2O molecules [1], produces H_2 utilizable in various applications [2], such as clean fuel for vehicles [3] and feedstock in chemical synthesis [4]. By converting electricity into chemical energy, water electrolysis also offers a promising approach to consuming fluctuating renewable sources [5], while alleviating the considerable greenhouse gas (GHG) emissions and related climate issues induced by substantially exploited fossil fuels [6].

In spite of the abovementioned advantages, traditional nonrenewable coal gasification and steam methane reforming (SMR) offer greatly cheaper approaches to H_2 production, and merely 4% of global H_2 production is currently accomplished by water electrolysis [7]. Compared with other sources, such as equipment, labor, and maintenance, the cost of electricity consumed to conduct electrolysis reactions is the dominant expense origin for large-scale systems. Therefore, improving the intrinsic electrolysis performance to realize increased energy efficiency and decreased specific energy consumption becomes a critical requirement for boosting green H_2 production by water electrolysis.

Among numerous operational parameters, temperature has the most pronounced influence on electrolysis performance, and elevated temperature has arisen as an emerging trend for research and development (R & D) of water electrolysis, which concurrently proposes challenges for stability and great opportunities for the prudent coupling of electrochemical and thermal characteristics [8]. Herein, we first present fundamental insights into electrolysis reactions, introducing basic working principles, thermodynamics, realistic operation features of electrolysis cells, and the influences of temperature variation. Furthermore, discussions about existing electrolysis technologies of alkaline electrolysis cell (AEC), polymer electrolyte membrane electrolysis cell (PEMEC) working at lower temperatures, with emphases on elevated temperature AEC (ET-AEC) and ET-PEMEC as critical research frontiers, as well as protonic ceramic electrolysis cell (PCEC) and solid oxide electrolysis cell (SOEC) operating at further increased temperature are sequentially introduced.

12.2 Fundamentals of Water Electrolysis

By the integration of cathodic hydrogen evolution reaction (HER) and anodic oxygen evolution reaction (OER), as well as ionic conduction in the medially placed electrolyte, the global splitting reaction of H_2O molecule with the generation of H_2 and O_2 is unveiled in Equation 12.1 [1], while the expressions of HER and OER are determined by the type of ion transferred through the electrolyte:

$$H_2O \rightarrow H_2 + 0.5O_2 \tag{12.1}$$

12.2.1 Various Electrolysis Cells and Working Principles

As a currently commercialized electrolysis method [9], AECs typically work at 25°C to 85°C and use 20 to 35 wt% KOH solution as an electrolyte for conducting OH^-, and a separator is adopted to segregate produced H_2 and O_2. The electrode half-reactions are shown in Equations 12.2 and 12.3 [1], and the working principles are depicted in Figure 12.1a.

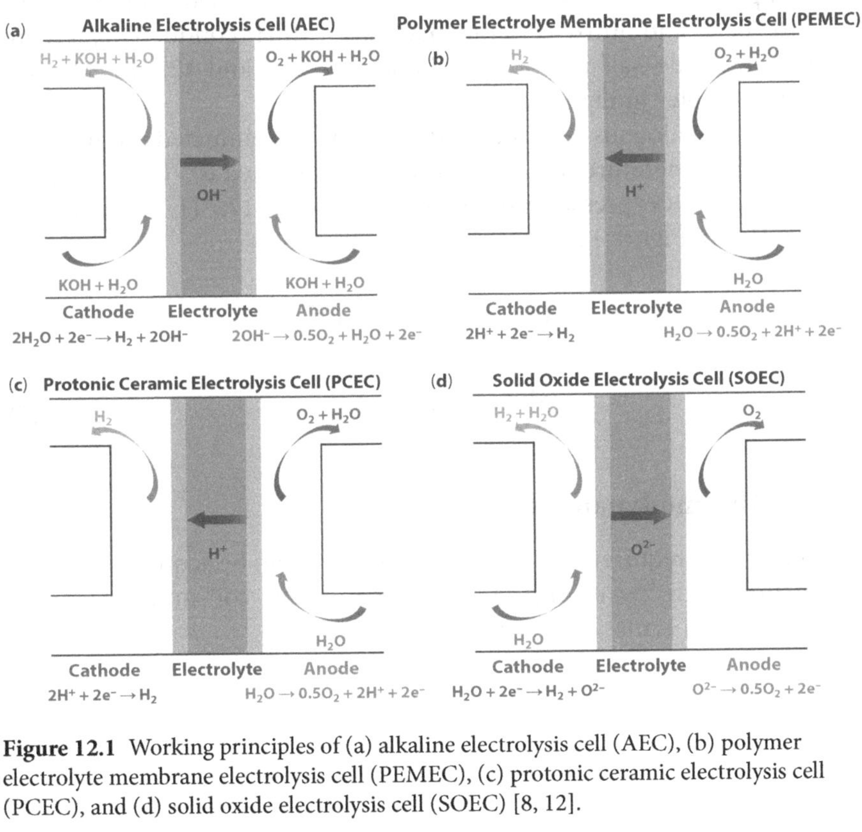

Figure 12.1 Working principles of (a) alkaline electrolysis cell (AEC), (b) polymer electrolyte membrane electrolysis cell (PEMEC), (c) protonic ceramic electrolysis cell (PCEC), and (d) solid oxide electrolysis cell (SOEC) [8, 12].

$$2H_2O + 2e^- \rightarrow H_2 + 2OH^- \quad (12.2)$$

$$2OH^- \rightarrow 0.5O_2 + H_2O + 2e^- \quad (12.3)$$

Meanwhile, PEMECs similarly work at 25°C to 85°C, whereas their electrolytes are substituted by polymer membranes with H^+ conductivity (e.g., Nafion) [10]. Consequently, expressions for HER and OER are replaced by half-reactions shown in Equations 12.4 and 12.5 [1], and feed water is supplied to the anode (Figure 12.1b).

$$2H^+ + 2e^- \rightarrow H_2 \quad (12.4)$$

$$H_2O \rightarrow 0.5O_2 + 2H^+ + 2e^- \quad (12.5)$$

In comparison, PCECs operate at an obviously increased temperature of 450°C to 700°C and utilize ceramic materials with H^+ conductivity, giving identical electrode half-reactions in Equations 12.4 and 12.5 [1, 11] and working principles in Figure 12.1c.

Relying on analogous ceramics and variation of conducted ion into O^{2-}, the working temperature of SOECs is further upgraded to 700°C, proposing HER and OER expressions in Equations 12.6 and 12.7 [1, 11], as well as cell structure in Figure 12.1d.

$$H_2O + 2e^- \rightarrow H_2 + O^{2-} \quad (12.6)$$

$$O^{2-} \rightarrow 0.5O_2 + 2e^- \quad (12.7)$$

12.2.2 Thermodynamics

The total energy requirement of water electrolysis can be acquired through the reaction enthalpy change ΔH, which includes consumed electrical energy ΔG and thermal energy $T\Delta S$ [1, 8, 11]:

$$\Delta H = \Delta G + T\Delta S \quad (12.8)$$

Moreover, the thermoneutral voltage V_{tn} is acquired [1, 8, 11]:

$$V_{tn} = \frac{\Delta H}{nF} \tag{12.9}$$

where n means transferred electrons and equals 2 for reaction in Equation 12.1, F refers to the Faraday's constant of 96485 $C{\cdot}mol^{-1}$. As a point representing the input power precisely equals the entire energy requirement, V_{tn} symbolizes where an electrolysis cell works without heat input/output, while additional heat generation/consumption is induced when operating at a voltage higher/lower than V_{tn}.

As the required electrical energy, ΔG analogously gives the reversible voltage V_{rev}, representing the theoretically minimum voltage to propel the electrolysis reaction [1, 8, 11]:

$$V_{rev} = \frac{\Delta G}{nF} \tag{12.10}$$

12.2.3 Realistic Operation and Influences of Temperature Variation

Nevertheless, the applied cell voltage V_{cell} during realistic operation is always higher than V_{rev}, as various overpotentials are involved [1, 8, 11]:

$$V_{cell} = V_{rev} + \eta_{act} + \eta_{ohm} + \eta_{mass} \tag{12.11}$$

In the above expression, the activation overpotential η_{act} is generated by the electronic transfer between electrodes and chemical species, and its relationship with current density j is described by the Butler-Volmer equation, reflecting the influences of electrode catalytic activity and structure [1, 8, 13]:

$$j = j_0\left[exp\left(\frac{\alpha nF\eta_{act}}{RT}\right) - exp\left(\frac{(1-\alpha)nF\eta_{act}}{RT}\right)\right] \tag{12.12}$$

where j_0 is electrode exchange current density, α is the charge transfer coefficient of the corresponding process, T is operational temperature, and the value of universal gas constant R is 8.315 J·mol^{-1}· K^{-1}.

Meanwhile, Ohm's law instructs the determination of ohmic overpotential [1, 8, 11]:

$$\eta_{ohm} = I_{cell}R_{cell} = I_{cell}(R_{electrode} + R_{electrolyte}) \tag{12.13}$$

where I_{cell} and R_{cell} are the current and resistance of the electrolysis cell, and $R_{electrode}$ and $R_{electrolyte}$ are the resistance of the electrode and electrolyte. Typically, $R_{electrolyte}$ contributes to most of R_{cell}, and η_{ohm} is thus majorly induced by ionic conduction in electrolytes.

When operating at an extremely increased current density, moreover, the influences of reactant depletion are pronounced and propose an enlarged mass transport overpotential η_{mass}, while Fick's law and Stefan-Maxwell equation can be employed to describe reactant/product diffusion [8].

With a lower heating value (LHV) or higher heating value (HHV) of produced H_2 as the criterion, electrolysis efficiency is given by Equations 12.14 and 12.15, and V_{cell} is the sole influential factor [1, 8, 11]:

$$\eta_{LHV} = \frac{\dot{n}_{H_2}LHV_{H_2}}{I_{cell}V_{cell}} = \frac{I_{cell}}{nF}\frac{LHV_{H_2}}{I_{cell}V_{cell}} = \frac{LHV_{H_2}}{nF}\frac{1}{V_{cell}} = \frac{1.229V}{V_{cell}} \tag{12.14}$$

$$\eta_{HHV} = \frac{\dot{n}_{H_2}HHV_{H_2}}{I_{cell}V_{cell}} = \frac{I_{cell}}{nF}\frac{HHV_{H_2}}{I_{cell}V_{cell}} = \frac{HHV_{H_2}}{nF}\frac{1}{V_{cell}} = \frac{1.481V}{V_{cell}} \tag{12.15}$$

Among various parameters, variation in temperature has the most pronounced influence on the electrolysis performance. As shown in Figure 12.2, ΔH and V_{tn} unveil weak dependence on temperature except water evaporation at 99.98°C and consequent changes, whereas $T\Delta S$ is drastically increased, and ΔG and V_{rev} are reduced with elevating temperature (e.g., 1.229 V at 25°C, 1.034 V at 600°C and 0.860 V at 1200°C). Relying on the integration of this phenomenon to decrease V_{rev}, facilitated reaction kinetics to dwindle η_{act}, as well as promoted ionic conduction to reduce η_{ohm}, an effectively decreased V_{cell} and improved energy efficiency can be accomplished at elevated temperatures.

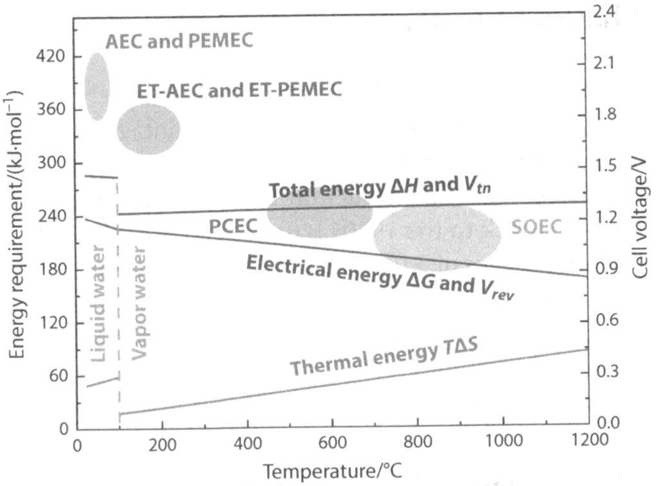

Figure 12.2 Water electrolysis thermodynamics at atmospheric pressure with working temperature and cell voltage distribution of (ET-)AEC, (ET-)PEMEC, PCEC, and SOEC [8, 12].

Therefore, a map of working temperature and cell voltage is drawn by the current electrolysis technologies, as displayed in Figure 12.2. With a lower temperature of 25°C to 85°C, AECs and PEMECs normally operate at 1.8 to 2.2 V [10, 14], while ET-AEC and ET-PEMEC at 100°C to 250°C have been proposed as a promising approach to reducing V_{cell} to 1.6 to 1.9 V [8, 15]. In contrast, PCECs at an obviously higher temperature of 450°C to 700°C possess V_{cell} of 1.1 to 1.4 V, which is decreased to 0.9 to 1.3 V for SOECs at >700°C [8, 16].

From the perspective of thermal behavior, the comparison between V_{cell} and V_{tn} also proposes insights into energy management in corresponding stacks and systems. As highly exothermic operation of AECs and PEMECs are induced by their V_{cell} obviously higher than V_{tn}, as presented in Figure 12.2, a substantial parasitic power is consumed to remove the excessive heat, which can be alleviated by the reduced V_{cell} of ET-AECs and ET-PEMECs [17–19]. In contrast, V_{cell} variation for PCECs in the vicinity of V_{tn} illustrates their feasibility to realize flexible endo/exothermic transition, while the operation of SOECs is typically restricted in endothermic mode by greatly decreased V_{cell}, and high-value heat is required for maintaining stable operation [16].

12.3 Alkaline Electrolysis Cell (AEC)

With KOH/NaOH solution as electrolytes and a relatively lower temperature of 25°C to 85°C, AEC possesses various advantages, such as low cost and prolonged durability, and thus becomes the most mature electrolysis technique at present. The construction of hundred-MW level AEC systems can be dated back to the 1920s [20], and large-scale AEC equipment with H_2 production rates of >1000 $Nm^3{\cdot}h^{-1}$ has been broadly developed [8, 9]. However, the higher specific energy consumption of traditional AECs leads to uncompetitive H_2 production costs and forbids their further progress, generating many studies focusing on performance improvement, and elevating working temperature is a promising approach.

12.3.1 Electrocatalysts (Electrodes) and Electrolytes

12.3.1.1 HER Electrocatalysts (Cathodes)

Specifically, alkaline media of AECs enable the utilization of low-cost transition metals, while Ni and Ni-based alloys are the mainstream HER and OER electrocatalysts based on their sufficient catalytic activity and outstanding stability [21, 22], whereas the applications of platinum group metals (PGMs) are forbidden by their scarcity and obviously higher expense.

Among various binary Ni-based alloys, Ni/Mo exhibits an excellent HER activity for the atomic synergistic interactions [22] and is typically fabricated on substrates such as Ni sheet, mild steel (MS), and Ni foam by the electrodeposition [23, 24] and hydrothermal [25–27] method. By generating MoO_2 cuboids on Ni foam, as-prepared $MoNi_4$ exhibited a considerably reduced HER overpotential of 15 mV at 10 $mA{\cdot}cm^{-2}$ and a Tafel slope of only 30 $mV{\cdot}dec^{-1}$, while its stability was validated by a 30-h test at 200 $mA{\cdot}cm^{-2}$ [25]. Electrodeposited Ni/Mo on an MS substrate also showed a diminished HER overpotential of 180 mV at 300 $mA{\cdot}cm^{-2}$ and 80°C in 6 M KOH solution [28], and the values of other alloys with replacing Mo by Zn, Co, W, Fe, and Cr lie in 200 to 400 mV [29].

Broadening Ni-based alloys into ternary materials, Ni/Mo/Fe revealed an overpotential of 187 mV at 300 $mA{\cdot}cm^{-2}$ and 80°C [28], while Ni/Mo/Cu, Ni/Mo/Zn, Ni/Mo/W, Ni/Mo/Co, and Ni/Mo/Cr have been proposed as HER electrocatalysts for AECs. Moreover, metal sulfides [30–32], phosphides [33, 34], nitrides [35, 36], carbides [37, 38], and selenides [39, 40] have been evaluated, and NiS_x with a hierarchically porous structure illustrated a 60-mV HER overpotential at 10 $mA{\cdot}cm^{-2}$ in 1 M KOH solution [30].

12.3.1.2 OER Electrocatalysts (Anodes)

In fundamentals, the OER overpotential of a four-electron reaction is typically higher than HER for its sluggish kinetics, while PGMs and their oxides, such as IrO_2 and RuO_2, with excellent activity are subject to their scarcity and higher cost [8], and substantial studies have been conducted to develop transition metal–based (normally Ni, Co, and Fe) materials as substitute OER electrocatalysts.

By leaching Al from Ni/Al alloy to generate a highly porous structure [41], Raney Ni was initially developed in 1948 and has become the most mature OER electrocatalyst in industrial applications. A considerably decreased OER overpotential of 190 mV was also demonstrated by Raney Co at 1000 mA·cm^{-2} and 160°C, while the incorporation of Co was adopted to fabricate Raney Ni/Co with various element compositions [42]. At 90°C and in concentrated 40 wt% KOH solution, synergistic effects with Ni also reduced the overpotential of 55 at% Fe electrode to 390 mV at 1000 mA·cm^{-2} [43].

Meanwhile, metal oxides including spinels, perovskites, as well as layered double hydroxides (LDHs) have revealed adequate catalytic activity with insufficient electronic conductivity. As a consequence, integration of metal chalcogenides (sulfides [44] and selenides [45]), pnictides (nitrides [46] and phosphides [47]) as conductive supports, and *in situ* constructed oxide layers to promote OER appears as a promising solution [21].

12.3.1.3 Electrolytes: Porous Diaphragms, Anion Exchange Membranes (AEMs), and Ion-Solvating Membranes (ISMs) as Separators

In addition to employing 20 to 35 wt% KOH solution as a liquid electrolyte, AECs concurrently use separators arranged between electrodes to prevent direct mixing of produced H_2 and O_2. Benefiting from sufficient stability in alkaline media, porous diaphragms are broadly utilized, and acquire OH^- conductivity by absorbing alkaline solution into their internal structures, as shown in Figure 12.3a. The initially adopted diaphragm was asbestos, whereas its applications were banned owing to the health threats [48], and polyphenylene sulfide (PPS) fabric is currently the mainstream diaphragm in industrial scenarios [18, 19, 49, 50].

However, the intrinsic hydrophobicity of organic materials proposes difficulty for the infiltration of liquid electrolytes and brings an enlarged area-specific resistance (ASR) improving η_{ohm} [51]. Therefore, incorporating composite coatings onto the PPS support has been utilized to enhance

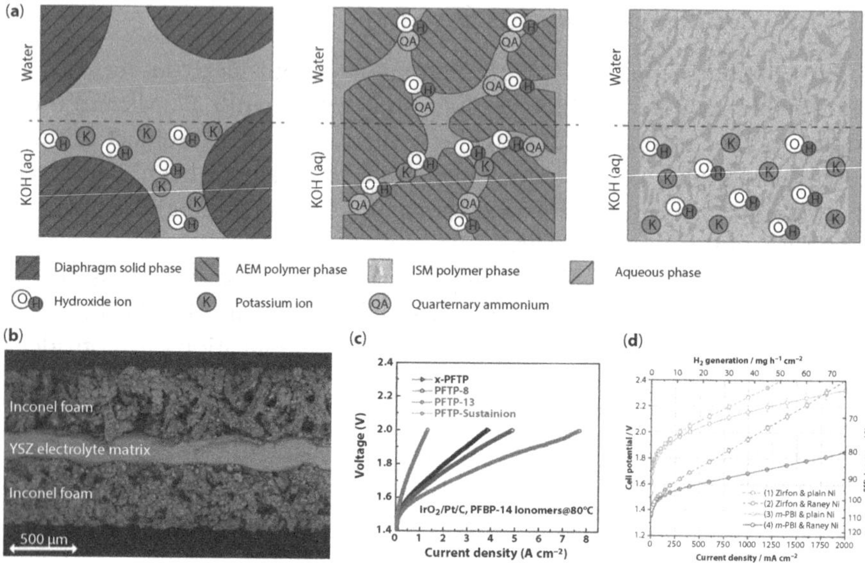

Figure 12.3 (a) Schematic illustration of OH^- conduction mechanisms in porous diaphragms, AEMs, and ISMs (Reprinted with permission [53]. Copyright 2023 American Chemical Society under CC BY 4.0). (b) SEM microstructure of YSZ diaphragm (Reprinted with permission [62]. Copyright 2016 The Electrochemical Society under CC BY 4.0). (c) Polarization curves of an AEMEC with PFAP-based and other AEMs at 80°C (Reprinted with permission [65]. Copyright 2021 Royal Society of Chemistry). (d) Polarization curves of AECs with Zirfon diaphragm and *m*-PBI ISM at 80°C in 24 wt% KOH solution (Reprinted with permission [67]. Copyright 2019 Royal Society of Chemistry under CC BY NC 4.0).

hydrophilicity, and well-known commercial Zirfon diaphragm possesses a hydrophilic layer composed of 15 wt% polysulfone (PSU) combined with 85 wt% ZrO_2, which exhibited an ASR of ~0.3 $\Omega\cdot cm^2$ at 25°C and 0.12 $\Omega\cdot cm^2$ at 80°C in 30 wt% KOH solution [48, 52, 53]. Varying the PSU/ZrO_2 weight ratio [54] and selecting novel fillers, such as CeO_2 [55] and cellulose nanocrystals (CNCs) [56], have been employed to optimize the properties of diaphragm materials.

For operation at elevated temperatures, KOH solution with a higher concentration of 30 to 45 wt% is adopted to acquire an improved OH^- conductivity [57], while porous ceramics including $SrTiO_3$ [58, 59], $BaTiO_3$ [60, 61], as well as yttria-stabilized zirconia (YSZ) [62] have been selected as diaphragms for their superior thermal stability compared with polymers. Nevertheless, as ceramics are typically prepared by cold press and sintering [8], a universal problem is their great brittleness, and an improved thickness (e.g., 1300 μm for $SrTiO_3$) [59] over Zirfon-type diaphragms (ca. 500 μm) [48, 54] is required to provide an adequate mechanical strength, which prolongs the distance for ion migration and brings an increased ASR. In comparison, a YSZ diaphragm with a considerably reduced thickness of 180 μm, as shown in Figure 12.3b, was fabricated by tape casting and realized an ASR of merely 0.058 $\Omega \cdot cm^2$ at 200°C in 45 wt% KOH solution [62].

Meanwhile, enabling current composite diaphragm operation at higher temperatures through stability enhancement is another method, whereas a critical challenge is polar groups in hydrophilic polymers degrade severely in alkaline media [52]. In this regard, as the practical systems typically work at improved pressures of >7 bar [18, 19, 49, 50], the electrolyte absorption may be facilitated and demand for hydrophilicity can thus be loosened, demonstrating hydrophobic materials with greatly superior thermal stability and alkaline resistance (e.g., polytetrafluoroethylene, PTFE) are potential materials for diaphragms at elevated temperatures.

For porous diaphragms, a higher thickness of >200 μm and an enlarged ASR limit the boosting of the overall performance for AECs, while developing alternative membranes possessing dense, thin structures and intrinsic OH^- conductivity has emerged in recent years, which can be categorized into anion exchange membranes (AEMs) and ion-solvating membranes (ISMs) [53]. Fundamentally, OH^- migration in AEMs relies on functional groups (e.g., quarternary ammonium) [63], as displayed in Figure 12.3a [53], offering an advantage of operation in less-concentrated alkaline media (typically in 1 M KOH solution) or even pure water for corrosion alleviation. Furthermore, the greatly reduced thickness of <100 μm empowers the AEMs to realize reduced ASRs [64], while an extraordinary performance of 7.68 $A \cdot cm^{-2}$ at 2.0 V was accomplished by an AEMEC with a poly (fluorenyl-*co*-aryl piperidinium) (PFAP) membrane in 1 M KOH solution, as unveiled in Figure 12.3c [65]. Analogous to the combination between porous diaphragms and AEMs, ISMs possess thin and dense structures, whereas require 20 to 30 wt% KOH solution to generate an internal homogeneous state for conducting OH-, as shown in Figure 12.3a [66, 67], and Figure 12.3d further illustrates that an excellent performance of 1.7 $A \cdot cm^{-2}$ at 1.8 V was realized by an AEC with a 40 μm *m*- (polybenzimidazole) PBI ISM at 80°C in a 24 wt% KOH solution [67]. However, the stability of AEMs and ISMs requires improvement as existing studies are generally limited to hundreds of hours, and their degradation mechanisms in long-term operation, especially at elevated temperatures, need to be evaluated.

12.3.2 Performance and Stability

For currently commercial large-scale AEC systems for green H_2 production, a low temperature of 60°C to 85°C is normally adopted [18, 19, 49, 50] and a prolonged lifetime of 10 to 15 years has been universally realized [9, 14]. Nevertheless, as summarized in Figure 12.4a, their unsatisfactory performance of 0.3 to 0.5 $A{\cdot}cm^{-2}$ at 1.8 to 2.3 V causes an enlarged reaction-specific energy consumption of 4.4-5.3 $kWh{\cdot}Nm^{-3}$ [8], leading to an obvious electricity expense.

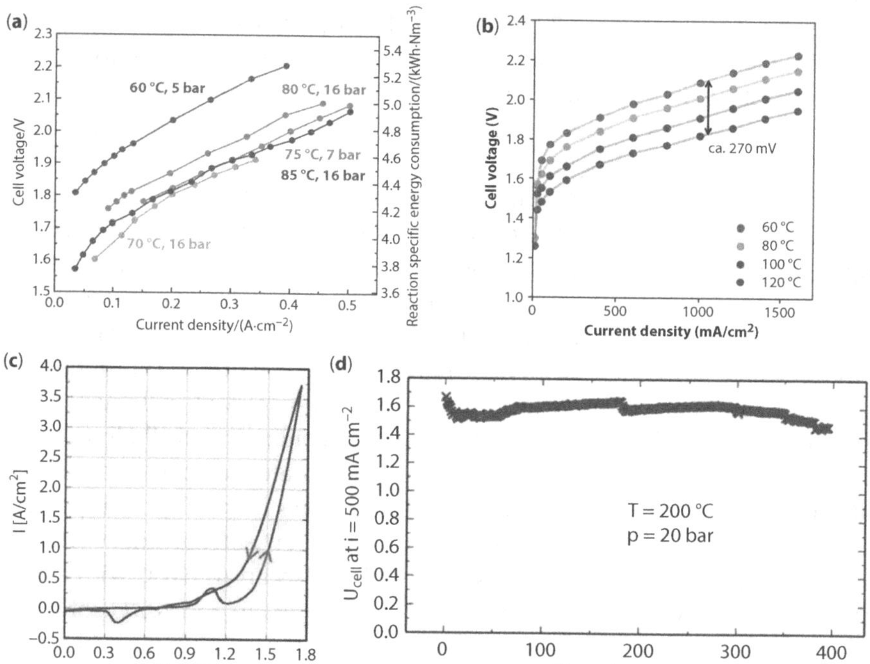

Figure 12.4 (a) Performance of conventional AEC systems operating at 60°C to 85°C [18, 19, 49, 50, 71]. (b) Polarization curves of an ET-AEC with activated cathode and plain nickel anode at 60°C to 120°C in 11 M KOH solution (Reprinted with permission [52]. Copyright 2023 American Chemical Society under CC BY NC ND 4.0). (c) Cyclic voltammogram and (d) 400-h stability test at 0.5 $A{\cdot}cm^{-2}$ of a foam-based ET-AEC with Inconel cathode, YSZ diaphragm, and Ni/PTFE/Ag anode at 200°C at 20 bar in 45 wt% KOH solution (Reprinted with permission [62]. Copyright 2016 The Electrochemical Society under CC BY 4.0).

In this aspect, ET-AECs with a temperature increase from 100°C to 250°C have been proposed to pursue improved performance, and several cell-level experiments have been conducted [52, 58, 59, 62]. Fixing the current density at a large value of 1 A·cm^{-2}, a remarkable voltage decrease rate of 4.5 mV·°C^{-1} from 60°C to 120°C, was revealed by an ET-AEC with activated cathode and plain Nickel anode in 11 M KOH solution, as presented in Figure 12.4b [52]. Integrating the enhanced reaction thermodynamics, electrode kinetics, and promoted ionic conduction, Figure 12.4c demonstrates an extraordinary performance of 3.75 A·cm^{-2} at 1.75 V illustrated by a foam-based ET-AEC consisting of Inconel cathode, YSZ diaphragm, and Ni/PTFE/Ag anode while adopting 45 wt% KOH solution at 200°C and 20 bar [62]. The performance enhancement could also be attributed to variations of underlying reaction mechanisms, as unexpected changes in Tafel slopes were observed on electrodes with elevating temperatures from <90°C to 120°C to 230°C [68, 69].

However, stability and related degradation phenomena of ET-AECs have not been thoroughly evaluated, while long-term experiments were generally limited to hundreds of hours, and Figure 12.4d depicts a 400-h test at 0.5 A·cm^{-2} and 200°C at 20 bar [62]. Accompanied by improved performance, elevated temperature and adoption of more concentrated alkaline solution concurrently promote the material corrosion, which causes an obvious weight wastage and uneven stress distribution to induce crackings [70]. In contrast to Ti and Zr entirely dissolved in 50 wt% KOH solution at 220°C after 100 h, Ni and Ni-based alloys as electrodes generally exhibited excellent stability at 100°C to 180°C [60, 70].

For composite porous diaphragms, their durability at elevated temperatures is affected by the detachment of hydrophilic coatings from substrates, restricting applicable temperature of <110°C [52]. The long-term stability of ceramic diaphragms also requires further study, and the incorporation of organic binders is needed to overcome the brittleness and provide sufficient mechanical strength. In addition, adopting polymers without hydrophilic treatment while exploiting an improved pressure to promote liquid electrolyte absorption enables to pursuit of enhanced stability, and stability of PTFE in extremely concentrated 50 wt% KOH solution at 260°C was probed, while a great tensile strength loss was observed on PBI in 30 wt% KOH solution at 80°C, and sulfone groups in PSU limits its operation at <125°C [51].

12.3.3 Challenges and Opportunities for AEC Stacks and Systems

Profiting from the prolonged lifetime and relatively lower cost, large-scale AEC systems with production rates of >500 $Nm^3 \cdot h^{-1}$ have been broadly constructed [9], and Figure 12.5a presents a typical diagram of industrial AEC system [19], consisting of an electrolysis stack and balance of plant (BOP) modules, including heat exchangers, gas-liquid separators, and pumps. Nevertheless, the higher specific energy consumption and narrow load range are major drawbacks of AEC systems, offering obstacles to efficient H_2 production and effective connections with renewable sources.

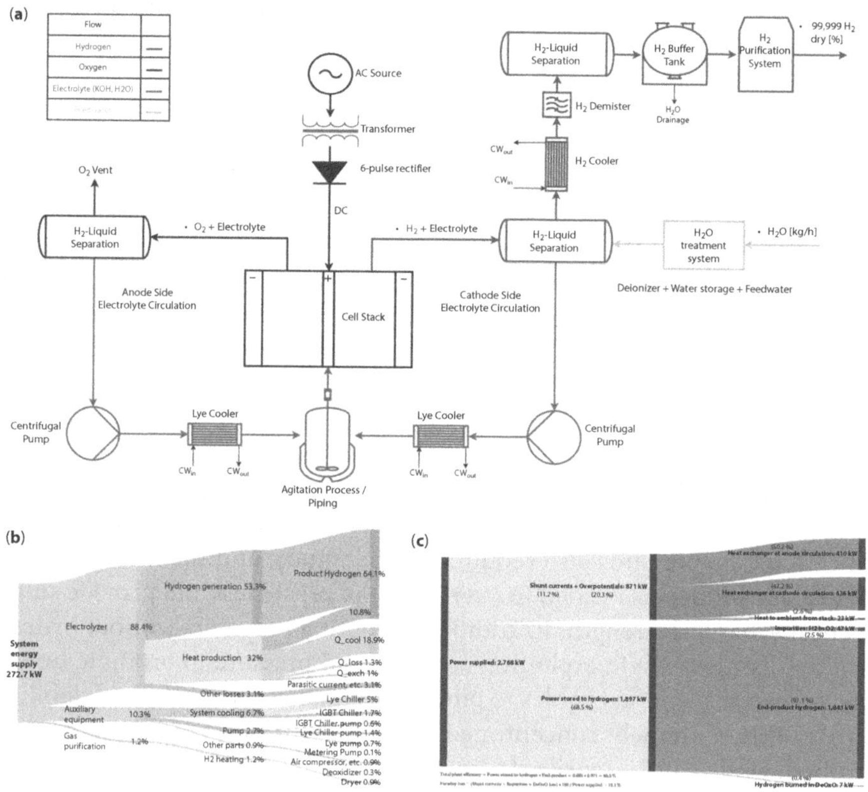

Figure 12.5 (a) Flow diagram of an industrial AEC system (Reprinted with permission [19]. Copyright 2022 Elsevier under CC BY 4.0). Energy flow of (b) a 250-kW AEC system under full-load operation at 85°C (Reprinted with permission [18]. Copyright 2023 Elsevier) and (c) a 3-MW AEC system at 70°C (Reprinted with permission [19]. Copyright 2022 Elsevier under CC BY 4.0).

12.3.3.1 Toward Thermoneutral Operation and Flexible Endo/Exothermic Transition

In addition to the substantial power consumed to propel electrolysis reaction, the unsatisfactory performance and relatively higher voltage of 1.8 to 2.3 V for conventional AEC systems concurrently lead to the generation of substantial excessive heat. As presented in Figure 12.5, b and c, 18.9% of the total power for a 250-kW AEC system existed as excessive heat under full-load operation at 85°C [18], and the portion was further improved to 30.63% for a 3-MW system at 70°C [19]. Therefore, additional cooling devices, including an industrial chiller and cooling water pump, are required to stabilize system temperature, and subsequent parasitic power typically accounts for >50% of BOP power [18].

The thermoneutral operation, originally symbolizing an electrolysis cell works without producing or absorbing additional heat, can be broadened into a system-level concept for eliminating excessive heat and parasitic cooling power, which has not been realized by conventional AEC systems due to the great mismatch between realistic and thermoneutral voltage but is feasible for ET-AEC systems.

Apart from the reduced applied voltage of 1.6 to 1.9 V [52, 58, 59, 62], elevated temperature concurrently improves the additional heat consumption for preheating feed water and extra water evaporation to reach a vapor partial pressure in equilibrium with liquid electrolyte, which broadens reaction-level V_{tn} into the thermoneutral voltage for an adiabatic electrolysis system $V_{tn, sys, adi}$ [17, 72]. In comparison with ~1.48 V of V_{tn}, the value of $V_{tn, sys, adi}$ at 10 bar was improved from 1.5 V at 80°C to 1.76 V at 160°C [17]. Integrated with promoted heat dissipation from components, such as electrolysis stack and gas-liquid separators [73], to the surrounding environment, a dissipation voltage V_{dis} increases $V_{tn, sys, adi}$ into a realistic system thermoneutral voltage $V_{tn, sys}$, which requires determined by experiments and can match with the reduced V_{cell}.

Therefore, ET-AEC systems possess the potential to accomplish highly efficient H_2 production through thermoneutral operation, while flexible endo/exothermic transition can also be achieved. In endothermic mode, the relatively lower temperature of 100°C to 150°C in contrast to other electrolysis cells, such as PCECs at 450°C to 700°C and SOECs at >700°C, loosens the requirement for input thermal energy, and other facilities including concentrated solar power (CSP) plants [74] and nuclear reactors [75] can be utilized as external heat sources. As the upper temperature limit for exploitable heat is also increased, exothermic ET-AEC systems can achieve the combined high-value H_2 and heat supply for heat-consuming sites, such as community households.

12.3.3.2 Broadening Load Range and Couplings with Renewable Sources

Another universal limitation for AEC systems is the relatively narrow load range with a lower limit of 20 to 25% and an upper limit of 100 to 110% [8, 9], restricting the couplings with fluctuating renewable sources. For operation at small current density, the reduction of gas production rates improves the relative influences of gas crossover, and anodic hydrogen to oxygen (HTO) of <2 vol% becomes a security criterion determining the lower limit of system operation, as the lower explosive limit (LEL) of H_2 in O_2 is 4 vol% [8]. Considering the system-level gas crossover is majorly induced by the mixing of circulated lye with dissolved H_2 and O_2 and is facilitated by improved gas solubility at increased pressure [76], adopting partial lye mixing with dynamically adjusting mixing proportion according to real-time gas analysis results [77], and appropriately reducing operational pressure [78] can be utilized as solutions.

To avoid severe material degradation at large current densities, overload operation is normally not permitted for AEC systems [8]. Herein, incorporating small-scale and compacted equipment, such as flow batteries and PEMECs (i.e., mixed electrolysis), offers an approach to consuming electricity beyond the upper limit, which also resolves the tardy response speed of AEC systems resulting from the bulky structures and large thermal capacity.

12.4 Polymer Electrolyte Membrane Electrolysis Cell (PEMEC)

Similarly, PEMECs work at a lower temperature of 25°C to 85°C, but their electrolytes vary into polymer membranes with intrinsic H^+ conductivity, while the catalyst layers (CLs) are fabricated by spraying electrocatalysts on the membranes or gas diffusion layers (GDLs), generating membrane electrode assemblies (MEAs) with compact catalyst coated on membrane (CCM) or GDL (CCG) structure [8]. At present, PEMECs have gradually stepped into the commercialization progress for their superior performance in contrast to AECs [9, 79], whereas further broadened applications are impeded by relatively higher expense induced by the dependence on PGMs as electrocatalysts, and ET-PEMECs have also been proposed to pursue an improved performance [80–87].

12.4.1 Electrocatalysts (Electrodes) and Electrolytes

12.4.1.1 HER Electrocatalysts (Cathodes)

As a highly acidic environment forbids the utilization of transition metals, such as Ni, Fe, and Co, to avoid severe corrosion and dissolution, electrodes for PEMECs strongly rely on PGMs and their oxides [88, 89], and Pt supported by carbon is currently the mostly used HER electrocatalyst [89]. Although a decreased Pt loading of 0.5 $mg{\cdot}cm^{-2}$ is typically adopted, substantial studies have been conducted to excavate substitutions to avoid the higher cost.

Fundamentally, combining Pt with non-precious metals including Ni [90], Co [91], and Fe [92] is a gentle and effective strategy to reduce Pt loading while maintaining an acceptable catalytic activity. At a decreased Pt loading of 5 wt% and merely 0.93 wt% in its surface regions, a binary Pt/Co catalyst supported by porous carbon unveiled an HER overpotential of 50 mV at 10 $mA{\cdot}cm^{-2}$ and Tafel slope of 27.1 $mV{\cdot}dec^{-1}$, which were superior compared with commercial 20 wt% Pt/C [91].

Moreover, integrating transition metals to form sulfides [93, 94], phosphides [95, 96], and carbides [97, 98] has been adopted to completely eliminate the dependence on PGMs. Among this group, MoS_2 arises as a representative dichalcogenide attracting numerous attention, as it possesses a special structure with highly active sites arranged in edge area and poorly conductive bulk [89, 99]. By engineering an unsaturated sulfur edge, the intrinsic catalytic activity of a MoS_2-C hybrid catalyst was boosted by optimizing free energy to adsorb hydrogen, revealing an HER overpotential of 136 mV at 10 $mA{\cdot}cm^{-2}$ in the acidic environment [100].

12.4.1.2 OER Electrocatalysts (Anodes)

In contrast, precious metal oxides, including IrO_2 and RuO_2, have been broadly employed as OER electrocatalysts in acidic media for their outstanding catalytic activity [101], whereas weakened stability and obvious weight loss induced by dissolution and aggregation were observed at positive potential [102, 103]. Doping Ru was validated as an effective approach to alleviating dissolution, while a satisfactory performance of 2.6 $A{\cdot}cm^{-2}$ at 1.8 V was achieved by a PEMEC with 1.5 $mg{\cdot}cm^{-2}$ $IrRuO_x$ anode and

0.5 mg·cm^{-2} Pt/C cathode at 90°C [104]. Meanwhile, Ir- and Ru-based perovskites [105–107] and pyrochlores [108, 109] have unveiled outstanding OER performance in an acidic environment, and a stable overpotential of 270 to 290 mV during the 30-h operation at 10 mA·cm^{-2} was exhibited by $IrO_x/SrIrO_3$ catalyst, which was ascribed to Sr leaching for generating highly active surface layers [107].

For OER electrocatalysts free of PGMs, Co and Mn oxides normally unveiled unstable features in acidic media [88], whereas improved interface quality between Co_3O_4 with substrates, such as carbon paper (CP) [110] and fluorine-doped tin oxide (FTO) [111], brought enhanced durability, and stabilizing of MnO_2 was also achieved with covering Mn sites by stable TiO_2 film [112]. Other materials including sulfides [113] and alloys [114] have been developed and measured, while the catalytic performance and stability of these non-noble electrocatalysts require further improvement.

12.4.1.3 *Electrolytes: Polymer Electrolyte Membranes (PEMs)*

Relying on the hopping (Grotthuss mechanism) and diffusion (vehicular mechanism) of protons through internal structures [115], polymer electrolyte membranes (PEMs) obtain H^+ conductivity, while Nafion series membranes with poly(perfluorosulfonic) acid (PFSA) groups on a tetrafluoroethylene (TFE) backbone are mainstream materials in PEMECs [15, 115], which generally reveal conductivity of ~100 mS·cm^{-1} [10].

As proton migration in Nafion-based membranes greatly relies on the intensity of the hydrogen bond network, an obvious reduction of H^+ conductivity is induced by a considerable loss of water content when the operating temperature is elevated to >100°C, limiting their applications in ET-PEMECs [10]. As a consequence, hydrophilic fillers including oxides [116–118] and metal-organic frameworks (MOFs) [119, 120] have been used to increase the water retention ability of Nafion-based membranes, while an improved conductivity of 280 mS·cm^{-1} was unveiled by a Nafion/ZIF-8 (Zn-based MOF)/graphene oxide (GO) membrane at 120°C and 40% relative humidity (RH) [119]. With relatively shorter side chains, the commercial Aquivion is another PFSA-based membrane, and its higher glass transition temperature (T_g) of 127°C over Nafion (67°C) indicates great applicability in ET-PEMECs [121].

As a widely used material for ET-PEM fuel cells (FCs), PBI has been proposed as an alternative membrane for ET-PEMECs based on its adequate thermal stability at >100°C. Although pure PBI exhibits poor conductivity [15], doping acids enable combinations with N-H sites in imidazole rings to promote proton hopping in as-generated hydrogen bond networks, and phosphoric acid (PA) is broadly adopted for its moderate vapor pressure and satisfactory stability at elevated temperature [122].

According to the polymer structure, every PBI unit can integrate with two PA molecules, while the excessive amount exists as free acid anions and water, which contribute the major portion of H^+ conductivity, and acid doping level (ADL, PA mole number for every mole PBI unit) thus obviously affects the performance of PA-doped PBI membranes [8]. Incorporating fillers, such as oxides [123–125], perovskites [126–128], and GOs [129, 130], effectively enhances the PA affinity and proposes improved ADLs (typically >10), and a remarkable conductivity of 310 $mS{\cdot}cm^{-1}$ was accomplished by a PA-doped PBI/Al_2O_3/SiO_2 membrane at 150°C under anhydrous condition [125].

Nevertheless, substantial PA absorption concurrently weakens the interaction between polymer chains and causes a considerably decreased tensile stress whereas increased elongation at break. In this regard, cross-linking appears as an approach to providing adequate mechanical strength, whereas ADLs are also reduced owing to the narrowed interchain space, causing a trade-off between stability and performance. Structural modifications of PBI backbones [131] or side chains [132] have also been used to improve ADLs of cross-linked PBI membranes.

Furthermore, various organic materials have also been proposed as alternative PEMs at elevated temperatures. With sulfonic acid groups and water molecules acting as proton conductors, sulfonated polyether ether ketone (SPEEK) acquires H^+ conductivity [15], which was improved to 198.5 $mS{\cdot}cm^{-1}$ at 170°C and 100% RH by introducing PBI to form acid-base interactions between sulfonic acid and nitrogen-containing groups, while a satisfactory tensile stress of 45.1 MPa was retained by the ionic cross-linking [133]. Applications as ET-PEMs were also evaluated on materials including cross-linked sulfonated poly (phenylene sulfone) (CSPPSU) [134], chitosan (CS) [135], as well as other PA-doped polymers [136–138].

12.4.2 Performance and Stability

With a similar working temperature of 60°C to 85°C, commercial PEMECs at present typically operate at 1.8 to 2.2 V, while an increased current density range of 1 to 2 $A{\cdot}cm^{-2}$ is enabled through the reduced resistance [8, 14]. Although an adequately prolonged durability of >10 years has been reached, large-scale applications of PEMECs are forbidden by their great dependence on PGMs, which can be analogously alleviated by elevating temperature to improve performance with intrinsically enhanced reaction thermodynamics and kinetics.

Presently, several research studies concentrating on ET-PEMECs operating at 100°C to 150°C have been conducted [80–87, 134, 139, 140], while utilizing an improved pressure to maintain liquid water [80–87] typically offered superior performance over steam electrolysis at atmospheric pressure [134, 139, 140]. Based on the test system shown in Figure 12.6a, a remarkable performance of 3 $A{\cdot}cm^{-2}$ at 1.86 V was realized by an

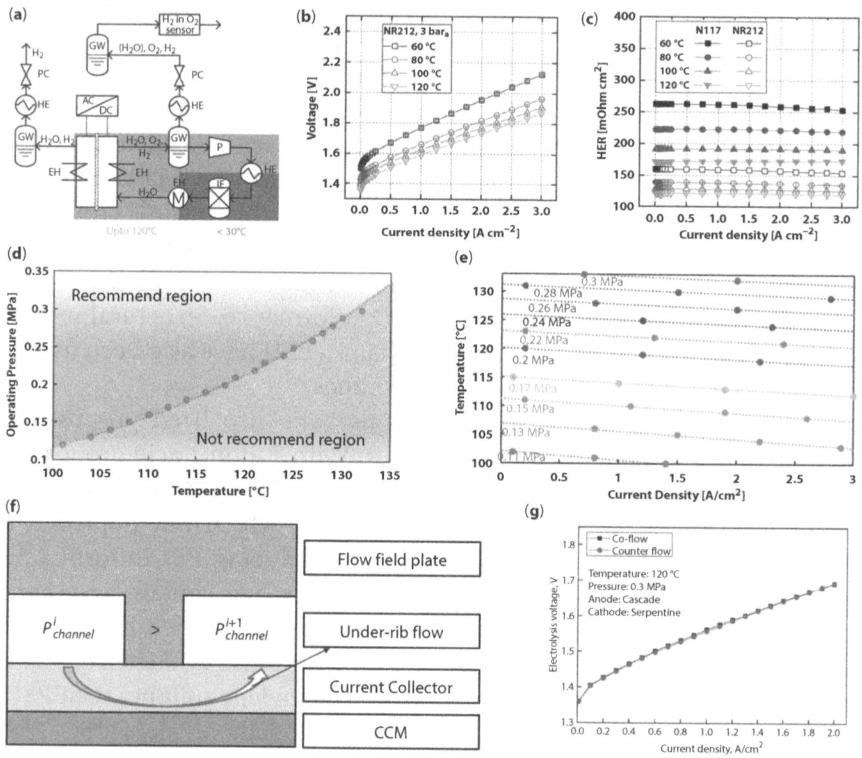

Figure 12.6 (a) Schematic illustration of a test bench for ET-PEMECs at up to 120°C, (b) polarization curves of an ET-PEMEC adopting Nafion NR212 membrane at 120°C at 3 bar, (c) high-frequency resistance (HFR) variation of ET-PEMECs with Nafion 117 and NR212 membrane at 60°C to 120°C at 3 bar (Reprinted with permission [83]. Copyright 2021 Elsevier under CC BY NC ND 4.0). Distribution of recommended pressure at (d) 2 $A{\cdot}cm^{-2}$ and (e) 0 to 3 $A{\cdot}cm^{-2}$ for efficient removal of anodic O_2 with temperature varying between 100°C to 135°C (Reprinted with permission [141]. Copyright 2019 The Electrochemical Society under CC BY 4.0). (f) Schematic illustration for the generation of under-rib flow in serpentine and cascade flow fields and (g) polarization curves of ET-PEMECs with co-flow and counterflow between cascade anodic flow field and serpentine cathodic flow field at 120°C at 3 bar (Reprinted with permission [80]. Copyright 2018 Elsevier).

ET-PEMEC at 120°C at 3 bar, and an obvious ohmic resistance reduction (decreased high-frequency resistance, HFR) was unveiled with increasing temperature from 60°C to 120°C, as depicted in Figure 12.6, b and c [83].

In spite of enhanced thermodynamics and kinetics at elevated temperatures, the practical realization of improved performance depends on the formation of gas-liquid two-phase flow at higher pressures. As steam electrolysis weakens the water retention capacity of PEMs, the liquid water supply for facilitated migration is required to avoid a severe decrease in hydration degree and H^+ conductivity, which improves the utilization of CLs and avoids an obvious η_{mass} for reactant depletion [8]. Meanwhile, considering the supplied water and anodic produced O_2 transport in opposite directions, a critical pressure exists for effective product removal, and its value is increased with elevating temperature and current density. As shown in Figure 12.6d, the recommended pressure at 2 $A{\cdot}cm^{-2}$ varied from 1.2 bar at 101°C to 2.9 bar at 130°C, and a map depicting its distribution at 0 to 3 $A{\cdot}cm^{-2}$ and 100°C to 135°C is presented in Figure 12.6e, giving instructions for selection of appropriate operational parameters [141]. As water is typically supplied to anodes in PEMECs (Figure 12.1b), a decreased membrane conductivity is also induced by cathodic water depletion. By exploiting the inter-channel pressure difference in serpentine and cascade flow fields shown in Figure 12.6f, the under-rib flow was generated to enhance the water delivery toward the cathode side, and Figure 12.6g demonstrates a performance of 2 $A{\cdot}cm^{-2}$ at 1.69 V was achieved at 120°C at 3 bar [80].

Currently, studies focusing on the stability and degradation issues of conventional PEMECs at ≤85°C have been extended into ~5000 hours, and their voltage degradation rates in the evaluated range of 1 to 2 $A{\cdot}cm^{-2}$ have been suppressed into <30 $\mu V{\cdot}h^{-1}$ [142–144]. As depicted in Figure 12.7a, excellent stability was achieved by a PEMEC with 0.08 $mg{\cdot}cm^{-2}$ Ir anode and 0.3 $mg{\cdot}cm^{-2}$ Pt cathode at 1.8 $A{\cdot}cm^{-2}$ and 80°C, which illustrated a voltage degradation rate of 27 $\mu V{\cdot}h^{-1}$ during >4500-h test [144]. Nevertheless, as elevated temperature promotes the material degradation, stability measurements of ET-PEMECs were restricted to a 300-h test at 2 $A{\cdot}cm^{-2}$ and 100°C, whereas Figure 12.7b demonstrates a substantially increased voltage degradation rate of 158 $\mu V{\cdot}h^{-1}$ [145].

As the pivotal part of PEMECs, the stability of MEAs is first challenged by elevated working temperature. Considering the PFSA groups in Nafion-based membranes, attack from radicals induces a severe polymer structure decomposition, while fluorine release rate (FRR) becomes a criterion to determine as-generated degradation, which revealed obviously higher values at 100°C over 60°C according to Figure 12.7c [145]. As kinetics for bubble nucleation and growth are concurrently facilitated, pronounced

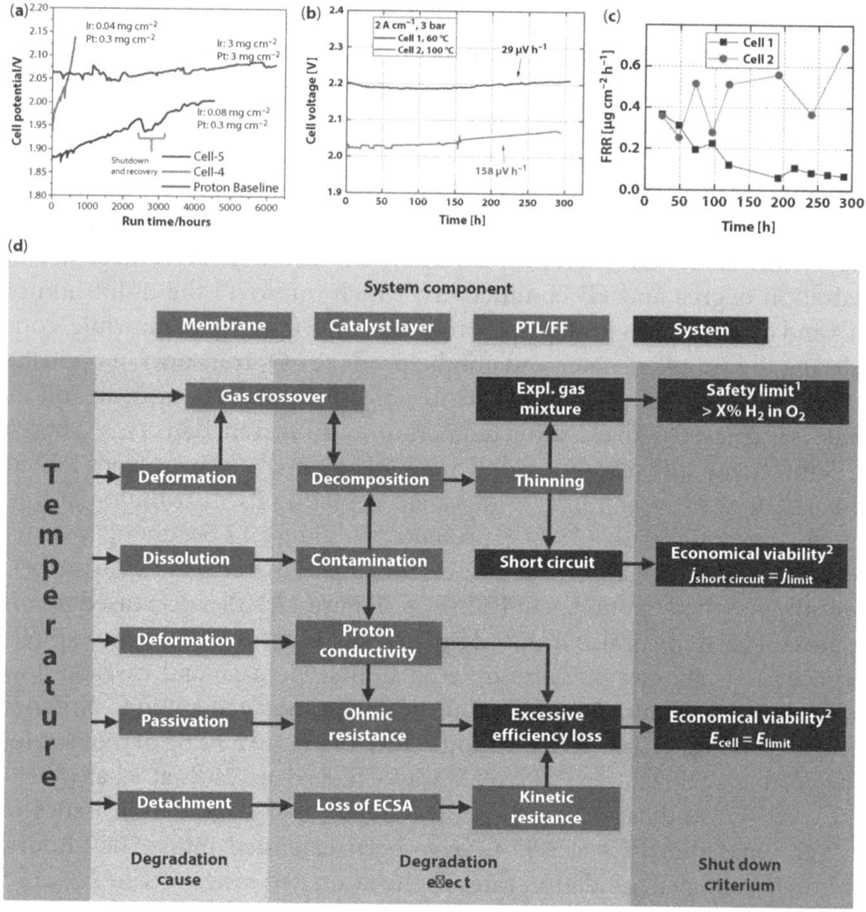

Figure 12.7 (a) Voltage evolution of PEMECs with 0.04 and 0.08 mg·cm^{-2} anodic Ir loading at 1.8 A·cm^{-2} and 80°C, and commercial PEMEC from Proton OnSite at 1.8 A·cm^{-2} and 50°C, with H_2 pressure all controlled at 27.6 bar (Reprinted with permission [144]. Copyright 2018 Elsevier). (b) Voltage and (c) fluorine release rate (FRR) of a PEMEC at 60°C and an ET-PEMEC at 100°C in a 300-h degradation measurement at 2 A·cm^{-2} and 3 bar. (d) Schematic illustration of degradation issues for ET-PEMEC components (Reprinted with permission [145]. Copyright 2021 The Electrochemical Society under CC BY NC ND 4.0).

impulse on CLs leads to dissolution and detachment of catalyst particles, as well as the electrochemical surface area (ECSA) reduction and subsequently weakened catalytic activity. The dissolved catalysts reversely facilitate the formation of radicals to decompose membranes and lead to a vicious circle for overall performance degradation of ET-PEMECs.

In addition, passivation and dissolution of GDLs and flow fields are also promoted at elevated temperatures, and Figure 12.7d gives an overview of degradation issues covering the entire ET-PEMECs, whereas more detailed studies are required to realize a prolonged lifetime.

12.4.3 Challenges and Opportunities for PEMEC Stacks and Systems

Profiting from excellent performance and acceptable durability, PEMECs have also realized the progress for industrialization, whereas their applications are restricted to small-scale systems with gas production rates of <500 $Nm^3 \cdot h^{-1}$, as higher construction costs are induced by the utilization of PGM-based electrocatalysts [8, 9].

However, highly compact structures bring intrinsically rapid response speeds to PEMECs [146–148] and offer enhanced adaptability with variable renewable sources [149, 150] and other industrial processes, such as methanol production [151]. Considering the mismatch between swift electrolysis reaction with relatively slower heat and mass transport, the generation of localized hot pots and reactant depletion adjacent to CLs require avoidance during highly dynamic operation [147]. In addition, dense structures of PEMs effectively forbid the gas crossover resulting from penetration driven by pressure difference, while PEMEC systems thus acquire advantages including decreased lower load limits of <10%, and increased applicable pressures of 50 to 80 bar [9].

From the perspective of thermal management, ET-PEMEC systems similarly need an enlarged amount of heat for preheating feed water and compensating heat dissipation, while the thermoneutral system operation can be realized by exploiting heat generation of electrolysis reaction and utilized to achieve optimized system-level energy/exergy efficiency [152]. Nevertheless, research concentrating on ET-PEMEC systems is in its infant stage and is currently limited to rare simulation studies [153].

12.5 Protonic Ceramic Electrolysis Cell (PCEC)

When the working temperature is further increased into an obviously higher range of >450°C, the electrolysis performance can be greatly enhanced, and the utilized materials generally transform into ceramics with sufficient thermal stability. Basically, ceramic-based electrolysis technologies can be classified into PCECs with H^+ conductivity at 450°C to

700°C, and SOECs relying on O^{2-} migration at >700°C. As the considerably elevated temperature also proposes great challenges to material stability, current R & D concerning PCECs are restricted to cell- and stack-level measurements [154], while area specific polarization resistance (ASR_p) and ohmic resistance (ASR_o) are major factors affecting the performance of PCECs [155].

12.5.1 Electrodes and Electrolytes

12.5.1.1 Hydrogen Electrodes (Cathodes)

Benefiting from considerably enhanced reaction kinetics, PCECs eliminate the dependence on PGMs, while Ni becomes the fundamental material for excellent thermal stability. To broaden active sites for an effective reaction from the electrode-electrolyte interface into the entire structure, Ni-based cermets by blending with ceramics acquiring protonic conductivity have been broadly adopted to fabricate hydrogen electrodes for PCECs [156], which are typically utilized as the supporting layer, and combine with an ultrathin electrolyte layer for a reduced ASR_o [8]. According to the electrode half-reactions in Equations 12.4 and 12.5, H_2O is supplied to anodes of PCECs, while the oxidation and migration of Ni can be avoided by anhydrous conditions.

With outstanding redox stability and electronic-ionic mixed conductivity, perovskites involving $La_{0.7}Sr_{0.3}V_{0.9}O_{3-\delta}$ (LSV) [157], $La_{0.8}Sr_{0.2}Cr_{0.5}Mn_{0.5}O_{3-\delta}$ (LSCM) [158], $La_{0.5}Sr_{0.5}Fe_{0.9}Mo_{0.1}O_{3-\delta}$ (LSFM) [159], and $SrFe_{0.75}Mo_{0.25}O_{3-\delta}$ (SFM) [160] have been selected as alternative cathodes for Ni-based cermets, whereas they suffer from inadequate catalytic activity and inferior performance.

12.5.1.2 Oxygen Electrodes (Anodes)

In comparison, the highly humidified and oxidative anodic environment greatly strengthens the stability demand for materials such as oxygen electrodes. Inspired by their initial applications in SOECs, mixed oxygen ion and electron conductors (MIECs), including $La_{0.6}Sr_{0.4}Co_{0.2}Fe_{0.8}O_{3-\delta}$ (LSCF) [161] and $Ba_{0.5}La_{0.5}CoO_{3-\delta}$ (BLC) [162], were developed as anodes for PCECs, whereas inadequate performance was induced by weakened catalytic activity at relatively lower temperature and active sites limited in electrode-electrolyte interface, as illustrated by the reduced current density of merely 0.2 A·cm^{-2} at 1.5 V and 600°C [161].

Similar to the preparation of Ni-based cermets, mixing MIECs with proton-conducting ceramics has been utilized to fabricate triple-conducting oxides (TCOs) with integrated $H^+/O^{2-}/e^-$ conductivity and thus effectively enhanced kinetics [163]. Additionally, the triple-conducting characteristic was also illustrated by double perovskites of $PrBa_{0.5}Sr_{0.5}Co_{1.5}Fe_{0.5}O_{5+\delta}$ (PBSCF) [164] and $BaGd_{0.8}La_{0.2}Co_2O_{6-\delta}$ (BGLC) [165], Ruddlesden-Popper (R-P) oxides of $Pr_2NiO_{4+\delta}$ (PNO) [166] and $La_{1.2}Sr_{0.8}NiO_{4-\delta}$ (LSN) [167], as well as misfit-layered $Ca_3Co_4O_{9-\delta}$ (CCO) [168] and its derivation $Gd_{0.3}Ca_{2.7}Co_{3.82}Cu_{0.18}O_{9-\delta}$ (GCCCO) [169]. Remarkably, an extraordinary performance of 9.35 $A{\cdot}cm^{-2}$ at 1.5 V was accomplished by a PCEC at 700°C, which was attributed to the special needle-type grain morphology of GCCCO electrode and considerably reduced ASR_p of merely 0.045 $\Omega{\cdot}cm^2$ [169].

Particularly, $PrBaCo_2O_{6-\delta}$ (PBC) with *in situ* exsolved $BaCo_2O_{3-\delta}$ (BCO) nanoparticles unveiled an excellent catalytic activity, whereas relatively higher cost of Pr and thermal expansion mismatch with other layers forbid its applications as bulk electrodes [170]. By coating the PBC/BCO catalyst onto the LSCF substrate, nevertheless, a substantial ASR_p reduction from 1.09 to 0.043 $\Omega{\cdot}cm^2$ at 600°C was realized by promoted O^{2-}/H^+ migration and H_2O/O_2 dissociation [171].

12.5.1.3 Electrolytes

Presently, mainstream electrolytes of PCECs include Ba-doped zirconate and cerate perovskites, whereas obvious boundary resistance and reduced ionic conductivity are induced during the sintering processes of $BaZrO_3$-based electrolytes [172], while the stability of $BaCeO_3$-based electrolytes is greatly weakened by their reactions with anodic humidity. Consequently, the composite $BaCe_{1-x}Zr_xO_3$-based electrolytes have been proposed to realize the compatibility between performance and stability [173].

Moreover, doping with Y and Yb to generate $BaCe_wZr_xY_yYb_zO_{3-\delta}$ (BCZYYb, $w + x + y + z = 1$) electrolytes has been widely employed to pursue further improved conductivity and durability, while the conductivity of BCZYYb7111 (i.e., $BaCe_{0.7}Zr_{0.1}Y_{0.1}Yb_{0.1}O_{3-\delta}$) was around 20 $mS{\cdot}cm^{-1}$ at 600°C [174], enabling a satisfactory PCEC performance of 4.42 $A{\cdot}cm^{-2}$ at 1.4 V at 700°C, and exhibiting 1000-h stability at 600°C [175].

Meanwhile, other proton conductors such as Hf-doped $BaHf_xCe_{0.8-x}Y_{0.1}Yb_{0.1}O_{3-\delta}$ (BHCYYb) [176], $La_2Ce_2O_7$ (LCO) [177], $Ba_7Nb_4MoO_{20}$ [178], $Ba_3CaNb_2O_9$, and $LaNbO_4$ [179] have also been regarded as potential electrolytes for PCECs.

12.5.2 Performance and Stability

Similarly, elevating temperature has the most pronounced effects on improving the performance of PCECs, and a remarkable example of 9.35 $A{\cdot}cm^{-2}$ at 1.5 V was accomplished at 700°C under 20% humidity, as demonstrated in Figure 12.8a [169]. Nevertheless, as the non-negligible electronic leakage of $BaCe_{1-x}Zr_xO_3$-based electrolytes induces parasitic oxidation, elevated working temperature promotes its competition with H_2 production and proposes an obviously decreased Faraday efficiency, which becomes another criterion for evaluating the performance of PCECs [180]. As illustrated in Figure 12.8b, although PCECs operating at 500°C typically unveiled Faraday efficiency of >95%, the values are reduced to 60~90% at 550°C to 600°C, and even 40~70% at 700°C to 800°C [179]. Through decreasing the electron hole concentration, Figure 12.8c indicates utilizing a higher operational pressure offers an improved Faraday efficiency, whereas strict demand for material strength and sealing structures are also proposed [165].

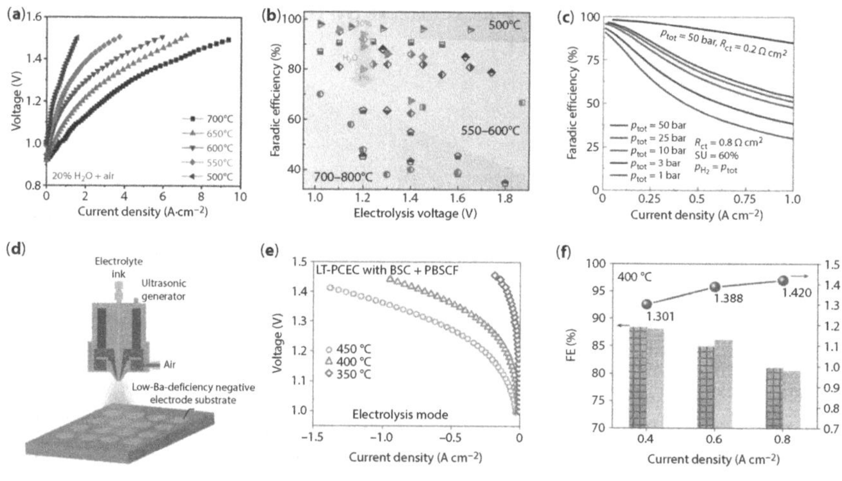

Figure 12.8 (a) Polarization curves of PCECs with GCCCO anodes at 500°C to 700°C and 20% H_2O (Reprinted with permission [169]. Copyright 2021 Royal Society of Chemistry). (b) Faraday efficiency distribution of PCECs at 500°C to 800°C (Reprinted with permission [179]. Copyright 2022 Taylor & Francis). (c) Influence of pressure on Faraday efficiency of PCECs (Reprinted with permission [165]. Copyright 2019 Springer). (d) Schematic illustration of ultrasonic spraying for fabricating ultrathin BCZYYb4411 electrolyte film, (e) polarization curves at 350°C to 450°C and (f) Faraday efficiency at 400°C of the corresponding low temperature (LT-) PCECs (Reprinted with permission [155]. Copyright 2023 Springer).

Therefore, lowering the working temperature is an attractive approach to pursuing an increased Faraday efficiency, whereas the weakened thermodynamics and reaction kinetics require overcomed to avoid obvious performance reduction. By ultrasonic spraying shown in Figure 12.8d, an ultrathin (~3 μm) BCZYYb4411 electrolyte with *in situ* formed $Ba_{0.62}Sr_{0.38}CoO_{3-\delta}$-$Pr_{1.44}Ba_{0.11}Sr_{0.45}Co_{1.32}Fe_{0.68}O_{6-\delta}$ (BSC+PBSCF) positive electrode was fabricated, while decreased ASR_o and ASR_p empowered the lowering of operating temperature to <450°C [155]. As presented in Figure 12.8, e and f, a compatibility between acceptable performance of 1.28 $A{\cdot}cm^{-2}$ at 1.4 V at 450°C with excellent Faraday efficiency of >80% at 400°C was realized, whereas further decreased temperature of 350°C brought a greatly weakened performance of 0.18 $A{\cdot}cm^{-2}$ at 1.46 V.

Another advantage brought by lowered operating temperature is alleviated material degradation and subsequent potential for realizing a prolonged lifetime. Presently, measurements of PCECs are yet restricted in a relatively short time and Figure 12.9a gives a 1833-h PCEC measurement at 1 $A{\cdot}cm^{-2}$ and 600°C unveiling a voltage degradation rate of 3.3 $\%{\cdot}k\ h^{-1}$, which is currently the longest evaluated durability [170]. In this regard, H_2O content in supplied air is a dominant factor affecting material stability. Owing to the reactions with H_2O, decomposition of $BaCe_{1-x}Zr_xO_3$-based electrolytes is normally observed when operating in highly humidified media [181], and incorporating a protective layer on the anode side was adopted for stability enhancement [177]. Exploiting higher reaction Gibbs free energy between $BaHfO_3$ and H_2O, replacing Zr with Hf was used to fabricate $BaHf_{0.1}Ce_{0.7}Yb_{0.2}O_{3-\delta}$ electrolyte with improved stability [182]. As anode materials typically involve alkaline earth metals, such as Ba and Sr, reactive with H_2O [183], humidity increase in the anodic environment also induces obvious cation segregation and leads to pronounced degradation of electrode activity and overall performance, as presented in Figure 12.9b [184]. Introducing elements with improved basicity [185] or developing anodes entirely free of alkaline earth elements [186, 187] have been adopted as solutions, and a 220-h stable operation with a reduced ASR_p of 0.05 $\Omega{\cdot}cm^2$ at 500°C was illustrated by a PCEC utilizing a 3D-$PrNi_{0.5}Co_{0.5}O_{3-\delta}$ (PNC) anode [186]. In comparison, the H_2O-free cathodic environment in PCECs effectively prevents obvious degradation of hydrogen electrodes, whereas the stability of widely used Ni-based cermets during longer operation time, such as >2000 h, requires further studied [188].

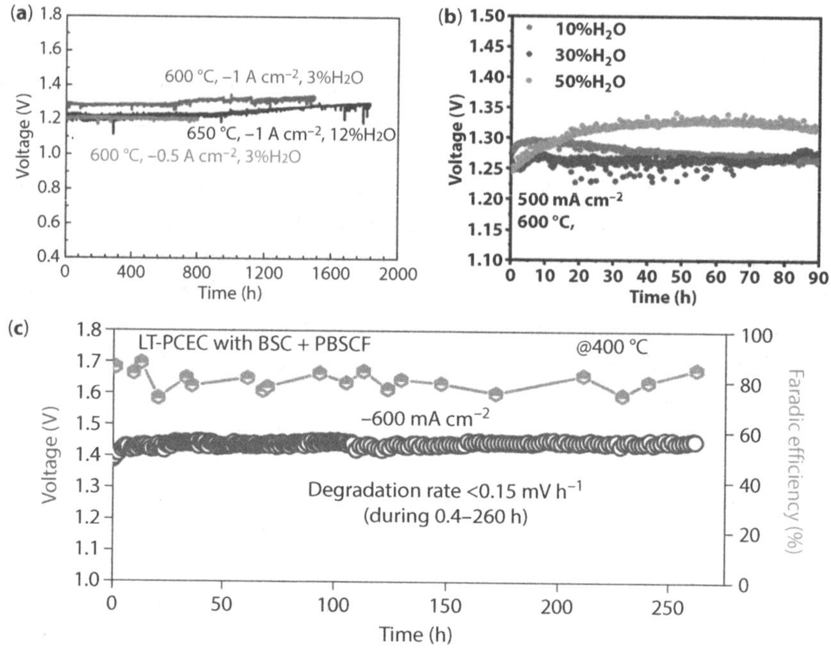

Figure 12.9 (a) 1833-h stability test of a PCEC at 1 A·cm^{-2} and 650°C under 12% H_2O (Reprinted with permission [170]. Copyright 2021 American Chemical Society). (b) Degradation of PCECs adopting PBSCF electrodes at 0.5 A·cm^{-2} and 600°C under 10 to 50% H_2O (Reprinted with permission [184]. Copyright 2022 American Chemical Society). (c) Voltage and Faraday efficiency evolution of an LT-PCEC during 260-h test at 0.6 A·cm^{-2} and 400°C (Reprinted with permission [155]. Copyright 2023 Springer).

In addition to focusing on the optimizations of electrode and electrolyte materials, appropriately lowering operation temperature is an appropriate method to mitigate degradation, and 400°C to 450°C is a promising range for simultaneously enhanced performance and stability. Although several research studies on this scale have been conducted [155, 189, 190], attention was concentrated on FC mode [189, 190], whereas stability and degradation issues especially under highly humidified media in electrolysis cell (EC) mode have not been thoroughly evaluated. As a representative, Figure 12.9c gives the voltage and Faraday efficiency evolution of an LT-PCEC with BSC+PBSCF positive electrode at 0.6 A·cm^{-2} and 400°C during the 260-h test, which unveiled a voltage degradation rate of <0.15 mV·h^{-1} [155].

12.5.3 Challenges and Opportunities for PCEC Stacks and Systems

Considering the inadequate material durability, the current R & D of PCECs are still in the infant stage for cell-level measurements, and difficulties exist for their scale-up and practical applications. Moreover, the incorporation of auxiliary components in PCEC stacks and systems, such as metallic current collectors and interconnects, as well as nonmetallic sealing structures, brings extra Cr- [191, 192] and Si-induced [183] degradation sources while further challenging electrode and electrolyte stability.

Nevertheless, voltage variation around V_{tn} enables PCECs to realize flexible transition between endothermic and exothermic mode, and their operating temperature range harmonious with chemical synthesis empowers the energy and mass couplings with various processes, such as methanol [193] and methane [194] production. By absorbing external heat provided by high-temperature steam from nuclear reactors, a considerably reduced specific energy consumption for H_2 production of 3.467 kWh·Nm^{-3} was achieved by a PCEC system [195], while other facilities, such as solar thermal plants [196], can also be utilized as external heat sources.

12.6 Solid Oxide Electrolysis Cell (SOEC)

With the highest operating temperature of >700°C, SOECs generally adopt ceramic materials analogous to PCECs whereas conducted ion becomes O^{2-} and possesses the best performance among various electrolysis technologies, which proposes great potential for accomplishing highly efficient and low-cost green H_2 production. However, the severest material degradation phenomena have also been exhibited by SOECs, restricting their applications as large-scale systems.

12.6.1 Electrodes and Electrolytes

12.6.1.1 Hydrogen Electrodes (Cathodes)

With expanded active sites in the bulk structures to enhance performance, Ni-based cermets are also widely used cathodes in SOECs, and Ni/YSZ is the representative material showing 0.18 Ω·cm^2 ASR_p at 900°C [197]. Relying on the integrated ionic and electronic conductivity to facilitate charge transfer, gadolinium-doped ceria (GDC) has been introduced for preparing the Ni/GDC composite cathodes, whereas Ni-based cermets suffer from obvious degradation in humidified cathodic media of SOECs [198].

Therefore, perovskites with excellent redox stability, such as $La_{0.6}Sr_{0.4}Ga_{0.3}Fe_{0.7}O_3$ (LSGF) [199], $La_{0.4}Sr_{0.4}TiO_3$ (LSTO) [200], and $Sr_2Fe_{1.5}Mo_{0.5}O_{3-\delta}$ (SFM) [201] have been utilized to replace the Ni-based cermets, while they generally showed insufficient catalytic activity and suffered from comparatively higher ASR_p. Performance improvement has been broadly accomplished by loading catalysts, and a reduced ASR_p of 0.3 $\Omega{\cdot}cm^2$ was demonstrated by a SOEC with $La_{0.43}Ca_{0.37}Ni_{0.06}Ti_{0.94}O_{3-\delta}$ (LCNT) cathode at 900°C, which also mitigated nanoparticle agglomeration by exsolving doped Ni into the lattice structure [202].

12.6.1.2 Oxygen Electrodes (Anodes)

Analogously, oxygen electrodes of SOECs typically rely on perovskites, while $La_{1-x}Sr_xMnO_{3+\delta}$ (LSM) [203] and LSCF [204] are the most adopted materials. Compared with the inadequate conductivity of LSM, the broadened active sites of LSCF offer an enhanced catalytic activity, which unveiled a reduced ASR_p of 0.16 $\Omega{\cdot}cm^2$ at 800°C [205]. Moreover, incorporating catalyst nanoparticles through infiltration has been validated effective for improved performance, and an ASR_p reduction from 0.08 to 0.06 $\Omega{\cdot}cm^2$ was illustrated by a SOEC at 800°C with adding $La_{0.6}Sr_{0.4}CoO_{3-\delta}$ (LSC) into LCSF anode [206].

With interstitial O^{2-} in the lattice structures to provide adequate catalytic activity, R-P oxides are another group of high-performance oxygen electrodes for SOECs [207]. Among first-order R-P oxides of $Ln_2NiO_{4+\delta}$ where Ln = La, Nd, and Pr, performance equivalent to LSCF of 0.1 $\Omega{\cdot}cm^2$ at 700°C has been reached by $Ln_{1.5}Pr_{0.5}NiO_{4+\delta}$ [208], while further enhanced activity was illustrated by Ni-based higher-order R-P oxides in FC mode. By incorporating 50 wt% GDC catalyst into $La_3Ni_2O_7$ cathode, an extremely decreased ASR_p of 0.03 $\Omega{\cdot}cm^2$ was exhibited by a SOFC at 700°C [209]. In contrast, applications of higher-order Fe- and Co-based R-P oxides are limited by their inadequate stability, especially under humidified media [210], and large thermal expansion coefficients [211].

12.6.1.3 Electrolytes

Benefiting from the outstanding stability and performance, as well as relatively lower cost, YSZ unveiling 90 $mS{\cdot}cm^{-1}$ conductivity at 1000°C [212] is currently a state-of-the-art electrolyte of SOECs, whereas its applications in the intermediate range of 600°C to 800°C are restricted by decreased performance of <30 mS·cm^{-1}. In comparison, the reduced ion radius of Sc^{3+} facilitates O^{2-} migration in scandia-stabilized zirconia (ScSZ) and brings an increased conductivity of 140 mS·cm^{-1} at 780°C [213], while ScSZ suffers from its obviously higher expense.

In addition to ZrO_2-based oxides, $La_{0.8}Sr_{0.2}Ga_{0.8}Mg_{0.2}O_{3-\delta}$ (LSGM) has been regarded as a promising electrolyte of SOECs for much higher conductivity of 170 mS·cm^{-1} at 800°C [214], and an ASR_o of 0.27 Ω·cm^2 was shown by a 300-μm LSGM electrolyte at 900°C [215]. However, an enlarged thickness is required to overcome the weakness of LSGM electrolytes, restricting further reduced ASR_o and improved performance [216].

12.6.2 Performance and Stability

Operating at the mostly increased range of >700°C, SOECs generally exhibit outstanding performance among various electrolysis technologies surpassing 1 to 2 A·cm^{-2} at 1.3 V. As presented in Figure 12.10a, insufficient performance of 1.41 A·cm^{-2} at 1.45 V was exhibited by a SOEC with nanocomposite $Sm_{0.5}Sr_{0.5}CoO_{3-\delta}$-$Ce_{0.8}Sm_{0.2}O_{1.9}$ (SSC-SDC) electrodes at 650°C, which was greatly enhanced to 4.08 A·cm^{-2} at 1.3 V at 800°C, while Figure 12.10b indicates a stable and decreased ASR_p of <0.05 Ω·cm^2 was maintained during a 2000-h measurement at 750°C [217]. As a representative, a SOEC equipping micro/nanochannel oxygen electrode (MC-SOEC) demonstrated an extraordinary performance of 5.96 A·cm^{-2} at 1.3 V at 800°C, whereas an obvious decrease of current density was also unveiled during a short time scale of <200 h, as depicted in Figure 12.10c [218]. Considering the pronounced effects of temperature elevation and existing SOECs generally operate at ≤900°C [219, 220], the further improved temperature of >1000°C proposes a novel concept of ET-SOEC, which possesses the feasibility to pursue an extremely decreased voltage of 0.9 to 1.1 V for efficient H_2 production, and has not been reached in current studies.

In contrast, adopting a higher operational pressure [221] and cathodic steam content [222] can be used to facilitate reaction kinetics, whereas their effects are relatively slight, and strengthened requirements for sealing structures and promoted degradation should be also considered. Introducing fuels, such as methanol [223], methane [224], and their mixtures [225], to anodic gas can be utilized to reduce voltage by consuming the produced O_2, and Figure 12.10d demonstrates the voltage of a methane-based solid oxide fuel–assisted electrolysis cell (SOFEC) could be reduced to 0.8 to 1.2 V [224].

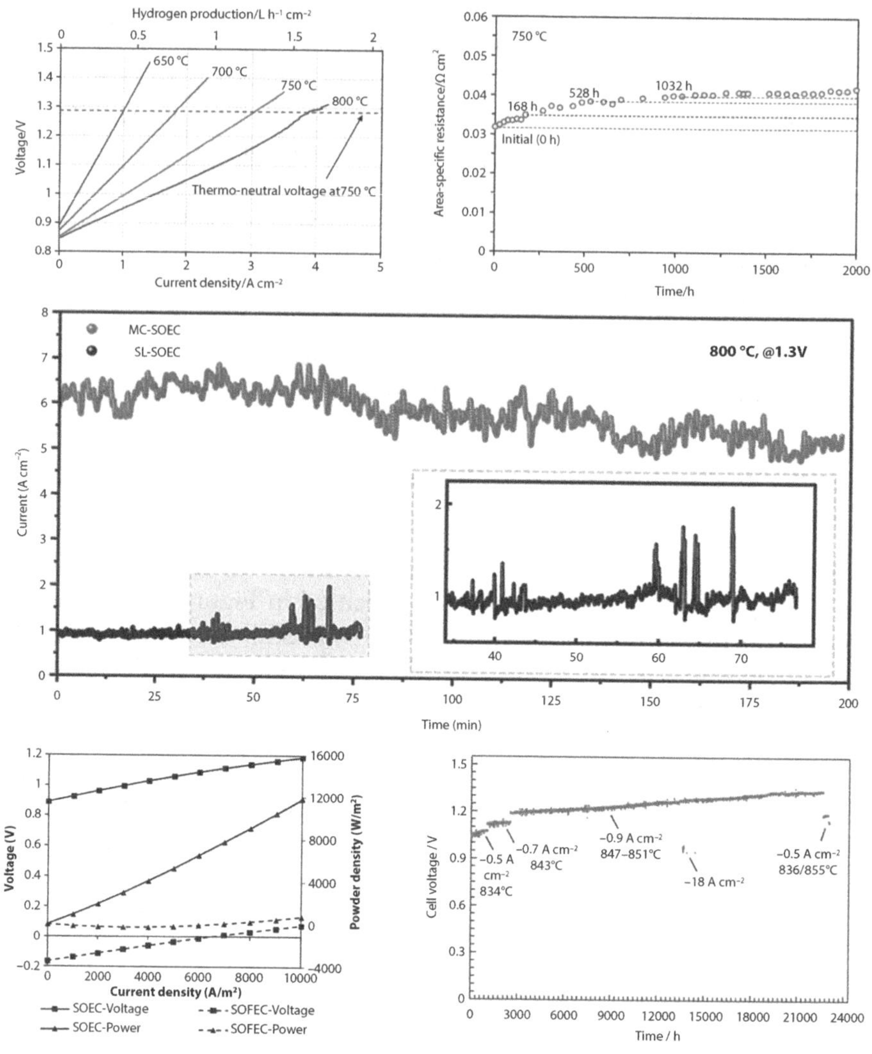

Figure 12.10 (a) Polarization curves of SOECs with SSC-SDC nanocomposite electrodes at 650°C to 800°C under 50% H_2O, and (b) ASR_p evolution during a 2000-h open circuit test at 750°C (Reprinted with permission [217]. Copyright 2019 Springer under CC BY 4.0). (c) Current density evolution of an MC-SOEC and SL-SOEC (SOEC with a sponge-like anode) at 1.3 V and 800°C (Reprinted with permission [218]. Copyright 2022 Wiley). (d) Polarization and power density curves of SOEC and methane-based SOFEC (Reprinted with permission [224]. Copyright 2016 Elsevier). (e) Voltage evolution of a SOEC during 23,000-h test at 834°C to 855°C (Reprinted with permission [226]. Copyright 2017 Elsevier).

However, the critical challenge for the progress of SOECs is severe material degradation under harsh conditions, and presently, the longest reported lifetime for SOECs is a 23,000-h measurement at 834°C to 855°C presented in Figure 12.10e [226], which still cannot satisfy the stability demand for industrial applications. Structural destruction is the major cause for reduced ionic conductivity and mechanical strength of widely adopted electrolytes, such as YSZ [227, 228], whereas degradation issues for electrodes are more complicated.

For broadly adopted LSM anodes, severe delamination is generated by grain destruction [229] and the formation of detrimental phases [230] under a highly oxidative environment, leading to an increased polarization resistance and potential for cell failure. As electrode delamination can be alleviated by a reduction of oxygen partial pressure, incorporating GDC nanoparticles [231] and tuning structures [232] can be used for stability reinforcement. Analogously, the degradation of LSCF anodes is induced by the Sr diffusion driven by an electric field, which results in oxygen vacancy depletion and detrimental phases, such as $SrZrO_3$ [233]. Similar phenomena were also observed on other perovskites including LSC [234] and $Ba_{0.5}Sr_{0.5}Co_{0.8}Fe_{0.2}O_{3-\delta}$ (BSCF) [235], while an enhanced delamination resistance was exhibited by R-P oxides for adequate oxygen mobility to reduce oxygen partial pressure [236], whereas their insufficient chemical stability at elevated temperatures of >750°C requires solved [237, 238].

Regarding Ni-based cermets as cathodes, their degradation majorly results from the destruction of microstructures. Under increased cathodic humidity, promoted degradation is induced by the reoxidation of Ni [239, 240], which was alleviated by increasing operational current density to accelerate the transformation of H_2O into H_2 [241]. Furthermore, causes for obviously decreased catalytic activity during long-term operation include particle coarsening and agglomeration [242, 243], as well as migration away from electrode-electrolyte interface [244, 245] generating Ni-depleting zones [244] or blocked active sites [246].

In contrast, perovskite-based hydrogen electrodes benefit from excellent redox stability [216], and tolerance to the contaminants in supplied gas, such as SO_2 [247] and CO_2 [248], whereas their instability at >800°C remains a major problem. Exsolving nanoparticles proposes enhanced durability by generating tight interactions with lattice structures [249], while a reduced voltage degradation rate of 48 $mV{\cdot}k\ h^{-1}$ during a 1000-h measurement at 0.67 $A{\cdot}cm^{-2}$ and 860°C was unveiled by a SOEC with $La_{0.65}Sr_{0.3}Cr_{0.85}Ni_{0.15}O_{3-\delta}$ (LSCrN) cathode, with coarsening and agglomeration of exsolved particles proposing additional degradation sources [250].

12.6.3 Challenges and Opportunities for SOEC Stacks and Systems

For extraordinary electrolysis performance and great potential to accomplish low-cost green H_2 production, numerous research studies focusing on R & D of SOECs have been conducted in recent years, while a prolonged durability of >20,000 h with a decreased voltage degradation rate of <1%·k h^{-1} has been realized [16]. The cost of SOEC stacks and systems has also been effectively reduced to 25 to 40 €·kW^{-1}, which is comparable to current AEC and PEMEC systems [251].

However, the current lifetime of SOECs yet cannot fulfill the demands for industrial applications, and the existing systems are comparatively small-scale equipment with H_2 production rates of <10 $Nm^3 \cdot h^{-1}$ [9]. Considering the usage of auxiliary components, including interconnects, current collectors, and sealing layers, in SOEC stacks, attendant elements of metallic Fe and Cr as well as nonmetallic B and Si also cause the formation of detrimental products, such as $SrCrO_4$, $LaBO_3$ [252] and $Si(OH)_4$ [253, 254], depositing on electrode surfaces and generating additional degradation. Analogous to the situation for PCECs, an appropriately lowered temperature appears as a promising strategy to achieve improved stability while maintaining a slightly but acceptable performance (e.g., 1 to 2 $A \cdot cm^{-2}$ at 1.4 V at ~700°C), which can effectively facilitate the broadening of SOECs.

Furthermore, the particular capacity of SOECs to flexibly transform between EC and FC mode proposes a widened load range of −100% to 100% [9], offering enhanced adaptability to the fluctuating renewable sources [255], and the ability to act as a bridge for realizing large-scale power storage [256]. When further considering the system-level thermal management, the operation of SOECs under V_{tn} enables the integration with various processes, such as the chemical synthesis of methanol [257] and ammonia, while external heat can be utilized to preheat the supplied gases and realize an improved overall efficiency, whereas stringent demand for input thermal energy is proposed by extremely elevated temperature.

12.7 Summary

Collectively, water electrolysis is an attractive technology for large-scale green H_2 generation, elimination of dependence on fossil fuels, and more sustainable developments for human society in the future. Substantial efforts to promote the progress of this field are continuously ongoing, while various electrolysis technologies confront their specific challenges and opportunities.

For traditional AECs and PEMECs working at 25°C to 85°C, adequate durability empowers their industrialization, whereas relatively higher specific energy consumption and dependence on PGMs are induced by inadequate performance, leading to uncompetitive costs restricting further expansion. Herein, ET-AECs and ET-PEMECs with increased temperature of 100°C to 250°C have been proposed and validated as effective solutions, while their stability and degradation during operation at a longer time scale should be more thoroughly evaluated.

In contrast, outstanding performance has been exhibited by PCECs at 450°C to 700°C and SOECs at >700°C for enhanced thermodynamics and reaction kinetics, but their obviously higher operating temperatures considerably promote material degradation and bring lifetime insufficient for industrial applications, limiting the R & D in cell- and stack-level experiments. In this aspect, an appropriately lowered temperature (e.g., 400°C to 450°C for PCECs [155] and 550°C to 600°C for SOECs) proposes a route to accomplish a performance-stability compatibility and offers attendant advantages such as improved Faraday efficiency for PCECs.

Moreover, electrolysis cells, and especially their systems, are the integration of the electrochemical reaction with thermal processes, which can also be flexibly adjusted by adopting different operation parameters (e.g., varying V_{cell} around V_{tn} to the realize reaction-level endo/exothermic transition). As a consequence, optimizations in electrolysis systems by prudent thermal management, such as using heat generated by reaction for thermoneutral operation and maximized efficiency of ET-AEC [17, 72] and ET-PEMEC [152] systems, as well as connections with other facilities *via* energy-mass couplings, like absorbing heat from nuclear reactions to maintain PCEC stack operation [195], shed light on the more intelligent applications of electrolysis technologies. Although scarce attention has been concentrated here at present, we strongly believe this proposes instructions for future studies.

References

1. Bard, A.J., Faulkner, L.R., White, H.S., *Electrochemical methods: fundamentals and applications*, John Wiley & Sons, New York, 2022.
2. Guan, D., Wang, B., Zhang, J., Shi, R., Jiao, K., Li, L., Wang, Y., Xie, B., Zhang, Q., Yu, J., Zhu, Y., Shao, Z., Ni, M., Hydrogen society: from present to future. *Energy Environ. Sci.*, 16, 4926–4943, 2023.

3. Aminudin, M.A., Kamarudin, S.K., Lim, B.H., Majilan, E.H., Masdar, M.S., Shaari, N., An overview: Current progress on hydrogen fuel cell vehicles. *Int. J. Hydrog. Energy*, 48, 4371–4388, 2023.
4. Guerra, O.J., Eichman, J., Kurtz, J., Hodge, B.-M., Cost Competitiveness of Electrolytic Hydrogen. *Joule*, 3, 2425–2443, 2019.
5. Sinsel, S.R., Riemke, R.L., Hoffmann, V.H., Challenges and solution technologies for the integration of variable renewable energy sources-a review. *Renew. Energy*, 145, 2271–2285, 2020.
6. Gür, T.M., Carbon Dioxide Emissions, Capture, Storage and Utilization: Review of Materials, Processes and Technologies. *Prog. Energy Combust. Sci.*, 89, 100965, 2022.
7. Younas, M., Shafique, S., Hafeez, A., Javed, F., Rehman, F., An Overview of Hydrogen Production: Current Status, Potential, and Challenges. *Fuel*, 316, 123317, 2022.
8. Zhang, W., Liu, M., Gu, X., Shi, Y., Deng, Z., Cai, N., Water Electrolysis toward Elevated Temperature: Advances, Challenges and Frontiers. *Chem. Rev.*, 123, 7119–7192, 2023.
9. Buttler, A. and Spliethoff, H., Current status of water electrolysis for energy storage, grid balancing and sector coupling via power-to-gas and power-to-liquids: A review. *Renew. Sustain. Energy Rev.*, 82, 2440–2454, 2018.
10. Liu, R.T., Xu, Z.L., Li, F.M., Chen, F.Y., Yu, J.Y., Yan, Y., Chen, Y., Xia, B.Y., Recent advances in proton exchange membrane water electrolysis. *Chem. Soc Rev.*, 52, 5652–5683, 2023.
11. Shi, Y., Cai, N., Cao, T., Zhang, J., *High-Temperature Electrochemical Energy Conversion and Storage: Fundamentals and Applications*, CRC Press, Taylor & Francis Group, London, 2017.
12. Chatenet, M., Pollet, B.G., Dekel, D.R., Dionigi, F., Deseure, J., Millet, P., Braatz, R.D., Bazant, M.Z., Eikerling, M., Staffell, I., Balcombe, P., Shao-Horn, Y., Schäfer, H., Water electrolysis: from textbook knowledge to the latest scientific strategies and industrial developments. *Chem. Soc. Rev.*, 51, 4583–4762, 2022.
13. Olivier, P., Bourasseau, C., Bouamama, P.B., Low-temperature electrolysis system modelling: A review. *Renew. Sustain. Energy Rev.*, 78, 280–300, 2017.
14. David, M., Ocampo-Martínez, C., Sánchez-Peña, R., Advances in alkaline water electrolyzers: A review. *J. Energy Storage*, 23, 392–403, 2019.
15. Sun, X., Simonsen, S., Norby, T., Chatzitakis, A., Composite Membranes for High Temperature PEM Fuel Cells and Electrolysers: A Critical Review. *Membranes*, 9, 83, 2019.
16. Hauch, A., Kungas, R., Blennow, P., Hansen, A.B., Hansen, J.B., Mathieser, B.V., Mogensen, M.B., Recent advances in solid oxide cell technology for electrolysis. *Science*, 370, eaba6118, 2020.
17. Lohmann-Richters, F.P., Renz, S., Lehnert, W., Müller, M., Carmo, M., Review-Challenges and Opportunities for Increased Current Density in Alkaline

Electrolysis by Increasing the Operating Temperature. *J. Electrochem. Soc.*, 168, 114501, 2021.
18. Ding, S., Guo, B., Hu, S., Tian, Z., Gu, J., Zhang, T., Yang, F., Ouyang, M., Experimental and modeling study on energy flow of 250 kW alkaline water electrolysis system under steady state conditions and cold start process. *Fuel*, 350, 128799, 2023.
19. Sakas, G., Ibáñez-Rioja, A., Ruuskanen, V., Kosonen, A., Ahola, J., Bergmann, O., Dynamic energy and mass balance model for an industrial alkaline water electrolyzer plant process. *Int. J. Hydrogen Energy*, 47, 4328–4345, 2022.
20. Zeng, K. and Zhang, D., Recent progress in alkaline water electrolysis for hydrogen production and applications. *Prog. Energy Combust. Sci.*, 36, 307–326, 2010.
21. Suen, N.-T., Hung, S.-F., Quan, Q., Zhang, N., Xu, Y.-J., Chen, H.M., Electrocatalysis for the oxygen evolution reaction: recent development and future perspectives. *Chem. Soc. Rev.*, 46, 337–365, 2011.
22. Ďurovič, M., Hnát, J., Bouzek, K., Electrocatalysts for the hydrogen evolution reaction in alkaline and neutral media-A comparative review. *J. Power Sources*, 493, 229708, 2021.
23. González-Buch, C., Herraiz-Cardona, I., Ortega, E., García-Antón, J., Pérez-Herranz, V., Study of the catalytic activity of 3D macroporous Ni and NiMo cathodes for hydrogen production by alkaline water electrolysis. *J. Appl. Electrochem.*, 46, 791–803, 2016.
24. Bao, F., Kemppainen, E., Dorbandt, I., Bors, R., Xi, F., Schlatmann, R., van de Krol, R., Calnan, S., Understanding the Hydrogen Evolution Reaction Kinetics of Electrodeposited Nickel-Molybdenum in Acidic, Near-Neutral, and Alkaline Conditions. *ChemElectroChem*, 8, 195–208, 2021.
25. Zhang, J., Wang, T., Liu, P., Liao, Z., Liu, S., Zhuang, X., Chen, M., Zschech, E., Feng, X., Efficient hydrogen production on MoNi4 electrocatalysts with fast water dissociation kinetics. *Nat. Commun.*, 8, 15437, 2017.
26. Wang, Y., Wang, Y., Bai, J., Lau, W.-M., Efficient Self-Supported Bifunctional NiMo Alloy Electrocatalysts for Water Splitting in Alkaline Environment. *ChemistrySelect*, 7, e202200468, 2022.
27. Fang, M., Gao, W., Dong, G., Xia, Z., Yip, S.P., Qin, Y., Qu, Y., Ho, J.C., Hierarchical NiMo-based 3D electrocatalysts for highly-efficient hydrogen evolution in alkaline conditions. *Nano Energy*, 27, 247–254, 2016.
28. Raj, I.A. and Venkatesan, V.K., Characterization of nickel-molybdenum and nickel-molybdenum-iron alloy coatings as cathodes for alkaline water electrolysers. *Int. J. Hydrogen Energy*, 13, 215–223, 1988.
29. Raj, I.A. and Vasu, K.I., Transition metal-based hydrogen electrodes in alkaline solution - electrocatalysis on nickel based binary alloy coatings. *J. Appl. Electrochem.*, 20, 32–38, 1990.
30. You, B. and Sun, Y., Hierarchically Porous Nickel Sulfide Multifunctional Superstructures. *Adv. Energy Mater.*, 6, 1502333, 2016.

31. Xing, Z., Yang, X., Asiri, A.M., Sun, X., Three-Dimensional Structures of MoS_2@Ni Core/Shell Nanosheets Array toward Synergetic Electrocatalytic Water Splitting. *ACS Appl. Mater. Interfaces*, 8, 14521–14526, 2016.
32. Anantharaj, S., Sugime, H., Noda, S., Surface amorphized nickel hydroxy sulphide for efficient hydrogen evolution reaction in alkaline medium. *Chem. Eng. J.*, 408, 127275, 2021.
33. Jin, L., Xia, H., Huang, Z., Lv, C., Wang, J., Humphrey, M.G., Zhang, C., Phase separation synthesis of trinickel monophosphide porous hollow nanospheres for efficient hydrogen evolution. *J. Mater. Chem. A*, 4, 10925–10932, 2016.
34. Tian, J., Liu, Q., Asiri, A.M., Sun, X., Self-Supported Nanoporous Cobalt Phosphide Nanowire Arrays: An Efficient 3D Hydrogen-Evolving Cathode over the Wide Range of pH 0–14. *J. Am. Chem. Soc.*, 136, 7587–7590, 2014.
35. Xing, Z., Li, Q., Wang, D., Yang, X., Sun, X., Self-supported nickel nitride as an efficient high-performance three-dimensional cathode for the alkaline hydrogen evolution reaction. *Electrochim. Acta*, 191, 841–845, 2016.
36. Lang, X., Qadeer, M.A., Shen, G., Zhang, R., Yang, S., An, J., Pan, L., Zou, J.-J., A Co-Mo_2N composite on a nitrogen-doped carbon matrix with hydrogen evolution activity comparable to that of Pt/C in alkaline media. *J. Mater. Chem. A*, 7, 20579–20583, 2019.
37. Chen, Z., Gong, W., Cong, S., Wang, Z., Song, G., Pan, T., Tang, X., Chen, J., Lu, W., Zhao, Z., Eutectoid-structured WC/W_2C heterostructures: A new platform for long-term alkaline hydrogen evolution reaction at low overpotentials. *Nano Energy*, 68, 104335, 2020.
38. Fan, X., Zhou, H., Guo, X., WC Nanocrystals Grown on Vertically Aligned Carbon Nanotubes: An Efficient and Stable Electrocatalyst for Hydrogen Evolution Reaction. *ACS Nano*, 9, 5125–5134, 2015.
39. Zhou, Q., Zhao, G., Rui, K., Chen, Y., Xu, X., Dou, S.X., Sun, W., Engineering additional edge sites on molybdenum dichalcogenides toward accelerated alkaline hydrogen evolution kinetics. *Nanoscale*, 11, 717–724, 2019.
40. Zhao, X., Zhao, Y., Huang, B., Cai, W., Sui, J., Yang, Z., Wang, H.-E., $MoSe_2$ nanoplatelets with enriched active edge sites for superior sodium-ion storage and enhanced alkaline hydrogen evolution activity. *Chem. Eng. J.*, 382, 123047, 2020.
41. Kim, J.-E., Bae, K.-K., Park, C.-S., Jeong, S.-U., Baik, K.-H., Kim, J.-W., Kang, K.-S., Lee, K.-B., Kim, Y.-H., Electrochemical characterization of Raney nickel electrodes prepared by atmospheric plasma spraying for alkaline water electrolysis. *J. Ind. Eng. Chem.*, 70, 160–168, 2019.
42. Wendt, H. and Plzak, V., Electrocatalytic and Thermal Activation of Anodic Oxygen- And Cathodic Hydrogen-Evolution in Alkaline Water Electrolysis. *Electrochim. Acta*, 28, 27–34, 1983.
43. Potvin, E. and Brossard, L., Electrocatalytic activity of Ni-Fe anodes for alkaline water electrolysis. *Mater. Chem. Phys.*, 31, 311–318, 1992.

44. Chen, J.S., Ren, J., Shalom, M., Fellinger, T., Antonietti, M., Stainless Steel Mesh-Supported NiS Nanosheet Array as Highly Efficient Catalyst for Oxygen Evolution Reaction. *ACS Appl. Mater. Interfaces*, 8, 5509–5516, 2016.
45. Gao, M.-R., Cao, X., Gao, Q., Xu, Y.-F., Zheng, Y.-R., Jiang, J., Yu, S.-H., Nitrogen-Doped Graphene Supported $CoSe_2$ Nanobelt Composite Catalyst for Efficient Water Oxidation. *ACS Nano*, 8, 3970–3978, 2014.
46. Chen, P., Xu, K., Tong, Y., Li, X., Tao, S., Fang, Z., Chu, W., Wu, X., Wu, C., Cobalt nitrides as a class of metallic electrocatalysts for the oxygen evolution reaction. *Inorg. Chem. Front.*, 3, 236–242, 2016.
47. Stern, L.-A., Feng, L., Song, F., Hu, X., Ni_2P as a Janus catalyst for water splitting: the oxygen evolution activity of Ni_2P nanoparticles. *Energy Environ. Sci.*, 8, 2347–2351, 2015.
48. Brauns, J., Schönebeck, J., Kraglund, M.R., Aili, D., Hnát, J., Itka, J., Mues, W., Jensen, J.O., Bouzek, K., Turek, T., Evaluation of Diaphragms and Membranes as Separators for Alkaline Water Electrolysis. *J. Electrochem. Soc.*, 168, 14510, 2021.
49. Li, Y., Zhang, T., Ma, J., Deng, X., Gu, J., Yang, F., Ouyang, M., Study the effect of lye flow rate, temperature, system pressure and different current density on energy consumption in catalyst test and 500W commercial alkaline water electrolysis. *Mater. Today Phys.*, 22, 100606, 2022.
50. Ding, S., Guo, B., Hu, S., Gu, J., Yang, F., Li, Y., Dang, J., Liu, B., Ma, J., Analysis of the effect of characteristic parameters and operating conditions on exergy efficiency of alkaline water electrolyzer. *J. Power Sources*, 537, 231532, 2022.
51. Renaud, R. and LeRoy, R.L., Separator materials for use in alkaline water electrolysers. *Int. J. Hydrogen Energy*, 7, 155–166, 1982.
52. Ehlers, J.C., Feidenhans L, A.A., Therkildsen, K.T., Larrazábal, G.O., Affordable Green Hydrogen from Alkaline Water Electrolysis: Key Research Needs from an Industrial Perspective. *ACS Energy Lett.*, 8, 1502–1509, 2023.
53. Aili, D., Kraglund, M.R., Rajappan, S.C., Serhiichuk, D., Xia, Y., Deimede, V., Kallitsis, J., Bae, C., Jannasch, P., Henkensmeier, D., Jensen, J.O., Electrode Separators for the Next-Generation Alkaline Water Electrolyzers. *ACS Energy Lett.*, 8, 1900–1910, 2023.
54. Lee, H., II, Dung, D.T., Kim, J., Pak, J.H., Kim, S.K., Cho, H.S., Cho, W.C., Kim, C.H., The synthesis of a Zirfon-type porous separator with reduced gas crossover for alkaline electrolyzer. *Int. J. Energy Res.*, 44, 1875–1885, 2019.
55. Lee, J., Lee, C.S., Lee, J., Kim, S.-K., Cho, H.-S., Kim, M.J., Cho, W., Joo, J., Kim, C.-H., Cerium Oxide-Polysulfone Composite Separator for an Advanced Alkaline Electrolyzer. *Polymers*, 12, 2821, 2020.
56. Lee, J.W., Lee, J.H., Lee, C.S., Cho, H.-S., Kim, M.J., Kim, S.-K., Joo, J.H., Cho, W.-C., Kim, C.-H., Cellulose nanocrystals-blended zirconia/polysulfone composite separator for alkaline electrolyzer at low electrolyte contents. *Chem. Eng. J.*, 428, 131149, 2022.

57. Allebrod, F., Chatzichristodoulou, C., Mollerup, P.L., Mogensen, M.B., Electrical conductivity measurements of aqueous and immobilized potassium hydroxide. *Int. J. Hydrogen Energy*, 37, 16505–16514, 2012.
58. Allebrod, F., Chatzichristodoulou, C., Mogensen, M.B., Alkaline electrolysis cell at high temperature and pressure of 250 °C and 42 bar. *J. Power Sources*, 229, 22–31, 2013.
59. Allebrod, F., Chatzichristodoulou, C., Mogensen, M.B., Cobalt and molybdenum activated electrodes in foam based alkaline electrolysis cells at 150-250 °C and 40 bar. *J. Power Sources*, 255, 394–403, 2014.
60. Wendt, H. and Hofmann, H., Ceramic diaphragms for advanced alkaline water electrolysis. *J. Appl. Electrochem.*, 19, 605–610, 1989.
61. Wendt, H. and Hofmann, H., Cermet diaphragms and integrated electrode-diaphragm units for advanced alkaline water electrolysis. *Int. J. Hydrogen Energy*, 10, 375–381, 1985.
62. Chatzichristodoulou, C., Allebrod, F., Mogensen, M.B., High Temperature Alkaline Electrolysis Cells with Metal Foam Based Gas Diffusion Electrodes. *J. Electrochem. Soc.*, 163, F3036–F3040, 2016.
63. Li, D., Park, E.J., Zhu, W., Shi, Q., Zhou, Y., Tian, H., Lin, Y., Serov, A., Zulevi, B., Baca, E.D., Fujimoto, C., Chung, H.T., Kim, Y.S., Highly quaternized polystyrene ionomers for high performance anion exchange membrane water electrolysers. *Nat. Energy*, 5, 378–385, 2020.
64. Abbasi, R., Setzler, B.P., Lin, S., Wang, J., Zhao, Y., Xu, H., Pivovar, B., Tian, B., Chen, X., Wu, G., Yan, Y., A Roadmap to Low-Cost Hydrogen with Hydroxide Exchange Membrane Electrolyzers. *Adv. Mater.*, 31, 1805876, 2019.
65. Chen, N., Paek, S.Y., Lee, J.Y., Park, J.H., Lee, S.Y., Lee, Y.M., High-performance anion exchange membrane water electrolyzers with a current density of 7.68 A cm^{-2} and a durability of 1000 hours. *Energy Environ. Sci.*, 14, 6338–6348, 2021.
66. Trisno, M.L.A., Dayan, A., Lee, S.J., Egert, F., Gerle, M., Kraglund, M.R., Jensen, J.O., Aili, D., Roznowska, A., Michalak, A., Park, H.S., Razmjooei, F., Ansar, S.-A., Henkensmeier, D., Reinforced gel-state polybenzimidazole hydrogen separators for alkaline water electrolysis. *Energy Environ. Sci.*, 15, 4362–4375, 2022.
67. Kraglund, M.R., Carmo, M., Schiller, G., Ansar, S.A., Aili, D., Christensen, E., Jensen, J.O., Ion-solvating membranes as a new approach towards high rate alkaline electrolyzers. *Energy Environ. Sci.*, 12, 3313–3318, 2019.
68. Divisek, J. and Schmitz, H., A bipolar cell for advanced alkaline water electrolysis. *Int. J. Hydrogen Energy*, 7, 703–710, 1982.
69. Wendt, H. and Plzak, V., Electrocatalytic and thermal activation of anodic oxygen-and cathodic hydrogen-evolution in alkaline water electrolysis. *Electrochim. Acta*, 28, 27–34, 1983.
70. Gras, J.M. and Spiteri, P., Corrosion of stainless steels and nickel-based alloys for alkaline water electrolysis. *Int. J. Hydrogen Energy*, 18, 561–566, 1993.

71. Sánchez, M., Amores, E., Rodríguez, L., Clemente-Jul, C., Semi-empirical model and experimental validation for the performance evaluation of a 15 kW alkaline water electrolyzer. *Int. J. Hydrogen Energy*, 43, 20332–20345, 2018.
72. LeRoy, R.L., Bowen, C.T., LeRoy, D.J., The Thermodynamics of Aqueous Water Electrolysis. *J. Electrochem. Soc*, 127, 1954–1962, 1980.
73. Diéguez, P.M., Ursúa, A., Sanchis, P., Sopena, C., Guelbenzu, E., Gandía, L.M., Thermal performance of a commercial alkaline water electrolyzer: Experimental study and mathematical modeling. *Int. J. Hydrogen Energy*, 33, 7338–7354, 2008.
74. Gonzalo, A.P., Marugán, A.P., Márquez, F.P.G., A review of the application performances of concentrated solar power systems. *Appl. Energy*, 255, 113893, 2019.
75. Zhang, Y., Wang, C., Lan, Z., Wei, S., Chen, R., Tian, W., Su, G., Review of Thermal-Hydraulic Issues and Studies of Lead-based fast reactors. *Renew. Sustain. Energy Rev.*, 120, 109625, 2020.
76. Trinke, P., Haug, P., Brauns, J., Bensmann, B., Hanke-Rauschenbach, R., Turek, T., Hydrogen Crossover in PEM and Alkaline Water Electrolysis: Mechanisms, Direct Comparison and Mitigation Strategies. *J. Electrochem. Soc.*, 165, F502–F513, 2018.
77. Brauns, J. and Turek, T., Experimental evaluation of dynamic operating concepts for alkaline water electrolyzers powered by renewable energy. *Electrochim. Acta*, 404, 139715, 2022.
78. Haug, P., Koj, M., Turek, T., Influence of process conditions on gas purity in alkaline water electrolysis. *Int. J. Hydrogen Energy*, 42, 9406–9418, 2017.
79. Wang, Y., Pang, Y., Xu, H., Martinez, A., Chen, K.S., PEM Fuel cell and electrolysis cell technologies and hydrogen infrastructure development - a review. *Energy Environ. Sci.*, 15, 2288–2328, 2022.
80. Li, H., Nakajima, H., Inada, A., Ito, K., Effect of flow-field pattern and flow configuration on the performance of a polymer-electrolyte-membrane water electrolyzer at high temperature. *Int. J. Hydrogen Energy*, 43, 8600–8610, 2018.
81. Li, H., Inada, A., Fujigaya, T., Nakajima, H., Sasaki, K., Ito, K., Effects of operating conditions on performance of high-temperature polymer electrolyte water electrolyzer. *J. Power Sources*, 318, 192–199, 2016.
82. Antonucci, V., Di Blasi, A., Baglio, V., Ornelas, R., Matteucci, F., Ledesma-Garcia, J., Arriaga, L.G., Aricò, A.S., High temperature operation of a composite membrane-based solid polymer electrolyte water electrolyser. *Electrochim. Acta*, 53, 7350–7356, 2008.
83. Garbe, S., Futter, J., Schmidt, T.J., Gubler, L., Insight into elevated temperature and thin membrane application for high efficiency in polymer electrolyte water electrolysis. *Electrochim. Acta*, 377, 138046, 2021.
84. Natarajan, V. and Basu, S., Performance and degradation studies of RuO_2-Ta_2O_5 anode electrocatalyst for high temperature PBI based proton exchange

membrane water electrolyser. *Int. J. Hydrogen Energy*, 40, 16702–16713, 2015.

85. Xu, W., Scott, K., Basu, S., Performance of a high temperature polymer electrolyte membrane water electrolyser. *J. Power Sources*, 196, 8918–8924, 2011.
86. Lee, B.-S., Park, H.-Y., Choi, I., Cho, M.K., Kim, H.-J., Yoo, S.J., Henkensmeier, D., Kim, J.Y., Nam, S.W., Park, S., Lee, K.-Y., Jang, J.H., Polarization characteristics of a low catalyst loading PEM water electrolyzer operating at elevated temperature. *J. Power Sources*, 309, 127–134, 2016.
87. Baglio, V., Ornelas, R., Matteucci, F., Martina, F., Ciccarella, G., Zama, I., Arriaga, L.G., Antonucci, V., Aricò, A.S., Solid Polymer Electrolyte Water Electrolyser Based on Nafion-TiO_2 Composite Membrane for High Temperature Operation. *Fuel Cells*, 9, 247–252, 2009.
88. Lin, Y., Dong, Y., Wang, X., Chen, L., Electrocatalysts for the Oxygen Evolution Reaction in Acidic Media. *Adv. Mater.*, 35, 2210565, 2023.
89. Liu, F., Shi, C., Guo, X., He, Z., Pan, L., Huang, Z.F., Zhang, X., Zou, J.J., Rational Design of Better Hydrogen Evolution Electrocatalysts for Water Splitting: A Review. *Adv. Sci.*, 9, 2200307, 2022.
90. Yang, J., Ning, G., Yu, L., Wang, Y., Luan, C., Fan, A., Zhang, X., Liu, Y., Dong, Y., Dai, X., Morphology controllable synthesis of PtNi concave nanocubes enclosed by high-index facets supported on porous graphene for enhanced hydrogen evolution reaction. *J. Mater. Chem. A*, 7, 17790–17796, 2019.
91. Qin, Y., Han, X., Gadipelli, S., Guo, J., Wu, S., Kang, L., Callison, J., Guo, Z., *In situ* synthesized low-PtCo@porous carbon catalyst for highly efficient hydrogen evolution. *J. Mater. Chem. A*, 7, 6543–6551, 2019.
92. Zhong, X., Wang, L., Zhuang, Z., Chen, X., Zheng, J., Zhou, Y., Zhuang, G., Li, X., Wang, J., Double Nanoporous Structure with Nanoporous PtFe Embedded in Graphene Nanopores: Highly Efficient Bifunctional Electrocatalysts for Hydrogen Evolution and Oxygen Reduction. *Adv. Mater. Interfaces*, 4, 1601029, 2017.
93. Corrales-Sánchez, T., Ampurdanés, J., Urakawa, A., MoS_2-based materials as alternative cathode catalyst for PEM electrolysis. *Int. J. Hydrogen Energy*, 39, 20837–20843, 2014.
94. Bolar, S., Shit, S., Kumar, J.S., Murmu, N.C., Ganesh, R.S., Inokawa, H., Kuila, T., Optimization of active surface area of flower like MoS_2 using V-doping towards enhanced hydrogen evolution reaction in acidic and basic medium. *Appl. Catal. B*, 254, 432–442, 2019.
95. Cao, E., Chen, Z., Wu, H., Yu, P., Wang, Y., Xiao, F., Chen, S., Du, S., Xie, Y., Wu, Y., Ren, Z., Boron-Induced Electronic-Structure Reformation of CoP Nanoparticles Drives Enhanced pH-Universal Hydrogen Evolution. *Angew. Chem. Int. Ed.*, 59, 4154–4160, 2020.
96. Song, H., Wu, M., Tang, Z., Tse, J.S., Yang, B., Lu, S., Single Atom Ruthenium-Doped CoP/CDs Nanosheets via Splicing of Carbon-Dots for Robust Hydrogen Production. *Angew. Chem. Int. Ed.*, 60, 7234–7244, 2021.

97. Yao, M., Wang, B., Sun, B., Luo, L., Chen, Y., Wang, J., Wang, N., Komarneni, S., Niu, X., Hu, W., Rational design of self-supported Cu@WC core-shell mesoporous nanowires for pH-universal hydrogen evolution reaction. *Appl. Catal. B*, 280, 119451, 2021.
98. Gong, Q., Wang, Y., Hu, Q., Zhou, J., Feng, R., Duchesne, P.N., Zhang, P., Chen, F., Han, N., Li, Y., Jin, C., Li, Y., Lee, S.-T., Ultrasmall and phase-pure W_2C nanoparticles for efficient electrocatalytic and photoelectrochemical hydrogen evolution. *Nat. Commun.*, 7, 13216, 2016.
99. Yu, Y., Nam, G.-H., He, Q., Wu, X.-J., Zhang, K., Yang, Z., Chen, J., Ma, Q., Zhao, M., Liu, Z., Ran, F.-R., Wang, X., Li, H., Huang, X., Li, B., Xiong, Q., Zhang, Q., Liu, Z., Gu, L., Du, Y., Huang, W., Zhang, H., High phase-purity 1T′-MoS_2- and 1T′-$MoSe_2$-layered crystals. *Nat. Chem.*, 10, 638–643, 2018.
100. Xu, Q., Liu, Y., Jiang, H., Hu, Y., Liu, H., Li, C., Unsaturated Sulfur Edge Engineering of Strongly Coupled MoS_2 Nanosheet-Carbon Macroporous Hybrid Catalyst for Enhanced Hydrogen Generation. *Adv. Energy Mater.*, 9, 1802553, 2019.
101. Danilovic, N., Subbaraman, R., Chang, K.-C., Chang, S.H., Kang, Y.J., Snyder, J., Paulikas, A.P., Strmcnik, D., Kim, Y.-T., Myers, D., Stamenkovic, V.R., Markovic, N.M., Activity-Stability Trends for the Oxygen Evolution Reaction on Monometallic Oxides in Acidic Environments. *J. Phys. Chem. Lett.*, 5, 2474–2478, 2014.
102. Kasian, O., Geiger, S., Li, T., Grote, J.-P., Schweinar, K., Zhang, S., Scheu, C., Raabe, D., Cherevko, S., Gault, B., Mayrhofer, K.J.J., Degradation of iridium oxidesvia oxygen evolution from the lattice: correlating atomic scale structure with reaction mechanisms. *Energy Environ. Sci.*, 12, 3548–3555, 2019.
103. Povia, M., Abbott, D.F., Herranz, J., Heinritz, A., Lebedev, D., Kim, B.-J., Fabbri, E., Patru, A., Kohlbrecher, J., Schäublin, R., Nachtegaal, M., Copéret, C., Schmidt, T.J., Operando X-ray characterization of high surface area iridium oxides to decouple their activity losses for the oxygen evolution reaction. *Energy Environ. Sci.*, 12, 3038–3052, 2019.
104. Siracusano, S., Van Dijk, N., Payne-Johnson, E., Baglio, V., Aricò, A.S., Nanosized IrO_x and $IrRuO_x$ electrocatalysts for the O_2 evolution reaction in PEM water electrolysers. *Appl. Catal. B*, 164, 488–495, 2015.
105. Diaz-Morales, O., Raaijman, S., Kortlever, R., Kooyman, P.J., Wezendonk, T., Gascon, J., Fu, W.T., Koper, M.T.M., Iridium-based double perovskites for efficient water oxidation in acid media. *Nat. Commun.*, 7, 12363, 2016.
106. Miao, X., Zhang, L., Wu, L., Hu, Z., Shi, L., Zhou, S., Quadruple perovskite ruthenate as a highly efficient catalyst for acidic water oxidation. *Nat. Commun.*, 10, 3809, 2019.
107. Seitz, L.C., Dickens, C.F., Nishio, K., Hikita, Y., Montoya, J., Doyle, A., Kirk, C., Vojvodic, A., Hwang, H.Y., Norskov, J.K., Jaramillo, T.F., Menlo Park CA United SLAC National Accelerator Lab., and Frontier Research Centers EFRC Energy, A highly active and stable $IrO_x/SrIrO_3$ catalyst for the oxygen evolution reaction. *Science*, 353, 1011–1014, 2016.

108. Kim, J., Shih, P.-C., Tsao, K.-C., Pan, Y.-T., Yin, X., Sun, C.-J., Yang, H., High-Performance Pyrochlore-Type Yttrium Ruthenate Electrocatalyst for Oxygen Evolution Reaction in Acidic Media. *J. Am. Chem. Soc.*, 139, 12076–12083, 2017.
109. Sardar, K., Petrucco, E., Hiley, C.I., Sharman, J.D.B., Wells, P.P., Russell, A.E., Kashtiban, R.J., Sloan, J., Walton, R.I., Water-Splitting Electrocatalysis in Acid Conditions Using Ruthenate-Iridate Pyrochlores. *Angew. Chem. Int. Ed.*, 126, 11140–11144, 2014.
110. Yang, X., Li, H., Lu, A.-Y., Min, S., Idriss, Z., Hedhili, M.N., Huang, K.-W., Idriss, H., Li, L.-J., Highly acid-durable carbon coated Co_3O_4 nanoarrays as efficient oxygen evolution electrocatalysts. *Nano Energy*, 25, 42–50, 2016.
111. Mondschein, J.S., Callejas, J.F., Read, C.G., Chen, J.Y.C., Holder, C.F., Badding, C.K., Schaak, R.E., Crystalline Cobalt Oxide Films for Sustained Electrocatalytic Oxygen Evolution under Strongly Acidic Conditions. *Chem. Mater.*, 29, 950–957, 2017.
112. Frydendal, R., Paoli, E.A., Chorkendorff, I., Rossmeisl, J., Stephens, I.E.L., Toward an Active and Stable Catalyst for Oxygen Evolution in Acidic Media: Ti-Stabilized MnO_2. *Adv. Energy Mater.*, 5, 1500991, 2015.
113. Wu, J., Liu, M., Chatterjee, K., Hackenberg, K.P., Shen, J., Zou, X., Yan, Y., Gu, J., Yang, Y., Lou, J., Ajayan, P.M., Exfoliated 2D Transition Metal Disulfides for Enhanced Electrocatalysis of Oxygen Evolution Reaction in Acidic Medium. *Adv. Mater. Interfaces*, 3, 1500669, 2016.
114. Schäfer, H., Küpper, K., Schmidt, M., Müller-Buschbaum, K., Stangl, J., Daum, D., Steinhart, M., Schulz-Kölbel, C., Han, W., Wollschläger, J., Krupp, U., Hou, P., Liu, X., Steel-based electrocatalysts for efficient and durable oxygen evolution in acidic media. *Catal. Sci. Technol.*, 8, 2104–2116, 2018.
115. Pan, M., Pan, C., Li, C., Zhao, J., A review of membranes in proton exchange membrane fuel cells: Transport phenomena, performance and durability. *Renew. Sustain. Energy Rev.*, 141, 110771, 2021.
116. Zhang, B., Cao, Y., Jiang, S., Li, Z., He, G., Wu, H., Enhanced proton conductivity of Nafion nanohybrid membrane incorporated with phosphonic acid functionalized graphene oxide at elevated temperature and low humidity. *J. Membr. Sci.*, 518, 243–253, 2016.
117. Cozzi, D., de Bonis, C., D'Epifanio, A., Mecheri, B., Tavares, A.C., Licoccia, S., Organically functionalized titanium oxide/Nafion composite proton exchange membranes for fuel cells applications. *J. Power Sources*, 248, 1127–1132, 2014.
118. Vinothkannan, M., Kim, A.R., Kumar, G.G., Yoo, D.J., Sulfonated graphene oxide/Nafion composite membranes for high temperature and low humidity proton exchange membrane fuel cells. *RSC Adv.*, 8, 7494–7508, 2018.
119. Yang, L., Tang, B., Wu, P., Metal-organic framework-graphene oxide composites: a facile method to highly improve the proton conductivity of PEMs operated under low humidity. *J. Mater. Chem. A*, 3, 15838–15842, 2015.

120. Donnadio, A., Narducci, R., Casciola, M., Marmottini, F., Amato, R.D., Jazestani, M., Chiniforoshan, H., Costantino, F., Mixed Membrane Matrices Based on Nafion/UiO-66/SO_3H-UiO-66 Nano-MOFs: Revealing the Effect of Crystal Size, Sulfonation, and Filler Loading on the Mechanical and Conductivity Properties. *ACS Appl. Mater. Interfaces*, 9, 42239–42246, 2017.
121. Stassi, A., Gatto, I., Passalacqua, E., Antonucci, V., Arico, A.S., Merlo, L., Oldani, C., Pagano, E., Performance comparison of long and short-side chain perfluorosulfonic membranes for high temperature polymer electrolyte membrane fuel cell operation. *J. Power Sources*, 196, 8925–8930, 2011.
122. Xiao, T., Wang, R., Chang, Z., Fang, Z., Zhu, Z., Xu, C., Electrolyte membranes for intermediate temperature proton exchange membrane fuel cell. *Prog. Nat. Sci.*, 30, 743–750, 2020.
123. Lobato, J., Cañizares, P., Rodrigo, M.A., Úbeda, D., Pinar, F.J., A novel titanium PBI-based composite membrane for high temperature PEMFCs. *J. Membr. Sci.*, 369, 105–111, 2011.
124. Li, X., Ma, H., Wang, P., Liu, Z., Peng, J., Hu, W., Jiang, Z., Liu, B., Construction of High-Performance, High-Temperature Proton Exchange Membranes through Incorporating SiO_2 Nanoparticles into Novel Cross-linked Polybenzimidazole Networks. *ACS Appl. Mater. Interfaces*, 11, 30735–30746, 2019.
125. Seo, K., Seo, J., Nam, K.H., Han, H., Polybenzimidazole/inorganic composite membrane with advanced performance for high temperature polymer electrolyte membrane fuel cells. *Polym. Compos.*, 38, 87–95, 2017.
126. Hooshyari, K., Javanbakht, M., Shabanikia, A., Enhessari, M., Fabrication $BaZrO_3$/PBI-based nanocomposite as a new proton conducting membrane for high temperature proton exchange membrane fuel cells. *J. Power Sources*, 276, 62–72, 2015.
127. Muthuraja, P., Prakash, S., Shanmugam, V.M., Radhakrsihnan, S., Manisankar, P., Novel perovskite structured calcium titanate-PBI composite membranes for high-temperature PEM fuel cells: Synthesis and characterizations. *Int. J. Hydrogen Energy*, 43, 4763–4772, 2018.
128. Shabanikia, A., Javanbakht, M., Amoli, H.S., Hooshyari, K., Enhessari, M., Polybenzimidazole/strontium cerate nanocomposites with enhanced proton conductivity for proton exchange membrane fuel cells operating at high temperature. *Electrochim. Acta*, 154, 370–378, 2015.
129. Abouzari-Lotf, E., Zakeri, M., Nasef, M.M., Miyake, M., Mozarmnia, P., Bazilah, N.A., Emelin, N.F., Ahmad, A., Highly durable polybenzimidazole composite membranes with phosphonated graphene oxide for high temperature polymer electrolyte membrane fuel cells. *J. Power Sources*, 412, 238–245, 2019.
130. Yang, J., Liu, C., Gao, L., Wang, J., Xu, Y., He, R., Novel composite membranes of triazole modified graphene oxide and polybenzimidazole for high temperature polymer electrolyte membrane fuel cell applications. *RSC Adv.*, 5, 101049–101054, 2015.

131. Li, X., Ma, H., Shen, Y., Hu, W., Jiang, Z., Liu, B., Guiver, M.D., Dimensionally-stable phosphoric acid-doped polybenzimidazoles for high-temperature proton exchange membrane fuel cells. *J. Power Sources*, 336, 391–400, 2016.
132. Tian, X., Wang, S., Li, J., Liu, F., Wang, X., Chen, H., Wang, D., Ni, H., Wang, Z., Benzimidazole grafted polybenzimidazole cross-linked membranes with excellent PA stability for high-temperature proton exchange membrane applications. *Appl. Surf. Sci.*, 465, 332–339, 2019.
133. Song, M., Lu, X., Li, Z., Liu, G., Yin, X., Wang, Y., Compatible ionic cross-linking composite membranes based on SPEEK and PBI for high temperature proton exchange membranes. *Int. J. Hydrogen Energy*, 41, 12069–12081, 2016.
134. Kim, J. and Ohira, A., Crosslinked Sulfonated Polyphenylsulfone (CSPPSU) Membranes for Elevated-Temperature PEM Water Electrolysis. *Membranes*, 11, 861, 2021.
135. Bai, H., Li, Y., Zhang, H., Chen, H., Wu, W., Wang, J., Liu, J., Anhydrous proton exchange membranes comprising of chitosan and phosphorylated graphene oxide for elevated temperature fuel cells. *J. Membr. Sci.*, 495, 48–60, 2015.
136. Anahidzade, N., Abdolmaleki, A., Dinari, M., Tadavani, K.F., Zhiani, M., Metal-organic framework anchored sulfonated poly(ether sulfone) as a high temperature proton exchange membrane for fuel cells. *J. Membr. Sci.*, 565, 281–292, 2018.
137. Wu, W., Zou, G., Fang, X., Cong, C., Zhou, Q., Effect of Methylimidazole Groups on the Performance of Poly(phenylene oxide) Based Membrane for High-Temperature Proton Exchange Membrane Fuel Cells. *Ind. Eng. Chem. Res.*, 56, 10227–10234, 2017.
138. Wang, K., Yang, L., Wei, W., Zhang, L., Chang, G., Phosphoric acid-doped poly(ether sulfone benzotriazole) for high-temperature proton exchange membrane fuel cell applications. *J. Membr. Sci.*, 549, 23–27, 2018.
139. Li, H., Fujigaya, T., Nakajima, H., Inada, A., Ito, K., Optimum structural properties for an anode current collector used in a polymer electrolyte membrane water electrolyzer operated at the boiling point of water. *J. Power Sources*, 332, 16–23, 2016.
140. Kim, J.-D. and Ohira, A., Water Electrolysis Using a Porous IrO_2/Ti/IrO_2 Catalyst Electrode and Nafion Membranes at Elevated Temperatures. *Membranes*, 11, 330, 2021.
141. Kai, J., Saito, R., Terabaru, K., Li, H., Nakajima, H., Ito, K., Effect of Temperature on the Performance of Polymer Electrolyte Membrane Water Electrolysis: Numerical Analysis of Electrolysis Voltage Considering Gas/Liquid Two-Phase Flow. *J. Electrochem. Soc.*, 166, F246–F254, 2019.
142. Siracusano, S., Van Dijk, N., Backhouse, R., Merlo, L., Baglio, V., Aricò, A.S., Degradation issues of PEM electrolysis MEAs. *Renew. Energy*, 123, 52–57, 2018.

143. Lewinski, K.A., van der Vliet, D., Luopa, S.M., NSTF Advances for PEM Electrolysis - the Effect of Alloying on Activity of NSTF Electrolyzer Catalysts and Performance of NSTF Based PEM Electrolyzers. *ECS Trans.*, 69, 893–917, 2015.
144. Yu, H., Danilovic, N., Wang, Y., Willis, W., Poozhikunnath, A., Bonville, L., Capuano, C., Ayers, K., Maric, R., Nano-size IrO_x catalyst of high activity and stability in PEM water electrolyzer with ultra-low iridium loading. *Appl. Catal. B*, 239, 133–146, 2018.
145. Garbe, S., Futter, J., Agarwal, A., Tarik, M., Mularczyk, A.A., Schmidt, T.J., Gubler, L., Understanding Degradation Effects of Elevated Temperature Operating Conditions in Polymer Electrolyte Water Electrolyzers. *J. Electrochem. Soc.*, 168, 44515, 2021.
146. Zhao, D., He, Q., Yu, J., Guo, M., Fu, J., Li, X., Ni, M., A data-driven digital-twin model and control of high temperature proton exchange membrane electrolyzer cells. *Int. J. Hydrogen Energy*, 47, 8687–8699, 2022.
147. Zhao, D., He, Q., Yu, J., Jiang, J., Li, X., Ni, M., Dynamic behaviour and control strategy of high temperature proton exchange membrane electrolyzer cells (HT-PEMECs) for hydrogen production. *Int. J. Hydrogen Energy*, 45, 26613–26622, 2020.
148. Crespi, E., Guandalini, G., Mastropasqua, L., Campanari, S., Brouwer, J., Experimental and theoretical evaluation of a 60 kW PEM electrolysis system for flexible dynamic operation. *Energy Convers. Manage.*, 277, 116622, 2023.
149. Zhang, H. and Yuan, T., Optimization and economic evaluation of a PEM electrolysis system considering its degradation in variable-power operations. *Appl. Energy*, 324, 119760, 2022.
150. Egeland-Eriksen, T., Jensen, J.F., Ulleberg, Ø., Sartori, S., Simulating offshore hydrogen production via PEM electrolysis using real power production data from a 2.3 MW floating offshore wind turbine. *Int. J. Hydrogen Energy*, 48, 28712–28732, 2023.
151. de Fournas, N. and Wei, M., Techno-economic assessment of renewable methanol from biomass gasification and PEM electrolysis for decarbonization of the maritime sector in California. *Energy Convers. Manage.*, 257, 115440, 2022.
152. Bonanno, M., Müller, K., Bensmann, B., Hanke-Rauschenbach, R., Peach, R., Thiele, S., Evaluation of the Efficiency of an Elevated Temperature Proton Exchange Membrane Water Electrolysis System. *J. Electrochem. Soc.*, 168, 94504, 2021.
153. Nafchi, F.M., Baniasadi, E., Afshari, E., Javani, N., Performance assessment of a solar hydrogen and electricity production plant using high temperature PEM electrolyzer and energy storage. *Int. J. Hydrogen Energy*, 43, 5820–5831, 2018.
154. Hanif, M.B., Rauf, S., Ul Abadeen, Z., Khan, K., Tayyab, Z., Qayyum, S., Mosiałek, M., Shao, Z., Li, C.-X., Motola, M., Proton-conducting solid oxide

electrolysis cells: Relationship of composition-structure-property, their challenges, and prospects. *Matter*, 6, 1782–1830, 2023.
155. Liu, F., Deng, H., Diercks, D., Kumar, P., Abdul Jabbar, M.H., Gumeci, C., Furuya, Y., Dale, N., Oku, T., Usuda, M., Kazempoor, P., Fang, L., Chen, D., Liu, B., Duan, C., Lowering the operating temperature of protonic ceramic electrochemical cells to $<450\,°C$. *Nat. Energy*, 8, 1145–1157, 2023.
156. Hwang, S.H., Kim, S.K., Nam, J.-T., Park, J.-S., Fabrication of an electrolyte-supported protonic ceramic fuel cell with nano-sized powders of Ni-composite anode. *Int. J. Hydrogen Energy*, 46, 1076–1084, 2021.
157. Song, S.-H., Yoon, S.-E., Choi, J., Kim, B.-K., Park, J.-S., A high-performance ceramic composite anode for protonic ceramic fuel cells based on lanthanum strontium vanadate. *Int. J. Hydrogen Energy*, 39, 16534–16540, 2014.
158. Anggia, E., Shin, E.-K., Nam, J.-T., Park, J.-S., Fabrication of ceramic composite anode at low temperature for high performance protonic ceramic fuel cells. *Ceram. Int.*, 46, 236–242, 2020.
159. Wang, L., Fan, Y., Li, J., Shao, L., Xi, X., Fu, X.-Z., Luo, J.-L., $La_{0.5}Sr_{0.5}Fe_{0.9}Mo_{0.1}O_{3-\delta}$-$CeO_2$ anode catalyst for Co-Producing electricity and ethylene from ethane in proton-conducting solid oxide fuel cells. *Ceram. Int.*, 47, 24106–24114, 2021.
160. Wachowski, S., Li, Z., Polfus, J.M., Norby, T., Performance and stability in H_2S of $SrFe_{0.75}Mo_{0.25}O_{3-\delta}$ as electrode in proton ceramic fuel cells. *J. Eur. Ceram. Soc.*, 38, 163–171, 2018.
161. Toriumi, H., Zhu, C., Habazaki, H., Aoki, Y., Intermediate Temperature H+-SOEC with Proton Conducting $BaZr_{0.6}Ce_{0.2}Y_{0.2}O_{3-\delta}$ Electrolyte. *ECS Trans.*, 91, 2669, 2019.
162. Leonard, K., Deibert, W., Ivanova, M.E., Meulenberg, W.A., Ishihara, T., Matsumoto, H., Processing Ceramic Proton Conductor Membranes for Use in Steam Electrolysis. *Membranes*, 10, 339, 2020.
163. Yang, Y., Shi, N., Xie, Y., Li, X., Hu, X., Zhu, K., Huan, D., Peng, R., Xia, C., Lu, Y., K doping as a rational method to enhance the sluggish air-electrode reaction kinetics for proton-conducting solid oxide cells. *Electrochim. Acta*, 389, 138453, 2021.
164. Choi, S., Davenport, T.C., Haile, S.M., Protonic ceramic electrochemical cells for hydrogen production and electricity generation: exceptional reversibility, stability, and demonstrated faradaic efficiency. *Energy Environ. Sci.*, 12, 206–215, 2019.
165. Vøllestad, E., Strandbakke, R., Tarach, M., Catalán-Martínez, D., Fontaine, M.-L., Beeaff, D., Clark, D.R., Serra, J.M., Norby, T., Mixed proton and electron conducting double perovskite anodes for stable and efficient tubular proton ceramic electrolysers. *Nat. Mater.*, 18, 752–759, 2019.
166. Li, W., Guan, B., Ma, L., Hu, S., Zhang, N., Liu, X., High performing triple-conductive $Pr_2NiO_{4+\delta}$ anode for proton-conducting steam solid oxide electrolysis cell. *J. Mater. Chem. A*, 6, 18057–18066, 2018.

167. Sun, C., Yang, S., Lu, Y., Wen, J., Ye, X., Wen, Z., Tailoring a micro-nanostructured electrolyte-oxygen electrode interface for proton-conducting reversible solid oxide cells. *J. Power Sources*, 449, 227498, 2020.
168. Rolle, A., Mohamed, H.A.A., Huo, D., Capoen, E., Mentré, O., Vannier, R.-N., Daviero-Minaud, S., Boukamp, B.A., $Ca_3Co_4O_{9+\delta}$, a growing potential SOFC cathode material: Impact of the layer composition and thickness on the electrochemical properties. *Solid State Ion.*, 294, 21–30, 2016.
169. Saqib, M., Choi, I.-G., Bae, H., Park, K., Shin, J.-S., Kim, Y.-D., Lee, J.-I., Jo, M., Kim, Y.-C., Lee, K.-S., Song, S.-J., Wachsman, E.D., Park, J.-Y., Transition from perovskite to misfit-layered structure materials: a highly oxygen deficient and stable oxygen electrode catalyst. *Energy Environ. Sci.*, 14, 2472–2484, 2021.
170. Zhou, Y., Liu, E., Chen, Y., Liu, Y., Zhang, L., Zhang, W., Luo, Z., Kane, N., Zhao, B., Soule, L., Niu, Y., Ding, Y., Ding, H., Ding, D., Liu, M., An Active and Robust Air Electrode for Reversible Protonic Ceramic Electrochemical Cells. *ACS Energy Lett.*, 6, 1511–1520, 2021.
171. Niu, Y., Zhou, Y., Zhang, W., Zhang, Y., Evans, C., Luo, Z., Kane, N., Ding, Y., Chen, Y., Guo, X., Lv, W., Liu, M., Highly Active and Durable Air Electrodes for Reversible Protonic Ceramic Electrochemical Cells Enabled by an Efficient Bifunctional Catalyst. *Adv. Energy Mater.*, 12, 2103783, 2022.
172. Pandiyan, A., Uthayakumar, A., Subrayan, R., Cha, S.W., Moorthy, S.B.K., Review of solid oxide electrolysis cells: a clean energy strategy for hydrogen generation. *Nanomater. Energy*, 8, 2–22, 2019.
173. Tang, C., Akimoto, K., Wang, N., Fadillah, L., Kitano, S., Habazaki, H., Aoki, Y., The effect of an anode functional layer on the steam electrolysis performances of protonic solid oxide cells. *J. Mater. Chem. A*, 9, 14032–14042, 2021.
174. Rajendran, S., Thangavel, N.K., Ding, H., Ding, Y., Ding, D., Arava, L.M.R., Tri-Doped $BaCeO_3$-$BaZrO_3$ as a Chemically Stable Electrolyte with High Proton-Conductivity for Intermediate Temperature Solid Oxide Electrolysis Cells (SOECs). *ACS Appl. Mater. Interfaces*, 12, 38275–38284, 2020.
175. Shin, J.-S., Park, H., Park, K., Saqib, M., Jo, M., Kim, J.H., Lim, H.-T., Kim, M., Kim, J., Park, J.-Y., Activity of layered swedenborgite structured $Y_{0.8}Er_{0.2}BaCo_{3.2}Ga_{0.8}O_{7+\delta}$ for oxygen electrode reactions in at intermediate temperature reversible ceramic cells. *J. Mater. Chem. A*, 9, 607–621, 2021.
176. Murphy, R., Zhou, Y., Zhang, L., Soule, L., Zhang, W., Chen, Y., Liu, M., A New Family of Proton-Conducting Electrolytes for Reversible Solid Oxide Cells: $BaHf_xCe_{0.8-x}Y_{0.1}Yb_{0.1}O_{3-\delta}$. *Adv. Funct. Mater.*, 30, 2002265, 2020.
177. Li, W., Guan, B., Ma, L., Tian, H., Liu, X., Synergistic Coupling of Proton Conductors $BaZr_{0.1}Ce_{0.7}Y_{0.1}Yb_{0.1}O_{3-\delta}$ and $La_2Ce_2O_7$ to Create Chemical Stable, Interface Active Electrolyte for Steam Electrolysis Cells. *ACS Appl. Mater. Interfaces*, 11, 18323–18330, 2019.
178. Fop, S., McCombie, K.S., Wildman, E.J., Skakle, J.M.S., Irvine, J.T.S., Connor, P.A., Savaniu, C., Ritter, C., Mclaughlin, A.C., High oxide ion and proton

conductivity in a disordered hexagonal perovskite. *Nat. Mater.*, 19, 752–757, 2020.

179. Tian, H., Luo, Z., Song, Y., Zhou, Y., Gong, M., W.Li, Z., Liu, M., Liu, X., Protonic ceramic materials for clean and sustainable energy: advantages and challenges. *Int. Mater. Rev.*, 68, 272–300, 2023.
180. Herradon, C., Le, L., Meisel, C., Huang, J., Chmura, C., Kim, Y.D., Cadigan, C., Hayre, R.O., Sullivan, N.P., Proton-conducting ceramics for water electrolysis and hydrogen production at elevated pressure. *Front. Energy Res.*, 10, 1020960, 2022.
181. Fang, S., Brinkman, K.S., Chen, F., Hydrogen permeability and chemical stability of Ni-$BaZr_{0.1}Ce_{0.7}Y_{0.1}Yb_{0.1}O_{3-\delta}$ membrane in concentrated H_2O and CO_2. *J. Membr. Sci.*, 467, 85–92, 2014.
182. Luo, Z., Zhou, Y., Hu, X., Kane, N., Li, T., Zhang, W., Liu, Z., Ding, Y., Liu, Y., Liu, M., Critical role of acceptor dopants in designing highly stable and compatible proton-conducting electrolytes for reversible solid oxide cells. *Energy Environ. Sci.*, 15, 2992–3003, 2022.
183. Lei, L., Zhang, J., Yuan, Z., Liu, J., Ni, M., Chen, F., Progress Report on Proton Conducting Solid Oxide Electrolysis Cells. *Adv. Funct. Mater.*, 29, 1903805, 2019.
184. Pan, J., Ye, Y., Zhou, M., Sun, X., Chen, Y., Revealing the Impact of Steam Concentration on the Activity and Stability of Double-Perovskite Air Electrodes for Proton-Conducting Electrolysis Cells. *Energy Fuels*, 36, 12253–12260, 2022.
185. Wang, Z., Yang, W., Shafi, S.P., Bi, L., Wang, Z., Peng, R., Xia, C., Liu, W., Lu, Y., A high performance cathode for proton conducting solid oxide fuel cells. *J. Mater. Chem. A*, 3, 8405–8412, 2015.
186. Ding, H., Wu, W., Jiang, C., Ding, Y., Bian, W., Hu, B., Singh, P., Orme, C.J., Wang, L., Zhang, Y., Ding, D., Self-sustainable protonic ceramic electrochemical cells using a triple conducting electrode for hydrogen and power production. *Nat. Commun.*, 11, 1907, 2020.
187. Wang, W., Li, Y., Liu, Y., Tian, Y., Ma, B., Li, J., Pu, J., Chi, B., Ruddlesden-Popper-Structured $(Pr_{0.9}La_{0.1})_2(Ni_{0.8}Cu_{0.2})O_{4+\delta}$: An Effective Oxygen Electrode Material for Proton-Conducting Solid Oxide Electrolysis Cells. *ACS Sustain. Chem. Eng.*, 9, 10913–10919, 2021.
188. Duan, C., Kee, R., Zhu, H., Sullivan, N., Zhu, L., Bian, L., Jennings, D., Hayre, R.O., Highly efficient reversible protonic ceramic electrochemical cells for power generation and fuel production. *Nat. Energy*, 4, 230–240, 2019.
189. Bian, W., Wu, W., Wang, B., Tang, W., Zhou, M., Jin, C., Ding, H., Fan, W., Dong, Y., Li, J., Ding, D., Revitalizing interface in protonic ceramic cells by acid etch. *Nature*, 604, 479–485, 2022.
190. Duan, C., Tong, J., Shang, M., Nikodemski, S., Sanders, M., Ricote, S., Almansoori, A., O'Hayre, R., Readily processed protonic ceramic fuel cells with high performance at low temperatures. *Science*, 349, 1321–1326, 2015.

191. Zhao, L., Ding, D., Zhang, L., Gui, L., Wang, Z., Wan, Y., Wang, R., Ling, Y., He, B., The effect of Cr deposition and poisoning on $BaZr_{0.1}Ce_{0.7}Y_{0.2}O_{3-\delta}$ proton conducting electrolyte. *Int. J. Hydrogen Energy*, 39, 18379–18384, 2014.
192. Wang, R., Würth, M., Pal, U.B., Gopalan, S., Basu, S.N., Roles of humidity and cathodic current in chromium poisoning of Sr-doped $LaMnO_3$-based cathodes in solid oxide fuel cells. *J. Power Sources*, 360, 87–97, 2017.
193. Schwabe, F., Schwarze, L., Partmann, C., Lippmann, W., Hurtado, A., Concept, design, and energy analysis of an integrated power-to-methanol process utilizing a tubular proton-conducting solid oxide electrolysis cell. *Int. J. Hydrogen Energy*, 44, 12566–12575, 2019.
194. Chen, Y., Liu, M., Shi, Y., Cai, N., Methane Production and Electricity Generation in a Proton Conducting Tubular Reversible Solid Oxide Cell. *ECS Trans.*, 91, 2641–2651, 2019.
195. Milewski, J., Kupecki, J., Szczęśniak, A., Uzunow, N., Hydrogen production in solid oxide electrolyzers coupled with nuclear reactors. *Int. J. Hydrogen Energy*, 46, 35765–35776, 2021.
196. Ghosh, A., Roy, D., Ghosh, S., Proton conducting reversible SOFC integrated in a solar thermal power generation system. *J. Phys. Conf. Ser.*, 1240, 12112, 2019.
197. Brisse, A., Schefold, J., Zahid, M., High temperature water electrolysis in solid oxide cells. *Int. J. Hydrogen Energy*, 33, 5375–5382, 2008.
198. Lang, M., Raab, S., Lemcke, M.S., Bohn, C., Pysik, M., Long Term Behavior of Solid Oxide Electrolyser (SOEC) Stacks. *ECS Trans.*, 91, 2713–2725, 2019.
199. Glisenti, A., Bedon, A., Carollo, G., Savaniu, C., Irvine, J.T.S., Reversible, all-perovskite SOFCs based on La C.OMMAS.R.X.X.X gallates. *Int. J. Hydrogen Energy*, 45, 29155–29165, 2020.
200. Gan, Y., Qin, Q., Chen, S., Wang, Y., Dong, D., Xie, K., Wu, Y., Composite cathode $La_{0.4}Sr_{0.4}TiO_{3-\delta}$-$Ce_{0.8}Sm_{0.2}O_{2-\delta}$ impregnated with Ni for high-temperature steam electrolysis. *J. Power Sources*, 245, 245–255, 2014.
201. Bernadet, L., Moncasi, C., Torrell, M., Tarancón, A., High-performing electrolyte-supported symmetrical solid oxide electrolysis cells operating under steam electrolysis and co-electrolysis modes. *Int. J. Hydrogen Energy*, 45, 14208–14217, 2020.
202. Myung, J.-h., Neagu, D., Miller, D.N., Irvine, J.T.S., Switching on electrocatalytic activity in solid oxide cells. *Nature*, 537, 528–531, 2016.
203. Liang, M., Yu, B., Wen, M., Chen, J., Xu, J., Zhai, Y., Preparation of LSM-YSZ composite powder for anode of solid oxide electrolysis cell and its activation mechanism. *J. Power Sources*, 190, 341–345, 2009.
204. Lee, K.J., Seo, J.W., Yoon, J.S., Hwang, H.J., Electrode Performance in a Solid Oxide Electrolyzer Cell (SOEC). *ECS Trans.*, 57, 3255–3260, 2013.
205. Kim, S.J. and Choi, G.M., Stability of LSCF electrode with GDC interlayer in YSZ-based solid oxide electrolysis cell. *Solid State Ion.*, 262, 303–306, 2014.
206. Vibhu, V., Yildiz, S., Vinke, I.C., Eichel, R.A., Bassat, J.M., de Haart, L.G.J., High Performance LSC Infiltrated LSCF Oxygen Electrode for High

Temperature Steam Electrolysis Application. *J. Electrochem. Soc.*, 166, F102–F108, 2019.
207. Huan, Y., Chen, S., Zeng, R., Wei, T., Dong, D., Hu, X., Huang, Y., Intrinsic Effects of Ruddlesden-Popper-Based Bifunctional Catalysts for High-Temperature Oxygen Reduction and Evolution. *Adv. Energy Mater.*, 9, 1901573, 2019.
208. Vibhu, V., Bassat, J.-M., Flura, A., Nicollet, C., Grenier, J.-C., Rougier, A., Influence of La/Pr Ratio on the Ageing Properties of $La_{2-x}Pr_xNiO_{4+\delta}$ as Cathodes in IT-SOFCs. *ECS Trans.*, 68, 825, 2015.
209. Sharma, R.K., Burriel, M., Dessemond, L., Bassat, J.-M., Djurado, E., $La_{n+1}Ni_nO_{3n+1}$ (n = 2 and 3) phases and composites for solid oxide fuel cell cathodes: Facile synthesis and electrochemical properties. *J. Power Sources*, 325, 337–345, 2016.
210. Matvejeff, M., Lehtimäki, M., Hirasa, A., Huang, Y.H., Yamauchi, H., Karppinen, M., New Water-Containing Phase Derived from the $Sr_3Fe_2O_{7-\delta}$ Phase of the Ruddlesden-Popper Structure. *Chem. Mater.*, 17, 2775–2779, 2005.
211. Adachi, Y., Hatada, N., Uda, T., Sintering, Electrical Conductivity, Oxygen Nonstoichiometry, Thermal Expansion and Thermal Stability of Ruddlesden-Popper Type Cobaltite $La_4Co_3O_{10}$. *J. Electrochem. Soc.*, 163, F1084–F1090, 2016.
212. Ebbesen, S.D., Jensen, S.H., Hauch, A., Mogensen, M.B., High Temperature Electrolysis in Alkaline Cells, Solid Proton Conducting Cells, and Solid Oxide Cells. *Chem. Rev.*, 114, 10697–10734, 2014.
213. Wang, Y.-p., Gao, J.-t., Chen, W., Li, C.-x., Zhang, S.-l., Yang, G.-j., Li, C.-j., Development of ScSZ Electrolyte by Very Low Pressure Plasma Spraying for High-Performance Metal-Supported SOFCs. *J. Therm. Spray Technol.*, 29, 223–231, 2020.
214. Stevenson, J.W., Armstrong, T.R., Pederson, L.R., Li, J., Lewinsohn, C.A., Baskaran, S., Effect of A-site cation nonstoichiometry on the properties of doped lanthanum gallate. *Solid State Ion.*, 113, 571–583, 1998.
215. Hosoi, K., Hagiwara, H., Ida, S., Ishihara, T., $La_{0.8}Sr_{0.2}FeO_{3-\delta}$ as Fuel Electrode for Solid Oxide Reversible Cells Using $LaGaO_3$-Based Oxide Electrolyte. *J. Phys. Chem. C*, 120, 16110–16117, 2016.
216. Ye, L. and Xie, K., High-temperature electrocatalysis and key materials in solid oxide electrolysis cells. *J. Energy Chem.*, 54, 736–745, 2021.
217. Shimada, H., Yamaguchi, T., Kishimoto, H., Sumi, H., Yamaguchi, Y., Nomura, K., Fujishiro, Y., Nanocomposite electrodes for high current density over 3 A cm^{-2} in solid oxide electrolysis cells. *Nat. Commun.*, 10, 5432, 2019.
218. Cao, J., Li, Y., Zheng, Y., Wang, S., Zhang, W., Qin, X., Geng, G., Yu, B., A Novel Solid Oxide Electrolysis Cell with Micro-/Nano Channel Anode for Electrolysis at Ultra-High Current Density over 5 A cm^{-2}. *Adv. Energy Mater.*, 12, 2200899, 2022.

219. Ioannidou, E., Neofytidis, C., Sygellou, L., Niakolas, D.K., Au-doped Ni/GDC as an Improved Cathode Electrocatalyst for H_2O Electrolysis in SOECs. *Appl. Catal. B*, 236, 253–264, 2018.
220. Liu, Q., Yang, C., Dong, X., Chen, F., Perovskite $Sr_2Fe_{1.5}Mo_{0.5}O_{6-\delta}$ as electrode materials for symmetrical solid oxide electrolysis cells. *Int. J. Hydrogen Energy*, 35, 10039–10044, 2010.
221. Bernadet, L., Gousseau, G., Chatroux, A., Laurencin, J., Mauvy, F., Reytier, M., Influence of pressure on solid oxide electrolysis cells investigated by experimental and modeling approach. *Int. J. Hydrogen Energy*, 40, 12918–12928, 2015.
222. Kim, S.-D., Seo, D.-W., Dorai, A.K., Woo, S.-K., The effect of gas compositions on the performance and durability of solid oxide electrolysis cells. *Int. J. Hydrogen Energy*, 38, 6569–6576, 2013.
223. Liu, F., Wang, T., Li, J., Wei, T., Ye, Z., Dong, D., Chen, B., Ling, Y., Shao, Z., Elevated-temperature bio-ethanol-assisted water electrolysis for efficient hydrogen production. *Chem. Eng. J.*, 434, 134699, 2022.
224. Patcharavorachot, Y., Thongdee, S., Saebea, D., Authayanun, S., Arpornwichanop, A., Performance comparison of solid oxide steam electrolysis cells with/without the addition of methane. *Energy Convers. Manage.*, 120, 274–286, 2016.
225. Nielsen, A.S., Peppley, B.A., Burheim, O.S., Tuning transport mechanisms in fuel-assisted solid oxide electrolysis cells for enhanced performance and product selectivity: Thermodynamic and kinetic modeling. *Chem. Eng. J.*, 452, 139079, 2023.
226. Schefold, J., Brisse, A., Poepke, H., 23,000 h steam electrolysis with an electrolyte supported solid oxide cell. *Int. J. Hydrogen Energy*, 42, 13415–13426, 2017.
227. Tietz, F., Sebold, D., Brisse, A., Schefold, J., Degradation phenomena in a solid oxide electrolysis cell after 9000 h of operation. *J. Power Sources*, 223, 129–135, 2013.
228. Laguna-Bercero, M.A., Campana, R., Larrea, A., Kilner, J.A., Orera, V.M., Electrolyte degradation in anode supported microtubular yttria stabilized zirconia-based solid oxide steam electrolysis cells at high voltages of operation. *J. Power Sources*, 196, 8942–8947, 2011.
229. Chen, K. and Jiang, S.P., Failure mechanism of (La,Sr)MnO_3 oxygen electrodes of solid oxide electrolysis cells. *Int. J. Hydrogen Energy*, 36, 10541–10549, 2011.
230. Keane, M., Mahapatra, M.K., Verma, A., Singh, P., LSM-YSZ interactions and anode delamination in solid oxide electrolysis cells. *Int. J. Hydrogen Energy*, 37, 16776–16785, 2012.
231. Chen, K., Ai, N., Jiang, S.P., Performance and structural stability of $Gd_{0.2}Ce_{0.8}O_{1.9}$ infiltrated $La_{0.8}Sr_{0.2}MnO_3$ nano-structured oxygen electrodes of solid oxide electrolysis cells. *Int. J. Hydrogen Energy*, 39, 10349–10358, 2014.

232. Wu, T., Zhang, W., Li, Y., Zheng, Y., Yu, B., Chen, J., Sun, X., Micro-/Nanohoneycomb Solid Oxide Electrolysis Cell Anodes with Ultralarge Current Tolerance. *Adv. Energy Mater.*, 8, 1802203, 2018.
233. Laurencin, J., Hubert, M., Ferreira Sanchez, D., Pylypko, S., Morales, M., Morata, A., Morel, B., Montinaro, D., Lefebvre-Joud, F., Siebert, E., Degradation mechanism of $La_{0.6}Sr_{0.4}Co_{0.2}Fe_{0.8}O_{3-\delta}/Gd_{0.1}Ce_{0.9}O_{2-\delta}$ composite electrode operated under solid oxide electrolysis and fuel cell conditions. *Electrochim. Acta*, 241, 459–476, 2017.
234. Zheng, Y., Li, Q., Chen, T., Wu, W., Xu, C., Wang, W.G., Comparison of performance and degradation of large-scale solid oxide electrolysis cells in stack with different composite air electrodes. *Int. J. Hydrogen Energy*, 40, 2460–2472, 2015.
235. Kim-Lohsoontorn, P., Brett, D.J.L., Laosiripojana, N., Kim, Y.-M., Bae, J.-M., Performance of solid oxide electrolysis cells based on composite $La_{0.8}Sr_{0.2}MnO_{3-\delta}$-yttria stabilized zirconia and $Ba_{0.5}Sr_{0.5}Co_{0.8}Fe_{0.2}O_{3-\delta}$ oxygen electrodes. *Int. J. Hydrogen Energy*, 35, 3958–3966, 2010.
236. Khan, M.S., Xu, X., Knibbe, R., Zhu, Z., Air electrodes and related degradation mechanisms in solid oxide electrolysis and reversible solid oxide cells. *Renew. Sustain. Energy Rev.*, 143, 110918, 2021.
237. Montenegro-Hernández, A., Vega-Castillo, J., Mogni, L., Caneiro, A., Thermal stability of $Ln_2NiO_{4+\delta}$ (Ln: La, Pr, Nd) and their chemical compatibility with YSZ and CGO solid electrolytes. *Int. J. Hydrogen Energy*, 36, 15704–15714, 2011.
238. Tong, X., Zhou, F., Yang, S., Zhong, S., Wei, M., Liu, Y., Performance and stability of Ruddlesden-Popper $La_2NiO_{4+\delta}$ oxygen electrodes under solid oxide electrolysis cell operation conditions. *Ceram. Int.*, 43, 10927–10933, 2017.
239. Subotić, V., Futamura, S., Harrington, G.F., Matsuda, J., Natsukoshi, K., Sasaki, K., Towards understanding of oxygen electrode processes during solid oxide electrolysis operation to improve simultaneous fuel and oxygen generation. *J. Power Sources*, 492, 229600, 2021.
240. Königshofer, B., Pongratz, G., Nusev, G., Boškoski, P., Höber, M., Juričić, Đ., Kusnezoff, M., Trofimenko, N., Schröttner, H., Hochenauer, C., Subotić, V., Development of test protocols for solid oxide electrolysis cells operated under accelerated degradation conditions. *J. Power Sources*, 497, 229875, 2021.
241. Subotić, V. and Hochenauer, C., Analysis of solid oxide fuel and electrolysis cells operated in a real-system environment: State-of-the-health diagnostic, failure modes, degradation mitigation and performance regeneration. *Prog. Energy Combust. Sci.*, 93, 101011, 2022.
242. Hubert, M., Laurencin, J., Cloetens, P., Morel, B., Montinaro, D., Lefebvre-Joud, F., Impact of Nickel agglomeration on Solid Oxide Cell operated in fuel cell and electrolysis modes. *J. Power Sources*, 397, 240–251, 2018.

243. Frey, C.E., Fang, Q., Sebold, D., Blum, L., Menzler, N.H., A Detailed Post Mortem Analysis of Solid Oxide Electrolyzer Cells after Long-Term Stack Operation. *J. Electrochem. Soc.*, 165, F357–F364, 2018.
244. The, D., Grieshammer, S., Schroeder, M., Martin, M., Al Daroukh, M., Tietz, F., Schefold, J., Brisse, A., Microstructural comparison of solid oxide electrolyser cells operated for 6100 h and 9000 h. *J. Power Sources*, 275, 901–911, 2015.
245. Tao, Y., Ebbesen, S.D., Mogensen, M.B., Degradation of solid oxide cells during co-electrolysis of steam and carbon dioxide at high current densities. *J. Power Sources*, 328, 452–462, 2016.
246. Hauch, A., Ebbesen, S.D., Jensen, S.H., Mogensen, M., Solid Oxide Electrolysis Cells: Microstructure and Degradation of the Ni/Yttria-Stabilized Zirconia Electrode. *J. Electrochem. Soc.*, 155, B1184, 2008.
247. Lim, C.K., Liu, Q., Zhou, J., Sun, Q., Chan, S.H., Sulfur Tolerant LSCM-based Composite Cathode for High Temperature Electrolysis/Co-electrolysis of H_2O and CO_2. *Fuel Cells*, 17, 464–472, 2017.
248. Hu, B., Aphale, A.N., Liang, C., Heo, S.J., Uddin, M.A., Singh, P., Carbon Tolerant Double Site Doped Perovskite Cathodes for High-Temperature Electrolysis Cells. *ECS Trans.*, 78, 3257–3265, 2017.
249. Teng, Z., Xiao, Z., Yang, G., Guo, L., Yang, X., Ran, R., Wang, W., Zhou, W., Shao, Z., Efficient water splitting through solid oxide electrolysis cells with a new hydrogen electrode derived from A-site cation-deficient $La_{0.4}Sr_{0.55}Co_{0.2}Fe_{0.6}Nb_{0.2}O_{3-\delta}$ perovskite. *Mater. Today Energy*, 17, 100458, 2020.
250. Amaya-Dueñas, D.M., Riegraf, M., Nenning, A., Opitz, A.K., Costa, R., Friedrich, K.A., Operational Aspects of a Perovskite Chromite-Based Fuel Electrode in Solid Oxide Electrolysis Cells (SOEC). *ACS Appl. Energy Mater.*, 5, 8143–8156, 2022.
251. Geipel, C., Hauptmeier, K., Herbrig, K., Mittmann, F., Münch, M., Pötschke, M., Reichel, L., Strohbach, T., Seidel, T., Surrey, A., Walter, C., Stack Development and Industrial Scale-Up. *ECS Trans.*, 91, 123–132, 2019.
252. Chen, K., Hyodo, J., Ai, N., Ishihara, T., Jiang, S.P., Boron deposition and poisoning of $La_{0.8}Sr_{0.2}MnO_3$ oxygen electrodes of solid oxide electrolysis cells under accelerated operation conditions. *Int. J. Hydrogen Energy*, 41, 1419–1431, 2016.
253. Chatzichristodoulou, C., Chen, M., Hendriksen, P.V., Jacobsen, T., Mogensen, M.B., Understanding degradation of solid oxide electrolysis cells through modeling of electrochemical potential profiles. *Electrochim. Acta*, 189, 265–282, 2016.
254. Jacobson, N.S., Opila, E.J., Myers, D.L., Copland, E.H., Thermodynamics of gas phase species in the Si-O-H system. *J. Chem. Thermodyn.*, 37, 1130–1137, 2005.

255. Luo, Y., Shi, Y., Cai, N., Power-to-Gas Energy Storage by Reversible Solid Oxide Cell for Distributed Renewable Power Systems. *J. Energy Eng.*, 144, 4017079, 2018.
256. Jensen, S.H., Graves, C., Mogensen, M., Wendel, C., Braun, R., Hughes, G., Gao, Z., Barnett, S.A., Large-scale electricity storage utilizing reversible solid oxide cells combined with underground storage of CO_2 and CH_4. *Energy Environ. Sci.*, 8, 2471–2479, 2015.
257. Luo, Y., Shi, Y., Li, W., Cai, N., Synchronous enhancement of H_2O/CO_2 co-electrolysis and methanation for efficient one-step power-to-methane. *Energy Convers. Manage.*, 165, 127–136, 2018.

Index

Also of Interest

Check out these published titles in "Advances in Hydrogen Production and Storage" series from Scrivener Publishing

Towards Green Hydrogen Generation
Edited by Mehmet Sankir and Nurdan Demirci Sankir
Published 2024. ISBN 978-1-394-23408-0

Hydrogen Electrical Vehicles
Edited by Mehmet Sankir and Nurdan Demirci Sankir
Published 2023. ISBN 978-1-394-16638-1

Hydrogen Storage Technologies
Edited by Mehmet Sankir and Nurdan Demirci Sankir
Published 2018. ISBN 978-1-119-45988-0

Hydrogen Production Technologies
Edited by Mehmet Sankir and Nurdan Demirci Sankir
Published 2017. ISBN 978-1-119-28364-5

www.scrivenerpublishing.com

Printed in the USA/Agawam, MA
September 16, 2026

910661.154